LIGHT IN BIOLOGY AND MEDICINE, Volume 1

LIGHT IN BIOLOGY AND MEDICINE
Volume 1

Ron H. Douglas
The City University
London, United Kingdom

Johan Moan
The Norwegian Radium Hospital
Oslo, Norway

and

F. Dall'Acqua
University of Padua
Padua, Italy

PLENUM PRESS • NEW YORK AND LONDON

ISBN-13:978-1-4612-8043-9   e-ISBN-13:978-1-4613-0709-9
DOI: 10.1007/978-1-4613-0709-9

Proceedings of the Second Congress of the European Society for
Photobiology, held September 6-10, 1987, in Padua, Italy

A Division of Plenum Publishing Corporation
233 Spring Street, New York, N.Y. 10013

PREFACE

Almost all life depends on light for its survival. It is the ultimate
basis for the food we eat (photosynthesis), and many organisms make use of
it in basic sensory mechanisms for guiding their behaviour, be it through
the complex process of vision, or by the relatively more simple photosensitivity of microorganisms. Furthermore, light has profound implications
for the field of medicine, both as a cause of disease (ie UV damage of
DNA), and as a therapeutic agent (ie photodynamic therapy). These and other
processes are the basis for the science of photobiology, which could be
defined as the study of the effects of (visible and ultraviolet) light
(from both the sun and artificial sources) on living matter. By its very
nature, therefore, it is a multidisciplinary science involving branches of
biology, chemistry, physics and medicine.

This book contains a selection of papers which have been chosen to
highlight recent advances in the various disciplines that make up photobiology. Although no book on photobiology can hope to be comprehensive,
we hope that this volume includes a representative sample of much of what
is new in the field. It is, however, inevitable that some areas will be
better represented than others reflecting the biases of conference org-
anisers and editors.

The chapters that constitute this book represent a collection of many of
the invited lectures and selected contributed papers presented at the 2nd
Congress of the European Society for Photobiology, held at The University
of Padua on 6-10th September 1987. Unlike many conference proceedings,
all contributions have been subjected to peer review.

The European Society for Photobiology (ESP) is now in its second year
of existence having been formally founded at the 1st Congress of the ESP
held in Grenoble in September 1986. The lectures presented at this first
meeting can be found in 'From Photobiology to Photophysics' (eds A Favre,
R Tyrrell & J Cadet, Elsevier 1987). In future it is planned to hold such
meetings every two years, the next one being in 1989 in Budapest. The most
significant advance since the last conference is that the ESP now has an
official journal (The Journal of Photochemistry and Photobiology, B; Biology).

This book is intended as a record of the scientific activities of the
ESP and it is hoped that everybody with an interest in photobiology will
find something of value to them.

CONTENTS

PHOTODYNAMIC EFFECTS ON CELLS

CLINICAL ASPECTS OF PHOTODYNAMIC CANCER TREATMENT

PHOTOTHERAPY OF HYPERBILIRUBINEMIA

PHOTOTHERAPY

LASER ANGIOPLASTY

PHOTOCHEMISTRY

UV EFFECTS

PHOTOTOTOXICOLOGY

PHOTOIMMUNOLOGY

PIGMENTATION

VISUAL TRANSDUCTION

PHOTOSENSITIVITY IN MICROORGANISMS

PHOTOSYNTHESIS

PLANT PHOTOMORPHOGENESIS

LASER TIME-RESOLVED FLUORESCENCE STUDIES FOR INVESTIGATING

CHROMOPHORE-BIOSUBSTRATE INTERACTIONS

Alessandra Andreoni

C.E.Q.S.E.- C.N.R., Milano and
Department of Biology, 2nd Faculty of Medicine
Via Sergio Pansini, 5 - 80131 Napoli (Italy)

INTRODUCTION

Fluorimetric studies utilizing continuous wave (cw) techniques performed on chromophores that bind specifically to the biomacromolecule or to the cell organelle or compartment to be examined allowed a great deal of information to be gained, in the past, on biomolecular structures and kinetics (Udenfried, 1969). Flow cytometry, which has become such an important methodology for auto mated cytology, is just an example of the extent to which the fluorescence properties of Acridines, i.e. dyes exhibiting selective affinity for nucleic acids (Albert, 1966), have been exploited to determine, for instance, DNA and RNA contents of single cells or chromosomes aberrations or cell-cycle phases (Mullaney et al., 1974). These investigations basically take advantage of the fact that the optical properties - fluorescence, in particular - of chromophores are affected unequivocally by binding to biomacromolecules or biostructures (Andreoni, 1985). Moreover, the stronger and/or more disruptive is the interaction with the substrate, the more pronounced is, usually, the effect of the binding on the chromophore emission. This property is so general that variations observed both "in vitro" and "in vivo" in the fluorescence quantum yield or in the emission spectrum of fluorescent drugs are commonly considered to be indicative of the stability of their complexation (Geacintov, 1987) or of their binding to specific cell substances or sites (Goormaghtigh et al., 1980; Kessel et al., 1985).

Time-resolved fluorescence studies, which add information on the dynamics of the chromophore-biosubstrate interaction, are relevant to all fields mentioned above as the photophysical mechanisms responsible for the fluorescence quenching or enhancing induced upon binding can be more directly determined, in general, from the knowledge of dynamic than of steady-state fluorescence parameters (Ware, 1983). In fact, the result of an ideal time-resolved fluorescence experiment, carried out on a chromophore staining a substrate with a δ - function exciting pulse and a detection apparatus with infinitely fast prompt response and sensitivity independent of the detected wavelength, would be a signal of the form

$$S(t) = \sum_i \left[S_i(0) \times \exp\left(-k_{Fi} t\right) \right] \qquad (1)$$

in which t is time and the decay rate k_{Fi} of each exponential component is the overall (radiative plus non-radiative) decay rate of a chromophore excited state. In fact, a multiexponential decay is to be expected because of the co-

1

existence, at least, of free and bound molecules (Geacintov, 1987). The latter ones, then, may be bound to different binding sites or form a variety of complexes with the substrate and thus undergo different ground- and excited-state interactions leading to different decay rates k_{Fi} for the variously bound chromophores. As a consequence, we can write the fluorescence molecular response function $i(t)$, according to Birks (1970), as

$$i(t) = \sum_i [k^0_{Fi} \sigma_i(\nu) N_{Oi} \times \exp(-k_{Fi}t)] / \sum_m [\sigma_m(\nu) N_{Om}] \qquad (2)$$

in which we assumed different values for the absorption cross-section $\sigma(\nu)$ at the excitation frequency ν for each of the variously bound and for the unbound ground-state species, which are present at the concentration values N_{Oi}, as well as different radiative decay rates k^0_{Fi}. The fluorescence quantum yield ϕ_F is, by definition, the time integral of eq.(2), that is

$$\phi_F = \sum_i [k^0_{Fi} \sigma_i(\nu) N_{Oi} / k_{Fi}] / \sum_m [\sigma_m(\nu) N_{Om}] \qquad (3)$$

It is noteworthy that, if the time-resolved decay $S(t)$ is properly measured, also its time integral provides an evaluation of ϕ_F in arbitrary units. To this end, the excitation/detection apparatus should have a prompt response faster than the shortest decay time constant $\tau_i = 1/k_{Fi}$ in the multiexponential decay of eq.(2) to ensure that both the measured decay rates are not affected by instrumental integration and all decay components are actually detected, so that the pre-exponential factors $S_i(0)$ in eq.(1) are proportional to the amplitudes at $t = 0$ of the fluorescence molecular response $i(t)$, i.e., from eq.(2),

$$S_i(0) \propto k^0_{Fi} \sigma_i(\nu) N_{Oi} / \sum_m [\sigma_m(\nu) N_{Om}] \qquad (4)$$

Comparing eq.s (2) and (4) to eq.(3) makes obvious that, if the ground-state equilibrium population N_{Oi} and the corresponding absorption cross-sections $\sigma_i(\nu)$ are determined by other methods, a measurement of the fluorescence quantum yield ϕ_F gives only a weighted mean value for the ratio of the radiative to overall decay rates in which the relative populations of excited chromophores of each species (i.e. $\sigma_i(\nu) N_{Oi} / \sum_m [\sigma_m(\nu) N_{Om}]$) appear as the weights. On the other hand, a fluorescence decay measurement allows one to evaluate the overall decay rates k_{Fi} singularly as the rate constants of the decaying exponentials in eq.(2) and, from the pre-exponential factors $S_i(0)$, the corresponding radiative decay rates k^0_{Fi}. Note that, while the values of k_{Fi} are readily obtained in absolute terms, only relative values can be extracted from eq.(4) for the radiative rates k^0_{Fi}. A measurement of ϕ_F with cw techniques is necessary to calculate their absolute values through eq.s (3) and (4).

EXPERIMENTAL SET-UP

The most recent generation of mode-locked lasers including both ion and neodimium lasers allows pulses of about 100 ps widths to be generated in different spectral regions (Svelto, 1984). The output pulses - or harmonics - can be used to synchronously pump dye lasers that offer, with relatively trouble-free operation, the advantages of tunability and of pulse shortening down to about 5 ps. If one adopts the single-photon timing technique to detect the fluorescence decay, it turns out that the power available from these excitation sources is more than ample and one can tolerate the losses associated with frequency doubling even in the case of the dye laser. Unless the lifetimes to be measured are very short, the typical repetition rates of 80 to 100 MHz of these lasers may cause repumping of the fluorescent system before the decay is complete. This can be avoided by using acousto-optic modulators to sort the output pulses (1 out of 1,000 typically) or by dumping the dye-laser cavity. The set-up developed at C.E.Q.S.E. - C.N.R. and described here, utilizes as the excitation light either the output of a mode-locked Argon-ion laser, which can be tuned at 514.5, 364 or 351 nm, or that of a synchronously--pumped Rhodamine 6G laser or its second harmonics.

The fluorescence is detected either at 90° to the exciting beam, if the samples are solutions, or, when fluorescence decays in single cells or sub-cellular organelles are to be measured, through the same microscope objective that focuses the exciting beam onto the sample. In all cases a suitable cut-off filter and a 0.5 metre Jarrell-Ash monochromator are put in front of the detector, which is a Hamamatsu R1564U-01 microchannel plate photomultiplier. When the experiment does not require measurements at different wavelengths in the chromophore emission spectrum, the monochromator is removed and, usually, the photomultiplier substituted with a single-photon avalanche diode developed by Cova et al. (1981). The two detectors give comparable time resolutions (see Cova et al., 1983 and below). Most results presented here were obtained using the monochromator/photomultiplier combination. The output of the photomulti-plier is sent to a EG&G preamplifier (model ESN VT110) and to a home-made in-tegrating amplifier to meet the requirements on the input-pulse amplitude and time duration of the Canberra 1428A constant-fraction discriminator (CFD) used in the stop channel of our time-to-amplitude converter (TAC) designed by Ber-tolaccini and Cova (1974). A multichannel analyser (MCA) is used to measure the delay distribution of the detected single photons with respect to the reference start pulses synchronous with the laser pulses. The start pulses for the TAC are obtained in different ways depending on the type of excitation sourse adopted. When the exciting pulses are either the 514.5 nm pulses emitted by the Argon-ion laser or the foundamental or the frequency-doubled output of the dye laser, the start pulses are obtained using a beam splitter on the (visible) laser beam and an ordinary fast diode (HP 4087) with fast electronic circuitry which includes a home-made CFD. The full width at half maximum (FWHM) of the prompt response of the detection apparatus, shown in Fig. 1, to the dye-laser pulse at 580 nm is 45 ps; a similar FWHM duration was obtained for the response to the frequency-doubled pulse. The 514.5 nm pulse is detected with a FWHM of 120 ps. For experiments requiring the UV excitation wavelengths available with the Argon-ion laser - namely, 351 and 364 nm - due to the lack of fast diodes with high sensitivity in this spectral range, the start pulses for the TAC are derived from the synchronous output of the acousto-optic modulator, which is operated as a passive pulse picker on the laser beam. Figure 2 shows a typical response to the laser pulse at 364 nm with a FWHM value of 160 ps which reflects both the longer time duration of the laser pulse itself and the higher jitter between synchronism and excitation pul-

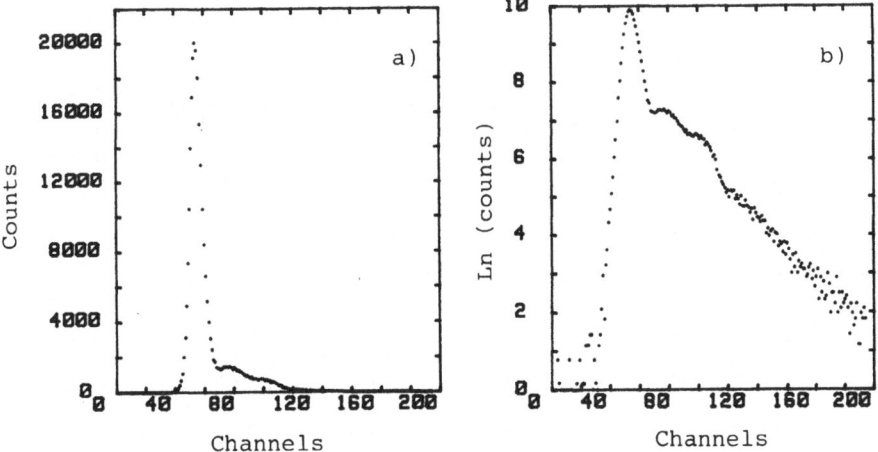

Fig. 1. Linear (a) and logarithmic (b) plots of the prompt response to the dye-laser pulse at 580 nm. Time scale: 5.08 ps/channel.

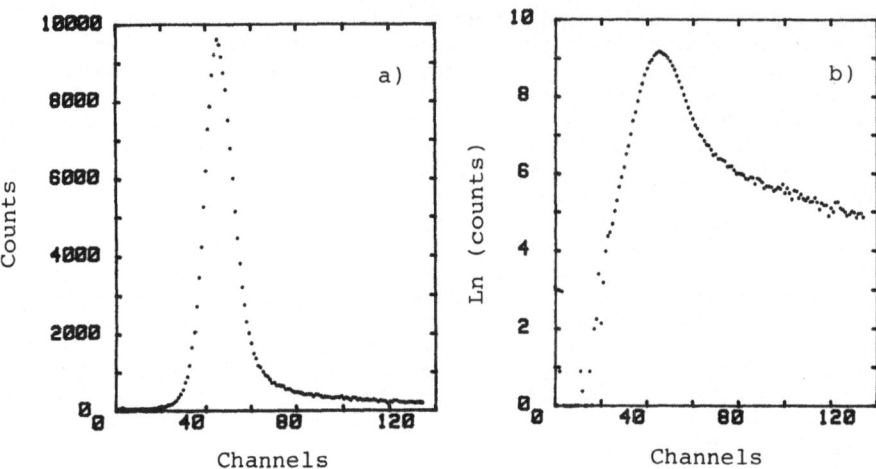

Fig. 2. Linear (a) and logarithmic (b) plots of the prompt response to the
Argon-ion laser pulse at 364 nm. Time scale: 10.79 ps/channel.

ses compared to the case of Fig. 1. In all cases the data collected by the
MCA are then transferred to a HP 9122 computer for numerical analysis. The
experimental decays are fitted to the multiexponential law of eq. (1) with i=3
at most using a non-linear least-square fitting program based on the "curfit"
method (Catterall and Duddelf, 1983). The prompt response to the excitation
pulse is not deconvoluted from the experimental fluorescence decay because, in
our experience, decay time constants faster than the FWHM of the prompt response are rather unreliable. Finally, corrections for the pulse pile-up errors,
which arise when the count rate for the fluorescence photons exceeds a few percent of the excitation repetition rate, are not applied. We prefer to accept
count rates of less than 2 % of the laser repetition rate, which is set typically at 40 KHz, and check it with a two-channel counter operated in the ratio
mode. The effective rate of collection of photons is still high enough to obtain good decay curves in a few minutes with an overall channel number allowing at least 10 channels to fall in one time constant of the fastest component
of the multiexponential decay. The chi-square value and the quality of the autocorrelation function of the weighted residuals are the parameters we use to
judge the goodness of the fit.

APPLICATIONS

     The use of sub-nanosecond laser pulses and of fast detection electronics
to study the time-resolved fluorescence of dyes interacting with biosubstrates
dates back to at least 1974, when Sacchi et al. reported the first measurement
of the excited singlet state lifetime of an extrinsic fluorophore in single
cells. Then a former version of the apparatus described was extensively used to
study the interaction of Acridines with DNA both in solutions (Andreoni et
al., 1978) and "in vitro" on histological preparations of bacteria, cells and
chromosomes (Bottiroli et al., 1979; Andreoni et al., 1979 a and b). These investigations were aimed at understanding and correlating two well established
properties of the Acridine fluorescence, namely: (i) the decrease in their
fluorescence quantum yield induced by binding to Guanine and, on the other
hand, its increase observed rather generally upon binding to DNA, which seemed
to be sensibly pronounced for DNAs relatively rich in Adenine-Thymine; (ii)
the peculiar banding pattern - i.e. regions of enhanced fluorescence over a
low-intensity background - exhibited by chromosomes stained by these dyes. The
time-resolved fluorescence study allowed us to establish the following main
points: (i) the binding to Guanine causes a shortening of the excited-state

4

lifetime down to ∿0.5 ns compared to the 4 ns lifetime for the free dye which parallels the behaviour of the quantum yield according to eq.(3); (ii) for the Acridines exhibiting increased fluorescence yield in the presence of Adenine- -Thymine pairs, we observed a corresponding lengthening of the fluorescence lifetime; (iii) the fluorescence decay in natural DNAs is non-exponential as a Förster non-radiative energy transfer mechanism becomes prominent that brings the excitation from Acridines with relatively high fluorescence yield and long lifetime (i.e. those intercalating sites formed by two Adenine-Thymine base pairs) towards those with low yield and short lifetime (i.e. the Acridines bound next to a Guanine); (iv) the banding pattern of chromosomes results from the concurrence of two facts contributing to be brightness of the fluorescen- ce bands. Actually both slower-decaying signals were always found when the la- ser was focused (with ∿0.3 μm spot diameter) onto a fluorescent band than when it was positioned elsewhere in any relatively dark remaining parts of the chro- mosome and higher peak signals, indicating higher local concentrations of dye, were detected in the bands than out of them. This led us to conclude that both the DNA base-pair composition and its accessibility to the staining molecules are important in determining the chromosome banding pattern.

Other classes of dye molecules of biomedical relevance are currently being investigated at C.E.Q.S.E. - C.N.R., namely: Porphyrins and related com- pounds that might be good candidates for tumor photochemotherapy utilizing red light (Dougherty et al., 1984), Furocoumarins that can be activated by UV-A light and cure skin deseases such as psoriasis (Wolff et al., 1974) and the well known antitumor drugs Anthracyclines (Di Marco et al., 1974). Different reasons led us to investigate the excited-state dynamics of these molecules with our apparatus. In the case of Porphyrins we found that the fluorescence lifetime was much dependent upon self-association. While most metal-free Por- phyrin monomers have a fluorescence decay time of ∿15.40 ns, their aggregates have shorter decay times and lower quantum yield values (Andreoni et al., 1982 and 1983). However, the fluorescence quantum yield of Hematoporphyrin (Hp) was found to decrease at increasing concentrations in buffer much faster if measur- ed with conventional cw techniques than if calculated as the time-integral of the decay $S(t)$ according to eq.(3). This proved that the complex self-associa- tion equilibrium of Hp in acquous solutions leads to the formation of aggrega- tes with excited-state lifetimes much below the time resolution of our appara- tus and/or with radiative decay rates $K_{F_i}^0$ becoming faster as the self-associa- tion proceeds (Andreoni et al., 1983). A similar result was recently reported by Kessel (1987) for the most aggregated chromatographic fractions of Hp "de- rivative". This effect is the actual drawback of the fluorescence imaging tech- niques that are being developed to monitor the drug distribution in tumor and normal tissues of patients injected with Porphyrins. On the other hand, as the effects of binding to serum proteins on the fluorescence decay time were found to be much less pronounced, cw fluorimetric techniques may be suitable for as- sessing the drug levels in blood or in serum.

Our studies on Furocoumarins were aimed at correlating their excited-sin- glet state lifetimes to the triplet quantum yield values obtained from flash- photolysis experiments (Bensasson et al., 1983). The mechanism generally ac- cepted to explain the photobiological activity of these drugs is photoreactivi- ty towards pyrimidine bases that leads to the formation of DNA adducts (Musajo et al., 1967). As Furocoumarins intercalate in duplex DNA, diffusional colli- sion is not required for photoreaction and thus the photoaddition might occurr, in principle, during the S_1 as well as during the T_1 lifetime. For most molec- ules examined, we have found a correspondence between short S_1 lifetimes and efficient conversion to the triplet state T_1 (Andreoni et al., 1985 and 1987). Measurements in the presence of DNA and after photobinding are in pro- gress to establish the nature of the primary steps of the photobiological activity of these drugs and to correlate their photophysical properties in the excited state with the chemical structures.

The extent and the strength of the interaction with DNA seems to play a role also in determining the antitumor activity of Anthracyclines (Di Marco et al., 1974). In particular, it has been suggested that the kinetics of the interaction rather that its equilibrium properties may correlate with the biological and pharmacological activity of these drugs (Chaires et al., 1985). Daunomycin, which exhibits high affinity towards DNA with a reported value for the apparent binding constant of 7×10^5 M^{-1}, is known to undergo fluorescence quenching upon intercalation of its chromophore moiety between DNA base pairs. To identify the mechanisms responsible for the quenching we recently carried out time-resolved fluorescence experiments (Malatesta and Andreoni, 1987) that are described in some detail here as they provide a rather exhaustive example of the potentials of this technique. Measurement were performed, using the photomultiplier/monochromator set-up, at different Daunomycin/DNA relative concentrations in TRIS buffer. The pulses at 364 nm emitted by the Argon-ion laser were chosen for the excitation because the addition of DNA causes a bathochromic shift and hypochromicity in the absorption spectrum of Daunomycin in

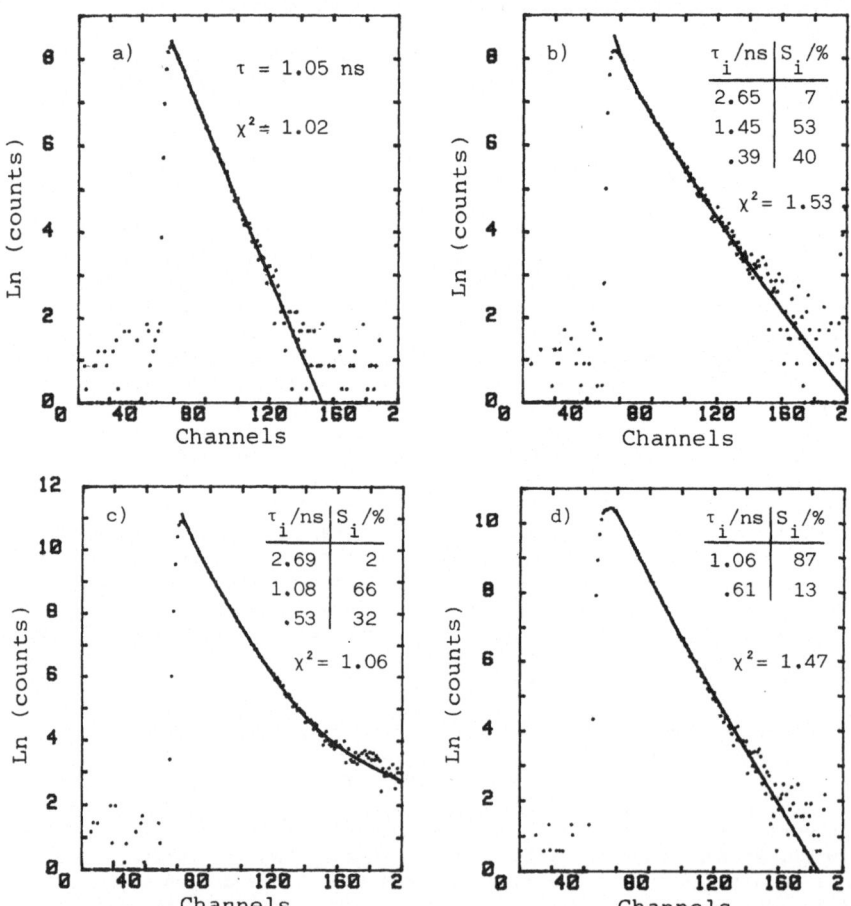

Fig. 3. Logarithmic plots of the experimental fluorescence decays (dots) of Daunomycin either in TRIS buffer (a) or bound to different amounts of calf thymus DNA: (b) r=.026; (c) r=.241; (d) r=.250. Time scale: 93.45 ps/channel. The full lines are the interpolating curves S(t) – see eq.(1) – whose time constants τ_i and pre-exponential factors S_i are listed in the inserts together with the corresponding chi-square values.

6

the wavelength range above 400 nm but leaves the spectrum unaltered at shorter wavelengths. While the fluorescence lifetime for Daunomycin in TRIS buffer was found to be 1.05 ns at all wavelengths observed in its emission spectrum (Fig. 3a), multiexponential decays were detected when calf thymus DNA was added. Figures 3 b, c and d show typical fluorescence decays for increasing values of the binding parameter r, i.e. of the ratio of occupied to total binding sites. The time constants and the relative pre-exponential factors are listed in the inserts, together with the chi-square values of the fittings. The knowledge of the equilibrium constants for the minimal binding scheme proposed for the interaction (Chaires et al., 1985) that is

$$\text{Dau + DNA} \rightleftharpoons \text{(Dau-DNA)}_{ext} \rightleftharpoons \text{(Dau-DNA)}_{int\ 1} \rightleftharpoons \text{(Dau-DNA)}_{int\ 2}$$

in which "ext" refers to the outside binding to the DNA backbone, whereas "int 1" and "int 2" are identified with two steps of the intercalation complex, allowed us to calculate the relative populations of Daunomycin engaged in each step of the binding process including the free drug for all values of r adopted in our study. As the absorption cross-section at the laser wavelength was independent of r, the correlation between pre-exponential factors and relative populations - $S_i(0)$ and N_{0i} in eq.(4) - was straightforward and the following conclusions could be drawn: (i) the final step "int 2" of the intercalation causes a shortening of the excited-state lifetime down to ~ 0.4 ns and a decrease in either the radiative lifetime or in σ at 364 nm by a factor of ~ 2 compared to the case of free Daunomycin; (ii) the externally-bound molecules "ext" contribute to the fluorescence decay with a time constant of ~ 2.67 ns. The lengthening of the excited-state lifetime may be due to the suppression of the non-radiative decay pathway operating for the free molecules represented by the excited-state proton exchange between the chromophore and the positively-changed ammonium group in the sugar moiety which is involved in the ionic bond with the DNA phosphate groups. On the basis of eq.(4) we calculate an increase by a factor of ~ 3 for the product $k_F^0 \sigma(\nu)$ for the "ext" molecules; (iii) a similar increase is found for Daunomycin in step "int 1" although the excited-state lifetime is rather similar in this case (~ 1.45 ns) to that of the free drug.

In summary the merits which, of course, are not limited to the field of applications reviewed here of pulse fluorometry compared to those of the cw techniques appear to be: (i) the straightforward measurement of the excited-state decay rates or, when the system fails to follow the simple kinetics of eq.(1), the immediate acquisition of the fluorescence decay law; (ii) the independent evaluation of steady-state and dynamic parameters of fluorescence as the former ones simply affect the initial amplitude of the signal or the pre-exponential factors of the decay law (see eq.(2)); (iii) the feasibility of investigations of phenomena occurring on a time scale as fast as 50 ps which are to be expected when short range and non diffusion-controlled interactions are possible.

ACKNOWLEDGEMENTS

The author is grateful to her previous or present colleagues G.Bottiroli, S.Cova, R.Cubeddu, G.Jori, V.Malatesta, C.A.Sacchi and T.G.Truscott whose work and ideas were the basis for this paper. The support and continuing encouragement of the director of C.E.Q.S.E. - C.N.R., Orazio Svelto, are also acknowledged. The author thanks the Hamamatsu Co. (Japan) for the loan of the photomultiplier tube which is in her opinion, when used with a suitable preamplifier, the best choice today for a single-photon detection system.

REFERENCES

Albert, A., 1966, "The Acridines", E.Arnold Ltd, London.
Andreoni, A., Sacchi, C.A., Svelto, O., Bottiroli, G., and Prenna, G., 1978,

Laser-induced fluorescence of acridine-DNA complexes, Sov. J. Quantum Electron. 8:1255.

Andreoni, A., Sacchi, C.A., and Svelto, O., 1979 a, Structural studies of biological molecules via laser-induced fluorescence: acridine-DNA complexes, in: "Chemical and Biochemical Applications of Lasers," C.Bradley-Moore, ed., Academic Press, New York.

Andreoni, A., Cova, S., Bottiroli, G., and Prenna, G., 1979 b, Fluorescence of complexes of quinacrine mustard with DNA. II: Dependence on the staining conditions, Photochem. Photobiol., 29:951.

Andreoni, A., Cubeddu, R., De Silvestri, S., Laporta, P., Jori, G., and Reddi, E., 1982, Hematoporphyrin derivative: experimental evidence for aggregated species, Chem. Phys. Letters, 88:33.

Andreoni, A., Cubeddu, R., De Silvestri, S., Jori, G., Laporta, P. and Reddi, E., 1983, Time-resolved fluorescence studies of hematoporphyrin in different solvent systems, Z. Natuforsch., 38c:83.

Andreoni, A., 1985, Time-resolved fluorescence of dyes of bio-medical relevance: influence of the environment, in: "Primary Photoprocesses in Biology and Medicine," R.V. Bensasson, G. Jori, E.J. Land and T.G. Truscott, ed.s, NATO-ASI Series A, Plenum Press, New York.

Andreoni, A., Cubeddu, R., Dall'Acqua, F., Knox, C.N., and Truscott, T.G., 1985, Fluorescence lifetimes of furocoumarins. Psoralens, Chem. Phys. Letters, 114:329.

Andreoni, A., Cubeddu, R., Knox, C.N., and Truscott, T.G., 1987, Fluorescence lifetimes of angular furocoumarins, Photochem. Photobiol., 46:169.

Bensasson, R.V., Land, E.J., and Truscott, T.G., 1983, "Flash Photolysis and Pulse Radiolysis," Pergamon Press, Oxford.

Bertolaccini, M., and Cova, S., 1974, The logic design of high precision time-to-pulse-height converters. Part 2: A converter design based on the use of integrated circuits, Nucl. Instrum. Methods, 121:557.

Birks, J.B., 1970, "Photophysics of Aromatic Molecules," Wiley-Interscience, London.

Bottiroli, G., Prenna, G., Andreoni, A., Sacchi, C.A., and Svelto, O., 1979, Fluorescence of complexes of quinacrine mustard with DNA. I: Influence of the DNA base composition on the decay time in bacteria, Photochem. Photobiol., 29:23.

Catterall, R., and Duddell, D.A., 1983, Beyond chi-square: evaluation of parametric models used in the analysis of data from fluorescence decay experiments, in: "Time-Resolved Fluorescence Spectroscopy in Biochemistry and Biology," R.B. Cundall and R.E. Dale, ed.s, NATO-ASI Series A, Plenum Press, New York.

Chaires, J.B., Dattagupta, N., and Crothers, D.M., 1985, Kinetics of the Daunomycin-DNA interaction, Biochemistry, 24:260.

Cova, S., Longoni, A., and Andreoni, A., 1981, Towards picosecond resolution with single-photon avalanche diodes, Rev. Sci. Instr., 52:408.

Cova, S., Longoni, A., Andreoni, A., and Cubeddu, R., 1983, A semiconductor detector for measuring ultraweak fluorescence decays with 70 ps FWHM resolution, IEEE J. Quantum Electron., QE-19:630.

Di Marco, A., Arcamone, F., and Zunino, F., 1974, Daunomycin (daunorubicin) and adriamycin and structural analogues: biological activity and mechanisms of action, in: "Antibiotics. Vol. 3: Mechanisms od Action of Antimicrobial and Antitumor Agents," J.W. Corcoran and F.E. Hahn, ed.s Springer-Verlag, Berlin, Heidelberg, New York.

Dougherty, T.J., Potter, W., and Weishaupt, K.R., 1984, The structure of the active component of hematoporphyrin derivative, in: "Porphyrin Localization and Treatment of Tumors," Alan R. Liss Inc., New York.

Geacintov, N.E., 1987, Principles and applications of fluorescence techniques in biophysical chemistry, Photochem. Photobiol., 45:547.

Goormaghtigh, E., Chatelain, P., Caspers, J., and Ruysschaert, J.M., 1980, Evidence of a complex between adriamycin derivatives and cardiolipin: possible role in cardiotoxicity, Biochem. Pharmac., 29:3003.

Kessel, D., Chang, C.K., and Musselman, B., 1985, Chemical, biological and biophysical studies on hematoporphyrin derivative, in: "Methods in Porphyrin Photosensitization," D. Kessel, ed., Plenum Press, New York.

Kessel, D., 1987, to be published in: "Photosensitization: Molecular, Cellular and Medical Aspects," G. Moreno, R.H. Pottier and T.G. Truscott, ed.s, NATO-ASI Series: Cell Biology, Springer-Verlag, Berlin, Heidelberg, New York.

Malatesta, V., and Andreoni, A., 1987, Dynamics of anthracyclines/DNA interaction: a laser time-resolved fluorescence study, Photochem. Photobiol., in press.

Mullaney, P.F., Steinkamp, J.A., Crissman, H.A., Crams, L.S., and Holm, D.M., 1974, Laser flow microphotometers for rapid analysis and sorting of individual mammalian cells, in: "Laser Applications in Medicine and Biology," Vol. 2, M.L. Wolbarsht, ed., Plenum Press, New York.

Musajo, L., Bordin, F., and Bevilacqua, R., 1967, Photoreactions at 3655 Å linking the 3-4 double bond of furocoumarins with pyrimidine bases, Photochem. Photobiol., 6:251.

Sacchi, C.A., Svelto, O., and Prenna, G., 1974, Pulsed tunable lasers in cytofluorometry, Histochem. J., 6:927.

Svelto, O., 1984, "Principles of Lasers," Plenum Press, New York.

Udenfried, S., 1969, "Fluorescence Assays in Biology and Medicine," Vol. II, Academic Press, LOndon, New York.

Ware, W.R., 1983, Kinetics of fluorescence decay: an overview, in: "Time-Resolved Fluorescence Spectroscopy in Biochemistry and Biology," R.B. Cundall and R.E. Dale, ed.s, NATO-ASI Series A, Plenum Press, New York.

Wolff, K., Fitzpatrick, T.B., Parrish, J.A., Gschnait, F., Cilchrest, B., Honigsmann, H., Pathak, M.A., and Tannenbaum, L., 1976, Photochemotherapy with orally administered methoxalen, Arch. Dermatol., 112:949.

APPLICATION OF POLARIZED LUMINESCENCE IN BIOLOGY AND MEDICINE

László Szalay, Gábor Laczkó and Péter Maróti

Department of Biophysics
József Attila University
Szeged, Hungary

Our purpose is to provide a brief report on the recent developments and trends in the field and to illustrate their manysided usefulness in diverse topics by many examples, mainly from the literature of the past two years. More details can be found in the books of Steiner (1983), Bayley and Dale (1985) and Lakowicz (1983).

EXPERIMENTAL METHODS

General Methods

The degree of steady-state polarization is relatively easy to measure and provides information on the spatial and (relative) angular distributions of the fluorophores. It is therefore adequate for characterizing the static features (structure, order, etc.) of a (biological) system. Rotational diffusion might be the most important depolarization process in experiments relating to biology, as it reflects the dynamics of the systems. Although basically appropriate for determination of rotational diffusion, the steady-state method has only limited value in investigations of dynamics, because it needs a knowledge of the fluorescence decay and a priori assumptions on the nature of the rotational diffusion. If these are not known, only time-resolved methods can help.

Lifetime-resolved anisotropy measurement is an intermediate possibility between the steady-state and the time-resolved methods (Lakowicz, 1983). It yields information on the time-dependent decay of anisotropy simply from determination of the steady-state anisotropies, r_s, while the lifetime of the fluorophore is varied. As r_s is the time average of $r(t)$ weighted by the fluorescence intensity decay function, if the latter is varied in a known way, then $r(t)$ can be reconstructed from the lifetime-dependent values of r_s. This is theoretically possible for any $r(t)$ decay, but the experimental uncertainities impose a limit on the number of parameters in the anisotropy decay function which can be recovered.

The maximum information on anisotropy decays is provided by time-resolved methods:
Time-domain (pulse) measurements have been greatly improved in recent years. The most widely used variant of this method is time-correlated single photon counting (O'Connor and Phillips, 1984). The use of picosecond pulses

of mode-locked lasers with repetition rates in the MHz region, time-to-amplitude converters in "reverse" mode and very fast microchannel plate photomultipliers has made the time-resolution of single photon counting much better and its traditionally slow data collection faster. An important recent advance in data analysis is the global approach (Knutson et al., 1983), which is especially relevant in the case of biological systems with many parameters to be recovered. Instead of comparing parameters of separate experiments, global analysis fits all experimental curves simultaneously. This greatly decreases the covariance between the parameters, thereby allowing the description of complex systems.

The development of the alternative technique, the frequency-domain method is perhaps even more impressive. It applies intensity-modulated excitation and measures the phase shift and the demodulation of the fluorescence signal caused by the response of the fluorophore. Anisotropy decays are determined from the phase angle difference and the ratio of the modulated amplitudes of the emission components polarized perpendicular and parallel relative to the vertically polarized excitation. The true frequency-domain instruments with (quasi-)continuously variable modulation frequency are able to resolve any type of intensity or anisotropy decay. The first practical frequency-domain instruments used cross-correlation detection, resulting in a narrow detection bandwidth, improving the signal-to-noise ratio (Gratton and Limkeman, 1983; Lakowicz and Maliwal, 1985). The harmonic content of the pulse train of picosecond mode-locked lasers is used as modulated excitation (Alcala et al., 1985). The best time-resolution with negligible color error has been achieved with microchannel plate photomultiplier applying external cross-correlation (Lakowicz et al., 1986). The maximum usable bandwidth is as high as 2 GHz, and lifetimes as short as 10 ps can be resolved. The data analysis is usually carried out by the method of nonlinear least squares (Lakowicz et al., 1984), and global analysis is applied in this field, too. As a result of this fast development, the frequency-domain method has become equivalent to the time-resolved one.

Special Methods

Some recent studies employing special methods related to polarized fluorescence, but not fitting into the classification above, are worth mentioning.

The laser T-jump technique with polarized fluorescence detection used by Genz and Holzwarth (1986) is a very potent method. The difference of the fluorescence components polarized parallel and perpendicular with respect to the vertically polarized continuous excitation light is detected after an iodine laser-generated T-jump. The T-jump causes a sudden increase in the cone angle of the wobbling motion of the probe, the relaxation of which is detected by the difference signal. The main phase transition in dipalmytoylphosphatidylcholine vesicles showed three relaxation signals in the microsecond and millisecond time ranges, supporting the authors' model of the main phase transition in these vesicles.

Fluorescence correlation spectroscopy (FCS) was applied by Kask et al. (1987) for measuring rotational correlation times in the nanosecond time range. This method has proved to be very useful in other fields (e.g. translational diffusion) in the time range from milliseconds to seconds. With polarized excitation and detection through a polarizer, FCS can be used for studying the rotational diffusion of the probe. Although the fluorescence dynamic depolarization method is inherently superior to FCS for determining rotational correlation times in the time range comparable to or shorter than the fluorescence lifetime (nanosecond range), FCS may be useful on a (much) longer time scale. The potential of FCS in this field has not been clarified yet. Its weakest side is the low signal-to-noise ratio. The classical time-resolved fluorescence depolarization method might well be superior even in the millisecond and longer time ranges if long-lived chromophores (also

phosphorence, if applicable) are applied. Another alternative is the fluorescence recovery after photobleaching technique with polarized bleaching
and detection.

Fluorescence-activated cell sorters are often employed in flow cytometry, using polarized fluorescence as a sorting parameter. Commercial
sorters can easily be modified to permit polarization measurements on single
cells (Dimitropoulos, 1986).

APPLICATIONS

The possibilities of applications are very wide and colorfully manysided. We enumerate some examples in headline form: membrane fluidity (also
in single cells), organization of membranes, orientation of membrane constituents, rotational mobility and lateral diffusion of membrane elements,
binding sites of molecules on cell membranes, cell division phenomena, interaction of cell constituens, behavior of cancer cells, immunoassay, etc.
Only two topics, dynamics of biological membranes and protein dynamics, will
be discussed in some detail as, in our opinion, the present, rapidly developing state of these fields makes their overview especially worthwhile.

Dynamics of Biological Membranes

The dynamics of the acyl side-chain region of membranes is frequently
investigated by measuring the fluorescence anisotropy of membrane-bound
fluorophores. This can reveal the rotational diffusion rate of the probe,
which, in turn, is indicative of the rates at which the acyl side-chains
undergo displacements. This "fluidity" of the lipid bilayer influences a
variety of membrane functions. The fluidity is highly dependent upon the
chemical composition of the bilayer, mostly on the cholesterol and sphingomyelin contents and the degree of saturation of the phospholipid acyl chains.
It also depends on the temperature and phase state.

The steady-state fluorescence polarization technique has been applied
for more than a decade to determine fluidity or to detect phase transitions.
In the early literature, quantitative data were obtained by comparing the
polarization in membranes with that in isotropic reference oils. Using the
Perrin equation, "microviscosity" values were calculated. However, time-
-resolved studies demonstrated that in membranes the anisotropies of rodlike
probes (e.g. 1,6-diphenyl-1,3,5-hexatriene, DPH) did not decay to zero, but
to a nonzero value, r_∞ . The rotation is thus hindered, distinctly different
from the motions in isotropic reference oils. The simple hydrodynamic analogy
cannot be applied. In spite of this, microviscosity is still calculated in
some recent papers (e.g. Kapitulnik et al., 1986). Instead of the microviscosity, we have now (at least) two independent quantities characterizing
the dynamics of lipid bilayers: the lipid order parameter (which can be determined from r_∞) and the rotational relaxation rate of the probe when it is
rotating unhindered between the hindrances.

These can easily be measured with time-resolved techniques, but it is
highly questionable to find them with the popular steady-state measurements.
Those who use the steady-state method usually calculate the lipid order
parameter or r_∞ based on semiempirical relations between the measured
steady-state anisotropy, r_s, and the limiting anisotropy, r_∞ . Van der Meer
et al. (1986) presented one of the most elaborate efforts to determine r_∞
from r_s. An extended Perrin equation was derived, stated to be applicable to
the restricted rotational diffusion, relating r_∞ to r_s. This equation contains a parameter, m, which expresses the difference between rotational diffusion in membranes and that in isotropic reference oils having the same r_s.
The value of m is calculated by a nonlinear least squares procedure so that
on the r_∞/r_0 vs. r_s/r_0 graph the theoretical curve fits best to the experi-

mental points taken from the literature (r_0 is the anisotropy in the absence of rotational diffusion). The equation is claimed to be in agreement with the literature data for a variety of artificial and biological membranes labelled with various probes, and to allow for an unambiguous interpretation of r_s data.

In our opinion, however, steady-state measurements are in general not suitable for the determination of r_∞ in unknown samples. Since r_s is the time-average of $r(t)$ weighted by the fluorescence intensity decay function, $I(t)$, the relation between r_s, r_0 (steady-state parameters) and r_∞ necessarily depends on the $I(t)$ function, even if the nature of the $r(t)$ function is a priori known. $I(t)$ frequently changes and can only be measured by time-resolved methods, and thus the physical information content of the steady-state experiments is simply not enough to determine r_∞. For the same reason, in the authors' equation expressing r_∞ with r_0, r_s and m, the values of r_∞ and r_0 do not depend on $I(t)$, but r_s does, and therefore m must also be dependent on $I(t)$. This was mentioned by the authors themselves, but was treated only as an error in m. They also stated that cholesterol and fatty acyl unsaturation affected the value of m. But if m is not an universal constant , then the estimation of r_∞ is always uncertain. Even in the authors' figure showing the r_∞/r_0 vs. r_s/r_0 function, significant differences exist between the best-fitting curve and the experimental points, amounting to 100% and 20% if r_s/r_0 = 0.2 and 0.5, respectively. The relative success of these semiempirical formulas follows from the observation that, for a given r_0, r_s is usually mainly determined by the hindrance and much less by the rotational diffusion rate. Striking examples illustrate how much care should be taken when interpreting steady-state data, and also show the possibilities of time-resolved methods.

Differential polarized phase fluorimetry studies of DPH showed that vanadyl ions increased the limiting anisotropy in crude plasma membranes of hamster, but the rotational relaxation time of DPH did not change (Amler et al., 1986). In contrast to many fluorescent probes, dehydroergosterol, a fluorescent cholesterol mimic, shows a decrease in r_s when the phospholipid matrix goes from the liquid-crystalline to the gel state (Chong and Thompson, 1986). The semiempirical laws correlating r_∞ and r_s would predict a decrease of the lipid order. Just the opposite is true. The decrease in anisotropy was found to be the result of a large increase in the fluorescence lifetime with decreasing temperature. In addition, it was also shown (Schroeder et al., 1987), that, with the time-resolved technique, dehydroergosterol was an excellent probe for examining the motions of cholesterol.

In conclusion, steady-state polarization measurements can be used in studying membrane dynamics, but great care must be exercised. If the fluorescence kinetic properties of the system are unknown, then we would prefer the simple use of the anisotropy values for characterizing the system instead of the order parameter, as the latter just cannot be determined correctly.

Polarized fluorescence techniques have yielded new results in membrane dynamics in many fields. Faucon and Lakowicz (1987) found that the hindered rotator model was not always adequate to account for the anisotropy decays of DPH. Frequency-domain fluorimetry allowed a distinction between the hindered and the two (or three) correlation time models.
There have been attempts to quantitatively correlate the composition of the membranes with the membrane fluidity. Van Blitterswijk et al. (1987) presented an expression describing the contributions of cholesterol, sphingomyelin and saturation of fatty acyl groups in various phospholipids to the order of membranes.
Popp-Snijders et al. (1986) suggested that the fluidity of the erythrocyte membrane was maintained by a change in the phospholipid class distribution, even when the fatty acid composition of the membrane was forced to change.

Most of the research has been devoted to the effects of treatment on fluidity, and the influence of fluidity on membrane functions.

Sire et al. (1986) found that mitochondria isolated from either galactosamine- or clofibrate- treated rats showed a decrease in lipid structural order, which was more drastic when both drugs were administered. These data could explain the altered kinetic properties of palmitoylcarnitine transferase I.

Charged anesthetics altered the order of the two monolayers of LM fibroblast plasma membranes selectively (Sweet et al., 1987). This helped in assigning a break point in the Arrhenius plots of DPH fluorescence near 31 ^{o}C to the inner monolayer of the LM plasma membrane.

The results of Kapitulnik et al. (1986) suggest that endogenous glucocorticoids play a major role in the perinatal fluidization of rat liver microsomal membrane.

Dolichol selectively fluidized and rigidified membrane areas detected with DPH and trans-parinaric acid, respectively (Wood et al., 1986). This demonstrated additional complexity in the biological membranes not reported in artificial systems.

The effect of blood-clotting factor Va on phospholipid vesicles indicated that the acidic phospholipid-dependent binding of factor Va to mixed vesicles had on ordering effect on the acyl chains in the outer layer, but left the bulk of the phospholipids unaltered (van de Waart et al., 1987).

Chester et al. (1986) suggested the existence of microheterogeneous regions in the bilayer which, when sufficiently modified, could affect enzyme functions, cell physiology, homeostasis and, ultimately, cell viability.

Effects of ethinyl estradiol on intestinal membrane structure and function in rabbit was associated with a significant decrease in membrane fluidity in the ileal microvillus membrane (Schwarz et al., 1986). However, no difference from the control was observed in microvillus membrane alkaline phosphatase activity following the estrogen treatment.

Barley plants having gradually diminished linolenic and increased linoleic acid contents were grown by application of a pyridazinone compound, SAN 9785 (Laskay and Lehoczki, 1986). The lowering effect of this treatment on the fluidity of the photosynthetic membrane was correlated with the decreasing photosynthetic activity.

The data presented by Brasitus et al. (1986) demonstrate that the fluidity of rat colonic brush-border membrane vesicles can influence Na^{+} - H^{+} exchange and osmotic water flow across these vesicles.

Protein Dynamics

Proteins are dynamic structures, and their functional properties depend upon these motions. For example, transient packing defects due to atomic motions play an essential role in the penetration of oxygen to the heme binding site in myoglobin and hemoglobin. Most of the efforts nowadays are concentrated on the unexplored field of ultrafast segmental motions in proteins. These motions have been predicted by molecular dynamics calculations. Simulations by Ichiye and Karplus (1983) showed that some of the tryptophans in lysozyme were highly mobile on the picosecond time scale, with a large anisotropy decay. Levy and Szabó (1982) predicted a fast, about 2 ps long decay in the tyrosine fluorescence anisotropy decay in pancreatic trypsin inhibitor. Evidence for these ultrafast motions has been found in fluorescence, NMR, IR, Raman and flash photolysis data. The fluorescence depolarization technique has the distinct advantage over NMR that the measured quantity depends directly on the decay of the time-correlation function. In the fluorescence anisotropy decay experiments, almost exclusively the intrinsic fluorophores of proteins, usually tryptophan and tyrosine, are used. Due to the complex motion of the probe and the frequent presence of more than one fluorescent amino acid residue, the anisotropy decay is very complex. The time range is also at the limit of the present time-resolved fluorimetry, and thus significant results can mostly be expected from the most sophisticated time- or frequency-domain methods. The fastest rotational correlation times provided by time-resolved fluorimetry are in the range of 10 picoseconds. These rotations were attributed to the local motions of proteins. However, the experimental values are about an order of magnitude slower than those predicted by the molecular dynamics calculations. The limited time-resolution of even the fastest instruments (about 10 ps) might well be a major factor in this discrepancy. This might be checked by comparing the value of r_0 measured in viscous solution with that obtained in proteins, but still there are doubts concerning the results.

Experiments relating to the relatively slow protein motions are listed first.

Qualitative conclusions can be drawn from steady-state measurements. Wasylewski et al. (1986) studied the role of Ca^{2+} ion in the conformational changes of staphylococcal metalloproteinase enzyme containing two tryptophans. Assuming that one of the tryptophans is totally exposed to iodine quenching, it was found that the mobility of the buried tryptophan was restricted, and removal of calcium increased its mobility.

The lifetime-resolved investigations of Sanyal et al. (1987) on human luteinizing hormone (hlH) could detect even the presence of fast, segmental motions in addition to the rotations of the whole protein. It was also shown that intersubunit motion need not be invoked for intact hlH, as had been suggested earlier.

The interaction of the fluorescent inhibitor N-dansyl-1,8-diaminooctane with the nucleotide site of pyridoxal kinase enzyme was examined by single photon counting (Kwok et al., 1986). The long rotational correlation time of the inhibitor trapped by the enzyme (42 ns) indicates that its mobility is restricted. The findings are consistent with the model that the inhibitor is immobilized by amino acids located at the catalytic site.

Hönes et al. (1986) investigated conformational changes of the substrate binding site of pig mitochondrial malate dehydrogenase enzyme with the frequency-domain technique. A substrate-analog dye, 8-hydroxypyrene-1,3,6-trisulfonate, was shown to be immobilized by the substrate-binding site, addition of NADH strengthened a hydrogen bond and further restricted the mobility, while $NAD^{\oplus}$ had the opposite effect.

In picosecond segmental protein dynamics, only the most advanced time-resolved techniques can provide significant results.

The mobilities of tryptophans in three homologous azurins were studied by Petrich et al. (1987) with single photon counting. Azurin Pae contains a single, buried tryptophyl residue in position W-48, azurin Afe also has a single tryptophyl residue, but on the surface, exposed to the solvent, in position W-118, and azurin Ade has two tryptophyl residues, buried W-48 and exposed W-118. The interiors of azurins Pae and Ade were not mobile enough to allow motion of the indole ring on the nanosecond time scale (single exponential decay with a rotational correlation time of about 5 ns), while the exposed tryptophans in azurins Afe and Ade showed considerable mobility on a few hundred picosecond time scale corresponding to an angular motion of 30°. The experiments may have failed to resolve a short component in Pae. W-118 could probably rotate about its χ_2 bonds without requiring motions of other residues.

The 2 GHz frequency-domain machine of Lakowicz et al. (1987) was shown to resolve rotational correlation times as short as 15 ps and the data suggested that even better resolution was possible. Tyrosine anisotropy decay experiments indicated segmental motions in different peptides and proteins in addition to their global rotations. Enkephalin displayed correlation times of 70 ps and 250 ps. In oxytocin and bovine pancreatic trypsin inhibitor (BPTI), the extended frequency range of the experiments made it possible to detect the shoulder above 300 MHz on the distribution of the differential phase angles as a function of the modulation frequency, which indicates the presence of picosecond rotations. Decay times of 30 ps and 450 ps were recovered for oxytocin, while BPTI also showed a short (40 ps) and a long (2.2 ns) component. The slower rotational rate in any of the studied molecules was approximately equal to the expected value for the rotation of the whole molecule based on hydrodynamic calculations. The fast components were attributed to the local torsional motions of the tyrosyl residues.

Nordlund et al. (1986) applied a signal averaging streak camera-optical multichannel analyzer system to resolve the anisotropy decay of lima bean trypsin inhibitor (LBTI) excited by the fourth harmonic of a Q-switched, mode-locked, single-pulse selected Nd:YAG laser. LBTI has a single tyrosine at position 69 (Tyr-69). This is an advantage over BPTI, as no energy transfer can take place. The anisotropy decay of LBTI has two components, with rotational correlation times of 40 ps and ≥ 3 ns. The slow process is consistent with the rotation of the whole protein, while the fast component, comparable to that of free tyrosine in aqueous solution, demonstrates the considerable freedom of tyrosine to move independently. As a consequence, the interior portion of LBTI around Tyr-69, consisting of a rigid, immobile backbone, is embedded in fluid, mobile amino acid side-chains.

These experiments provide a challenge for molecular dynamics calculations, and comparison of the experimental data with theory is now possible. The time-resolution of the fluorescence technique and its ability to resolve complex decays should be further improved to permit a more detailed description of protein motions.

REFERENCES

Alcala, J.R., Gratton, E., and Jameson, D.M., 1985, A multifrequency phase fluorometer using the harmonic content of a mode-locked laser, Anal. Instrum., 14:225.

Amler, E., Teisinger, J., Svobodová, J., and Vyskocil, F., 1986, Vanadyl ions increase the order parameter of plasma membranes without changing the rotational relaxation time, Biochim. Biophys. Acta, 863:18.

Bayley, P.M., and Dale, R.E., 1985, "Spectroscopy and the dynamics of molecular biological systems," Academic Press, London.

Brasitus, T.A., Dudeja, P.K., Worman, H.J., and Foster, E.S., 1986, The lipid fluidity of rat colonic brush-border membrane vesicles modulates Na^+-H^+ exchange and osmotic water permeability, Biochim. Biophys. Acta, 855:16.

Chester, D.W., Tourtellotte, M.E., Melchior, D.L., and Romano, A.H., 1986, The influence of saturated fatty acid modulation of bilayer physical state on cellular and membrane structure and function, Biochim. Biophys. Acta, 960:383.

Chong, L.-G., and Thompson, T.E., 1986, Depolarization of dehydroergosterol in phospholipid bilayers, Biochim. Biophys. Acta, 863:53.

Dimitropoulos, K., Rolland, J.M., and Nairn, R.C., 1986, Flow cytofluorimetry of fluorescein fluorescence polarization to assay lymphocyte activation, Biochem. Biophys. Res. Commun., 136:1021.

Faucon, J.F., and Lakowicz, J.R., 1987, Anisotropy decay of dephenylhexatriene in melittin-phospholipid complexes by multifrequency phase-modulation fluorometry, Arch.Biochem. Biophys., 252:245.

Genz, A., and Holzwarth, J.F., 1986, Dynamic fluorescence measurements on the main phase transition of dipalmytoylphosphatidylcholine vesicles, Eur. Biophys. J., 13:323.

Gratton, E., and Limkeman, M., 1983, A continuously variable frequency cross-correlation phase fluorometer with picosecond resolution, Biophys. J., 44:315.

Hönes, G., Hönes, J., and Hauser, M., 1986, Studies of enzyme-ligand complexes using dynamic fluorescence anisotropy I-II, Biol. Chem. Hoppe-Seyler, 367:95.

Ichiye, T., and Karplus, M., 1983, Fluorescence depolarization of tryptophan residues in proteins: a molecular dynamics study, Biochemistry, 22:2884.

Kapitulnik, J., Weil, E., and Rabinowitz, R., 1986, Glucocorticoids increase the fluidity of the fetal-rat liver microsomal membrane in the peri-natal period, Biochem. J., 239:41.

Kask, P., Piksarv, P., Mets, U., Pooga, M., and Lippmaa, E., 1987, Fluores-cence correlation spectroscopy in the nanosecond time range: rota-tional diffusion of bovine carbanic anhydrase B, Eur. Biophys. J., 14:257.

Knutson, J.R., Beechem, J.M., and Brand, L., 1983, Simultaneous analysis of multiple fluorescence decay curves: a global approach, Chem. Phys. Lett., 102:501.

Kwok, F., Kerry, J.A., and Churchich, J.E., 1986, Sheep brain pyridoxal kinase: fluorescence spectroscopy of the dimeric enzyme, Biochim. Biophys. Acta, 874:167.

Lakowicz, J.R., 1983, "Principles of fluorescence spectroscopy," Plenum, New York.

Lakowicz, J.R., Laczkó, G., Cherek, H., Gratton, E., and Limkeman, M., 1984, Analysis of fluorescence decay kinetics from variable-frequency phase shift and modulation data, Biophys. J., 46:463.

Lakowicz, J.R., and Maliwal, B., 1985, Construction and performance of a variable-frequency phase-modulation fluorometer, Biophys. Chem., 21:61.

Lakowicz, J.R., Laczkó, G., and Gryczynski, I., 1986, 2-GHz frequency--domain fluorometer, Rev. Sci. Instrum., 57:2499.

Lakowicz, J.R., Laczkó, G., and Gryczynski, I., 1987, Picosecond resolution of tyrosine fluorescence and anisotropy decays by 2-GHz frequency--domain fluorometry, Biochemistry, 26:82.

Laskay, G., and Lehoczki, E., 1986, Correlation between linolenic-acid defi-ciency in chloroplast membrane lipids and decreasing photosynthetic activity in barley, Biochim. Biophys. Acta, 849:77.

Levy, R.M., and Szabó, A., 1982, Initial fluorescence depolarization of tyrosine in proteins, J. Am. Chem. Soc., 104:2073.

Nordlund, T.M., Liu, X-Y., and Sommer, J.H., 1986, Fluorescence polarization decay of tyrosine in lima bean trypsin inhibitor, Proc. Natl. Acad. Sci. USA, 83:8977.

O'Connor, D.V., and Phillips, D., 1984, "Time-correlated single photon counting," Academic, London.

Petrich, J.W., Longworth, J.W., and Fleming, G.R., 1987, Internal motion and electron transfer in proteins: a picosecond fluorescence study of three homologous azurins, Biochemistry, 26:2711.

Popp-Snijders, C., Schouten, J.A., van Blitterswijk, W.J., and van der Veen, E.A., 1986, Changes in membrane lipid composition of human erythro-cytes after dietary supplementation of (n-3) polyunsaturated fatty acids. Maintenance of membrane fluidity. Biochim. Biophys. Acta, 854:31.

Sanyal, G., Charlesworth, M.C., Ryan, R.J., and Prendergast, F.G., 1987, Tryptophan fluorescence studies of subunit interaction and rotational dynamics of human luteinizing hormone, Biochemistry, 26:1860.

Schroeder, F., Barenholz, Y., Gratton, E., and Thompson, T.E., 1987, A flu-orescence study of dehydroergosterol in phosphatidylcholine bilayer vesicles, Biochemistry, 26:2441.

Schwarz, S.M., Watkins, J.B., Ling, S.C., Fayer, J.C., and Mone, M., 1986, Effects of ethinyl estradiol on intestinal membrane structure and function in the rabbit, Biochim. Biophys. Acta, 860:411.

Sire, O., Mangeney, M., Montagne, J., Bereziat, G., and Nordmann, J., 1986, Changes of fatty acid composition of phospholipids and lipid structur-al order in rat liver mitochondrial membrane subsequent to galactos-amine intoxication. Effect of clofibrate, Biochim. Biophys. Acta, 860:75.

Steiner, R.F., ed., 1983, "Excited states of biopolymers," Plenum, New York.

Sweet, W.D., Wood, W.G., and Schroeder, F., 1987, Charged anesthetics selectively alter plasma membrane order, Biochemistry, 26:2828.

Van Blitterswijk, W.J., van der Meer, B.W., and Hilkmann, H., 1987, Quantitative contributions of cholesterol and the individual classes of phospholipids and their degree of fatty acyl (un)saturation to membrane fluidity measured by fluorescence polarization, Biochemistry, 26:1746.

Van der Meer, B.W., van Hoeven, R.P., and van Blitterswijk, W.J., 1986, Steady-state fluorescence polarization data in membranes. Resolution into physical parameters by an extended Perrin equation for restricted rotation of fluorophores, Biochim. Biophys. Acta, 854:38.

Van de Waart, P., Visser, A.J.W.G., Hemker, H.C., and Lindhout, T., 1987, The effect of factor Va on lipid dynamics in mixed phospholipid vesicles as detected by steady-state and time-resolved fluorescence depolarization of diphenylhexatriene, Eur. J. Biochem., 164:337.

Wasylewski, Z., Stryjewski, W., Wasniowska, A., Potempa, J., and Baran, K., 1986, Effect of calcium binding on conformational changes of staphylococcal metalloproteinase measured by means of intrinsic protein fluorescence, Biochim. Biophys. Acta, 871:177.

Wood, W.G., Gorka, C., Williamson, L.S., Strong, R., Sun, A.Y., Sun, G.Y., and Schroeder, F., 1986, Dolichol alters dynamic and static properties of mouse synaptosomal plasma membranes, FEBS Lett., 205:25.

TIME-RESOLVED FLUORIMETRY USING SYNCHROTRON RADIATION AND MAXIMUM ENTROPY METHOD OF ANALYSIS

Jean-Claude Brochon* and Alastarir Livesey[+]

* LURE CNRS Bât.209D
  Université Paris-Sud 91405 Orsay Cedex, France

[+] M.R.C. Laboratory of Molecular Biology
  Hills Road and DAMTP Silver Street Cambridge U.K.

INTRODUCTION

Unpolarised pulse fluorescence is being used increasingly in biophysics to study the fluctuations and/or heterogeneity in the three-dimensional structure of macromolecules. Polarised pulse fluorescence also gives a direct physical measurement of the Brownian rotational constants in solution of the whole molecule, deformations and flexions of the structure and local motions of side-chains around their chemical bonds. Both these aspects can be related to biological activity.

Taking advantages of the properties of synchrotron radiation pulse and of a new powerfull method of data analysis (1,2) the pulse fluorimetry displays high sensitivity and gives for the first time an uncorrelated description of emission kinetics and molecular dynamics.

EXCITATION FLASH OF SYNCHROTRON RADIATION

Previously excitation light pulses at various wavelengths were provided by gas filled discharge lamps or lasers using different dyes. Synchrotron radiation (S.R.) is the light emitted by a charged particle moving at a relativistic energy in a circular path. This light displays usefull properties :

- The particles are gathered in bunches so the light is pulsed.

- S.R has continous wavelength spectrum from X-rays to IR

- The flash profile is independent of wavelength (3)

- The emission rate is in the MHz domain

- The intensity of the light is directly related to the number of particles stored in the machine which is smoothly decreasing with a lifetime of several hours. Thus there is no rapid intensity fluctuation.

- The width of S.R. pulse depends on machine parameters. Recently, pulses of few tens of picoseconds have been currently obtained (4,5), although with the stable broad flash (FWHM = 1.6 ns) from ACO we can measured a lifetime as short as 75 picoseconds (2).

MAXIMUM ENTROPY ANALYSIS OF PULSE-FLUORIMETRY

With a vertically polarised excitation the parallel I// and perpendicular I⊥ components of the fluorescence intensity at time t after the start of the excitation flash are

$$I_{//}(t) = \frac{1}{3} E_\lambda(t) * [\int_o^\infty \int_o^\infty \int_{-0.2}^{0.4} \gamma(\tau,\theta,A) e^{-t/\tau} (1+2Ae^{-t/\theta}) \, d\tau d\theta dA] \quad (1)$$

$$I_\perp(t) = \frac{1}{3} E_\lambda(t) * [\int_o^\infty \int_o^\infty \int_{-0.2}^{0.4} \gamma(\tau,\theta,A) e^{-t/\tau} (1 - Ae^{-t/\theta}) \, d\tau d\theta dA] \quad (2)$$

where E(t) is the temporal shape of the flash and $\gamma(\tau,\theta,A)$ are the number of fluorophores with fluorescence decay τ, rotation time θ, and initial amplitude of anisotropy A (related to angle between absorption and emission dipoles).   * denotes a convolution with time.

Since the measured data set (I// (t) & I⊥ (t)) is noisy and finite in extent there is strictly an infinite set of $\gamma(\tau,\theta,A)$ which agree with our data within the error bars. We then choose that distribution which maximises the Skilling-Jaynes entropy (6).

$$S= \int_o^\infty \int_o^\infty \int_{-0.2}^{0.4} \gamma(\tau,\theta,A) - m(\tau,\theta,A) - \gamma(\tau,\theta,A) \log \frac{\gamma(\tau,\theta,A)}{m(\tau,\theta,A)} \, d\tau d\theta dA \quad (3)$$

where $m(\tau,\theta,A)$ is a model (see below).

We have shown that this choice has many compelling features : it introduces the minimum correlations between τ, θ and A. The recovered $\gamma(\tau,\theta,A)$ is smooth, positive, unique and robust to noise. In order to ensure our recovered distribution agrees with our data, we maximise S subject to the following constraint

$$C = \sum_{k=1}^{M} \frac{(I_k{}^{calc} - I_k{}^{obs})^2}{M \; \sigma_k{}^2} \qquad < 1.0 \qquad (4)$$

where I^{calc}, I^{obs} are the kth calculated and observed intensities, $\sigma^2{}_k$ is the variance of the kth point, and M is the total number of observations. The summation is over both the parallel and perpendicular components. Our programme automatically corrects for the amount of scattered light present in the data and any background.

The user may sometimes have some prior model (2,7) for the distribution but if not, it can be shown that the correct model is flat in log τ, log θ and cos λ space where

$$A = \frac{3 \cos^2 \lambda - 1}{5}$$

TOTAL FLUORESCENCE

If we are only interested in the fluorescence decay constants (τ) we can considerably simplify the data by summing the parallel and (twice the) perpendicular components.

$$T(t) = I//(t) + 2 \; I\perp(t) = E * \int_0^\infty \alpha(\tau) \; \exp(-t/\tau) \; d\tau \qquad (5)$$

where $\alpha(\tau)$ is the distribution of fluorescence decays given by

$$\alpha(\tau) = \int_0^\infty \int_{-0.2}^{0.4} \gamma(\tau,\theta,A) \; d\theta dA \qquad (6)$$

Note the variance of T(t) is : $\sigma_T{}^2 = \sigma_{//}{}^2 + 4\sigma_\perp{}^2$

T(t) can also be measured in one experiment by setting the polarizer at the "magic" angle of 54.75°.

The entropy is also independent of θ and A and is given by

$$S = \int \alpha(\tau) - m(\tau) - \alpha(\tau) \log\frac{\alpha(\tau)}{m(\tau)} \; d\tau \qquad (7)$$

where the model m is flat in log τ space if the user has no knowledge of the τ values expected.

APPLICATIONS OF M.E.M. ANALYSIS

Total fluorescence

To illustrate the one-dimensional analysis, the single cysteine of the catalytic subunits of ATCase (wild type from E. Coli) was labelled by 1,5 I-EADANS (8,9). The whole enzyme reconstituted (10) has a final molar ratio dye/protein of 0.1.

The measured fluorescence and flash curves are displayed in fig. 1a and the reconstruction of the spectrum is shown in fig.1b. Three significant (asymmetrical) peaks are clearly visible centred at 0.15, 2.4 and 7.7 ns with a fourth minor contribution at about 24 ns. The spectrum displays the heterogeneity and asymmetry contributions to the lifetimes, particularly at short decay times which accounts for 60% of the total area under the spectrum.

I-AEDANS is a specific reagent for the cysteine residue in proteins. The average ratio of label to molecules of ATCase was only 0.1 so we can safely assume that only one dye molecule was bound to each labelled ATCase.

The I-AEDANS dye has a long arm which enables the fluorescent part (naphatalene-sulfonic acid) to interact in many different ways with neighbouring residues. Similarly we can expect the enzyme to have many different conformations around the cystein residue. Maximum entropy gives a good view of the resulting heterogeneity of the fluorescent emission which, in turn, can be correlated with the conformational dynamics of the associated dye-enzyme system.

COMPUTER SIMULATION OF POLARISED FLUORESCENT DECAYS

In order to test maximum entropy's ability to analyse polarised fluorescent decays, we calculated the theoretical parallel and perpendicular fluorescent decay data at 240 pts (Δt = 0.08 ns) from a "mixture" of two fluorophores with τ_1 = 1.5199, θ_1 = 1.2844, A1 = 0.4 and τ_2 : 5.337, θ_2 = 7.008, A2 = -0.2. The species were present in a ratio of 1:2. The peak counts were 213,000 and 143,000 for parallel and perpendicular components. Random Gaussian noise has been added to the points with a variance equal to the number of "photons" detected.

In Figures 2 we present contour plots of 2 cross-sections through $\gamma(\tau,\theta,A)$. When A=0 we cannot (as expected) determine any θ value. Nevertheless, the two peaks are well separated and their initial amplitudes of rotation are well determined. The τ value can be localised quite accurately but we remain less certain of the θ value (strictly we cannot differentiate between the broad distribution determined and any sharper structure centred on θ = 1.28 or 7.0 respectively). This we would expect since θ is essentially determined from the difference between I// and I⊥, whereas τ is determined from its sum, thus τ is better conditioned with respect to the data than θ.

24

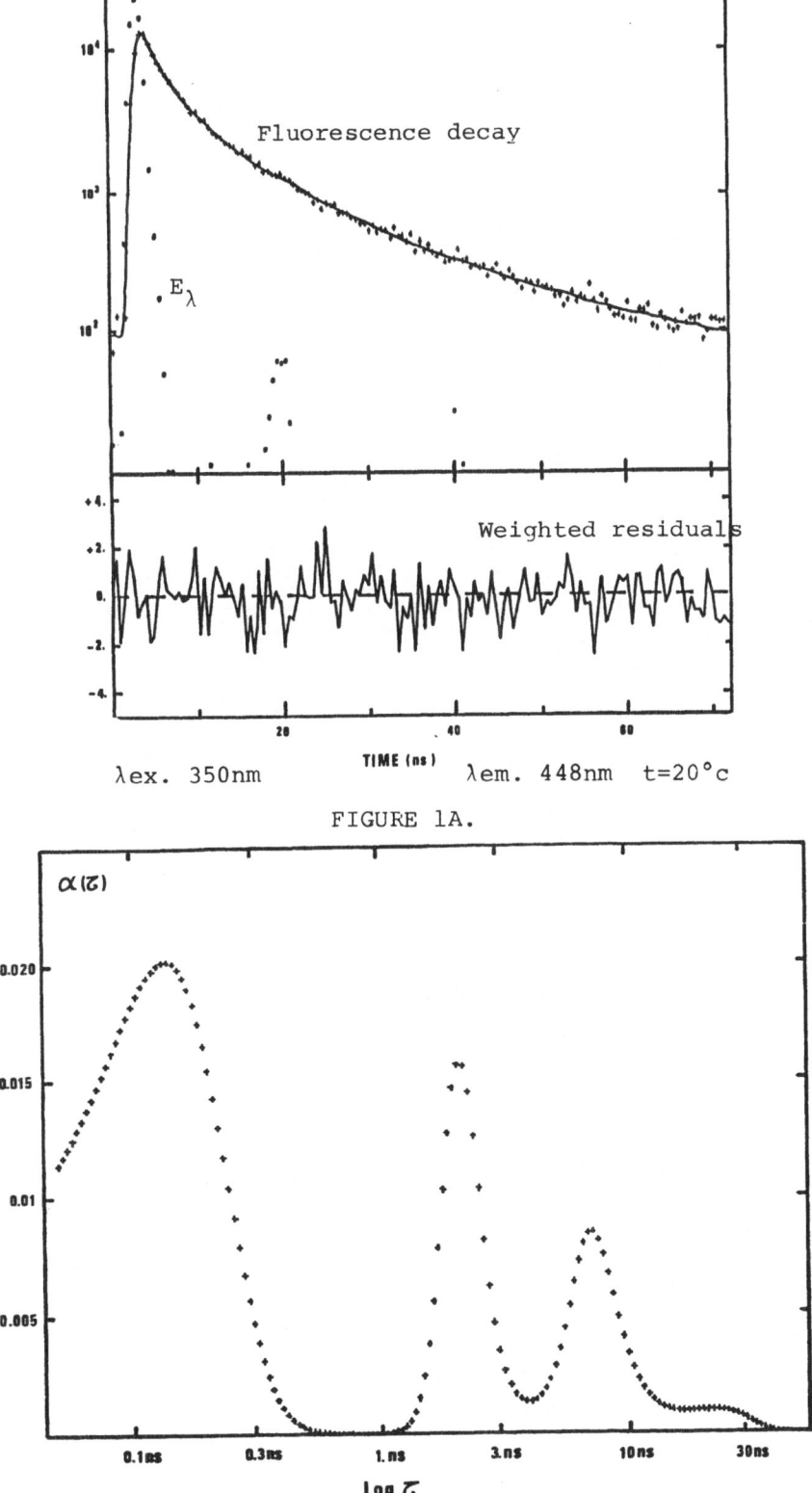

FIGURE 1A.

FIGURE 1B. Spectrum of α(τ) on 150 points

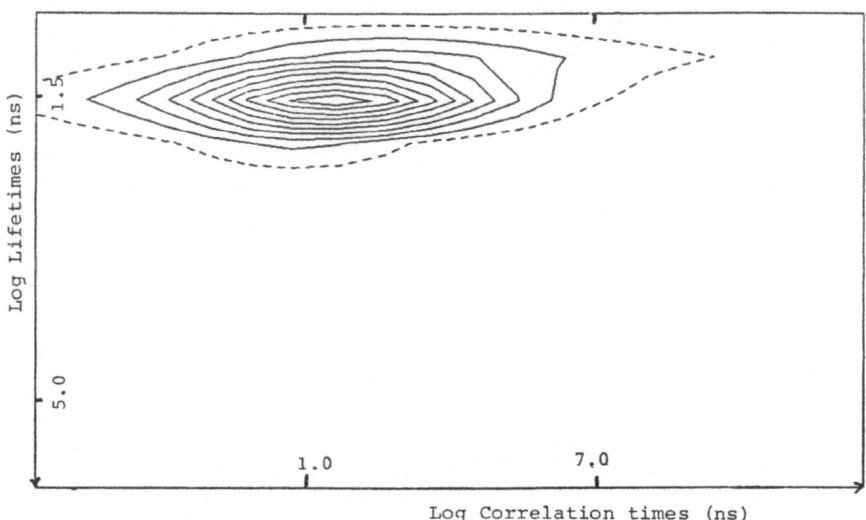

FIGURE 2A. Contour plot of cross-section through $\gamma(\tau,\theta,A)$ for A=0.4

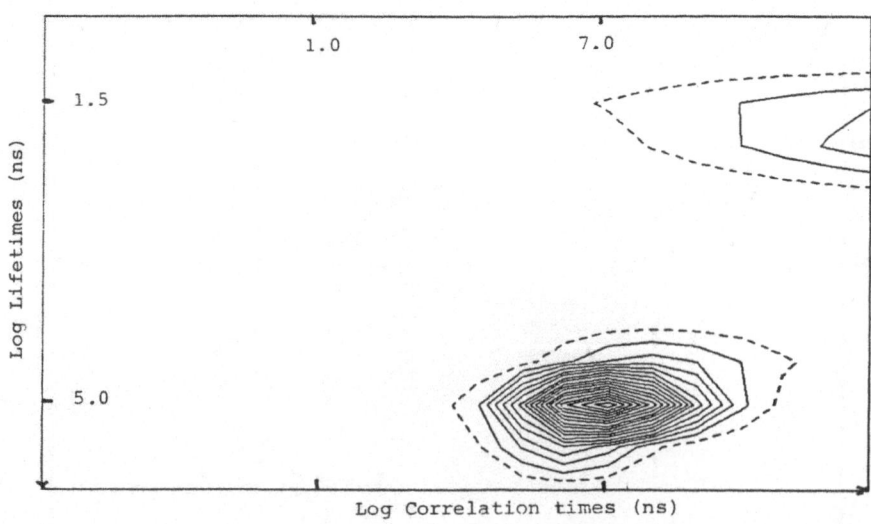

FIGURE 2B. Contour plot of cross-section through $\gamma(\tau,\theta,A)$ for A=-0.2

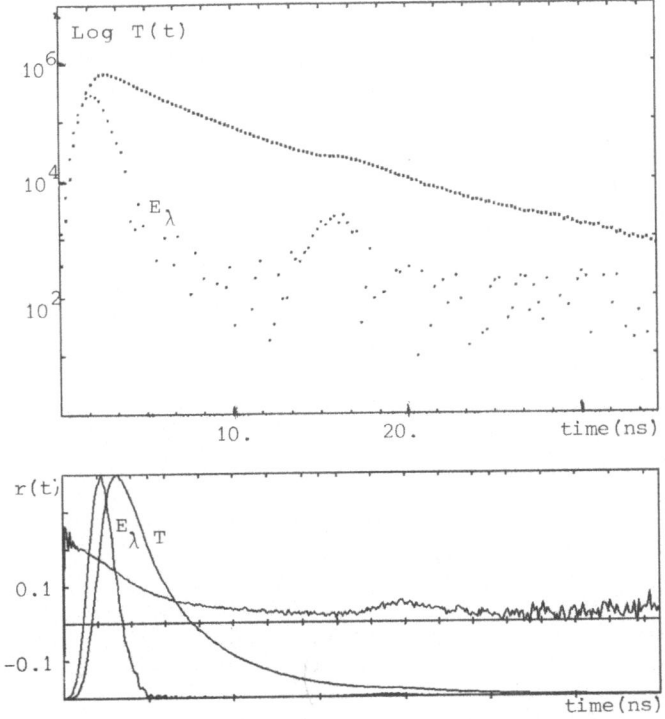

FIGURE 3A. Upper part: Fluorescence decay of Apocytochrome and
flash profile E_λ. Peak counts on T(t) was 671299.
Lower part: Plot of the anisotropy decay in linear scale.

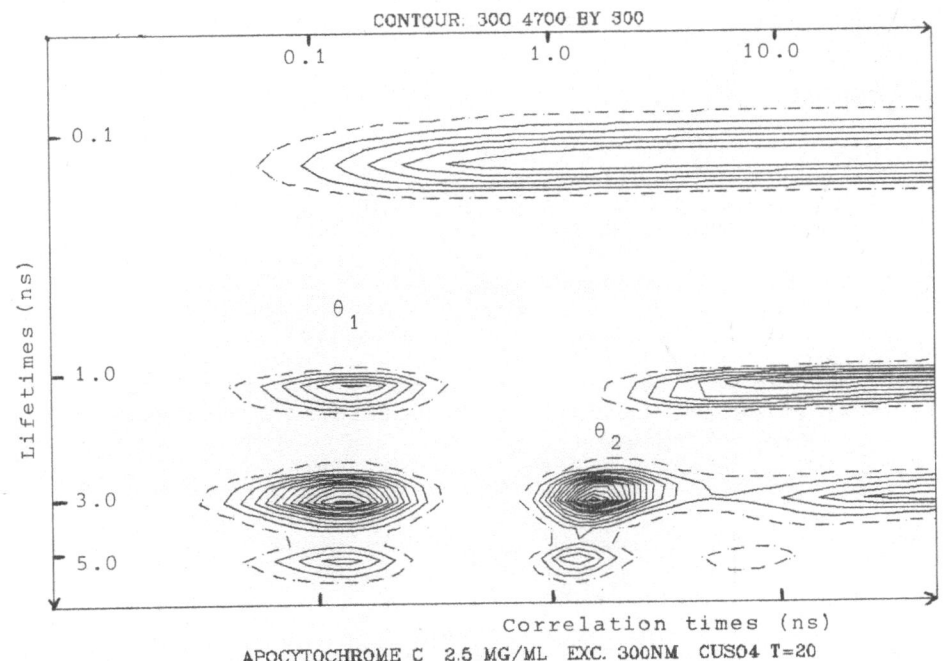

APOCYTOCHROME C 2.5 MG/ML EXC. 300NM CUSO4 T=20

FIGURE 3B. Contour plot of a section through $\gamma(\tau, \theta, A)$ for A=0.258

POLARISED FLUORESCENCE OF APOCYTOCHROME C PROTEIN

Data were measured at the synchrotron on a single tryptophane residue of apocytochrome in aqueous buffer at 20°C at the concentration of 2.5 mg/ml. The data sets are shown in fig.3a together with the anisotropy profile. The contour plot of a section through $\gamma(\tau,\theta,A)$ for A = 0.258 is presented in figure 3b. We can clearly distinguish four lifetime components centered at 0.15, 1.2, 3.14 and 5.4 ns as found previously in a one-dimensional analysis of T(t) (11).

Along the θ axis we can see two high peaks from lifetime τ=3.1 ns centered at θ_1=0.14 ns and θ_2 = 1.4 ns which reflect the major contribution to the depolarisation process. An identical minor contribution is given by the longest τ of 5.4 ns.

On the contrary the shortest lifetime (τ=0.15) does not play any role in the fast motion (θ_1=0.14 ns). Its corresponding θ value remains uncertain due to the weak contribution to the signal of a such short lifetime.

The intermediate τ of 1.2 ns monitors the fast flexibility and the overall motion of the protein (MW=11 900) but is clearly not involved in the intermediate flexibility (θ_2).

The long tail component of correlation times has probably no physical meaning and could be related to an anisotropy of noise on I// and I⊥.

In conclusion this example fully demonstrates the ability of Maximum Entropy Method of analysis to resolve both structural heterogeneity and complex protein dynamics from pulse fluorimetry data.

REFERENCES

1. A.K. Livesey, M. Delaye, P. Licinio and J.C. Brochon (1987) Faraday discuss. Chem. Soc. Vol. 83 paper 14 London
2. A.K. Livesey and J.C. Brochon (1987) Biophys. J. in press
3. C. Benard and M. Rousseau (1974) J. Opt. Soc. Am. 64, 1433.
4. R. Rigler, O. Kristensen, J. Roslund, P. Thyberg, K. Oba, and M. Eriksson (1987) Physica Scripta in press
5. P. Marin et als (1987) private communication:Super ACO, Orsay
6. E.T. Jaynes, (1983), Collected works. Papers on Probability Statistics and Statistical Physics Ed. R.D. Rosenkrantz (D. Reidel, Dordrecht, Holland)
7. A. Mac Lachlan, personal communication.
8. E.N. Hudson and G. Weber (1973) Biochemistry 12 : 4154
9. F. Merola and J.C. Brochon (1986) Eur. J.Biophys. 13:291
10. P. Tauc (1987) Thèse d'état Université de Paris Sud Orsay
11. M. Vincent, J.C. Brochon, W. Jordi, B. de Kruijff, and J. Gallay, in preparation

Acknowledgements

We are indebted to the technical staff at LURE for running the machines during beam-time sessions. P. TAUC and G. HERVE kindly supplied the labelled ATCase-AEDANS. We gratefully acknowledge M. VINCENT and J. GALLAY for measurements of polarized fluorescence decays of apocytochrome protein.

TIME-RESOLVED FLUORESCENCE SPECTROSCOPY OF ADENOSINE

Jean-Pierre Ballini[1,3],Malcolm Daniels[2,3] and Paul Vigny[1,3]

[1] L.P.C.B C.N.R.S.(UA198), Univ. Pierre et Marie Curie (Paris VI)
Institut Curie 11, rue Pierre et Marie Curie, 75231 Paris Cedex 05
France
[2] Chem. Dept. and Radiation Center, OSU, Corvallis, OR 97331, USA
[3] L.U.R.E., C.N.R.S. and Univ. Paris-Sud, Bat. 209C, 91405 Orsay
France

INTRODUCTION

The fluorescence from adenosine (ADO) and adenosine–5'– monophosphate (AMP) in neutral aqueous solution at room temperature has been little studied despite its obvious importance for understanding excited state behavior and photochemistry of many polymeric nucleic acid species. In contrast to this there have been many studies on adenine (Ade) and AMP in acid solution and at low temperature. The origin for such disparate treatment is found simply in the fact that the fluorescence quantum efficiency of ADO and AMP at room temperature is so low ($\approx 4 \times 10^{-5}$) , while in acidic solution ϕ_f is around twenty times greater and at 77K the increase is by three orders of magnitude.

The work reported here has been carried out on the fluorescence lifetime facility which is coupled to the UV/visible port of the ACO synchrotron at LURE, Orsay (France). An entire class of measurements which we term 'time-resolved excitation spectra', requiring repeated scanning over a range of excitation wavelength in the UV from 230 nm to 357 nm, needs a continuously tunable exciting source such as synchrotron.

The starting point for the present work was our observation[1] of an unexpected 'tail' during the measurement of the decay profile of the fluorescence from adenosine (Fig.1), whereas a lifetime of ≈ 1 psec was anticipated. With the 1.76 nsec exciting pulse of the LURE synchrotron, the fluorescence profile should then be that of the exciting pulse. Accordingly we have carried out the extensive series of measurements reported here utilizing variable-wavelength time-correlated single photon counting detection in conjunction with variable wavelength 1.76 nsec pulsed excitation at 13.6 MHz. This technique is particularly suitable for carrying out time-delayed measurements and as well as the conventional time-delayed emission spectra, we have relied heavily on the corresponding time-delayed excitation spectra, a much less common procedure, to elucidate the behavior of adenosine at room temperature. With the aim of aiding the interpretation of neutral adenosine we have carried out similar measurements on ^{6}N,^{6}N–dimethyladenosine (6DMA), ^{6}N–methyladenosine (6MA), 1-methyladenosine (1MA), and on protonated-adenosine (ADO H$^+$). Detailed results concerning these compounds will be published elsewhere.

EXPERIMENTAL

All experiments have been carried out using the pulsed synchrotron excitation source together with gated time-correlated photon counting. The general synchrotron facility at LURE has been completely described earlier[2]. A measure of the overall sensitivity of the facility may be gained by noting that while most work in fluorescence decays is with systems for which the product of emission quantum yield ($\phi_f \approx 0.1$) and absorbance (A$\approx$0.1) is $\approx 10^{-2}$, the work we report here (and in references 1 and 3) is for systems for which $\phi_f A \approx 10^{-4}$.

Time-resolved spectroscopy

The channel resolution over 128 ch. is 2 nm for a scan range from 280 nm to 536 nm in emission, or over 64 ch. from 220 nm to 347 nm in excitation. Optical resolution has been either 16 nm or 8 nm, for either excitation or emission spectra, depending on the compound. The dwell times are $\approx$1sec/ch./scan . No smoothing has been carried out on the results presented here. Correction for the decay of the synchrotron beam is taken into account between recordings of sample and buffer. The geometry of the beams within the 1cm cuvet has been verified in order to correct the emission and the excitation spectra for the absorbance of the solution, using the procedure of Vigny & Duquesne[4] (OD has always been kept $\leq$ 1). Excitation spectra are corrected for the wavelength variation of exciting beam intensity using solid sodium salicylate as a quantum counter.

The time-delays and windows used in this work are illustrated in Fig.1 for the fluorescence from adenosine.

The following time-windows have been commonly used.

- early time-window,   0 $\rightarrow$ 2.35 nsec
- late time-window,   4.70 $\rightarrow$ $\leq$36.0 nsec

Decay kinetics

Recording of one decay is always carried out using a sequence of four measurements. (i) an exciting pulse profile (preflash) at the emitting wavelength of our sample, with a constant geometry of the beams within the cuvet and onto the photomultiplier, and with the same counting rate of events (obtained by aperturing the exciting beam before the double monochromator). (ii) decay profile of our sample. (iii) decay profile of the buffer in the same conditions. (iv) another exciting pulse profile (postflash) identical with (i) in order to verify the stability of the experiment.

Decay-time analysis; we use the same treatment concerning the absorbance of the solution and the decay of the synchrotron beam as for spectra. Then analysis has been carried out on a Vax 780 by re-convolution from the excitation pulse profile assuming a multi-exponential model. Fitting is by non-linear least squares minimisation using the Marquardt algorithm, the goodness of fit being judged graphically by randomness of the plot of the weighted residuals and of the autocorrelation function of the weighted residuals, and by the value of χ^2. Each decay determination always consists of at least two data sets because of the pre- and post-flashes, and the computation is carried out a number of times (5 to 10), varying the starting values, in order to check the stability of the analysis. For sets of multicomponent decays for which the same decay constants are anticipated, only the proportion of each varying, global analysis of all the sets leads to a considerable reduction in number of independent parameters[5,6].

Absorption spectra have been measured with a Cary 118CX spectrophotometer. To be sure of the purity of adenosine, liquid chromatography separation has been carried out on a Beckman model 344 HPLC system, using an Altex Ultrasphere-ODS column (25x0.46 cm; mean particle size 5µm) at 35°C, eluting with a gradient of 2.5 % to

12.5 % methanol/water at a flow rate of 1.5 ml/min, monitoring by absorbance at 254 nm on a Shimadzu detector. No evidence of more than one peak has been found.
Adenosine (ADO) was obtained from Calbiochem and from Boehinger, solutions were usually prepared in neutral buffer with phosphate 10^{-2}M, pH7.2

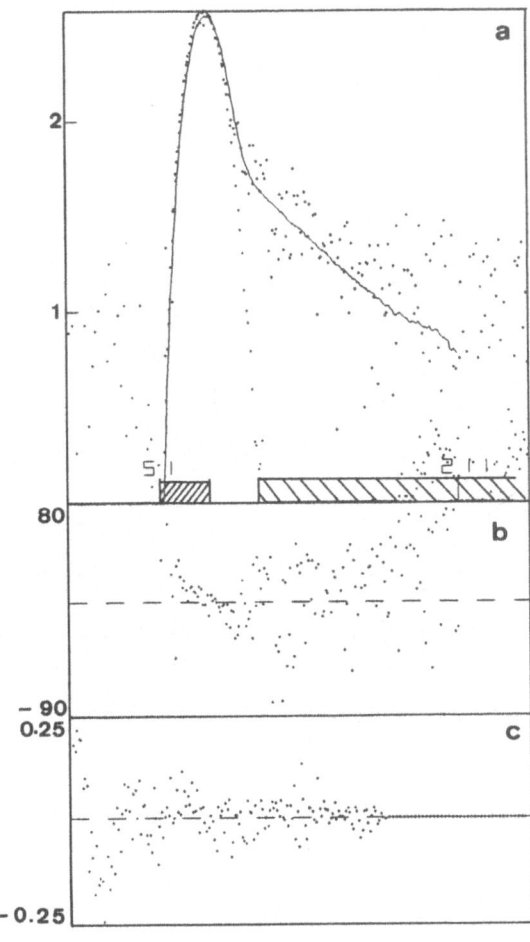

Fig. 1

+   Decay profile for adenosine pH 7.2 excited at 275 nm and monitored at 350 nm.
        log(counts) versus channel number (time).
●   exciting pulse profile
        continuous line is the fitting by 2 exponentials.
/////   early time window within the leading edge of the excitation.
\\\\\   late time window after the excitation.
1b   weighted residuals.
1c   autocorrelation function of the weighted residuals.

RESULTS

Time-delayed emission spectra

As shown in Fig.1, Adenosine decay indicates that, excited at 275 and monitored at 350 nm , most of the emission is much faster than nanosecond, as expected, together with a much smaller proportion of longer-lived component with τ in the nsec range. To determine the emission spectrum of the fast component with the minimum of interference from slower-decaying component(s) the emission spectrum has been determined using the early time-window, with the results shown in Fig.2. This result is closely similar to that observed for ADO at 77K in glycol/water[7] . No other species is obviously present and we shall refer to this spectrum as originating from the '320' species.

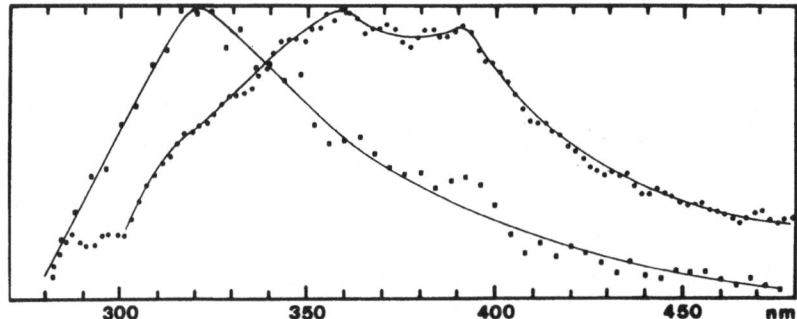

Fig. 2 Time-delayed emission spectra for adenosine pH 7.2

■   early time-window (λ_{exc}=250nm)

●   late time-window (λ_{exc}=270nm)

To discriminate against the '320' species and to favor the slower-decaying species responsible for the 'tail' of Fig.1 is the reason for using the late time-window. There is considerable gain of intensity at 360 nm, and beyond to 500 nm while little of the '320' remains. It is important to note that the intensity at 360 nm is very close to that at 390 nm and a slight change in window can lead to a reversed intensity ratio. This suggests that the emissions at 360 nm and 390 nm may originate from different species. To help clarify the number of emitting species and their relationships, excitation spectra have been determined, as well as decay profiles at different wavelengths.

Time-delayed excitation spectra

These are shown in Fig.3. For the early time-window an excitation spectrum is obtained which superimposes very nicely with the absorption spectrum of adenosine, peaking at 258 nm .

The late time-window profile monitored at 400 nm peaks $\approx$ 269 nm while monitored at 360 nm the peak is broader and extends to $\approx$281nm with a mean of the values around 276 nm. In interpreting data such as this it must be noted that such time-delayed spectra do not represent single species but are the result of the overlapping in wavelength and time of several emitting species which have been partially resolved by appropriate selection of λex, λem,delay time and dwell time.

The behavior in the late time window indicates the existence of two species having lifetimes in the nsec range, the first emitting most strongly around 360 nm and having an excitation peak $\approx$ 276 nm, the second emitting primarily at longer wavelengths $\approx$ 390 nm and having an excitation peak $\approx$ 269 nm. Our understanding of this system at this stage may be summarized as follows

	λ_{ex}(max)	λ_{em}(max)	τ
Species I	258 nm	320 nm	< 100 psec
Species II	$\approx$ 269 nm	$\approx$ 390 nm	$\approx$ 4 nsec
Species III	$\approx$ 276 nm	$\approx$ 360 nm	$\approx$ 4 nsec

Further evidence has come from investigating the decay profiles corresponding to this classification.

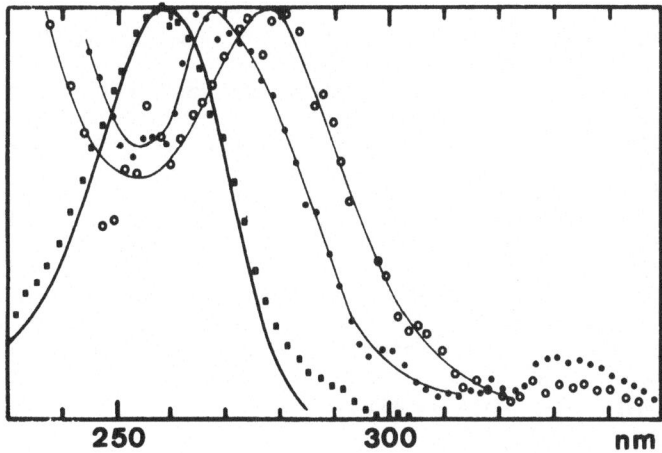

Fig. 3 Time-delayed excitation spectra for neutral adenosine.

■   early time-window excitation monitored at λ_{em}=340nm

●   late time-window excitation monitored at λ_{em}=400nm

○   late time-window excitation monitored at λ_{exc}=360nm

Absorption spectrum is continuous line.

<u>Wavelength-resolved decay profiles</u>

Typical decay profile is shown in Fig.1 . Excitation at 260 nm and monitoring at 320 nm gives mostly (84%) of a very fast emission which can be characterized as having $\tau < 100$ ps, also a small amount (16%) of a longer-lived component with a lifetime of 3.4 ns. Changing the excitation to 275 nm while monitoring at 330 nm changes the proportion of longer-lived component to 56% without significantly changing the lifetime (3.7 ns). These results are consistent with the idea that there are two emitting species in this region, a very fast one with an excitation maximum at $\approx$ 260 nm and another with a much longer 3.7 ns lifetime preferentially excited between 270 nm and 280 nm. Exciting at 275 nm and monitoring at 390 nm or 350 nm gives 58% or 55% of a nsec component with a lifetime of 4.1 ns (global value) and this is also consistent with time-delayed spectra suggesting there are two species resulting from 275 nm excitation.

While the emission from neutral adenosine peaking at 320 nm and decaying much faster than the exciting pulse can be immediately assigned as the 'normal' or anticipated fluorescence from adenosine on the basis of the agreement of its excitation spectrum with the absorption spectrum of adenosine and previous work[1,3] the nature of the species responsible for the longer lived emissions in the range 350 nm – 400 nm is not obvious. From the similarity of this emission with the fluorescence from adenosine in acidic solution ($\lambda_{max} \approx$ 390 nm) it occurred to us that we may be observing emission from protonated adenosine even in neutral solution. To provide a better basis for such an assignment we have investigated in detail the fluorescence behavior of adenosine in acid solution. Since tautomeric forms could be also involved, a detailed study of ^{6}N,^{6}N–dimethyladenosine (6DMA) having a fixed tautomer system has been also carried out. These studies will be published elsewhere.

DISCUSSION

The problems we face in understanding this system can be stated as follows :
1) from time-delayed emission spectra there are three emitting states ('320','350','390'). The '320' is a fast decaying state with lifetime <100 ps. Kinetic studies of the '350' and '390' states show only one lifetime around 4 ns, so that either they have the same lifetime, or they cannot be resolved under our conditions of low counts (low S/N ratio).
2) from time-delayed excitation spectra there are three absorbing species present, '258','269' and '276', one of which ('258') coincides with the absorption spectrum, the '269' being correlated with the '390' emission and the '276' with the '350'.
Because of limitations of space, we only summarize the similar results which we have obtained for 6DMA and for ADO in acid solution. 6DMA exhibits two emission spectra, one of which coincides with the absorption spectrum. ADO in acid solution has a single emission/excitation pair but the excitation spectrum differs from the absorption spectrum. The problem then is to identify the absorbing species in each case, to relate them to the emitting states, and to relate 6DMA and ADO H$^+$ to ADO.

<u>Multicomponent spectra</u>

A conceptual framework for understanding our results would be provided if adenosine exists in three tautomeric forms, which have distinct fluorescence spectra. Tautomerisation in adenosine is due to the two 'mobile' H atoms of the extra cyclic amino group, and to eliminate the problem due to tautomerism is the reason for investigating 6DMA. If the phenomena of multiple excitation/emission spectra are due to the existence of tautomers, then 6DMA should show just one emission spectrum and an excitation spectrum superimposing on the absorption spectrum. This is not the case. 6DMA shows two emission and excitation spectra which correlate with two of the adenosine excitation/emission pairs. Consequently tautomerism cannot be source of the multiple spectra of adenosine.

It appears that the behavior of ADO can be understood as a combination of the behavior of 6DMA and of ADO H$^+$ .Tautomerisation as the origin of the dual excitation/emission of 6DMA can be ruled out a priori. Protonation of 6DMA has no effect on its emission spectrum ($\lambda_{max} \approx 358$ nm) or quantum yield (which is as low as adenosine in neutral solution) and the excitation and absorption spectra essentially coincide[8]. Consequently protonation of 6DMA must also be ruled out. In view of this we suggest that the two excitation spectra originate in two rotational conformers of the exocyclic $-\bar{N}(CH_3)_2$ group, which may be coplanar with or orthogonal to the plane of the heterocyclic ring. Such a situation has been shown to be responsible for the well-known dual fluorescence of p–(dimethylamino)benzonitrile in which excitation gives rise to a twisted internal charge transfer state[9,11] with a large excited state dipole, a nsec lifetime and strong solvation.

To be effective this proposal concerning rotational conformers implies a considerable role of the amino group in the electronic transitions which is supported by the strong effects of amino group methylation on the absorption spectra. Similar effects are known in the spectroscopy of anilines[12] and have been discussed by Kasha and Rawls[13] in term of partial charge transfer to the ring, described as $a_\pi^* \leftarrow 1$.

If the $-NH_2$ group of ADO has the same conformational behavior as the $-N(CH_3)_2$ group of 6DMA, then two of the three excitation/emission pairs can be accounted for. This seems to be the case. However there remains another excitation/emission pair (269/390). This pair has almost identical characteristics with the excitation/emission pair of protonated adenosine, which we now consider.

The outstanding features about the fluorescence of adenosine in acid solution are (i) the positions of, and shift between the excitation and absorption spectra are unchanged from the 269/390 excitation/emission couple in neutral solution (ii) the emission is almost entirely long-wavelength with a much larger Stoke's shift than for the other excitation/emission couples of neutral solution, and (iii) has an overall efficiency an order of magnitude greater, but still only $\approx 8 \times 10^{-4}$. Items (i) and (iii) tell us that the behavior of the 269/390 excitation/emission of neutral adenosine is unchanged in nature and is simply enhanced in the protonated forms. As this effect does not exist in acidic 6DMA [8], it cannot be due to any protonated 6–dimethylamino tautomer.

To keep the same tautomeric structures after protonation is only possible if some protonation takes place at N7; this is in agreement with the experimental conclusions of Knighton[8]. The role of N7 is supported by the strong and completely depolarized fluorescence very similar to protonated adenosine which has been reported to occur from 7–ethyladenosine in neutral solution[8]. This substitution is likely to favor the out of plane rotamer of the amino group[14,15]. We suggest that the 269/390 may originate from a perpendicular $-NH2$ conformer strongly solvated in the ground state, this interaction being enhanced when protonation occurs (probably at N7).

The differences in the behavior of 6DMA and ADO may be traced to the non-polar nature of the methyl groups of 6DMA which cannot be involved in hydrogen-bonded solvent interactions. In contrast to this, the amino group of ADO is strongly involved in H–bonding with the solvent; its conformational behavior can thus be influenced by solvation at N7 and changes in that solvation due to protonation. The strong role of protic solvent in stabilizing twisted internal charge transfer transitions to which our model is closely analogous is well known[15,18].

Acknowledgements

We thank Dr. Jean-Claude Brochon who is in charge of the UV fluorescence facilities at LURE, for the accessibility and the reliability of the experimental set–up.
This work has been supported in part by PHS grant GM 30474.
We thank Ding-guo Ho for computational assistance.

References

1. J. P. Ballini, M. Daniels and P. Vigny, J. Lumines, 27, 389 (1982).
2. P. M. Guyon, C. Depautex and G. Morel, Rev. Sci. Inst., 47, 1349 (1976).
3. J. P. Ballini, P. Vigny and M. Daniels, Biophys. Chem., 18, 61 (1983).
4. P. Vigny and M. Duquesne, Photochem. and Photobiol., 20, 15 (1974).
5. J. R. Knutson, J. M. Beecham and L. Brand, Chem. Phys. Lett., 102, 501 (1983).
6. S. L. Frye, Jaeju Ko, and A. M. Halpern, Photochem. and Photobiol., 40, 555, (1984)
7. J. W. Longworth, R. O. Rahn and R. G. J. Schulman, J. Chem. Phys., 45, 2930 (1966).
8. W. B. Knighton, G. O. Giskaas, and P. R. Callis, J. Phys. Chem., 86, 49 (1982).
9. Z. R. Grabowski, K. Rotkiewicz and A. Siemiaczuk, D. J. Cowley and W. Baumann, Nouv. J. Chim., 3, 443 (1979).
10. D. Huppert, S. D. Rand, P. M. Rentzepis, P. F. Barbara, W. S. Struve and Z. R. Grabowski, J. Chem. Phys., 75, 5714 (1982).
11. Y. Wang, M. McAuliffe, F. Novak, and K. B. Eisenthal, J. Phys. Chem., 85, 3736, (1981).
12. J. N. Murrel, in: "Theory of Electronic Spectra of Organic Molecules", Methuen, London (1963).
13. M. Kasha and H. R. Rawls, Photochem. and Photobiol., 7, 561 (1968).
14. J. Smagowicz, K. Berens and K. L. Wierzchowxki, p24 in: "Excited States of Biological Molecules", J. B. Birks Ed., John Wiley & Sons.
15. C. Cazeau-Dubroca, A. Peirigua, S. Ait Lyazidi and G. Nouchi, Chem. Phys. Lett., 98, 51 (1983).
16. Y. Wang and K. B. Eisenthal, J. Chem. Phys., 77, 6076 (1982).
17. J. Hicks, M. Vandersall, Z. Babarogic and K. B. Eisenthal, Chem. Phys. Lett., 116, 18 (1985).
18. D. W. Anthon and J. H. Clark, J. Phys. Chem., 91, 3530 (1987).

MEASUREMENT OF BIOLOGICAL EFFECTIVE IRRADIANCE AND IRRADIATION

Dieter Kockott
Vogelsbergst. 27, D-6450 Hanau 7, W. Germany

Jürgen Kochmann
Geneststr. 6, D-1000 Berlin 62, W. Germany

STARTING POINT

Each radiation-exposed object absorbs a greater or lesser part of the impinging radiation, part of which is converted into heat and part of which is responsible for photobiological/photochemical processes. In this lecture, only the photobiological effective part of the radiation will be considered. Each photobiological effect is characterized by its relative spectral action spectrum $s(\lambda)_{biol, rel}$. This indicates which wavelength range of the spectrum causes each particular photobiological effect. As an example, figure 1 shows the relative spectral action spectrum of the UV-erythema $s(\lambda)_{er, rel}$ [1] and of psoriasis phototherapy $s(\lambda)_{ps, rel}$ [2].

One value which is important for the description of experimental conditions and also of photomedical treatment is the photobiological effective irradiance E_{biol} within the photobiological effective spectral range. In order to measure this one should have a receiver of exactly the same relative spectral sensitivity as the biological object for each particular photobiological effect. As there are a large number of different photobiological effects, the ideal receiver should have an adjustable relative spectral sensitivity. Such a receiver, however, is not yet on the market. Thus, one tries to manage by measuring the irradiance in the UV-A or UV-B wavelength range. As an example, figure 2 shows the relative spectral sensitivity in UV-A and UV-B of a commercial radiation meter. The biological effective irradiance E_{biol} is determined by

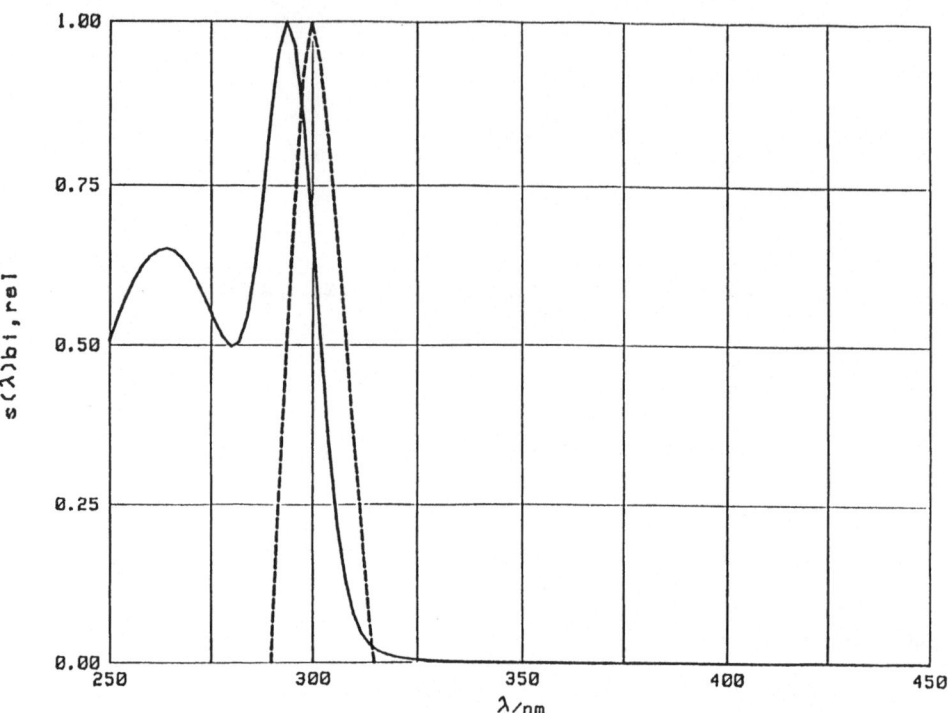

Fig 1. Relative spectral action function $s(\lambda)_{rel}$
---- for the phototherapy of psoriasis
—— of UV erythema

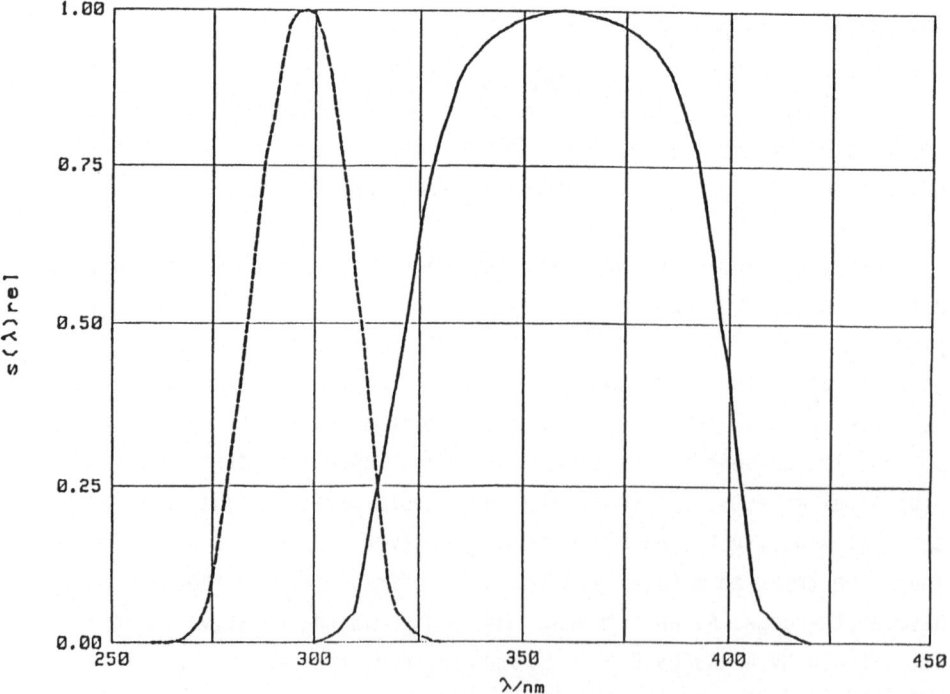

Fig 2. Relative spectral sensitivity $s(\lambda)_{rel}$
---- UV-B receiver
—— UV-A receiver

(1) $E_{biol} = K \int S(\lambda) \cdot s(\lambda)_{biol, \, rel} \, d\lambda$

$S(\lambda)$ relative spectral energy distribution of the radiation employed

$s(\lambda)_{biol, \, rel}$ relative spectral action function of each particular photobiological effect

K constant.

Accordingly, the irradiance measured by a UV-A or UV-B receiver of relative spectral sensitivity $s(\lambda)_{A, \, rel}$ or $s(\lambda)_{B, \, rel}$ is

(2) $E_{UV-A} = K \int S(\lambda) \cdot s(\lambda)_{A, \, rel} \, d\lambda$

(3) $E_{UV-B} = K \int S(\lambda) \cdot s(\lambda)_{B, \, rel} \, d\lambda$

Figure 3 shows the function $S(\lambda) \cdot s(\lambda)_{biol, \, rel}$ as well as the functions $S(\lambda) \cdot s(\lambda)_{A, \, rel}$ and $S(\lambda) \cdot s(\lambda)_{B, \, rel}$ at a given relative spectral energy distribution.

The areas below the functions are a measurement for each respective irradiance. One realizes that the irradiances in UV-A or UV-B deviate from the biological effective irradiance. If the relative spectral energy distribution is constant, i.e. if one and the same radiation source is used, the measured irradiances, e.g. in UV-B, can be considered as a measure for the biological effective irradiance. Therefore, when comparing two experiments performed with the same radiation source and the same radiation meter one should indicate the irradiance in UV-B as a relative value. Of course, when radiation sources with a different relative spectral energy distribution are used, the measurement in UV-A or UV-B is still physically correct. However, these measurements cannot be used to describe the photobiological effects. Here we surely have one of the reasons for the widely varying figures quoted for the irradiance or radiant exposure in UV-A or UV-B necessary to reach a certain photobiological effect.

PROBLEM

In order to measure the biological effective irradiance or the biological effective radiant exposure exactly, a receiver is required whose relative spectral sensitivity is equivalent to the sensitivity of the photobiological effect in question. In principle, it is possible to use the technique of the diode array, but this is still very expensive at

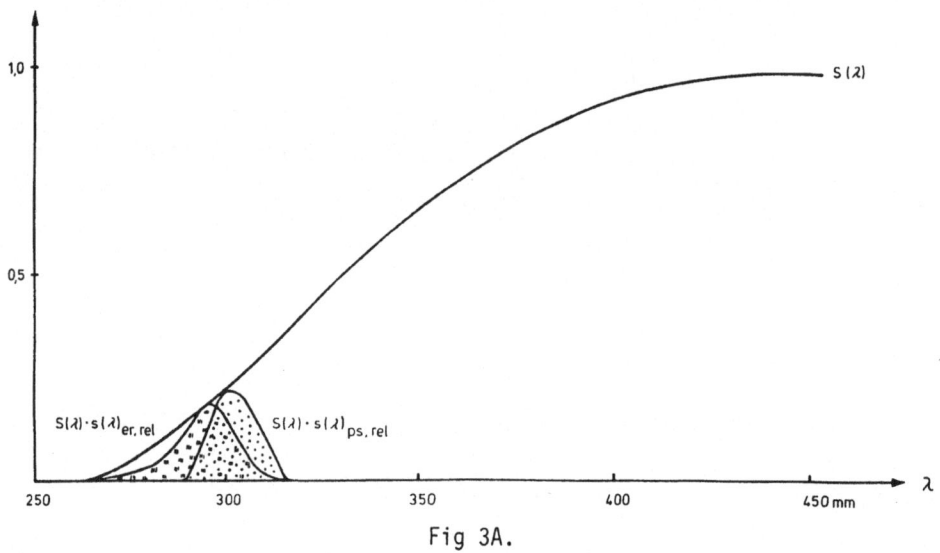

Fig 3A.

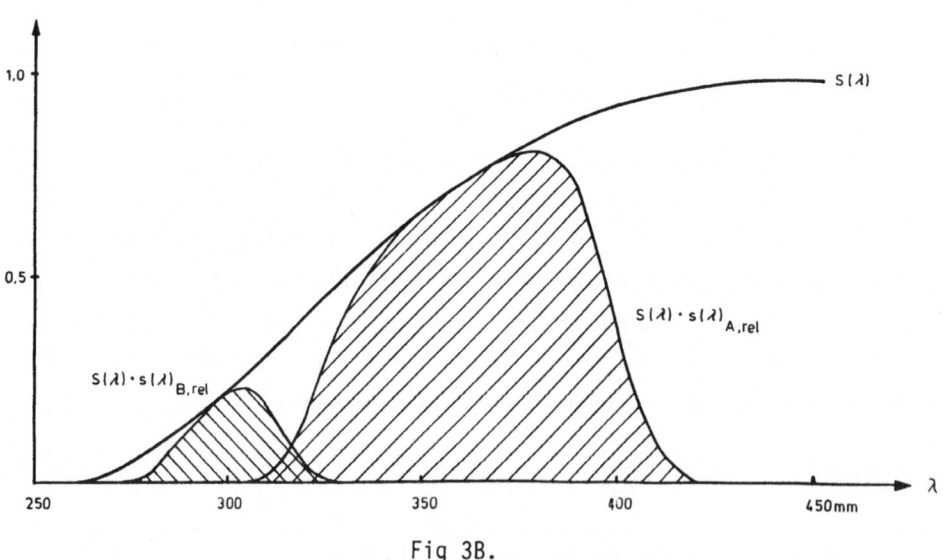

Fig 3B.

present. The problem can also be solved, however, by calculation, using correcting factors.

SOLUTION

Following equations (1) and (2), the biological effective irradiance E_{biol} results in

$$(4) \quad E_{biol} = \frac{E_{UV-A}}{\int S(\lambda) \cdot s(\lambda)_{A, rel} \, d\lambda} \cdot \int S(\lambda) \cdot s(\lambda)_{biol, rel} \, d\lambda$$

and by analogy

$$(5) \quad E_{biol} = \frac{E_{UV-B}}{\int S(\lambda) \cdot s(\lambda)_{B, rel} \, d\lambda} \cdot \int S(\lambda) \cdot s(\lambda)_{biol, rel} \, d\lambda$$

This means that the biological effective irradiance E_{biol} is calculable from the irradiance measured in UV-A (or UV-B) if the relative spectral energy distribution of the radiation employed and the relative spectral sensitivity of the radiation meter in UV-A (or UV-B) and also the relative spectral action function of the photobiological effect in question are known. These functions can be obtained from the relevant literature or from the producers of radiation equipment and radiation meters. In principle, it does not matter whether the irradiance measured in UV-A is used for the calculation or that measured in UV-B.

In order to minimize the influence of errors in measurement it is advisable to use the measurement range that is nearest in spectrum to the particular photobiological effect. Therefore, the measurement in UV-A will serve as a basis for, e.g., the calculation of the biological effective irradiance for direct pigmentation, while, on the other hand, the measurement in UV-B will serve for the calculation of the biological effective irradiance for UV-erythema or the phototherapy of psoriasis.

Our programmable pocket calculator performs the calculation. The calculator is pre-programmed and can be free-programmed as follows:

- 2 pre-programmed and 1 free-programmed relative spectral energy distributions $S(\lambda)$ of a radiation device
- 2 pre-programmed and 1 free-programmed relative spectral sensitivities $s(\lambda)_{rel}$ of the radiation meter used
- 2 pre-programmed and 1 free-programmed relative spectral action functions $s(\lambda)_{biol, rel}$ of a photobiological effect.

Using the pre-programmed functions, the calculator is able to perform any routine task; when the free-programmed functions are used, each individual problem can be solved. The calculation is performed in 2-nm-steps and covers the spectral range of 250 to 450 nm.

The pre-programmed and the free-programmed functions can be combined with each other.

Example

relative spectral energy distribution
for radiation device   A
                                                B
                                                C

relative spectral sensitivity
for radiation meter   X
                                                 Y
                                                Z

relative spectral action function
for the photobiological effect   UV-erythema
                                                 psoriasis phototherapy
                                                 direct pigmentation

An object is exposed to radiation device B. The radiation meter X measures an irradiance in UV-B of 5 W/m² on the object. Then the biological effective irradiance relating to UV-erythema, psoriasis phototherapy and direct pigmentation can be recalled by pressing the appropriate button, e.g.

$$E_{er} = \quad 1 \ W/m^2$$

$$E_{ps} = \quad 2 \ W/m^2$$

$$E_{pi} = \quad 250 \ W/m^2$$

With this calculator it is possible to calculate the biological effective irradiance E_{biol} or the biological effective radiant exposure $H_{biol} = \int E_{biol} \, dt$. These photobiologically relevant physical values make it possible to compare the test results of various radiation devices and radiation meters. This is the case not only for the current tests or photomedical treatment, but also with data presented in scientific literature, provided the necessary information is available.

References

(1) Report CIE/TC 6-10
(2) J. Krochmann, Licht-Forschung 6 (1984) 1, 41/42

THE OPTICAL ABSORPTION AND SCATTERING PROPERTIES OF TISSUES IN THE VISIBLE AND NEAR-INFRARED WAVELENGTH RANGE

Brian C. Wilson, Michael S. Patterson, Stephen T. Flock
and J. David Moulton

Hamilton Regional Cancer Centre and McMaster University
Hamilton, Ontario, Canada   L8V 1C3

INTRODUCTION

The development of diagnostic and therapeutic photomedicine has generated a need to determine the optical properties of tissues in the U.V., visible and infrared regions of the spectrum. In this paper we will review the experimental techniques and resulting data on the optical properties of mammalian tissues. These will include recent results from this laboratory as well as a summary of the work of other groups. Measurements are most abundant at around 630 nm, the wavelength of greatest current interest for clinical photodynamic therapy. We will use these data as the reference values for examining the wavelength-dependence of the optical properties.

EXPERIMENTAL TECHNIQUES

A. Reflectance and Transmittance Methods

Most of the early work comprised measurements of diffuse reflectance, R_d, or transmittance, T_d, through tissue slices. This is particularly useful for skin where it is possible to physically separate the different layers; stratum corneum, epidermis and dermis.[1,2] Some transmittance spectra have been obtained by placing tissue samples in standard spectrophotometers.[3] These studies demonstrated that the optical properties are strongly wavelength dependent, at least in the U.V. and visible. A number of workers applied approximate models of light propagation to these R&T measurements in order to estimate the separate absorption and scattering contributions.[1,4,5] We note that these measurements depend on whether diffuse or collimated irradiation is used, so that values are not always comparable between studies.

B. Interstitial Optical Fiber Detector Methods

The introduction of optical fiber detectors by Doiron and co-workers[6] in the late 1970's lead to a new approach in tissue optics. Single-strand optical fibers a few hundred microns in diameter can be placed interstitially to map out the distribution of light energy fluence in bulk, intact tissue. By measuring the depth-dependence of energy fluence within tissue, the effective attenuation coefficient, μ_{eff}, or its inverse the effective penetration depth, δ_{eff}, can be determined. (For surface irradiation by a collimated beam the energy fluence far from the surface varies with depth, x, as exp(- $\mu_{eff}.x$): corresponding relationships apply for interstitial source irradiations.[6]) The technique has been used for animal and human tissues, both in vitro and in vivo.[7-11] The spatial distribution may depend strongly on, for example, the irradiation beam diameter,[9] so that intercomparison of data from different studies must be made with caution.

By more detailed measurements of both the spatial and angular distribution of the energy radiance, and by using detector fibers with isotropic response to measure the absolute energy fluence within tissue,[9,10] it has been possible to determine separately the absorption coefficient of tissues, μ_a, and the reduced scattering coefficient μ_s'. These are, respectively, the probability of absorption per infinitesmal path length, and $\mu_s(1-g)$ where the scattering coefficient μ_s is the probability of scatter per infinitesimal path length and g is the mean cosine of the scattering angle (g=0 for isotropic scattering, ->1 as the scattering becomes forward peaked). Diffusion theory has commonly been used as the propagation model to derive the fundamental optical properties from interstitial measurements.[6,7,9-11]

C. Direct Methods

The separate coefficients μ_a and μ_s, and the scattering angular distribution or phase function, $S(\theta)$, can be measured by direct (i.e. model-independent) methods using optically thin tissue samples in which the effects of multiple scattering can be eliminated. This has been done to some extent in skin,[2,5] and we have recently reported[12] values for the total attenuation coefficient, $\mu = \mu_a + \mu_s$, and $S(\theta)$ measured in this way for a range of soft tissues at 633 nm. Below we will present preliminary data on μ_a for these tissues.

D. The Added Absorber Method

Originally introduced by Profio and Sarnaik[13] and extended by Wilson et al,[11] this technique is based on measuring the changes in μ_{eff} resulting from the addition of known amounts of absorber to the tissue. For scatter-dominated cases, such as most tissues above about 600 nm, diffusion theory may then be applied to yield values for μ_a and μ_s'. At shorter wavelengths or for heavily pigmented tissues, more complex modelling is required. Both situations will be illustrated by examining the wavelength-dependence of these two coefficients from preliminary measurements carried out in this laboratory.

PUBLISHED RESULTS AND RECENT MEASUREMENTS

Effective Attenuation Coefficient

The approximate range of values of μ_{eff} measured by method B in human and animal tissues at 630 nm (including both in vitro and in vivo data) are: red muscle 0.2-0.7, white muscle 0.1-0.5, brain 0.2-1.2, fat 0.3, liver 0.8-2.8, and tumor 0.2-0.8 mm^{-1}. The wide ranges may be due to real differences between tissue samples, differences in the irradiation and detection geometries between experiments, or differences in tissue handling and preparation in vitro. There have been few systematic studies to determine the importance of each of these factors.

In order to examine the wavelength dependence of μ_{eff}, we have recalculated the in vivo data of Wilson et al,[8] normalizing the values to the value in each tissue at 630 nm, as shown in Figure 1. Also included are the normalized extinction coefficients for hemoglobin and oxyhemoglobin.[14] The data above 800 nm for muscle are taken from the in vitro spectrophotometric results of Bolin et al,[3] matching to the in vivo value at 800 nm. At the shorter visible wavelengths the tissue curves clearly show the contribution from HbO_2 absorption. However, this is 'suppressed' as compared to the pure HbO_2 spectrum due to the absorption of other pigments, the tissue scattering*, and possibly incomplete oxygenation. Least suppression occurs in brain, which has relatively little pigmentation but high scattering, while the heavily pigmented liver strongly suppresses the blood signal.

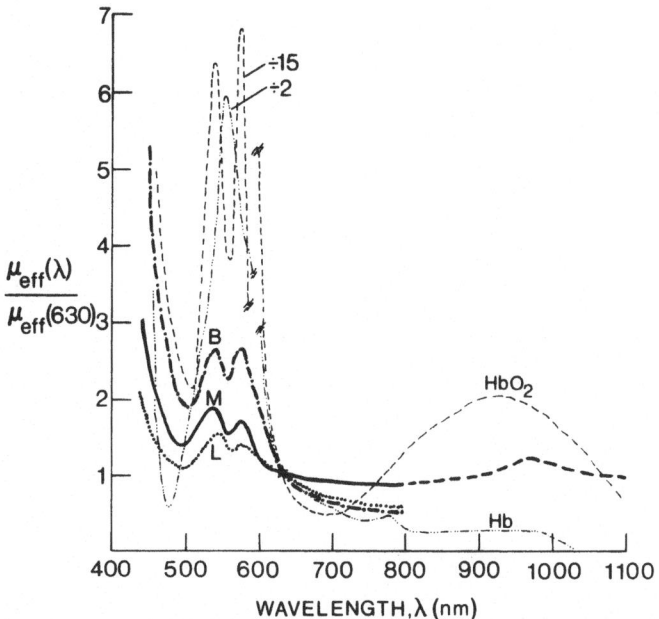

Figure 1. Effective attenuation coefficient in vivo vs. wavelength in brain (B), muscle (M) and liver (L), normalized to the value at 630 nm, adapted from ref 7. Muscle in vitro above 800 nm adapted from ref 3. The hemoglobin and oxyhemoglobin spectra[14] show the normalized extinction coefficients, scaled down at the shorter wavelengths by 2 and 15, respectively.

*The effect of scattering can be seen, for example, in diffusion theory, where $\mu_{eff} \doteq \sqrt{\mu_a}$ for $\mu_a \ll \mu_s'$, whereas for a pure absorber $\mu_{eff} = \mu_a$.

Table 1. Absorption and Scattering Properties of tissues at 630 nm.
DI - direct (method C)   IN - indirect (method D)
CO - combined DI and IN data: $\mu_s = \mu_s'/(1-g)$.
Published data[11,12], plus preliminary values(p)
from this laboratory.

TISSUE	μ (mm^{-1})	μ_a (mm^{-1})	μ_s (mm^{-1})	μ_s' (mm^{-1})	g
BRAIN DI	95+10				0.940+0.029
(pig) IN		0.026		5.7	
CO			95+46		
MUSCLE DI	33+4	0.16+0.10(p)			0.954+0.016
(red) IN		0.15		0.7	
CO			15+8		
MUSCLE DI	35+4	0.09+0.02(p)			0.965+0.004
(white) IN		0.012		0.8	
CO			23+3		
LIVER DI		0.48+0.02(p)			
IN		0.27		1.7	
FAT DI	38+7				0.77+0.06
(pig)					
TUMOR DI	63+11				0.64+0.07
(VX2)					

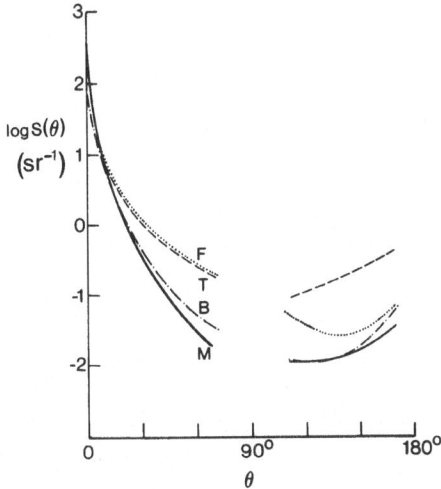

Figure 2. Scattering phase functions for animal red muscle (M), brain (B), fat (F) and VX2 tumor (T), measured directly in vitro (adapted from ref. 12). Note the high forward peak ($\theta = 0°$) and the small backward peak ($\theta = 180°$) in all tissues. The corresponding mean cosine values (g) are given in Table 1, calculated from $S(\theta)$ as in ref. 12.

Absorption and Scatter at 630 nm

Table 1 summarizes published and preliminary values of μ_a, μ_s, μ_s' and g for animal tissues measured in this laboratory at 630 nm. The phase function plots for 4 of these tissues are summarized in Figure 2. The striking features of these results are that scatter takes place predominantly in the forward direction, and that the ratio of scatter to absorption coefficients is very high (>100 for muscle, >1000 for brain). Even for the heavily pigmented liver, if one assumes a typical g value of >0.9, the scatter-to-absorption ratio would be >35.

Wavelength Dependence of Absorption and Scatter

Preliminary data for the wavelength dependence of μ_a and the reduced scattering coefficient μ_s' are shown in Figure 3 for pig brain and bovine muscle. These results were obtained using the added absorber method described above(D). Diffusion theory was applied, as in Wilson et al,[11] to derive the coefficients at wavelengths above 600 nm. A numerical solution of the radiation transport equation, based on the work of van de Hulst,[15] was used at shorter wavelengths where the absorption is large enough to invalidate simple diffusion theory. The absorption spectrum shows most structure in these tissues, with evidence of both the hemoglobin absorption band around 550 nm, and the water absorption peak at around 970 nm. The latter corresponds to the small peak seen in μ_{eff} for muscle in Figure 1. As in the case of μ_{eff}, the 550 nm peak is much less pronounced than in the extinction coefficient of oxyhemoglobin. This suggests that there is a significant contribution to absorption from other chromophores, particularly in muscle.

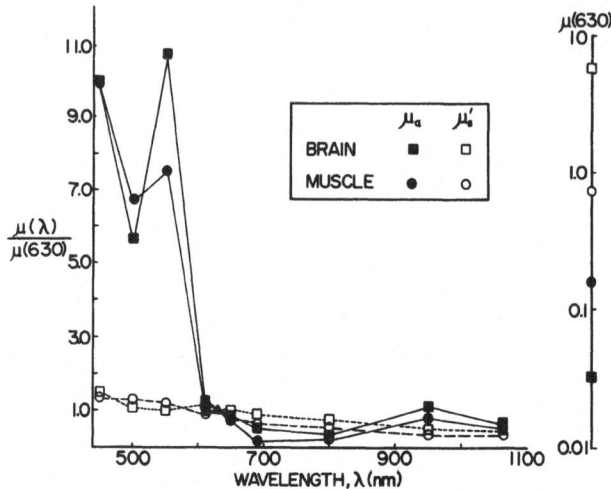

Figure 3. Absorption and reduced scatter spectra for pig brain and bovine muscle, each normalized to the values at 630 nm. The scale on the right shows the absolute values at 630 nm (Table 1).

The scatter spectrum shows a monotonic decrease in μ_s' with increasing wavelength, as has also been seen in skin components.[1] We have not yet measured the phase functions to determine how g varies with wavelength. However, the g-values in human epidermis in vitro derived from the work of Bruls and van der Leun[2] varied only from 0.826 to 0.868 between 302 and 546 nm. Further, assuming that Mie theory is an approximate description of scatter in tissues in the visible,[12] the wavelength dependence of g should not be very strong. Thus, the wavelength dependence of μ_s is probably similar to that of μ_s' in Figure 3.*

Relationship of In Vitro to In Vivo Values and Effects of Tissue Handling In Vitro

A critical question in all the in vitro work done to date is whether the optical properties measured are close to the true in vivo values. Changes in μ_a could arise from biochemical changes in the chromophores or loss of chromophores. Indirect evidence of blood deoxygenation and drainage has been obtained in various tissues from changes in μ_{eff} between in vivo and immediately post mortem.[8] We have also observed small changes in μ_{eff} occurring at 630 nm in tissues held at room temperature,[11] and Crilly[16] has reported a 20% increase in diffuse transmittance through a 1.5 mm thick section of adipose tissue from 22 to 37°C. The scattering may also be affected by tissue handling, particularly freezing where the formation of ice crystals may

---

* We note, however, that the preliminary values in Figure 3 were derived using a fit to the model of van de Hulst with g = 0. The assumption is certainly not correct, and work is in progress on models which will allow more appropriate g values, such as those given in Table 1, in the fitting of the added absorber data. It is likely that μ_a is relatively insensitive to the assumed g value, but μ_s' may be quite sensitive. The shape of the μ_s' spectra below 600 nm in Figure 3 should, therefore, be regarded as tentative.

cause cell rupture. Preuss et al[17] have reported changes in diffuse transmittance through optically thick tissue samples following freezing and thawing. We have investigated these effects in the added absorber experiments and the results are shown in Figure 4. In method C where very thin (<100 μm) samples were used, blood loss could occur during preparation of the sections, although, as seen in Table 1, the μ_a values were comparable to within a factor of 2 with those obtained by the added absorber method where blood loss was not a problem. However, further studies are required to investigate systematically the effects of tissue handling and preparation on the optical properties.

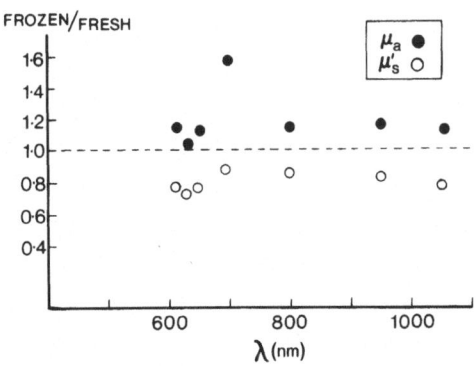

Figure 4. Ratio of the optical coefficients in red muscle after slow freezing and thawing to the values in fresh tissue.

CONCLUSIONS

The optical properties of soft tissues at 630 nm are now known in general terms, and are characterized by small mean free path, very high albedo, and highly forward-directed scattering. These characteristics have been mainly determined by in vitro studies. The corresponding in vivo properties are not known, although there is indirect evidence that the tissue handling and preparation in vitro do not invalidate these general conclusions. Nevertheless, methods are required to make measurements in vivo so that the validity of in vitro results can be checked in detail and so that accurate values may be obtained for target tissues in individual patients.

The preliminary studies reported here for the wavelength dependence of the reduced scattering coefficients show a monotonic decrease with increasing wavelength, while the absorption spectra show structure relating to the absorption characteristics of chromophores, in particular (oxy)hemoglobin. Further work is in progress to determine the spectral shape of μ_a, μ_s and g in detail. Using the known extinction spectra of possible tissue chromophores-hemoglobin, melanin, etc - as "fingerprints", it may then be possible to estimate the relative contribution of each to the total tissue absorption coefficient, and thereby to construct detailed quantitative models for light propagation in specific tissues.

ACKNOWLEDGEMENTS

This work was supported by the National Cancer Institute of Canada.

REFERENCES

1. Wan, S. Anderson, R.R. and Parrish, J.A., Analytic modelling for the optical properties of skin with in vitro and in vivo applications, Photochem. Photobiol. 34:493 (1981)
2. Bruls, W.A.G. and van der Leun, J.C., Forward scattering properties of human epidermal layers, Photochem. Photobiol. 40:231 (1984)
3. Bolin, F.P., Preuss, L.E. and Cain, M.W., A comparison of spectral transmittance for several mammalian tissues: effects at PRT frequencies, in "Porphyrin Localization and Treatment of Tumors", D.R. Doiron and C.J. Gomer, eds. Liss, N.Y., 211 (1984)
4. Wilksch, P.A., Jacka, F. and Blake, A.J. Studies of light propagation through tissues, in "Porphyrin Localization and Treatment of Tumors", D.R. Doiron and C.J. Gomer, eds., Liss, N.Y., 149 (1984)
5. van Gemert, M.J.C. and Star, W.M., Relations between the Kubelka-Munk and the transport equation models for anisotropic scattering, Lasers Life Sci., in press
6. Wilson, B.C. and Patterson, M.S. The physics of photodynamic therapy, Phys. Med. Biol. 31: 327 (1986)
7. Doiron, D.R., Svaasand, L.O. and Profio, A.E. Light dosimetry in tissue: application to photoradiation therapy, in "Porphyrin Photosensitization", D. Kessel and T.J. Dougherty, eds., Plenum, N.Y., 63 (1983)
8. Wilson, B.C., Jeeves, W.P. and Lowe, D.M., In vivo and post mortem measurements of the attenuation spectra of light in mammalian tissues, Photochem. Photobiol. 42:153 (1985)
9. Marijnisen, J.P.A. and Star, W.M. Phantom measurements for ligh dosimetry using isotropic and small angle detectors, in "Porphyrin Localization and Treatment of Tumors", D.R. Doiron and C.J. Gomer, eds., Liss, N.Y., 133 (1984)
10. Marijnisen, J.P.A. and Star, W.M. Light intensity measurements in optical phantoms and in vivo during HPD-photoradiation therapy, using a miniature light detector with isotropic response, in "Photodynamic Therapy of Tumors and other Diseases", G. Jori and C. Perria, eds., Edizioni Libreria Progetto, Padova, 388 (1985)
11. Wilson, B.C., Patterson, M.S. and Burns, D.M., The effect of photosensitizer concentration in tissue on the penetration depth of photoactivating light, Lasers Med. Sci. 1:235 (1986)
12. Flock, S.T., Wilson, B.C. and Patterson, M.S. Total attenuation coefficients and scattering phase functions of tissues and phantom materials at 633 nanometers, Med. Phys. 14: 835 (1987)
13. Profio, A.E. and Sarnaik, J., Fluorescence of HPD for tumor detection and dosimetry in photoradiation therapy, in "Porphyrin Localization and Treatment of Tumors", D.R. Doiron and C.J. Gomer, eds., Liss, N.Y., 163 (1984)
14. Van Assendelft, W.W. "Spectrophotometry of Hemoglobin Derivatives", Royal Vangorcum, Netherlands (1970)
15. van de Hulst, H.C. "Multiple Light Scattering Tables, Formulas and Applications", Academic, N.Y. (1980)
16. Crilly, R., A study of the optical properties of tissues in the near infra-red, Med. Phys. 13:603 (1986)
17. Preuss, L.E., Bolin, F.P. and Taylor, R.C. Environmental effects on light transport in tumors in vitro (400-1100 nm), Photochem. Photobiol. 42S:32S (1986)

PHOTOSENSITIZATION IN A LIGHT SCATTERING MEDIUM

Leonard I. Grossweiner

Biophysics Laboratory, Physics Department
Illinois Institute of Technology
Chicago, IL 60616, U.S.A.

INTRODUCTION

In essentially all photosensitized processes in biology, the chromophores are embedded in a turbid tissue matrix. When collimated light enters tissue, light scattering leads to a loss of the initial directionality and spreading, accompanied by internal reflections at layer interfaces and energy absorption by chromophores. As a consequence, conventional measurements of transmission and reflection are inapplicable owing to the loss of collimation. The determination and analysis of optical parameters in turbid media require special experimental and analytical methods. Current research in tissue optics has been directed towards calculating the macroscopic optical constants from diffuse transmission and reflection measurements. Approximational theories of light propagation in turbid media have been employed, including the early two-flux model of Kubelka and Munk (1931), multi-flux models (Welch et al., 1987), various forms of the diffusion approximation to radiative transfer (Ishimaru, 1978), and Monte Carlo simulation (Bonner et al., 1987). The results have led to radiant flux profiles within samples of specified geometries (e.g., Welch et al., 1987), absorption spectra of weak chromophores (e.g., Norris and Butler, 1961), and dosimetry models for phototherapy procedures (e.g., Doiron et al., 1983; Pratesi at al., 1984; Grossweiner 1986; Profio and Doiron, 1987). Relatively little attention has been given to the effects of light scattering on the rate of a photosensitized chemical reaction in a turbid medium, which might provide a link to biological photosensitization in tissue. In recent work, Grossweiner and Messina (1987) investigated the photoinactivation of a colorless enzyme, subtilisin Carlsberg, by methylene blue in the presence of polystyrene microspheres. The results were analyzed with a simplified form of the diffusion approximation applicable to optically dense, highly scattering layers. The present paper is an extension of that work, in which the same data are correlated with other theoretical models and compared. The results show that the simpler models leading to analytical expressions for the power absorption are adequate for this dye-scatterer system.

METHODS

Experimental Data

A full description of the experimental methods leading to the data employed in this paper has been published (Grossweiner and Messina, 1987). Briefly, polystyrene microspheres ranging in size from 0.5 to 5 μm were suspended in pH 7.4 phosphate buffer containing methylene blue and 20 μM subtilisin Carlsberg. The dye and microsphere sphere concentrations were varied to give at least 90 % attentuation in 2 cm depth. The 1/e attenuation depth of each suspension was measured using the light source and photomultiplier detection system of a modified a Farrand manual spectrofluorimeter. The sample was contained in a 1 x 4 x 4 cm Pyrex cuvette and illuminated at the front face with a broad beam of collimated light. The flux profile at 665 nm was determined in 1 mm steps with a 200 μm flat-ended optical fiber traversed parallel to the 4 cm optical path and perpendicular to the beam. The attenuation coefficient was obtained by non-linear least squares fitting of the profiles to Eq. (5), starting 3 to 5 mm from the front surface. The same suspensions were irradiated in a cylindrical 2 x 2 cm Pyrex cuvette under oxygen saturation with a 200 W Hg-Xe arc followed by a narrow-band 650 nm filter. The rate of enzyme inactivation was determined with a spectrophotometric assay using Nα-benzoyl-L-arginine ethyl ester as the substrate. Table 1 summarizes the results, where δ is the 1/e attenuation

Table 1. Optical Constants of Polystyrene Microsphere-Methylene Blue Suspensions at 650 nm[*]

Sphere Diameter	# δ	$ k	@ s'	& κ
(μm)	(cm)	(1/cm)	(1/cm)	
5.0	0.53	0.26	4.56	0.19
5.0	0.58	0.25	5.96	0.24
5.0	0.67	0.25	2.97	0.22
5.0	0.74	0.36	1.69	0.27
5.0	0.80	0.44	1.18	0.31
2.0	0.46	1.01	1.56	0.35
2.0	0.51	0.41	7.63	0.23
2.0	0.54	0.28	4.08	0.20
2.0	0.59	0.48	1.99	0.28
2.0	0.64	0.65	1.25	0.33
2.0	0.70	0.26	2.62	0.22
1.1	0.42	0.24	7.87	0.15
1.1	0.43	0.14	1.29	0.11
1.1	0.44	0.37	4.65	0.21
1.1	0.46	0.59	6.08	0.27
1.1	0.65	0.29	2.72	0.23
1.1	0.68	0.48	1.50	0.31
0.5	0.45	0.84	1.96	0.32
0.5	0.46	0.26	6.06	0.32
0.5	0.40	0.18	1.16	0.14
0.5	1.19	0.20	1.18	0.20

* Data taken from Grossweiner and Messina (1987).
# δ is the measured value of 1/e penetration depth.
$ k is the linear absorption coefficient of methylene blue at 650 nm.
@ s' is the apparent scattering coefficient.
& κ is the measured value of absorbed power on a relative scale.

depth corrected to 650 nm, k is the linear absorption coefficient of the dye and κ is the first-order rate constant for enzyme inactivation. The apparent attenuation coefficient for multiple scattering was calculated with the diffusion approximation: $s' = 1/3k\delta^2$, according to Eq. (3). Completely absorbing methylene blue solutions without microspheres led to $\kappa = 0.50$, corresponding to a photoinactivation quantum yield of 0.00116.

Calculation of Power Absorption with the Diffusion Approximation

When light scattering dominates over absorption, the complicated integro-differential equation of radiative transfer (Chandrasekhar, 1950) reduces to a diffusion equation for photons, analogous to the diffusion of material particles through a percolative medium (Morse and Feshbach, 1953). The diffusion approximation leads to profiles of the diffuse energy fluence rate or "space irradiance" for the assumed geometry and boundary conditions. The power absorption in a slab of unit cross-sectional area exposed to uniform illumination is given by:

$$P = k \int_0^L \phi(x) \, dx \qquad (1)$$

where ϕ is the space irradiance. The linear absorption coefficient was calculated from the measured extinction coefficient of the dye. The direct use of the scattering coefficient was avoided by expressing the results in terms of δ, which involves both k and s. This procedure is advantageous for multiple scattering conditions where s cannot be measured directly. The predicted values of P (expressed as fractional absorption) were compared to the data in Table 1 using linear regression.

The approximate form of the diffusion equation derived by Ishimaru (1978) for a collimated beam incident on a slab of particles can be expressed as:

$$d^2\phi(x)/dx^2 - 3ks\phi(x) = -Q_o\exp[-(s + k)x] \qquad (2)$$

where $Q_o = (3/4\pi)I[k\gamma + s(s + k)g]$, and $\gamma = s(1 - g) + k$. The parameter γ is the transport coefficient, g is the mean cosine of the scattering angle or anisotropy factor, and I is the incident irradiance. The least complicated case obtains by ignoring the right side of Eq. (2), which is valid for optically dense layers, except within a depth the order of $1/(s + k)$ from the illuminated surface. The resultant homogeneous diffusion equation can be expressed as:

$$d^2\phi/dx^2 - \phi/\delta^2 = 0 \qquad (3)$$

where $\delta = \sqrt{(1/3sk)}$. Following Svaasand and Ellingsen (1983), an approximate front surface boundary condition applicable to Eq. (3) is:

$$I = (1/4)\phi(x = 0) + (1/2)j(x = 0) \qquad (4)$$

where j is the radiant energy flow vector. The magnitude of j is given by: $j = -(1/3s)d\phi/dx$. Grossweiner and Messina (1987) assumed that the flux was uniformly isotropic at the rear surface, i.e., $d\phi/dx = 0$ at $x = L$. The corresponding solution to Eq. (3) is:

$$\phi(x) = \frac{4I\cosh[(L - x)/\delta]}{\cosh(L/\delta) + 2k\delta\sinh(L/\delta)} \qquad (5)$$

The fractional power absorption for this case is:

$$P = \frac{4k\delta}{\coth(L/\delta) + 2k\delta} \tag{6}$$

An alternative rear boundary condition was employed by Hemenger (1977), based on the assumption that the exiting flux extrapolates linearly to zero outside the interface. This condition can be expressed as $[d\phi/dx]_{x\,=\,L} = - s\phi(L)/\Delta$, where $\Delta = 0.7104$ (Morse and Feshbach, 1953). The solution to Eq. (3) with this rear boundary condition is:

$$\phi(x) = \frac{4I[e^{(L - x)/\delta} + \Gamma e^{-(L - x)/\delta}]}{(1 + 2k\delta)e^{L/\delta} + \Gamma(1 - 2k\delta)e^{-L/\delta}} \tag{7}$$

where $\Gamma = [(1 + s\delta/\Delta)/(1 - s\delta/\Delta)]$. The fractional absorbed power is:

$$P = \frac{4k\delta[(e^{L/\delta} - 1) + \Gamma(1 - e^{-L/\delta})]}{(1 + 2k\delta)e^{L/\delta} + \Gamma(1 - 2k\delta)e^{-L/\delta}} \tag{8}$$

More general solutions of Eq. (2) retain the source term on the right side, in which case s' = s(1-g). A recent analysis of Jacques and Prahl (1987) employs the boundary condition of Groenhuis et al. (1983), in which the total diffuse flux within the tissue at the air-tissue interface is equated to the reflected part of the diffuse flux directed outward. This condition requires a value for the internal reflection coefficient (r) at each interface. For uniformly diffuse flux, r depends only on the relative refractive index at the air-tissue interface (n). Specifically, $r = 1 - (1 - R')/n^2$, where R' is the regular or "specular" reflection coefficient for uniformly diffuse incident light (Kortum, 1969). However, r is lower for partially collimated flux and cannot be calculated exactly. The unknown value of g is expected to be close to unity for large scatterers. A practical procedure is to solve Eq. (2) with r and g as parameters in order to evaluate the sensitivity of the flux profiles to these quantities. The computer program of Jacques and Prahl (1987) based on Eq. (2) was employed in the present work.

Calculation of the Power Absorption with the Kubelka-Munk Model

The Kubelka-Munk model treats uniformly diffuse flux within a layer as if there are forward-directed and backward-directed fluxes, each subject to independent back-scattering and absorption events. The total flux ignoring specular reflection at the interfaces is given by:

$$E(x) = \frac{2I[\sinh\alpha(L - x) + \beta\cosh\alpha(L - X)]}{(1 + \beta^2)\sinh\alpha L + 2\beta\cosh\alpha L} \tag{9}$$

where $\alpha = \sqrt{K(K + 2S)}$ and $\beta = \sqrt{(K/K + 2S)}$. S and K are the Kubelka-Munk scattering and absorption parameters, which are approximately related to the true coefficients by: $k \simeq 2K$ and $s(1 - g) \simeq (4/3)S + (1/6)K$ (Groenhuis et al., 1983). The integration of KE(x) over the slab gives:

$$P = \frac{\cosh(L/\delta) + K\delta\sinh(L/\delta) - 1}{\cosh(L/\delta) + (1/2)(1/K\delta + K\delta)\sinh(L/\delta)} \tag{10}$$

where $\delta = 1/a$.

Calculation of Power Absorption with the Modified Beer-Lambert Law

A frequently used approximation for turbid media extends the Beer-Lambert law to include light scattering:

$$P = [k/(k + s)][1 - e^{-(k + s)L}] \tag{11}$$

Equation (11) was employed to test the validity of this approximation for the data in Table 1.

RESULTS

The theoretical models were compared to the experimental data by means of the linear correlation between κ and P. A good correlation indicates that theory is <u>consistent</u> with the data on a relative basis. However, it does not prove the theory is quantitatively correct or applicable to a broader range of parameters than those examined. Plots of κ vs P for the analytical solutions to Eq. (1) are shown in Fig. 1. The data for the different microsphere sizes fall on the same line, indicating that the scatterer size per se is not a determinant of the absorbed power, except to the extent that it affects the macroscopic coefficients. The presence of the microspheres reduced the quantum yield of the reaction from 0.00116 by a factor $\kappa/0.50$. The lower efficiency is due mostly to energy loss by diffuse reflection. A summary of the linear

Table 2. Correlation of Predicted Power Absorption with Measured Values

Model	Basis	# g	\$ r	Linear Regression[*] a	b	ρ	@ P_{ave}
Kubelka-Munk	Eq.(10)	-	-	- 0.1582	2.976	0.966	0.58
Beer-Lambert	Eq.(11)	-	-	- 0.1583	1.194	0.824	0.24
diff approx	Eq.(6)	-	-	- 0.1716	3.154	0.982	0.59
" "	Eq.(8)	-	-	- 0.1985	3.132	0.964	0.58
" "	&	0	0	0.1982	1.674	0.852	0.67
" "	&	0.5	0	0.2759	1.696	0.923	0.68
" "	&	0.8	0	0.3463	1.666	0.941	0.73
" "	&	0	0.2	0.2614	1.601	0.854	0.71
" "	&	0.5	0.2	0.3420	1.583	0.932	0.75
" "	&	0.8	0.2	0.4018	1.548	0.926	0.74
" "	&	0	0.5	0.4091	1.359	0.871	0.72
" "	&	0.5	0.5	0.4754	1.324	0.949	0.80
" "	&	0.8	0.5	0.5013	1.325	0.872	0.77

* P = a + bκ; ρ is the correlation coefficient.
\# g is the anisotropy factor.
\$ r is the internal reflection coefficient at front interface.
@ P_{ave} is the mean absorbed fractional power absorption for the range of measurements.
\$ Values calculated with computer program of Jacques and Prahl (1987).

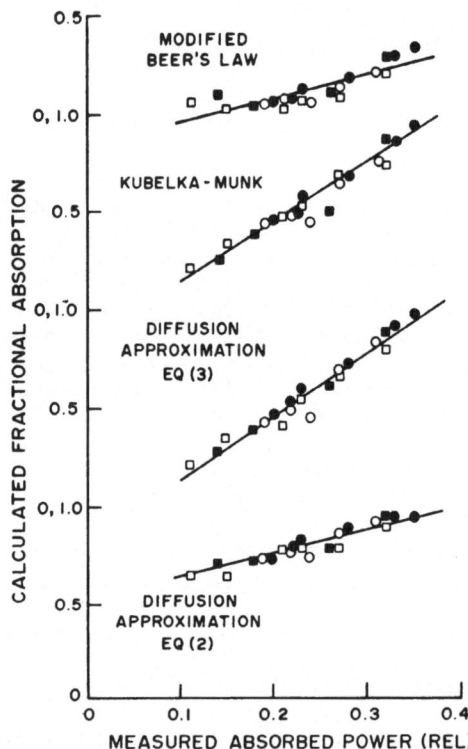

Figure 1. Correlation between measured and calculated power absorptions. The microsphere diameters are: ○ 5.0 μm, ● 2.0 μm, □ 1.1 μm, ■ 0.5μm. The value 0.5 on the horizontal axis corresponds to a photoinactivation quantum efficiency of 0.00116. The diffusion approximation of Eq. (2) was calculated with g = 0.5 and r = 0.5. The regression coefficients are given in Table 2.

regression coefficients in Table 2 shows that the Kubelka-Munk model and the two simplified forms of the diffusion approximation lead to similar regression coefficients. The last column in Table 2 is the mean of the highest and lowest values of the absorbed power predicted by the model. The modified Beer-Lambert law approximation predicts significantly smaller absorption values with greater scatter. The correlation with the solution to Eq. (2) derived by Jacques and Prahl (1987) was tested with their computer program using the "similarity principle". According to this procedure, as suggested by Jacques and Prahl (private communication), s , k, and g are transformed as: $k' = k$, $s' = s(1 - g)$, $g' = 0$. The analysis was carried out with $g = 0$, .5, and 0.8 and $r = 0$., 0.2, and 0.5. In general, the correlations show smaller values of the slope and larger intercepts than the Kubelka-Munk and the simplified diffusion approximations (Table 2). There is no obvious physical interpretation to this result, because a and b are correlation coefficients. However, the mean absorbed power was higher than predicted by the simpler theories. The fits for $g = 0.8$, $r = 0.0$ and $g = 0.5$, $r = 0.5$ (Fig. 1) gave the best statistics, although other combinations were not very different.

DISCUSSION

The present study differs from the usual case in which the objective is to calculate s, k, and g from measurements of diffuse reflectance (R) and transmission (T) on representative tissue samples. Instead, s was deduced from measurements of δ, under conditions where k is known. (In neutron diffusion physics, this approach is referred to as the "poisoned moderator" method.) The models leading to Eqs. (6), (8), and (10) assume uniformly diffuse flux, i.e., $g = 0$. The analytical methods have the advantages of few parameters and ease of computation. The best correlation was found with Eq. (6), which is applicable to highly turbid, optically dense systems. Equation (8) differs from Eq. (6) only in the rear boundary condition. Since most of the light absorption occurred in the frontal regions of the layer, the integrated absorptions are almost identical. However, the two solutions give different flux profiles near the rear face, which may be relevant for other applications. The Kubelka-Munk method led to almost the same results. There appears to be little justification for the approximation of Eq. (11), based on the additivity of k and s. The computer program BHMIE of Bohren and Huffman (1983) was used to calculate the values of g according to Mie theory for non-absorbing spheres. This program leads to the angular dependence of the S_{11} Stokes parameter. The average value of $\cos\theta$ was calculated by numerical integration, leading to 0.85 for 0.5 μm spheres, 0.94 for 1.1 μm spheres, and close to unity for 2 and 5 μm spheres. This limited range of g had no apparent effect on power absorption for the strongly scattering systems employed. It is not evident why the "similarity principle" is required for the calculation of R and T with the solution to Eq. (2) of Jacques and Prahl (1987), which is a point that should be clarified.

ACKNOWLEDGEMENTS

This work was supported by ONR Contract No. N00014-86-C-0188 and NIH Grant No. GM 20117. The assistance of E. W. Grossweiner of the Wenske Laser Center, Ravenswood Hospital Medical Center, Chicago, with the computer calculations is much appreciated.

REFERENCES

Bohren, C. F. and Huffman, D. R., 1983, "Absorption and Scattering of Light by Small Particles", John Wiley & Sons, New York.

Bonner, R. F., Nossal, R., Havlin, S., and Weiss, G. H., 1987, Model for photon migration in turbid biological media, J. Opt. Soc. Am. A, 4:423

Chandrasekhar, S., 1950, "Radiative Transfer", Clarendon Press, Oxford.

Doiron, D. R., Svaasand, L. O., and Profio, A. E., 1983, Light dosimetry in tissue: application to photoradiation therapy, in: "Porphyrin Photosensitization", D. Kessel and T. J. Dougherty, eds., Plenum Press, New York.

Groenhuis, R. A. J., Ten Bosch, J. J., and Ferwerda, H. A., 1983, Scattering and absorption of turbid materials determined from reflection measurements. 2: Measuring method and calibration, Applied Optics, 22:2463.

Grossweiner, L. I., 1986, Optical dosimetry in photodynamic therapy, Lasers Surg. Med, 6:462.

Grossweiner L. I., and Messina, J. W. 1987, Investigation of power absorption by methylene blue in a light scattering medium with an internal photodynamic actinometer, Photochem. Photobiol., 45:617.

Hemenger, R.P., 1977, Optical properties of turbid media with specularly reflecting boundaries: applications to biological problems, Appl. Optics, 16:2007.

Ishimaru, A., 1978, "Wave Propagation and Scattering In Random Media", Vol. 1, Academic Press, New York.

Jacques, S. L., and Prahl, S. A., 1987, Modeling optical and thermal distributions in tissue during laser irradiation, Lasers Surg. Med., 6:494.

Kortum, G., 1969, "Reflectance Spectroscopy", Springer-Verlag, Berlin, Heidelberg, New York.

Kubelka, P., and Munk, F., 1931, Ein Beitrag zur Optik der Farbanstriche, Z. Tech. Phys., 12:593.

Morse, P. M. and Feshbach, H., 1953, "Methods of Theoretical Physics", McGraw-Hill Book Co., New York.

Norris, K. H. and Butler, W. L., 1961. Techniques for obtaining absorption spectra on intact biological samples. IRE Trans. on Bio-Med. Appl., July, 1961:153.

Pratesi, R., Ronchi, L., Cecchi, G., Sbrana, G., Migliorini, M. G., Vecchi, C., and Donzelli, G., 1984, Skin optics and phototherapy of jaundice, Photochem. Photobiol., 40:77.

Svaasand, L. O. and Ellingsen, R., 1983, Optical properties of human brain, Photochem. Photobiol., 38:29.3

Welch, A. J., Yoon, G., and van Gemert, M. J. C., 1987, Practical models for light distribution in laser-irradiated tissue, Lasers Surg. Med., 6:488.

60

OPTICAL DOSIMETRY IN PHOTODYNAMIC THERAPY

Leonard I. Grossweiner

Physics Department, Illinois Institute of Technology
and Wenske Laser Center, Ravenswood Hospital Medical
Center, Chicago, Illinois, U.S.A.

INTRODUCTION

Photodynamic therapy (PDT) is an experimental treatment for malignant tumors utilizing the combined action of a tumor-localizing drug and
light. Consequently, a useful treatment plan must determine a drug dose
and light dose that maximize the probability of tumor eradication. The
drug dose is specified by the injection dose, which introduces uncertain-
ties about the concentration of photosensitizer in the tumor at the time
of light exposure. Recent dosimetry theories for PDT with hematoporphyrin derivative (HPD) and dihematoporphyrin ether (DHE) assume that the
drug concentration in the tumor is proportional to the injection dose
(Grossweiner, 1986; Potter et al., 1987; Grossweiner et al., 1987). The
specification of the light dose depends on the method of light delivery.
For front surface (FS) delivery from an external light source, the light
dose is specified by the incident fluence. For interstitial delivery
with optical fibers, the light dose is specified as energy output per
unit length for cylindrical diffusing tips (CI) and total energy output
for spherical diffusing tips (SI). The relationship between the incident
light dose and the energy absorbed by the photosensitizer must either be
measured or calculated with a tissue model. In an early approach of
Doiron et al. (1983), energy absorption by HPD in tissue at a given depth
was calculated with a modified form of the Beer-Lambert law. The total
attenuation coefficient was taken as the sum of the linear absorption
coefficient of HPD and the attenuation coefficient of the tissue. The
results lead to useful insights about optical flux profiles at different
wavelengths, although they are not directly applicable to PDT dosimetry.
Profio and Sarnaik (1984) showed that analysis of HPD fluorescence with
the diffusion approximation provides a potentially useful assay for the
tissue drug concentration. The diffusion approximation was employed to
calculate the optical flux profiles. The more general application of the
diffusion approximation to tissue optics was treated by Svaasand and
Ellingsen (1983) for FS and SI delivery. Potter et al. (1987) proposed a
theory of PDT dosimetry emphasizing photobleaching of the sensitizer by
the therapeutic light. The model is based on the concept of "photodynamic dose", defined as the integral of the sensitizer concentration
over the range of incident light dose. The analysis showed that the drug
dose and light dose are not "reciprocal" quantities, i.e., halving the
drug dose requires more than twice the light dose for the same biological
response. An independent analysis of Grossweiner (1986) is based on the

assumption that tumor eradication requires a minimum absorbed energy by the sensitizer, the value of which was estimated from clinical data for PDT of superficial tumors. The diffusion approximation was employed to calculate the incident light dose that provides the required absorbed energy in those tumor regions most distant from the source. The model is applicable to idealized planar, cylindrical, and spherical geometries, and incorporates a correction for sensitizer photobleaching. Dosimetry tables were calculated for head and neck squamous cell carcinoma ·(HNSCC) and compared retrospectively with the results of a clinical trial in which FS and CI deliveries were employed (Grossweiner et al., 1987). The present paper describes a more general approach, plus numerical estimations that lead to some additional insights about PDT kinetics.

OPTICAL DOSIMETRY MODEL

The dosimetry model is based on the calculation of the energy absorption rate by a photosensitizer in tissue with the relationship:

$$\partial E(r,t)/\partial t = \alpha c(r,t)\phi(r) \tag{1}$$

where E (J/cm^3) is the absorbed energy density, α $[cm(\mu g/g)]^{-1}$ is the absorption coefficient of the tumor-localized photosensitizer, c ($\mu g/g$) is the tissue concentration of photosensitizer, ϕ (W/cm^2) is the diffuse energy fluence rate or "space irradiance" at position r, and t (s) is the exposure time. Defining E^* as the minimum absorbed energy per unit volume for tumor eradication, the PDT dosimetry condition is:

$$E(d,t) = E^* \tag{2}$$

where d is the maximum distance from the light source to the tumor margin. It is assumed that the rate of sensitizer photobleaching is proportional to the rate of energy absorption:

$$- \partial c/\partial t = k(\partial E/\partial t) \tag{3}$$

where k $[(\mu g/g)/(J/cm^3)]$ is the specific photobleaching rate constant. Integration of Eq. (3), substitution in Eq. (1), and integration of Eq. (1) gives:

$$E(r,t) = (c_o/k)[1 - \exp(-k\alpha\phi(r)t] \tag{4}$$

where c_o is the initial sensitizer concentration, which is assumed to be uniformly distributed. It is evident from Eq. (4) that the maximum absorbed energy cannot exceed c_o/k.

The value of E^* may be estimated from PDT trials on superficial tumors. Svaasand and Ellingsen (1983) derived the following approximate solution for the space irradiance in an optically thick, planar layer exposed at uniform irradiance (I):

$$\phi(x) = 2I(1 + R)\exp(-x/\delta) \tag{5}$$

where R is the diffuse reflectance and δ is the 1/e optical penetration depth. As a simplifying assumption, the thickness of a superficial tumor is taken as equal to δ, typically 1-3 mm. Since Eq. (2) must apply equally for superficial and thicker tumors, the corresponding right sides of Eq. (4) may be equated, leading to:

62

$$(It) = (It)' \ [(1 + R')/(1 + R)]\exp(d/\delta - 1) \tag{6}$$

where the primed parameters refer to a superficial tumor and the unprimed to a thick tumor. Equation (6) provides a convenient method for the estimation of PDT light dose for a thick tumor by comparing it to a superficial tumor with comparable values of c_o and k, which need not be known. Alternatively, the value of E^*/c_o can be calculated with Eqs. (2), (4), and (5) using the best available parameters.

Equivalent solutions can be obtained for idealized cylindrical and spherical geometries. The result for an infinitely long cylinder of radius a located in an infinite volume of an isotropic medium is (Grossweiner et al., 1987):

$$\phi(r) = \frac{2J/\pi a}{K_o(a/\delta) + (2D/\delta)K_1(a/\delta)} \ K_o(r/\delta) \tag{7}$$

where J (W/cm) is output power per unit length of fiber, r (cm) is the radial distance from the fiber axis, and D (cm) is the light scattering diffusion constant of the tissue. [The value of D can be calculated from the diffuse reflectance of a semi-infinite planar with the relationship: $R = (1 - 2D/\delta)/(1 + 2D/\delta)$; Profio and Sarnaik, 1984]. The functions K_o and K_1 are modified Bessel functions of the second kind, which are tabulated in standard references. For a spherical source of radius a located in an infinite volume of isotropic medium the result is:

$$\phi(r) = \frac{(P/\pi a^2)}{[1 + 2D/a + 2D/\delta]} \ (a/r)\exp[-(r - a)/\delta)] \tag{8}$$

where P is the total output power.

The PDT dose is calculated by setting x = d in Eq. (5) or r = d in Eqs. (7) and (8), leading to the general form:

$$\phi(d) = \left.\begin{array}{c} I \\ J \\ P \end{array}\right\} x \ F(D,d/\delta,...) \tag{9}$$

where F represents the right side of each equation except for the appropriate incident power term. Substituting Eqs. (4) and (9) in Eq. (2) and rearranging leads to:

$$\left.\begin{array}{c} I \\ J \\ P \end{array}\right\} x \ t = -(1/k\alpha F)\ln(1 - kE^*/c_o) \tag{10}$$

where the left side is the PDT dose. When the photobleaching rate constant k is negligible, Eq. (10) reduces to:

$$\left.\begin{array}{c} I \\ J \\ P \end{array}\right\} x \ t = E^*/\alpha c_o F \tag{11}$$

Equation (10) can be employed to determine the light dose for different values of the drug dose. If t is the exposure time at drug dose c_o, the exposure time t_1 for drug dose c_1 is given by:

$$t_1/t = [\ln(1 - kE^*/c_1)]/[\ln(1 - kE^*/c_o)] \tag{12}$$

This result does not depend on the functional form of ϕ and should be valid for any geometry.

APPLICATIONS OF THE DOSIMETRY MODEL

PDT is currently being carried out with pre-determined protocols based on empirical dosimetry. In principle, analytical models can provide a rational approach to PDT dosimetry. However, models are based on simplifying assumptions and require values of parameters which usually are not known for a given tumor. The present model assumes that the tumor and immediately surrounding tissues are equivalent to an optically dense, randomly scattering medium with uniform values of the optical constants. In actuality, tumor tissue is inhomogeneous and variegated in color, with discernible boundaries. It is likely that the photosensitizer distribution in the tumor is not uniform, as assumed, because the drug is delivered by the vasculature. The model also assumes that the optical constants of the tissue remain constant during the treatment, which may not be valid when diffuse hemorrhaging or drainage occur. Notwithstanding these potential sources of error, the model provide some useful insights about PDT dosimetry, and leads to numerical results based on ab initio parameters that are consistent with clinical practice.

PDT of Thick Tumors

Equation (6) scales the PDT dose for a thick tumor treated by FS delivery to a thin tumor, by assuming they have the same values of c_o and k. For comparable values of the diffuse reflectance, each increase of d/δ by unity increases the PDT dose by a factor of e, e.g., for $\delta = 2$ mm, a 6 mm thick tumor should require a 7.4 fold higher incident dose than a 2 mm tumor. Svaasand (1984) noted that eradication of superficial tumors requires approximately 12 J/cm^2 when treated at 625-640 nm after injection of HPD at 2.5 to 5 mg/kg. Taking this dose as W, a 6 mm tumor should require 90 J/cm^2. This value is comparable to the FS dose employed for PDT of HNSCC reported by Grossweiner et al. (1987), e.g., tumors from 6 to 8 mm thick were treated at 75 to 100 J/cm^2 at 24 h post-injection 2 mg/kg DHE. Dougherty (1987) noted that typical depths of necrosis in skin lesions induced by PDT are in the 3 to 5 mm range with a DHE dose of 2.0 mg/kg and light doses from 20-50 J/cm^2. The prediction of Eq. (6) for $\delta = 2$ mm and d = 4 mm is 33 J/cm^2. Similar considerations show that a relatively opaque tumor (e.g., $\delta = 0.05$ mm) requires 250 J/cm^2 for 2 mm thickness and thicker tumors of this opacity cannot be treated. It may be concluded that highly pigmented tumors cannot be treated by PDT unless they are very thin.

Estimation of Parameters

Potter et al. (1987) measured the apparent rate constant (β) for DHE photobleaching at 630 nm in a variety of human tumors using an in vivo fluorescence assay, which led to $\beta = 0.036 \pm 0.006$ $l/(J/cm^2)$. The present analysis utilizes the 1st order rate constant in terms of the absorbed dose and not the incident dose as specified by β. An approximate relationship between the two quantities follows from Eqs. (1) and (3) by assuming $x = \delta$ for a superficial tumor:

$$k \simeq \beta e / 2\alpha (1 + R) \tag{13}$$

Taking α = 0.0048/cm(μg/g) for DHE in tissue at 630 nm (Profio et al., 1985) and R = 0.5 as typical of human skin leads to k = 6.5 (μg/g)/(J/cm^3). Substitution in Eqs. (4) and (5) with x = δ gives E^*/c_o = 0.054 (J/cm^3)/(μg/g). Based on 600 daltons as the formula weight of an HPD monomer unit, this value of E^*/c_o is equivalent to an overall PDT quantum efficiency of 0.0071 at 630 nm. This quantum efficiency is comparable to values reported for the photodynamic inactivation of enzymes by photosensitizing dyes (Grossweiner, 1976). If the putative action mechanism of singlet oxygen attack on tumor tissue membranes is correct, PDT appears to be a very efficient process, considering that singlet oxygen is an unstable intermediate and probably must react at multiple targets in order to effect tumor eradication.

Non-Reciprocity Between Light Dose and Drug Dose

The light dose required to compensate for changes in drug dose can be estimated with Eq. (12). The results in Table 1 for k = 6.5 and E^*/c_o = 0.054 show that a lower drug dose requires a higher than reciprocal light dose, and vice-versa for a higher drug dose.

Table 1. Non-Reciprocity of Light Dose and Drug Dose[a]

c_1/c_o	1/2	2/3	1	3/2	2
t_1/t	2.8	1.7	1	0.62	0.45

a) calculated for k = 6.5 and E^*/c_o = 0.054.

For these parameters, the lowest drug dose at which PDT is possible is 0.35 c_o. There is essentially no information about the concentration of DHE in human tumors. Assuming it is comparable to experimental rodent tumors, the order of 3-5 μg/g after interperitoneal injection of DHE at 5 mg/kg (Dougherty (1986), necrosis may not occur below about 2 μg/g. Potter et al. (1987) suggested that this DHE concentration occurs in normal tissue below 1 mg/kg injection dose. The same authors proposed that normal tissue can be spared in PDT by employing lower drug doses and increasing the light dose with a correction for photobleaching. The practicality of this procedure remains to be tested.

DISCUSSION

A dosimetry model for PDT based on the principle that tumor eradication requires a specific minimum energy absorption by the tumor-localized photosensitizer leads to values of the light dose in general agreement with clinical results. The model is not limited to PDT with HPD or DHE at 630 nm. The calculations can be carried out for other photosensitizers and wavelengths by employing the appropriate values of the key parameters: δ, α, c_o, D (or R), and k. A major problem at present is that literature values of these parameters may not be applicable to the specific tumor to be treated. The calculated doses are exponentially dependent on the value of δ and linearly dependent on the value of c_o. The measurement of both quantities in vivo with fiber optic surface probes is technically feasible. The availability of this information would provide a significant advance in the dosimetry, because the other parameters can be estimated to acceptable accuracy from in vitro measurements. The present model is similar in many respects to that of Potter et al.(1987). However, the latter was developed for FS exposures, whereas the present more general approach is applicable also to PDT with interstitial fibers. This aspect is especially important for thicker tumors, for which FS delivery leads to impractically long exposure times. Finally, it might be noted that controlled clinical trials of PDT mandate the randomized assignment of patients to dosimetry groups, in order to acquire the statistical data required to ascertain the efficacy and safety of the procedure. The potential value of dosimetry models will be realized at such a time that each tumor site can be treated under conditions which optimize the probability of achieving a complete response.

ACKNOWLEDGEMENT

This work was supported in part by NIH Grant No. GM 20117.

REFERENCES

Doiron, D. R., Svaasand, L. O. and Profio, A. E., 1983, Light dosimetry in tissue: application to photoradiation therapy, in: "Porphyrin Photosensitization" D. Kessel and T. J. Dougherty, eds., Pergamon Press, New York.

Dougherty, T. J., 1986, Photosensitization of malignant tumors, Seminars Surg. Oncol., 2:24-37.

Grossweiner, L. I., 1976, Photochemical inactivation of enzymes, Curr. Topics Radiat. Res. Quart., 11:141-199.

Grossweiner, L. I., 1986, Optical dosimetry in photodynamic therapy, Lasers Surg. Med., 6:462-466.

Grossweiner, L. I., Hill, J. H. and Lobraico, R. V., 1987, Photodynamic therapy of head and neck squamous cell carcinoma: Optical dosimetry and clinical trial, Photochem. Photobiol., 46:379-382.

Potter, W. R., Mang, T. S. and Dougherty, T. J., 1987, The theory of photodynamic therapy dosimetry: Consequences of photodestruction of sensitizer, Photochem. Photobiol., 46:97-101.

Profio, A. E. and Sarnaik, A. E., 1984, in: "Porphyrin Localization and Treatment of Tumors", D. R. Doiron and C. J. Gomer, eds., Alan R. Liss, Inc.,New York.

Profio, A. E., Wudl, L. R, and Sarnaik, J., 1985, Dosimetry methods in photoradiation therapy, in: "Methods in Porphyrin Photosensitization", D. Kessel, ed., Plenum Press, New York.

Svaasand, L. O. and Ellingsen, R., 1983, Optical Properties of human brain, Photochem. Photobiol., 38:293-299.

Svaasand, L. O., 1984, Thermal and optical dosimetry for photoradiation therapy of malignant tumors, in: "Porphyrins in Tumor Phototherapy", A. Andreoni and R. Cubeddu, eds., Plenum Press, New York.

SINGLET OXYGEN AND ELECTRON TRANSFER INDUCED OXYGENATION PATHWAYS IN

REACTIONS PHOTOSENSITIZED BY DICYANOANTHRACENE AND HYDROXY-ANTHRAQUINONES

Klaus Gollnick*, Albert Schnatterer, Gerald Utschick,
Uwe Paulmann, and Stephan Held

Institut für Organische Chemie der Universität München
Karlstrasse 23, D-8000 München 2, Fed. Rep. of Germany

INTRODUCTION

Some time ago, Schenck and Gollnick (1958) observed that ascaridole formation by hydroxy-anthraquinone-photosensitized oxygenation of α-terpinene, now recognized as a (4+2)-cycloaddition of singlet oxygen to the cyclic 1,3-diene system, was dependent on the number as well as on the positions of the OH groups in the anthraquinone moiety. Since naturally occurring quinones carrying hydroxy groups such as hypericin (for a recent review, see Duran and Song, 1986), cercosporin and dohistromin (Youngman and Elstner, 1984) are of importance in many biochemical processes, for example in photodynamic actions, it appears to be interesting to study the capability of model compounds such as anthraquinone and a series of its hydroxy derivatives as sensitizers of Type II (singlet oxygen) and Type I ((1) H-atom induced, and (2) electron transfer induced) photooxygenation reactions. Since the methods available for distinguishing between singlet oxygen and electron transfer induced oxygenation reactions have their severe drawbacks (Davidson et al., 1987), our approach is to apply simple, chemically well-defined systems which may be able to give a clear-cut answer with regard to the mechanisms involved. These results may be useful to photobiologists for their studies on phtooxygenations that proceed in the rather complicated biological systems. Although our only recently commenced studies on Type I and Type II photooxygenations sensitized by hydroxy-anthraquinones are far from being complete, first conclusions may be drawn from the following results.

RESULTS AND DISCUSSION

I. Phtosensitized Oxygenations of (+)-Limonene ($\underline{1}$) and (+)-α-Pinene ($\underline{8}$)

(+)-Limonene ($\underline{1}$) and (+)-α-pinene ($\underline{8}$) were irradiated in oxygen-saturated as well as in air-saturated solutions of CCl_4 and acetonitrile (MeCN) in the presence of typical singlet oxygen ($^1O_2$) producing sensitizers such as rose bengal (RB) (in MeCN) and tetraphenylporphin (TPP) (in CCl_4) as well as in the presence of 9,10-dicyanoanthracene (DCA), anthraquinone (AQ), and a series of hydroxy-AQs (the numbers indicate the positions of the OH groups).

Irrespective of the solvents used, products $\underline{2}$ through $\underline{7}$ from $\underline{1}$ and $\underline{9}$ from $\underline{8}$ were identified after reduction of the original hydroperoxide

Scheme I, A & B. Ene Reactions with Singlet Oxygen

A.

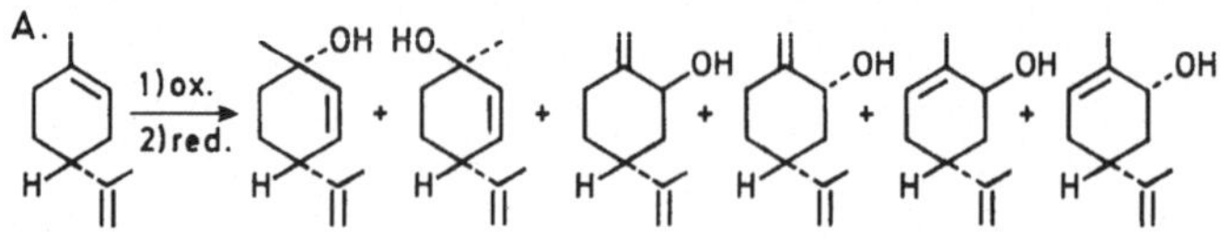

1) ox.
2) red.

(+)-1   (+)-2   (+)-3   (+)-4   (+)-5   (-)-6   (-)-7

$E_{ox} = + 1.87V$

$^1O_2$ [a]		39 ± 4 [c]	11 ± 1	19 ± 2	20 ± 2	8 ± 2	3 ± 1
Autox. [b]		17	18	0	0	34	31

a) sensitizers: AQ, hydroxy-AQs, and those reported by K. Gollnick and A.Schnatterer (1986) and lit. cited; b) Schenck et al. (1965); c) in %.

B.

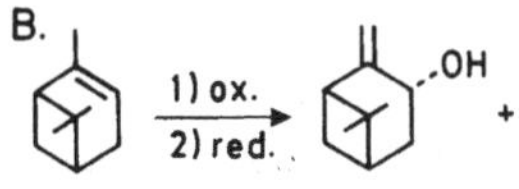

1) ox.
2) red.

(+)-8   (-)-9   + 7 further alcohols, aldehydes, ketones and epoxides [a]

$^1O_2$	100 %	----
Autox. [a]	14 %	86 %

a) Gollnick and Schenck (1964).

mixtures as singlet oxygen products. There is no indication for H-atom abstraction from (+)-1 by electronically excited sensitizers which should occur with allylic hydrogens at positions 3 and 6 of (+)-1 thus yielding 1:1-mixtures of (+)-2/(+)-3 and 1:1-mixtures of racemic 6/7 (Scheme 1, A). Similarly, (+)-8 yields (-)-pinocarveol (-)-9 exclusively with $^1O_2$, whereas H-atom abstraction induced oxygenation of 8 leads to mixtures of alcohols, aldehydes, ketones, and epoxides which contain some (-)-9 (about 15% in autoxidations, see Gollnick and Schenck, 1964). Singlet oxygen generated from $^1O_2$ by microwave discharge gave the same products with 1 (Gollnick and Schade, 1973) so that all the sensitizers used here (RB, TPP, DCA, AQ, and the hydroxy-AQs) are unequivocally shown to be efficient singlet oxygen generators. It is, therefore, not surprizing that cyclic 1,3-dienes such as α-terpinene and 2,5-dimethylfuran are easily oxygenated to ascaridole and 2,5-dimethylfuran endoperoxide, respectively (for 2,5-methylfuran/$^1O_2$ oxygenations, see Gollnick and Griesbeck, 1985, and lit. cited there), although (4+2)-cycloaddition products may be obtained by electron transfer induced oxygenations as well (Barton et al., 1972; Eriksen et al., 1977; Haynes, 1978; Tang et al., 1978). From the very small "half-value concentration of oxygen" ($4 \cdot 10^{-6}$ M for RB; $8 \cdot 10^{-5}$ M for 1,8-AQ, Gollnick and Franken, unpublished; Franken, 1969) it is, however, clear that $^1O_2$ is the oxygenating species produced from the triplet excited sensitizers leading to rate constants for ^{3}RB* (^{3} 1,8-AQ*) + $^3O_2$ ⟶ RB (1,8-AQ) + $^1O_2$ of $1.3 \cdot 10^9$ M^{-1} s^{-1} in accord with those obtained earlier for triplet xanthene dyes/$^3O_2$ interactions by flash photolysis experiments and cyclooctatetraene-quenched photooxygenations of 2,5-dimethylfuran in oxygen- and air-saturated solutions (Schenck and Gollnick, 1963; Gollnick and Schenck, 1964; Gollnick, 1968). It should further be noted that there

Scheme II, A & B. Ene-Reactions with Singlet Oxygen and Electron Transfer
Oxygenation that Requires Participation of Superoxide O_2^{-}

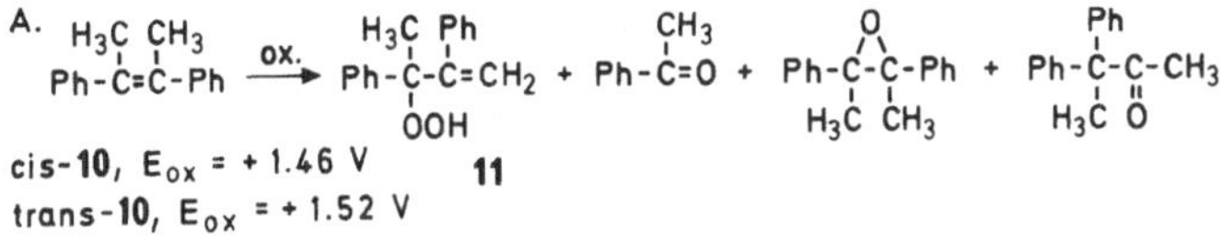

A.

$$
\begin{array}{c}
\underset{\substack{\text{H}_3\text{C} \;\; \text{CH}_3 \\ \text{Ph-C=C-Ph}}}{} \xrightarrow{\text{ox.}} \underset{\substack{\text{H}_3\text{C} \;\; \text{Ph} \\ \text{Ph-C-C=CH}_2 \\ \text{OOH}}}{} + \text{Ph-}\underset{\text{CH}_3}{\text{C}}\text{=O} + \text{Ph-}\underset{\substack{\text{H}_3\text{C} \;\; \text{CH}_3}}{\overset{\text{O}}{\text{C-C}}}\text{-Ph} + \text{Ph-}\underset{\substack{\text{H}_3\text{C} \;\; \text{O}}}{\text{C-C-CH}_3}
\end{array}
$$

cis-**10**, E_{ox} = + 1.46 V   **11**
trans-**10**, E_{ox} = + 1.52 V

$^1O_2$ 100 %   ---   ---   ---
$^1O_2$ + El.Tr./O_2^{-} a)   88 – 9 %   9 – 62 %   2 – 10 %   1 – 19 %

a) product distribution depends on reactions conditions such as initial
concentration of <u>10</u>, mediator conc., oxgen conc. (Gollnick, Schnatterer,1986)

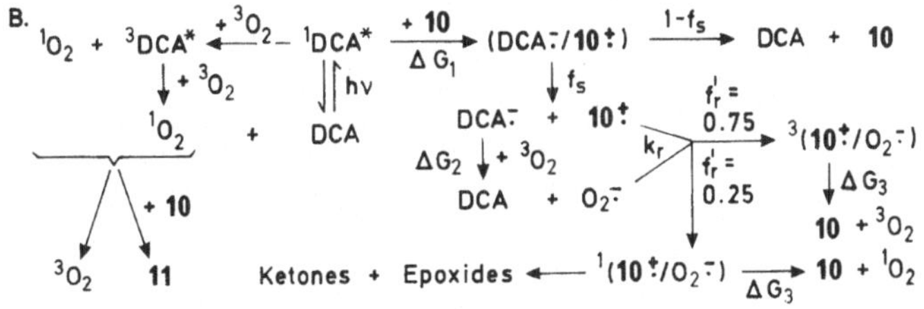

B.

$$\Delta G_1 = E_{ox}(\underline{10}) - E_{red}(sens) - e^2/\epsilon\, a - E_s(sens) \tag{1}$$

DCA: +1.46 + 0.89   – 0.06 – 2.88   = – 0.59 eV
1,8–AQ: +1.46 + 0.64   – 0.06 – 2.49   = – 0.45 eV

$$\Delta G_2 = E_{red}(sens) - E_{red}(O_2) \tag{2}$$

DCA: –0.89 + 0.78   = – 0.11 eV
1,8–AQ: –0.64 + 0.78   = + 0.14 eV

is obviously no electron transfer induced oxygenation proceeding, because
in this case one would expect the formation of acyclic carbonyl compounds
from the presumed dioxetane precursors formed via (2+2)-cycloaddition of
singlet oxygen to <u>1</u> and <u>8</u>, respectively.

II. Photosensitized Oxygenations of α,α'-Dimethylstilbenes (<u>10</u>)

Irradiations of cis-<u>10</u> and trans-<u>10</u> in oxygen-saturated as well as
in air-saturated CCl$_4$ or MeCN resulted in the exclusive formation of ene-
product <u>11</u> if RB in MeCN, TPP in CCl$_4$, AQ and the hydroxy-AQs in both solvents were used as sensitizers. The same result was obtained with DCA in
CCl$_4$. However, with DCA in MeCN, acetophenone, cis- and trans-epoxides of
10 and 3,3-diphenyl-2-butanone were formed in addition to allylic hydroper-
oxide <u>11</u>. Starting with $5 \cdot 10^{-2}$ M of cis-<u>10</u>, 12% of electron transfer induced oxygenation products were obtained in oxygen-saturated MeCN; in air-saturated solutions, these products increased to 49% (Scheme II, A).
Gollnick and Schnatterer (1986) have discussed these results at some length
showing that <u>11</u> is produced via the singlet oxygen pathway that competes
with the electron transfer induced oxygenation of <u>10</u> (Scheme II, B). The
latter oxygenation reaction gives rise to the formation of ketones and epoxides via the radical cation <u>10</u>$^{+}$ generated by electron transfer from <u>10</u> to
singlet excited DCA, ^{1}DCA*. The radical anion of DCA, DCA^{-}, produced by
this transfer, interacts with $^3O_2$ in a secondary electron transfer reaction

$$Ph_2C=CPh_2 \xrightarrow{\text{ox.}} Ph_2C=O + Ph_3C-OH + Ph_3C-\overset{\overset{\text{O}}{\|}}{C}-Ph$$

12

$E_{ox} = + 1.33V$

El.Tr./$O_2^{\overline{\cdot}}$ 85 %   10 %   5 %

$^1O_2$ No Reaction

to regenerate DCA and affords superoxide $O_2^{\overline{\cdot}}$ (Spada and Foote, 1980; Schaap
et al., 1980). This species is necessary for the production of ketones and
epoxides from 10, and DCA meets all the requirements for being a successful
sensitizer for the electron transfer induced photooxygenation of 10:
(1) the electron transfer from 10 to $^1DCA^*$ is exothermic; ΔG_1 , calculated
by using the Rehm-Weller equation (1) (Rehm and Weller, 1970) is negative
by about 0.6 eV in agreement with an almost diffusion-controlled electron
transfer step; (2) an appreciable amount of the radical cations 10^+ can
escape from the primary solvent cage it shares with $DCA^{\overline{\cdot}}$ thus avoiding
electron back-transfer; (3) electron transfer from $DCA^{\overline{\cdot}}$ to $^3O_2$ is exothermic; ΔG_2 calculated from equation (2) equals –0.11 eV by using a value
of –0.78 eV for the reduction potential of $^3O_2$ (Mattes and Farid, 1983).
As is shown for 1,8-AQ, electron transfer from 10 to singlet excited 1,8-AQ
should also occur in a nearly diffusion-controlled reaction (ΔG_1 = –0.45
eV). Since we have only recently begun to study the anthraquinones as photosensitizers for electron transfer oxygenations, we do not know yet which
electronically excited state (singlet, triplet, or both) is involved in
this step. Thus, the fact that only the singlet oxygen product 11 is obtained with hydroxy-AQ sensitizers could be due to a very inefficient electron transfer from 10 to either excited states of the hydroxy-AQs and/or
to a very inefficient escape of 10^+ from the common cage ($f_s \rightarrow 0$). But
this does not seem to be the reason because 1,8-AQ and all the other AQs do
sensitize the electron transfer oxygenation of 1,1-di-(p-anisyl)-ethylene
(13) (see Section IV). 1,8-$AQ^{\overline{\cdot}}$, however, should not (or only very inefficiently) be able to undergo an electron transfer with $^3O_2$ since this pro-
cess is endothermic (ΔG_2 = +0.14 eV) so that $O_2^{\overline{\cdot}}$ is not available for the
reaction with 10^+ to give the ketonic and epoxidic products.

III. Photosensitized Oxygenation of Tetraphenylethylene (12)

Our proposition that the absence of the electron transfer induced
oxygenation products from 10 in MeCN in case of AQ and hydroxy-AQs as sensitizers is due to the very inefficient production of superoxide is supported by the results obtained with tetraphenylethylene (12) as a substrate
(Scheme III). None of our sensitizers gives rise to any oxygenation of 12
in CCl_4 in accord with previous results which showed that 12 is inert towards $^1O_2$ (Eriksen et al., 1977) and that DCA in non-polar solvents such as
CCl_4 acts merely as a singlet oxygen sensitizer (Gollnick and Schnatterer,
1986). With the exception of DCA, none of our sensitizers oxygenated 12 to
any product in MeCN. With 12, electron transfer to the electronically excited AQ and hydroxy-AQs should even be better than with 10 since the oxidation potential of 12 (+1.33 V) is smaller than those of cis-10 (+1.46 V)
and trans-10 (+1.52 V). So again, it is the incapability of the anthraquinone radical anions to transfer an electron to oxygen that prevents the
oxygenation of 12 in MeCN which otherwise occurs so readily with DCA. It
should be noted, however, that according to the reduction potentials of AQ
and 1-AQ (Table 1), $O_2^{\overline{\cdot}}$ should be formed. That this is not (or only very
inefficiently) the case may be due to slightly too high reduction potentials reported for AQ (Takahashi et al., 1983) and 1-AQ (Kawenoki et al.,
1982), and/or to too low a reduction potential reported for $^3O_2$. Thus, for

Scheme IV, A. Electron Transfer Oxygenation Occurring as a Chain Reaction

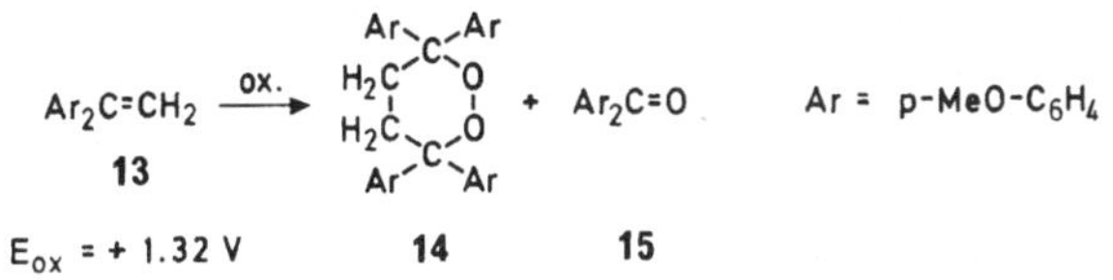

$$Ar_2C=CH_2 \xrightarrow{ox.} \begin{array}{c} Ar\ \ \ Ar \\ H_2C-C-O \\ | \ \ \ \ \ | \\ H_2C-C-O \\ Ar\ \ \ Ar \end{array} + Ar_2C=O \qquad Ar = p\text{-MeO-}C_6H_4$$

13   **14**   **15**

$E_{ox} = + 1.32$ V

Table 1. Electron Transfer Induced Photooxygenation of 13 in MeCN

Sens	E_{red} (V) [a]	E_S (eV) [a]	ΔG_1 (eV) [b]	14 (%)	15 (%)
DCA	– 0.89	2.88	– 0.63	97	3
1,4-AQ	– 0.65	2.35	– 0.44	71	29
1,8-AQ	– 0.64	2.49	– 0.59	58	42
AQ	– 0.94 [c]	2.73 [c]	– 0.53	57	43
1-AQ	– 1.00	2.50	– 0.24	47	53
2,6-AQ				45	55
1,2-AQ	– 0.72	2.37	– 0.39	39	61
RB				25	75

a) reduction potential and singlet energies after Kawenoki et al., 1982;   b) according to Rehm–Weller equ. (1);   c) reduction potential and triplet energy after Takahashi et al., 1983.

example, Eriksen et al. (1983) report a value of –0.87 V for AQ rather than a value of –0.94 V, and Sawyer et al. (1982) assume a value of –0.87 V for $^3O_2$ rather than a value of –0.78 V.

IV. Photosensitized Oxygenation of 1,1-Di-(p-anisyl)-ethylene (13)

Electron-donor substituted 1,1-diphenylethylenes such as 1,1-di-(p-anisyl)-ethylene (13) undergo 1,2-dioxane (14) formation in oxygen-saturated MeCN in high yields if irradiated in the presence of DCA (Gollnick and Schnatterer, 1984a, 1984b) or tetracyanoanthracene (TCA) (Mattes and Farid, 1986) (Scheme IV, A). With DCA as a sensitizer, 13 yields 97% of 14 and 3% of p,p'-dimethoxybenzophenone (15). A quantum yield of about eight clearly indicates that 14 is formed in a chain reaction thus requiring $^3O_2$ to participate in the oxidation step rather than $O_2^{\bar{}}$ (which may be involved in a chain termination step) (Scheme IV, B). As Table 1 shows, AQ and the hydroxy-AQs are also efficient sensitizers for the production of 14 as would be anticipated from the ΔG_1-values and the fact that $O_2^{\bar{}}$ is not required for this reaction. However, in contrast to DCA, AQ and the hydroxy-AQs give rise to the formation of substantial amounts of ketone 15, and the ratio of 14/15 is apparently dependent on the sensitizer used. Even RB appears to form 14 and 15, though in a very slow reaction. How ketone 15 is formed is not yet known. Since neither TCA nor AQ and the hydroxy-AQs are capable of producing $O_2^{\bar{}}$, 15 cannot be the result of superoxide addition to radical cation $13^{\pm}$. Mattes and Farid (1986) reject the possibility that ketone formation occurs via addition of $^3O_2$ to $13^{\pm}$ since such reactions are very slow and cannot compete with the almost diffusion-controlled addition of $13^{\pm}$ to olefin 13. They propose that ketone formation proceeds via reaction of $^3O_2$ with the dimeric radical cation, 13–$13^{\pm}$, in competition with the dioxane formation. Since, however, the ratio of 14/15 appears to be sensitizer-dependent, neither $^3O_2$ addition to $13^{\pm}$ nor $^3O_2$ addition to 13–$13^{\pm}$ should be involved in ketone formation. In order to explain the sensitizer-dependent ratio of 14/15, interaction of $^3O_2$ with the radical ion pairs (sens$^{\bar{}}$/$13^{\pm}$) to some unknown intermediate or singlet oxygen oxygenation of 13 may be invoked which, probably via an unstable 1,2-dioxetane, may then give rise to ketone formation.

$$\text{Sens}^* + \text{Ar}_2\text{C=CH}_2 \longrightarrow (\text{Sens}^{\bar{\cdot}}/\text{Ar}_2\overset{+}{\text{C}}\text{-}\overset{\cdot}{\text{C}}\text{H}_2) \longrightarrow \text{Sens} + \text{Ar}_2\text{C=CH}_2$$

$$\text{13} \qquad\qquad\qquad\qquad\qquad\qquad\qquad\qquad \text{13}$$

$$\Big\downarrow + {}^3\text{O}_2 \qquad\qquad \Big\downarrow + {}^3\text{O}_2$$

$$\text{Sens} + {}^1\text{O}_2 \xrightarrow[\text{?}]{+ \text{13}} \text{Ar}_2\underset{\overset{|}{\text{O-O}}}{\text{C-CH}_2} \qquad \text{Sens}^{\bar{\cdot}} + \text{13}^{\ddagger}$$

$$\text{15} + \text{CH}_2\text{O} \qquad\qquad \text{Sens} + \text{O}_2^{\bar{\cdot}} \qquad\qquad \Big\downarrow + \text{13}$$

$$\text{14} + \text{13}^{\ddagger} \xleftarrow[\text{2) } + \text{13}]{\text{1) } + {}^3\text{O}_2} \qquad \text{13-13}^{\ddagger}$$

If $^1\text{O}_2$ is involved, the dioxetane must be formed in a rather inefficient reaction with 13 since in CCl_4 oxygenation of 13 is very slow with all sensitizers yielding some 15; but we have still to determine whether other products are formed besides 15. 14, however, does not seem to be contained in the reaction mixture. If, in the case of AQ and the hydroxy-AQs as sensitizers singlet oxygen should turn out to be the oxidizing species that produces 15, this result would mean that electron back-transfer from (sens$^{\bar{\cdot}}$/13‡.) to give sens + 13 would be rather enhanced with sens = hydroxy-AQs in comparison with sens = DCA; i.e., the value of f_S = 0.06 for DCA (assumed to be the same as in case of 10) should be reduced for the hydroxy-AQs appreciably. Such an enhanced electron back-transfer may be expected since the exothermicity of this process given by

$$\Delta G_3 = E_{red}(\text{sens}) - E_{ox}\ (\underline{13}) \qquad\qquad (3)$$

decreases from −2.21 eV with DCA to −1.96 eV with 1,8-AQ as sens. Mattes and Farid (1983) have shown that such a relationship exists for the formation of solvent-separated radical ions in electron transfer reactions of various donors with ^{1}DCA* and explained these results in terms of the gap theory for radiationless decay and the Marcus "inverted region" of electron transfer (see also Schaap et al., 1984). However, within the series of AQs, ΔG_3 decreases from 1-AQ (−2.32 eV) to 1,4-AQ (−1.97 eV) so that one would expect less dioxane (14) formation with 1,4-AQ than with 1-AQ as a sensitizer. We are presently engaged with more quantitative work (fluorescence quenching, oxygen concentration effects, mediator (co-sensitizer) effects, etc.) in order to elucidate the mechanistic details of these reactions.

V. Photosensitized Oxygenation and Retro-di-π-methane Rearrangements of 1,1,2,3-Tetraarylcyclopropanes (16)

1,1,2,3-Tetraarylcyclopropanes 16a-f do not react with singlet oxygen in CCl_4 and MeCN. With DCA as a sensitizer, 16a-c give rise to quantitative formation of 3,3,4,5-tetraaryldioxolanes 17a-c in oxygen-saturated MeCN, whereas 16d-f reacted under the same conditions exclusively to the corresponding 1,1,3,3-tetraarylpropenes 18d-f from which the cyclopropanes were synthesized by a di-π-methane rearrangement (Gollnick and Paulmann, in press). (For conversion of 1,1,3,3-tetraphenylpropene to trans-1,1,2,3-tetraphenylcyclopropane, see Griffin et al., 1965). We have explained these results and the fact (1) that under N_2 cyclopropanes 16a-c are completely inert, and (2) that retro-di-π-methane rearrangements of

Scheme V. Electron Transfer Oxygenation that Requires Participation of Superoxide $O_2^{\overline{\cdot}}$ and Electron Transfer Induced "Retro-di-π-methane Rearrangement"

(Di-π-methane Rearrangement)

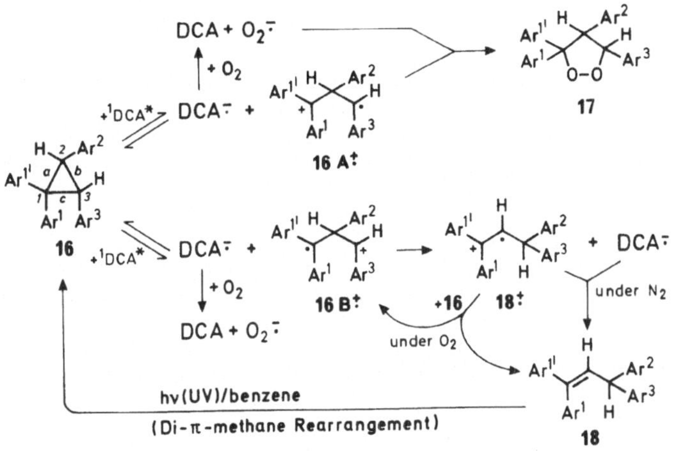

Table 2. DCA-Photosensitized Reactions of 1,1,2,3-Tetraarylcyclopropanes

$\underline{16}^a$	R^b in				$E_{ox}(V)^c$	in O_2-satur. MeCN^d			in N_2-sat.MeCN^d	
	Ar^1	$Ar^{1'}$	Ar^2	Ar^3		$t_{irr}(h)$	$\underline{17}(\%)$	$\underline{18}(\%)$	$t_{irr}(h)$	$\underline{18}(\%)$
$\underline{a}$	OMe	H	H	H	1.37	5	100^e	0^e	24	0^e
$\underline{b}$	OMe	OMe	H	H	1.28	1	100^e	0^e	24	0^e
$\underline{c}$	Me	Me	H	H	1.43	20	100^e	0^e	13	0^e
$\underline{d}$	H	H	H	OMe	1.31	1	0	95	13	25^f
$\underline{e}$	H	H	OMe	OMe	1.16	1	0	100	13	100^f
$\underline{f}$	H	H	Me	Me	1.42	1	0	45	13	60^f

a init.conc.: $2.5 \cdot 10^{-2}$ M;   b Ar = p-R-C_6H_4;   c irrev. ox. pot. vs. SCE in MeCN; scan rate 400 mV/s;   d no other products were observed; e no products $\underline{17}$ and/or $\underline{18}$ were obtained with 1-AQ, 1,4-AQ and 1,8-AQ; f product $\underline{18}$ was also obtained with 1-AQ, 1,4-AQ, and 1,8-AQ as sens.

cyclopropanes $\underline{16d}$-$\underline{f}$ occur much more readily in the presence of oxygen than under nitrogen (Table 2) by a mechanism represented in Scheme V. We propose that electron transfer from $\underline{16}$ to ^{1}DCA* occurs always under C-C bond cleavage between C-1 and C-3 giving rise to 1,3-radical cations $\underline{16A}^{\ddag}$ or $\underline{16B}^{\ddag}$. Which of the two radical cations is formed is proposed to depend on the stabilization of the cationic site by assuming that one p-anisyl or p-tolyl group stabilizes the cation better than even two phenyl groups. $\underline{16a}$-$\underline{c}$ should thus lead to $\underline{16A}^{\overset{+}{\cdot}}$, $\underline{16d}$-$\underline{f}$ should lead to $\underline{16B}^{\overset{+}{\cdot}}$. Only the latter carrying a secondary cationic site should rearrange to radical cation $\underline{18}^{\ddag}$

which suffers an electron back-transfer (yielding 18) from DCA$\bar{\cdot}$ present in the common solvent cage in N_2-saturated MeCN. In O_2-saturated MeCN, some of the DCA$\bar{\cdot}$ present in the common cage with 16B$\overset{+}{\cdot}$ may be trapped by O_2 thus producing a higher yield of "free" radical cation 16B$\overset{+}{\cdot}$ that undergoes rearrangement to "free" radical cation 18$\overset{+}{\cdot}$. The latter should react with cyclopropane 16 to give 18 and again a "free" radical cation 16B$\overset{+}{\cdot}$ in an almost diffusion-controlled electron transfer process since the oxidation potentials of 16 are always appreciably lower than those of the corresponding propenes 18 (e.g., E_{ox} (18b) = 1.38 V, E_{ox} (18e) = 1.49 V). Therefore, in the presence of O_2, retro-di-π-methane rearrangement should occur in a chain reaction, i.e. with a rate rather enhanced compared to that observed in N_2-saturated MeCN.

In accord with the results obtained with AQ and hydroxy-AQs as sensitizers for electron transfer induced oxygenations in Sections II through IV, cyclopropanes 16a-c being inert if irradiated in O_2-saturated MeCN in the presence of these sensitizers obviously need superoxide $O_2\bar{\cdot}$ rather than $^3O_2$ in order to produce dioxolanes 17a-c, in contrast to the results obtained for dioxolane formation with 1,2-diarylcyclopropanes (Mizuno et al., 1985) and 2,2-diphenyl-1-methylenecyclopropane (Takahashi et al., 1983). Cyclopropanes 16d-f are slowly converted to the corresponding propenes 18d-f if irradiated under the same conditions in the presence of hydroxy-AQs. This shows that the hydroxy-AQs are able to abstract an electron from the cyclopropanes as is anticipated from the oxidation and reduction potentials of the reactants. As with the 1,1-diarylethylenes, we are presently studying these reactions in more detail in order to elucidate the mechanistic pathways taken in hydroxy-AQ-photosensitized oxygenation reactions.

CONCLUSIONS

Besides the well-known singlet oxygen photosensitizers RB and TPP (and many others such as chlorophyll, hematoporphyrin, zinc-TPP, methylene blue, etc.), DCA, AQ and hydroxy-AQs are efficient singlet oxygen producers in non-polar and polar solvents. DCA was recognized as a singlet oxygen producing sensitizer independently by Steichen and Foote (1981), Santamaria (1981) and Schnatterer (1982). DCA, AQ and the hydroxy-AQs and even RB (though in a rather inefficient way) are sensitizing electron transfer induced photooxygenations in polar solvents such as MeCN, but they appear to be no (or only very inefficient) sensitizers for H-atom transfer induced photooxygenation reactions. However, whereas DCA sensitizes electron transfer induced photooxygenations that require superoxide $O_2\bar{\cdot}$ as well as those that proceed with $^3O_2$ in chain reactions, only the latter appear to be sensitized by AQ and the hydroxy-AQs due to the fact that the secondary electron transfer from the anthraquinone radical anions to oxygen should be endothermic and thus proceed inefficiently (if at all) in MeCN at room temperature.

REFERENCES

Barton, D.H.R., Leclerc, G., Magnus, P.D., and Menzies, I.D., 1972, An unusual synthesis of ergosterol acetate peroxide, J. Chem. Soc., Chem. Comm., 447.

Davidson, R.S., Goodwin, D., and Pratt, J., 1987, Problems associated with distinguishing between singlet oxygen and electron transfer photooxygenation reactions, Free Rad. Res. Comm., 2:313.

Duran, S. and Song, P.-S., 1986, Hypericin and its photodynamic action, Photochem. Photobiol., 43:677.

Eriksen, J., Foote, C.S., and Parker, T.L., 1977, Photosensitized oxygenation of alkenes, J. Am. Chem. Soc., 99:6455.

Eriksen, J., Lund, H., and Nyvad, A.I., 1983, Electron transfer fluorescence quenching of radical ions, Acta Chem. Scand. B, 37:459.

Franken, T., 1969, Beiträge zur Kenntnis der Typ II photosensibilisierten Oxygenierungsreaktionen in Lösung, Dissertation, University of Bonn.

Gollnick, K., 1968, Type II photooxygenation reactions in solutions, Adv. Photochem., 6:1.

Gollnick, K. and Schenck, G.O., 1964, Mechanism and stereoselectivity of photosensitized oxygen transfer reactions, Pure Appl. Chem., 9:507.

Gollnick, K. and Schade, G., 1973, Mikrowellenentladung von CO_2 : eine neue, ergiebige Quelle für Singulett-Sauerstoff, O_2 ($^1\Delta_g$), Tetrahedron Lett., 857.

Gollnick, K. and Schnatterer, A., 1984a, Formation of a 1,2-dioxane by electron transfer photooxygenation of 1,1-di(p-anisyl)ethylene, Tetrahedron Lett., 25:185.

Gollnick, K. and Schnatterer, A., 1984b, Formation of 1,2-dioxanes by electron transfer photooxygenation of 1,1-disubstituted ethylenes, Tetrahedron Lett., 25:2735.

Gollnick, K. and Griesbeck, A., 1985, Singlet oxygen photooxygenation of furans. Isolation and reactions of (4+2)-cycloaddition products (unsaturated sec.-ozonides), Tetrahedron, 41:2057.

Gollnick, K. and Schnatterer, A., 1986, 9,10-Dicyanoanthracene-sensitized photooxygenation of α, α'-dimethylstilbenes. Mechanism and kinetics of the competing singlet oxygen and electron transfer photooxygenation reactions, Photochem. Photobiol., 43:365.

Gollnick, K. and Paulmann, U., Electron transfer induced photooxygenation and "retro-di-π-methane rearrangement" of 1,1,2,3-tetraarylcyclopropanes, Tetrahedron, in press.

Gollnick, K. and Franken, T., manuscript in preparation.

Griffin, E.W., Marcantonio, A.F., Kristensen, H., Petterson, R.C., and Irving, C.S., 1965, Photocyclization of propenes to cyclopropanes. Novel phenyl and hydrogen migrations in π,π* systems, Tetrahedron Lett., 2951.

Haynes, R.K., 1978, Lewis-acid-catalyzed oxygenation of 1,1-bicyclohexenyl and α-terpinene. Reactions in dichloromethane and liquid sulphur dioxide, Austr. J. Chem., 31:131.

Kawenoki, I., Keita, B., Kossanyi, J., and Nadjo, L., 1982, Reversible electron transfer between the singlet excited state of variously substituted anthraquinones and electron donors. Free activation energy influence, Nouveau J. Chim., 6:387.

Mattes, S.L. and Farid, S., 1983, Photochemical electron transfer reactions of olefins and related compounds, Organic Photochemistry, 6:233.

Mattes, S.L. and Farid, S., 1986, Photochemical electron transfer reactions of 1,1-diarylethylenes, J. Am. Chem. Soc., 108:7356.

Mizuno, K., Kamiyama, N., Ichinose, N., and Otsuji, Y., 1985, Photooxygenation of 1,2-diarylcyclopropanes via electron transfer, Tetrahedron, 41:2207.

Rehm, D. and Weller, A., 1970, Kinetics of fluorescence quenching by electron and H-atom transfer, Isr. J. Chem., 8:259.

Santamaria, J., 1981, Photo-oxygenation de composes aromatiques sensibilisees par des accepteurs d'electron, Tetrahedron Lett., 22:4511.

Sawyer, D.T., Chiericato, G., Angelis, C.T., Nanni, E.J., and Tsuchija, T., Effects of media and electrode materials on the electrochemical reduction of dioxygen, 1982, Anal. Chem., 54:1720.

Schaap, A.P., Zaklika, K.A., Kaskar, B., and Fung, W.-M., 1980, Formation of 1,2-dioxetanes via 9,10-dicyanoanthracene-sensitized electron transfer processes, J. Am. Chem. Soc., 102:389.

Schaap, A.P., Siddiqui, S., Prasad, G., Palomino, E., and Lopez, L., 1984, Cosensitized electron transfer photooxygenation: the photochemical preparation of 1,2,4-trioxolanes, 1,2-dioxolanes and 1,2,4-dioxazolidines, J. Photochem., 25:167.

Schenck, G.O. and Gollnick, K., 1958, Cinetique et inhibition de reaction photosensibilisees en presence d'oxygene moleculaire, J. Chim. Physique, 55:892.

Schenck, G.O. and Gollnick, K., 1963, Über die schnellen Teilprozesse photosensibilisierter Substrat-Übertragungen; Untersuchungen über Chemismus und Kinetik der durch Xanthenfarbstoffe photosensibilisierten O_2-Übertragungen, Forschungsber. d. Landes Nordrhein-Westfalen, Nr. 1256, Westdeutscher Verlag, Köln und Opladen.

Schenck, G.O., Neumüller, O.A., Schroeter, S., and Ohloff, G., 1965, Zur Autoxidation des (+)-Limonens, Liebigs Ann. Chem., 687:26.

Schnatterer, A., 1982, Dimethylstilbene als Sonde für Singulett-Sauerstoff- und Elektronen-Transfer Photooxygenierungen, Diplomarbeit, University of München.

Spada, L.T. and Foote, C.S., 1980, Detection of radical-ion intermediates in the cyanoaromatic-sensitized photooxidation of trans- and cis-stilbene, J. Am. Chem. Soc., 102:391.

Steichen, D.S. and Foote, C.S., 1981, Indirect sensitized photooxygenation of aryl olefins, J. Am. Chem. Soc., 103:1855.

Tang, R., Yue, H.J., Wolf, J.F., and Mares, F., 1978, Oxygenation of cyclic dienes to endoperoxides, J. Am. Chem. Soc., 100:5248.

Takahashi, Y., Miyashi, T., and Mukai, T., 1983, Trimethylenemethane cation radical: photosensitized (electron transfer) generation and reactivity, J. Am. Chem. Soc., 105:6511.

Youngman, R.J. and Elstner, E.F., 1984, Photodynamic and reductive mechanisms of oxygen activation by the fungal phototoxins, cercosporin and dothistromin, p.501, in "Oxygen Radicals in Chemistry and Biology", Bors, W., Saran, M., and Tait, D., Eds., W. de Gruyter, Berlin.

MEDIATING EFFECTS OF DNA ON SOME PHOTOCHEMICAL PROCESSES

Stephen J. Atherton

Center for Fast Kinetics Research
University of Texas at Austin
Austin, Texas 78712

INTRODUCTION

The work which follows may be thought of in terms of photosensitization of simple one electron transfer reactions and of the way in which the DNA environment effects both their yields and rates. Electron transfer reactions in DNA have important biological consequences; for example, they are involved in the formation and repair of thymine dimers which are the major damaging species produced in the cell by UV light (Jorns, 1987). Also, abnormally high mobilities for electrons in DNA have been reported (Van Lith et al., 1986), a phenomenon which may be of considerable importance in the understanding of many biological processes. The following studies make no pretense of being actual biological processes which commonly occur in living systems. The reactions studied are, however, electron transfer reactions which are caused to take place 'in the DNA matrix, and as such are valid reflections of the way DNA mediates such processes.

The quenching of ethidium bromide (EB) excited singlet states by Cu^{2+} and by methylviologen (MV^{2+}) are described. In all the experiments the reactants are almost exclusively associated with DNA. Binding constants are $2.6 \times 10^6 M^{-1}$ for EB (Gaugain et al., 1978), $1.8 \times 10^5 M^{-1}$ for MV^{2+} (Fromherz and Rieger, 1986) and from ca. 10^3 to $10^5 M^{-1}$, depending on ionic strength and method of determination, for Cu^{2+} (Izatt et al., 1971).

Both the ground state absorption and fluorescence characteristics of EB are a function of the solvent. As the polarity of the solvent decreases, the absorption and emission shift to the red and the emission quantum yield increases. The lifetime of EB fluorescence is similarly a function of solvent, increasing from 1.8 ns in water to 10.6 ns in acetonitrile and 23 ns in aqueous DNA solutions (Burns, 1969). These solvent-induced changes in the photophysical parameters of EB are useful in the present study since they would serve to indicate any significant change in the physical environment of EB caused by the addition of Cu^{2+} or MV^{2+}.

EXPERIMENTAL

DNA was Calf Thymus Type 1 from Sigma and either used as received or purified by phenol extraction (Maniatis et al., 1982). Water was purified by passage through a Millipore filtration system. In all DNA solutions the supporting electrolyte (Na_2SO_4) concentration was always greater than $10^{-3}M$ to inhibit denaturation (Eichhorn, 1975). The pulse radiolysis and laser flash photolysis equipment has been described previously (Foyt, 1981; Atherton and Beaumont, 1987). For laser flash photolysis the excitation source was the 532 nm (ca. 11 ns pulse width) second harmonic of either a Quantel YG 481 or YG 581 Nd:YAG laser. Beam energies were always less than 100 mJ per cm^2.

RESULTS AND DISCUSSION

No changes in the ground state absorption spectrum of DNA-intercalated EB were observed on the addition of Cu^{2+} or MV^{2+}, under all conditions of concentration studied. The emission spectrum was similarly unchanged, except in intensity due to quenching of EB excited singlet states. This is evidence against significant changes in the environment of EB being caused by the addition of Cu^{2+} or MV^{2+}. It also argues against ground state complex formation between EB and either Cu^{2+} or MV^{2+}. Figure 1(a) shows the emission spectra of DNA-intercalated EB with and without the addition of 2 x $10^{-4}M$ Cu^{2+} as a function of salt (Na_2SO_4) concentration. Figure 1(b) is identical except that MV^{2+} is substituted for Cu^{2+}. Both Cu^{2+} and MV^{2+} show a very high efficiency of quenching which can only be interpreted in terms of both reactants being concentrated in the DNA matrix. The rate constants for quenching of EB excited singlet states by Cu^{2+} and MV^{2+} in water (no DNA) have been measured as 6.8 ± 0.5 x $10^9 M^{-1}s^{-1}$ (Atherton and Beaumont, 1984) and ca. $10^9 M^{-1}s^{-1}$ (Fromherz and Rieger, 1986) respectively. Thus if both reactants are in water, 0.2% and 0.04% of EB excited singlet states should be quenched by 2 x $10^{-4}M$, Cu^{2+}, and MV^{2+} respectively. If EB was bound and the quencher free in bulk solution, then these percentages may rise due to the increase in EB fluorescence lifetime; but it is inconceivable that they would approach the efficiency seen in Figure 1. Clearly the presence of DNA leads to large increases in the efficiency of fluorescence quenching. It is important to note that this increase in efficiency is no more than may be accounted for quite simply, in terms of concentration effects, kinetic salt effects, increase of EB fluorescence lifetime on binding to DNA etc. In fact, Fromherz and Rieger, 1986, after consideration of all the probable mechanisms of increased quenching by MV^{2+}, due to the presence of DNA, find that the efficiency should be ca. 100 times higher than observed. There is no evidence that the DNA changes the energetics of the reaction, for example.

There is also a large dependence of the quenching efficiency on salt concentration, for both MV^{2+} and Cu^{2+}. This may be explained for Cu^{2+} in terms of polyelectrolyte theory (Manning, 1977; 1978), which predicts that the charge fraction of a linear polyelectrolyte (e.g., DNA) is constant over a large range of ionic strength, at least up to 0.1 M. Thus as the salt concentration rises, Na^+ ions will replace Cu^{2+} ions at the helix, whilst maintaining the DNA charge fraction, leading to a decrease in EB fluorescence quenching. Polyelectrolyte theory only applies to binding caused by electrostatic interaction between the charged groups on the polymer and counterions. Since there is a possibility of hydrophobic interaction between MV^{2+} and DNA, it is difficult to know to what extent polyelectrolyte theory is applicable in this case. However, the salt dependencies of fluorescence quenching are of comparable magnitude for Cu^{2+} and MV^{2+}, and intercalative binding does not have a large dependence on ionic strength (Le Pecq and Paoletti, 1967; Armstrong et al., 1970). This is evidence against intercalation of MV^{2+} in DNA and supports binding of MV^{2+} to the exterior of DNA via electrostatic attraction.

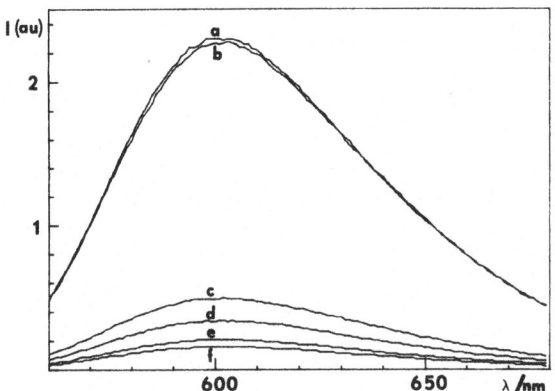

Fig. 1(a). Emission spectra of 3.6×10^{-4}M DNA and 4×10^{-5}M
EB with a. 10^{-2}M Na_2SO_4; b. 1.25×10^{-3}M Na_2SO_4;
c. 2×10^{-4}M $Cu^{2+} + 10^{-2}$M Na_2SO_4; d. 2×10^{-4}M
$Cu^{2+} + 5 \times 10^{-3}$M Na_2SO_4; e. 2×10^{-4}M $Cu^{2+} + 2.5$
$\times 10^{-3}$M Na_2SO_4; f. 2×10^{-4}M $Cu^2 + 1.25 \times 10^{-3}$M
Na_2SO_4. Excitation at 532 nm.

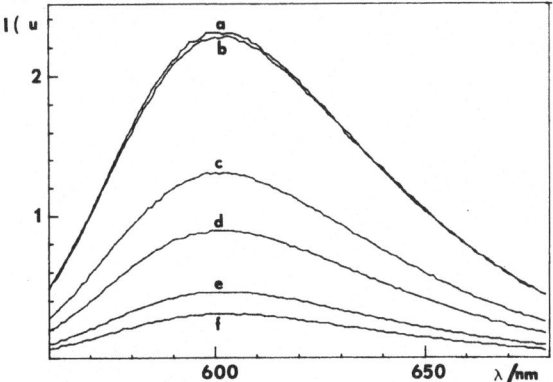

Fig. 1(b). Emission spectra of 3.6×10^{-4}M DNA and 4×10^{-5}M
EB with a. 10^{-2}M Na_2SO_4; b. 1.25×10^{-3}M Na_2SO_4;
c. 2×10^{-4}M $MV^{2+} + 10^{-2}$M Na_2SO_4; d. 2×10^{-4}M
$MV^{2+} + 5 \times 10^{-3}$M Na_2SO_4; e. 2×10^{-4}M $MV^{2+} + 2.5$
$\times 10^{-3}$M Na_2SO_4; f. 2×10^{-4}M $MV^{2+} + 1.25 \times 10^{-3}$M
Na_2SO_4. Excitation at 532 nm.

Mechanism of Quenching

The mechanism of quenching of excited singlet EB for both Cu^{2+} and MV^{2+} is via electron transfer from EB excited singlet state to Cu^{2+} or MV^{2+}.

$$EB^{1*} + Cu^{2+}(MV^{2+}) \xrightarrow{k_Q} EB^+ + Cu^+(MV^+) \tag{1}$$

This has been shown previously for MV^{2+} (Atherton and Beaumont, 1986; 1987; Fromherz and Rieger, 1986). In the case of Cu^{2+}, the products of Reaction 1 are oxidized EB (EB^+) and reduced Cu^{2+} (Cu^+). EB^+ may be produced independently by pulse radiolysis. The major reactive primary products found on pulse radiolysis of water are hydrated electrons (e_{aq}^-) and the OH radical. A minor yield of hydrogen atoms is also formed. In an aqueous solution containing an appropriate concentration of azide (N_3^-) and saturated with N_2O, the following sequence of reactions will occur:

$$H_2O \xrightarrow{\text{fast } e^-} e_{aq}^- + OH$$

$$e_{aq}^- + N_2O \xrightarrow{H_2O} OH + OH^- + N_2$$

$$OH + N_3^- \longrightarrow N_3^{\cdot} + OH^-$$

The azide radical ($N_3^{\cdot}$) is a well known oxidant which reacts exclusively via one electron transfer (Alfassi and Schuler, 1985). Figure 2 shows the absorption spectrum observed 20 μs after pulse radiolysis of an aqueous N_2O-saturated solution containing 5.8×10^{-4}M DNA, 4×10^{-5}M EB, and 10^{-2}M NaN$_3$. These absorptions are assigned to one electron oxidized EB (EB^+) formed via reaction of EB with $N_3^{\cdot}$. Figure 3 shows the absorption spectrum observed 800 ns after laser flash photolysis (532 nm, 11 ns pulse) of a N_2-saturated solution of 5.8×10^{-4}M DNA, 4×10^{-5}M EB, 10^{-3}M Cu^{2+}, and 5×10^{-3}M Na_2SO_4. The similarity of this spectrum to that in Figure 2 indicates that they are due to the same species, and that EB^+ is formed via Cu^{2+} quenching of EB excited singlet state. Photoionization of EB is ruled out

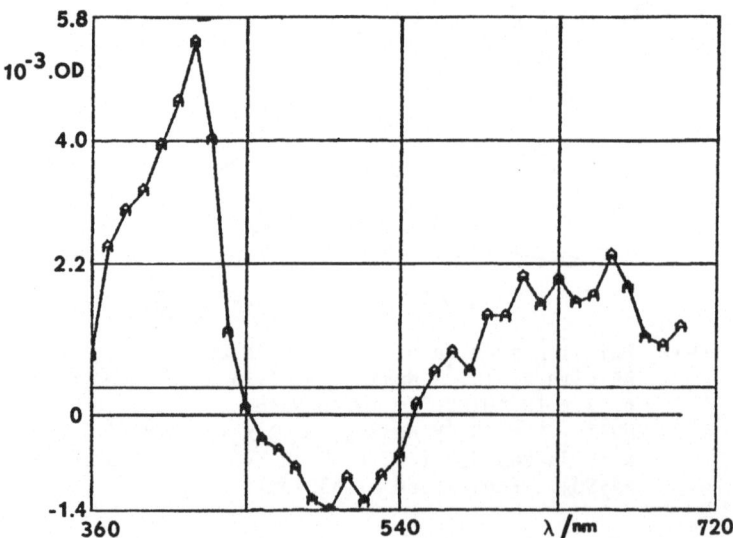

Fig. 2. Spectrum observed 20 μs after pulse radiolysis of N_2O-saturated 5.8×10^{-4}M DNA, 4×10^{-5}M EB, and 10^{-2}M NaN$_3$.

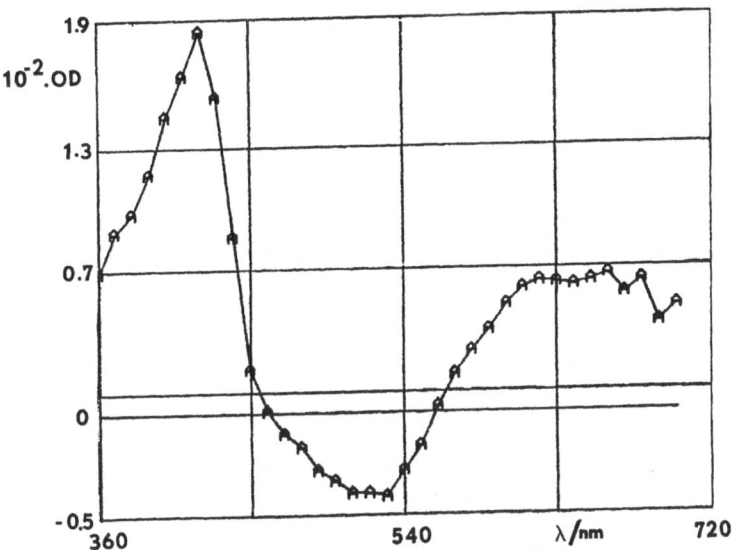

Fig. 3. Spectrum observed 800 ns after laser flash photolysis
(532 nm, 11 ns pulse) of N_2-saturated $5.8 \times 10^{-4}M$
DNA, $4 \times 10^{-5}M$ EB, $10^{-3}M$ Cu^{2+}, and $5 \times 10^{-3}M$ Na_2SO_4.

since under the same flash photolysis conditions, with a solution containing
no Cu^{2+}, neither the familiar absorption of the hydrated electron (λ_{max} =
720 nm) nor the absorptions shown in Figure 3 are observed.

Figure 4 (taken from Atherton and Beaumont, 1987) shows the absorption
spectra observed (A) 0.9 μs and (B) 27 μs after flash photolysis of aerated

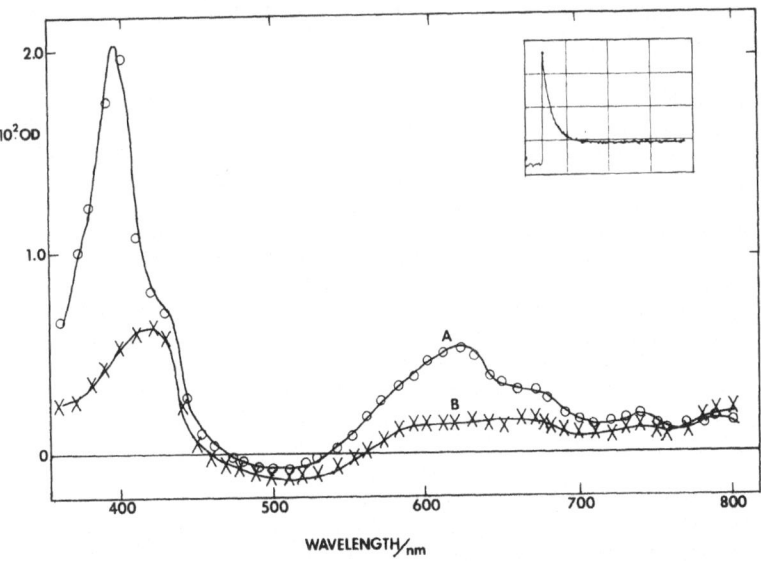

Fig. 4. Absorption spectra observed after 532 nm irradiation
of aerated $3.6 \times 10^{-4}M$ DNA, $4 \times 10^{-5}M$ EB, $2.5 \times 10^{-3}M$
Na_2SO_4, and $2 \times 10^{-4}M$ MV^{2+}: (A) 0.9 μs after the
pulse; (B) 27 μs after the pulse. Inset: Decay at 395
nm. Solid line is the best single-exponential fit to
a nonzero base level. One horizontal division is 26 μs.

3.6×10^{-4}M DNA, 4×10^{-5}M EB, 2×10^{-4}M MV^{2+}, and 2.5×10^{-3}M Na$_2$SO$_4$. Spectrum A shows the familiar features of reduced viologen (MV$^+$), a peak around 400 nm and a broad absorption around 600 nm, in addition to a slight shoulder at $420 \rightarrow 430$ nm, a bleaching in the region 470-500 nm, and an absorption into the red past 800 nm. In aerated solution, MV$^+$ decays by reaction with O$_2$ to yield MV^{2+} and O$_2^-$ with a rate constant of $5.6 \pm 0.4 \times 10^8$M^{-1}s^{-1}, close to the value of 6×10^8M^{-1}s^{-1} measured in DNA-free aqueous solution (Rodgers, 1984). This leaves the only visible absorbing species as EB$^+$ (Figure 4 spectrum B), which decays via second order kinetics with k = $1.2 \pm 0.3 \times 10^9$M^{-1}s^{-1}. In N$_2$-saturated solution, MV$^+$ and EB$^+$ decay by recombination; second order kinetics are observed with k = $5.6 \pm 1.5 \times 10^9$M^{-1}s^{-1}.

The rate constant for the reaction of MV$^+$ with O$_2$ is highly sensitive to the delectric constant of the medium (Rodgers, 1984), and the similarity of the values measured both in the presence and absence of DNA indicate that the MV$^+$ observed has escaped the DNA helix and is free in bulk solution. Further evidence for this is the observation of second order decay kinetics for MV$^+$ and EB$^+$ in N$_2$-saturated solution. If this reaction occurred without either reactant leaving the DNA helix, the kinetics should be first order. Thus after the initial quenching event, there is a competition between back electron transfer from MV$^+$ to EB$^+$ on the same helix where they were formed, and escape of MV$^+$ into bulk solution.

$$EB^{1*}_{DNA} + MV^{2+}_{DNA} \xrightarrow{k_Q} MV^+_{DNA} + EB^+_{DNA}$$

$$MV^{2+}_{DNA} + EB_{DNA} \xleftarrow{k_R} MV^+_{DNA} + EB^+_{DNA} \xrightarrow{k_{esc}} MV^+_F + EB^+_{DNA}$$

The subscripts DNA and F indicate species bound to DNA and free in solution respectively; EB^{1*} is EB excited singlet state; and k_Q, k_R, and k_{esc} are rate constants for quenching, recombination on the helix, and escape of MV$^+$ from the helix respectively. The fraction of MV$^+$ which escapes from the helix, F_{esc}, is given by:

$$F_{esc} = \frac{k_{esc}}{k_R + k_{esc}}$$

Values of F_{esc} may be calculated from the fraction of EB excited states quenched, F_Q, and the quantum yield of observed MV$^+$, Φ_{MV}. Table 1 (also taken from Atherton and Beaumont, 1987) gives values of F_Q, Φ_{MV}, and F_{esc} as a function of ionic strength, for solutions of 3.6×10^{-4}M DNA, 4×10^{-5}M EB, and 2×10^{-4}M MV^{2+}. F_Q was measured by integrating the areas under the steady state fluorescence spectra shown in Figure 1(b). The ratio of the area under the fluorescence curve in the presence of MV^{2+} to that in its absence is the fraction of EB excited singlet states which are not quenched, F_{NQ}. Obviously $F_Q = 1 - F_{NQ}$. Values for Φ_{MV} were measured using the charge transfer excited state of Ru(bipy)$_3^{2+}$ as an actinometer (for details see Atherton and Beaumont, 1987).

Table 1 shows that F_{esc} values are very small and exhibit an ionic strength dependence. Therefore, $k_R \gg k_{esc}$; and the ionic strength effect may be due to an increase in k_{esc} as the ionic strength rises. It has proved impossible to measure k_R since even at high (2×10^{-2}M) concentration of MV^{2+} k_R is still larger than k_Q, i.e., the quenching event remains the rate-determining step. A lower limit of 10^9s^{-1} is given for k_R.

In the case of quenching of EB excited singlet state by Cu^{2+} in the DNA system, preliminary results suggest different kinetics for the decay of EB$^+$ (Figure 3) than is the case for the MV^{2+}-EB-DNA system. The decay is now

Table 1. Quantum Yields and Helix Escape Yields of Reduced Viologen[a]

10^3 [Na$_2$SO$_4$]	F_Q	$10^2 \, \Phi_{MV}$	$10^2 \, F_{esc}$
10	0.43	0.88 ± 0.08	2.0
5	0.64	1.04 ± 0.15	1.6
2.5	0.78	1.13 ± 0.15	1.4
1.25	0.82	1.00 ± 0.02	1.2

[a]Symbols as described in text.

biphasic; there is an initial faster decay which is first order, k = 2.7 x $10^5 s^{-1}$, independent of beam energy; and it accounts for ca. 45% of the total absorption. The remaining ca. 55% of the total absorption decays over 100's of μs. There are also indications that the yield of EB$^+$ may be higher in the case of Cu^{2+}. This system is the subject of continuing study.

One possibility is that Cu$^+$ is more mobile around the DNA helix than MV$^+$ and may move away from its EB$^+$ partner whilst still remaining on the helix. About 55% of this Cu$^+$ escapes the helix whilst 45% re-encounters EB$^+$ on the same helix and is responsible for the initial first order decay. The remaining 55% must recombine via diffusion from bulk solution.

Thus far the mediating effects of DNA on these electron transfer reactions consist of bringing the reactants closer together and diminishing the mobility of both the reactants and products. Thus the initial electron transfer quenching reaction is much more efficient; and also back electron transfer, particularly in the case of MV$^+$ recombining with EB$^+$, is highly efficient.

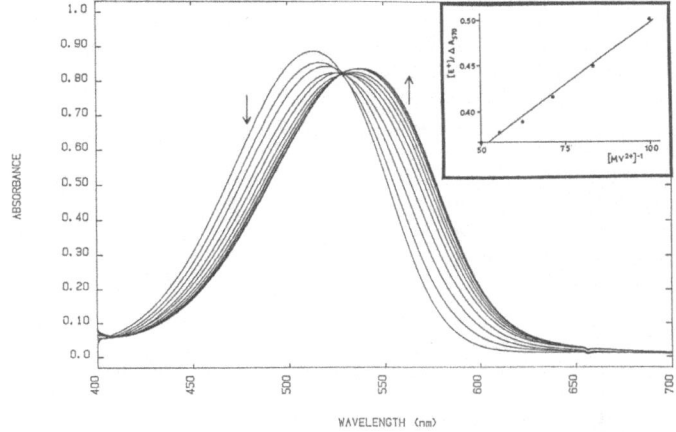

Fig. 5. Ground state absorption spectra of solutions of 5 x 10^{-2}M SDS, 1.5 x 10^{-4}M EB, and 2 x 10^{-3} to 2 x 10^{-2}M MV^{2+} in steps of 2 x 10^{-3}M. Arrows indicate increasing salt concentration. Inset: Benezi-Hildebrand Plot.

In the MV^{2+}-EB-DNA system, however, the opposite effect can be seen simultaneously. This is shown by the fact that, even under high concentrations of MV^{2+}, the EB fluorescence cannot be fully quenched. This may be due to the different nature of the binding sites for MV^{2+} and EB preventing sufficiently close approach. Further evidence for the separation of MV^{2+} and EB on the DNA helix comes from the observation that a charge transfer complex between MV^{2+} and EB can be induced by the presence of sodium dodecylsulphate (SDS) micelles. Figure 5 (taken from Atherton et al., 1987) shows the effect of increasing MV^{2+} concentration on the visible absorption spectrum of $1.5 \times 10^{-4}M$ EB in aqueous $5 \times 10^{-2}M$ SDS. The inset shows a Benesi-Hildebrand plot of these data; $K = 100 \pm 20$ M^{-1}, for the formation of the 1:1 EDA complex between MV^{2+} and EB. There is no sign of complex formation in pure water at these concentrations of EB and MV^{2+}, hence the major role of SDS in complex formation is probably to concentrate the reactants on the micelle surface. Although it is clear from the studies of the MV^{2+}-EB-DNA system that DNA has a large concentrating effect on EB and MV^{2+}, addition of even up to 0.1 M of MV^{2+} to solutions of $3.6 \times 10^{-4}M$ DNA, $4 \times 10^{-5}M$ EB, and $1.25 \times 10^{-3}M$ Na_2SO_4 fails to produce any sign of complex formation. The effect of DNA is to hold the reactants close together; however, DNA also prevents sufficiently close approach, with the appropriate orientation, to form the charge transfer complex.

ACKNOWLEDGEMENTS

Experiments were performed at the Center for Fast Kinetics Research, which is supported jointly by National Institutes of Health Grant RR00886 from the Biomedical Research Technology Program of the Division of Research Resources and by the University of Texas at Austin.

REFERENCES

Alfassi, Z. B., and Schuler, R. H., 1985, J. Phys. Chem., 89:3359.
Armstrong, R. W., Kuruczec, T., and Strauss, U. P., 1970, J. Am. Chem. Soc., 92:3174.
Atherton, S. J., and Beaumont, P. C., 1984, Photobiochem. Photobiophys., 8:103.
Atherton, S. J., and Beaumont, P. C., 1986, J. Phys. Chem., 90:2252.
Atherton, S. J., and Beaumont, P. C., 1987, J. Phys. Chem., 91:3933.
Atherton, S. J., Hubig, S. M., Callan, T. J., Duncanson, J. A., Snowden, P. T., and Rodgers, M. A. J., 1987, J. Phys. Chem., 91:3137.
Burns, V. W. F., 1969, Arch. Biochem. Biophys., 183:420.
Eichhorn, G. L., 1975, "Inorganic Biochemistry," Elsevier, Amsterdam, pp. 1210-1243.
Foyt, D. C., 1981, Comput. Chem., 5:49.
Fromherz, P., and Reiger, B., 1986, J. Am. Chem. Soc., 108: 5361.
Gaugain, B., Barbet, J., Capelle, N., Roques, B. P., and Le Pecq, J.-B., 1978, Biochemistry, 17:5078.
Izatt, R. M., Christensen, J. J., and Rytting, J. H., 1971, Chem. Rev., 71:349.
Jorns, M. S., 1987, J. Am. Chem. Soc., 109:3133.
Le Pecq, J.-B., and Paoletti, C. J., 1967, J. Mol. Biol., 27:87.
Maniatis, T., Fritsch, E. F., and Sambrook, J., 1982, "Molecular Cloning," Cold Spring Harbour Laboratory, New York, p. 458.
Manning, G. S., 1977, Biophys. Chem., 7:95.
Manning, G. S., 1978, Q. Rev. Biophys., 11:179.
Rodgers, M. A. J., 1984, Radiat. Phys. Chem., 23:245.
Van Lith, D., Warman, J. M., De Haas, M. P., and Hummel, A., 1986, J. Chem. Soc. Faraday I, 82:2933.

ACTION SPECTRA FOR PHOTOINACTIVATION OF CELLS IN THE PRESENCE OF TETRA-(3-HYDROXYPHENYL)PORPHYRIN, CHLORIN e_6 AND ALUMINIUM PHTHALOCYANINE TETRASULPHONATE

André Western and Johan Moan

Institute for Cancer Research
The Norwegian Radium Hospital
Montebello, 0310 Oslo 3, Norway

INTRODUCTION

Recently, a number of new photosensitizers have been proposed for use in photodynamic cancer therapy (PDT). Three of the most promising ones of these sensitizers are tetra(3-hydroxyphenyl)porphyrin (3THPP), chlorin e_6 (Chl e_6) and aluminium phthalocyanine tetrasulphonate (AlPCTS). 3THPP was recently shown to be a better tumorlocalizer and a more potent and selective tumor photosensitizer than the until now most widely used drugs HpD and DHE (Berenbaum et al., 1986; Peng et al., 1987; Moan et al., 1987). Chl e_6 has a similar structure as porphyrins and a significantly stronger absorbance of red light. AlPCTS and other phthalocyanines also have high absorbance of red light, are efficient photosensitizers and have been launched as future PDT sensitizers (see review by Spikes, 1986). In order to choose the most optimal wavelength for therapy it is necessary to know the action spectrum for cell inactivation. All these dyes tend to aggregate in aqueous media and one can therefore not expect that their action spectra for photoinactivation of cells and sensitization of tumors parallel their absorption spectra.

In the present work the absorption spectra-, the fluorescence excitation spectra- and the action spectra for photoinactivation of cells have been determined for the three sensitizers mentioned.

MATERIALS AND METHODS

Chemicals

3THPP, Chl e_6 and AlPCTS were obtained from Porphyrin Products, Logan, UT. Stock solutions of 3THPP and Chl e_6 were prepared in 0.05M NaOH. Stock solution of AlPCTS was prepared in PBS. Small aliquots of stock solution ($\sim$2%) were added directly to the cell culture medium. This addition did not result in any pH change of the medium.

Cells

Human cells of the line NHIK 3025 were cultivated in Minimum Essential Medium (MEM) supplied with 10% newborn calf serum (Gibco, Scotland) as described earlier (Christensen et al., 1983).

Action Spectra

Survival curves were registered by a method described elsewhere (Moan et al., 1984). About 10^6 cells were incubated in 25 cm² Falcon tissue culture flasks. 6 h later the medium was changed to MEM with 3% serum containing the photosensitizing drug. After an incubation period of 18 h at 37°C the cells were exposed to light in the presence of the drug. The light source was a 900 W Osram high pressure Xe lamp fitted to a Bausch & Lomb grating monochromator. The bandwidth of the exposure light was 20 nm as measured by means of a second monochromator (Jarrel Ash) with narrow slits ($\Delta\lambda \approx 1$ nm) and a UDT 11a detector (United Detector Technology, St. Monica, CA). This detector was calibrated at the wavelengths used for irradiation and thus also gave the absolute fluence rates at the position of the cells. One day after light exposure the cells were fixed in ethanol and stained with methylene blue.

Absorption- and Fluorescence Spectra

Cells for spectroscopic studies were incubated in 25 cm² Falcon tissue culture dishes as described above. After an incubation period of 18 h with the dyes, the dishes were washed 5 times in cold PBS and brought into suspension in 3 ml PBS by means of a cell scraper. The samples contained about 10^6 cells/ml, as determined by Bürker chamber counting. Absorption spectra were registered by means of a Perkin-Elmer Lamda 15 spectrophotometer equipped with an integrating sphere. Cuvettes with 0.5 cm optical path lengths and containing 1 ml of cell suspension each were used. In no cases was the absorbance of the cell sample (measured with PBS in the reference cuvette) larger than 0.1, ensuring that inner filter effects were of no importance in the fluorescence experiments, where the optical path length, both for excitation and emission, were 1.5 mm. Fluorescence measurements were carried out in parallel with the absorption measurements, using a Perkin-Elmer LS5 spectrofluorimeter. A 3x3 mm cuvette containing 0.3 ml cell suspension was used. When recording the excitation spectra a filter was used to reduce the amount of scattered light, notably in the excitation region 230-350 nm. It should be noted that the spectrofluorimeter does not give corrected excitation spectra at wavelength higher than about 640nm. Corrections in this region were made by use of another spectrofluorimeter (Perkin-Elmer MPF2).

RESULTS AND DISCUSSION

The absorption- and fluorescence excitation spectra of the three dyes in NHIK 3025 cells (18h incubation, 5 times washing in PBS, suspended in PBS) are shown in figures 1-3. In all cases the absorption- and fluorescence excitation spectra are different. The difference is largest in the case of AlPCTS but is significant also in the two other cases. Such differences are most likely due to the well-known fact that aggregated material of these dyes has a very low fluorescence quantum yield compared with that of monomers. Thus, in the case of AlPCTS, the absorption spectrum in cells (fig. 2) is obviously a superposition of typical phthalocyanine monomer- and aggregate- (or dimer) spectra (Wagner et al., 1987). The fluorescence excitation spectrum (fig. 2) closely resembles what is assumed to be the absorption spectrum of phthalocyanine monomers (Wagner et al., 1987). To the knowledge of the present authors the absorption spectra of 3THPP and Chl e_6 in aqueous solution have not been separated into monomer-/aggregate spectra so far.

The values for the points in the action spectra are normalized to fall on the absorption- and fluorescence excitation spectra at a given wavelength. For 3THPP, Chl e_6 and AlPCTS this wavelength is 420nm, 405nm and 675nm, respectively.

In all cases the action spectra for cell inactivation closely resembled the fluorescence excitation spectra (figures 1-3). This indicates that aggregated material (and probably also dimeric material which contributes to the absorption- but not to the fluorescence excitation spectra) has a low photobiological activity in the cells. A similar conclusion has been drawn in the case of hematoporphyrin derivative and Photofrin II bound to cells (Moan and Sommer, 1984). The data are also in agreement with the result of Wagner et al., (1987) which indicate that light absorption by dimers and aggregates of phthalocyanines does not result in any significant yield of singlet oxygen.

The present action spectra for cell inactivation sensitized by 3THPP and Chl e_6 seem to be the first ones published so far. The action spectrum for AlPCTS (fig. 2) is almost similar to the corresponding action spectrum for chloroaluminium phthalocyanine found by Ben Hur and Rosenthal (1986).

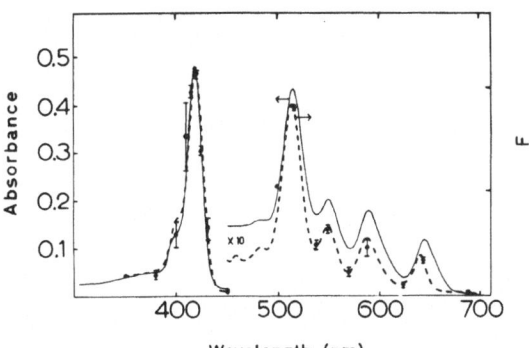

Wavelength (nm)

Fig 1. 3THPP in cells. Absorption spectrum (———) and fluorescence excitation spectrum (– – – –, λ_{em} = 646nm for excitation up to 600nm, and 720nm for excitation up to 700nm). F = Fluorescence intensity in relative units. The action spectrum for cell inactivation is given by the points where the bars correspond to error limits determined from at least 3 independent experiments. The values given are the inverse values of the relative mean numbers of incident photons needed to inactivate 50% of the cells. The fluence rate was measured by means of a calibrated UDT detector. Incubation conditions: 18h with 10 μg 3THPP/ml MEM containing 3% serum; light exposure in the presence of the dye.

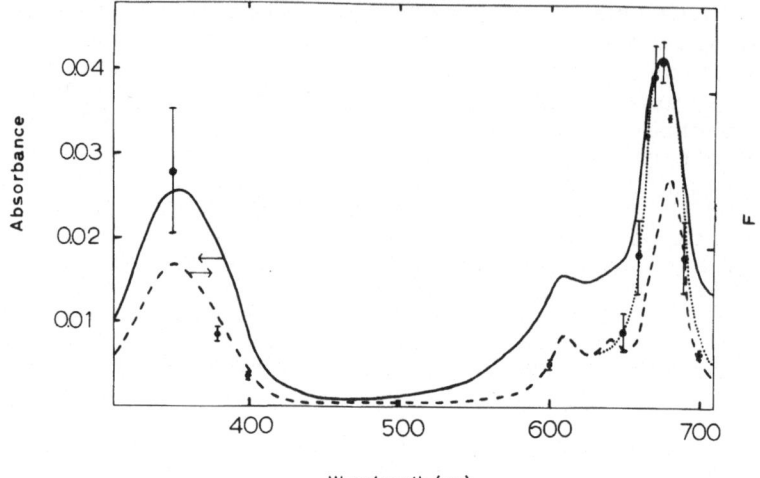

Fig 2. AlPCTS in cells. Absorption spectrum (———) and fluorescence excitation spectrum (————, λ_{cm} = 675nm for excitation up to 610nm, and 720nm for excitation up to 700nm). F = Fluorescence intensity in relative units. The action spectrum is given by the points and determined as described under fig. 1. Incubation conditions: 18h with 60 µg AlPCTS/ml MEM with 3% serum; light exposure in the presence of the dye.
The dotted line is the corrected fluorescence excitation spectrum. Relatively large error limits are associated with the point at 350 nm, supposedly due to the low fluence rate and hence long exposure times at this wavelength.

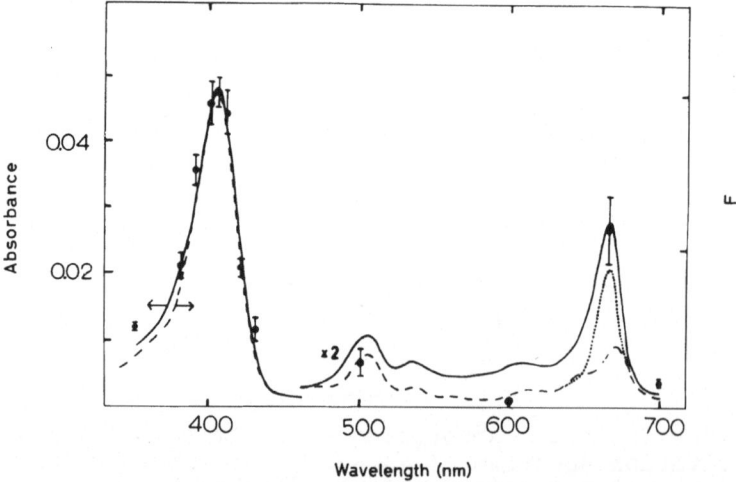

Fig 3. Chl e_6 in cells. Absorption spectrum (———) and fluorescence excitation spectrum (————, λ_{cm} = 668nm for excitation up to 650nm, and 720nm for excitation up to 700nm). F = Fluorescence intensity in relative units. The action spectrum for cell inactivation is given by the points which are determined as described under fig. 1. Incubation conditions: 18h with 6 µg Chl e_6/ml MEM with 3% serum; light exposure in the presence of the dye. The dotted line is the corrected fluorescence excitation spectrum.

ACKNOWLEDGMENT

The authors acknowledge the financial support of the Norwegian
Cancer Society (Landsforeningen Mot Kreft).

REFERENCES

Ben Hur, E. and Rosenthal,I., 1986, Action spectrum (600-700
    nm) for chloroaluminium phthalocyanine-induced photo-
    toxicity in Chinese hamster cells. Laser in Life Sci.
    1:79-86.
Berenbaum, M.C., Akande,S.L., Bonnett,R., Kaur,H., Ioannou,S.,
    White,R.D., and Winfield,U.J., 1986, Meso-tetra(hydroxy-
    phenyl)porphyrins, a new class of potent photosensitizers
    with favorable selectivity. Br.J.Cancer 54:717-725.
Christensen, T., Sandquist,T., Feren,K., Waksvik,H., and
    Moan,J., 1983, Retention and photodynamic effects of
    hematoporphyrin derivative in cells after prolonged culti-
    vation in the presence of porphyrin.
    Br.J.Cancer 48:35-43.
Moan, J., Høvik,B., and Sommer,S., 1984, A device to determine
    fluence response for photoinactivation of cells in vitro.
    Photobiochem.Photobiophys. 8:11-17.
Moan, J. and Sommer,S., 1984, Action spectra of hematoporphyrin
    derivative and Photofrin II with respect to sensitization
    of cells in vitro to photoinactivation. Photochem.Photo-
    biol. 40:631-634.
Moan, J., Peng,Q., Berg,K., Western,A., and Rimington,C., 1987,
    Photosensitizing efficiencies, tumor- and cellular uptake
    of different photosensitizing drugs relevant for
    photodynamic therapy of cancer. Photochem.Photobiol.
    46:713-722.
Peng, Q., Evensen,J.F., Rimington,C., and Moan,J., 1987, A com-
    parison of different photosensitizing dyes with respect
    to uptake in C3H-tumors and tissues of mice. Cancer Lett.
    36:1-10.
Spikes, J.D., 1986, Phthalocyanines as photosensitizers in bio-
    logical systems and for the photodynamic therapy of tu-
    mors. Photochem. Photobiol. 43:691-699.
Wagner, J.R., Ali, H., Langlois,R., Brasseur,N., and van Lier,
    J.E., 1987, Biological activities of phthalocyanines-
    VI. Photooxidation of L-tryptophan by selectively sulfo-
    nated gallium phthalocyanines: Singlet oxygen yields
    and effect of aggregation. Photochem.photobiol. 45:587-
    594.

MECHANISMS OF PHOTODYNAMIC DAMAGE INDUCED IN CELLULAR SYSTEMS

T.M.A.R. Dubbelman, J.P.J. Boegheim and J. Van Steveninck

Dept. Medical Biochemistry, Sylvius Laboratories
P.O. Box 9503
2300 RA Leiden, The Netherlands

Although it is well known that cells do not survive photodynamic treatment with photosensitizers, the actual mechanism of cell death is unknown in most cases and may also be different in different cells. Most cellular systems prove to be sensitive to photodynamic damage, because most constituents of cells, proteins, nucleic acids, lipids and coenzymes can in principle be photooxidized. The actual oxidation depends on many factors such as the intracellular localization of both the photosensitizer and the target molecules. An example is the intralysosomal enzyme β-glucuronidase, which in intact L929 cells is not inactivated at all, but when the cells are disrupted by sonication this enzyme proves to be quite sensitive to photoinactivation by HPD (Boegheim et al., 1987). The mechanism of inactivation of cellular systems has only been clarified in very few cases. Photooxidation of cysteine or histidine residues is often the reason for the inactivation of functional proteins. Examples are the inactivation of the enzymes glyceraldehyde-3-phosphate dehydrogenase and succinate dehydrogenase and the inhibition of some membrane transport systems.

Details of the inhibition of K^+-leakage from red blood cells induced by the SH specific sensitizer Cu-HPD, of anion transport in erythrocytes with deuteroporphyrin disulfonic acid as sensitizer and aminoisobutyric acid in L929 cells with HPD are discussed.

K^+-leakage from red blood cells induced by Cu-HPD

Cu-HPD is a stable complex that only acts as a photosensitizer with SH containing compounds such as dithiothreitol and cysteine, presumably via a type I reaction (Van Steveninck et al., 1985). Also in red blood cell membranes only SH groups are photooxidized. Illumination of intact erythrocytes in the presence of the Cu-HPD complex resulted in K^+-leakage followed by hemolysis (Fig. 1), most likely caused by photooxidation of some membrane SH groups. In order to localize these groups erythrocytes were pretreated with several SH reagents and its effect upon the photodynamically induced K^+-leakage was studied. As also shown by others (Brown et al., 1975) N-ethylmaleimide (NEM) is very reactive with many SH groups, but caused only a minor enhancement of passive K^+ permeability (5% after 120 min). The 17 kDa fragment, membrane spanning part, of the aniontransporter contains a SH group that does not react with NEM. It however reacts with 5,5'-dithio-bis(-2-nitrobenzoic acid) (DTNB). As expected, pretreatment with NEM induced a very slight increase in the K^+-

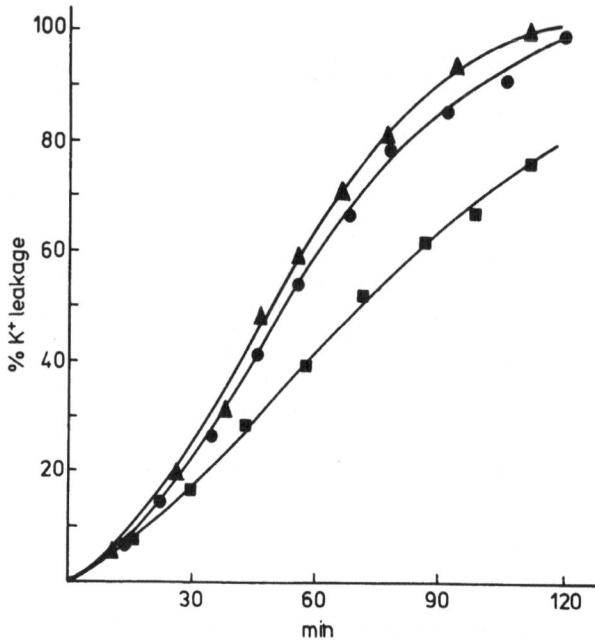

Fig. 1. K$^+$-leakage from red blood cells. ●-●, cells (10% suspension) illuminated in the presence of 30 μM Cu-HPD. ▲-▲, cells pretreated with NEM (10 mM, 45 min) and subsequently illuminated. ■-■, cells pretreated with NEM (10 mM, 45 min)plus DTNB (added after 45 min at 5 mM for another 30 min) followed by illumination.

leakage induced by Cu-HPD, but pretreatment with NEM plus DTNB (which itself causes only 10% leakage after 120 min) decreased the photodynamically induced K$^+$ leakage considerably (Fig. 1). Thus, these results show that the SH group located in the 17 kDa fragment of the anion transporter is involved in the induction of K$^+$-leakage by this SH specific photosensitizer.

Action of HPD on the transport system of aminoisobutyric acid

Aminoisobutyric acid (AIB) influx in L929 murine fibroblasts is inhibited by the photodynamic action of HPD in a dose dependent manner (Dubbelman and Van Steveninck, 1984). As shown in fig. 2, kinetic analysis reveales that the inhibition is characterized by an increase in the apparent Km value, while the apparent Vmax remains constant. The increased apparent Km cannot simply be interpreted as a decrease of the affinity of the carrier for its substrate because it is an active co-transport system with Na. The constant apparent Vmax however suggests that carrier molecules are not completely inactivated, as this would most likely have been reflected by a decrease in Vmax (Van den Broek and Van Steveninck, 1982).

Upon postincubation after photodynamic treatment (10 μg/ml HPD (1 h) and illumination (10 min)) AIB transport velocity is restored to its normal value in about 16 h. This can be caused either by synthesis of new transporters or by repair of damaged transport proteins, e.g. damaged SH groups. Treatment of the cells with anisomycin, a transfer inhibiting agent (Grollman, 1967) shows that restoration of photodynamically inhibited AIB transport is caused by protein synthesis.

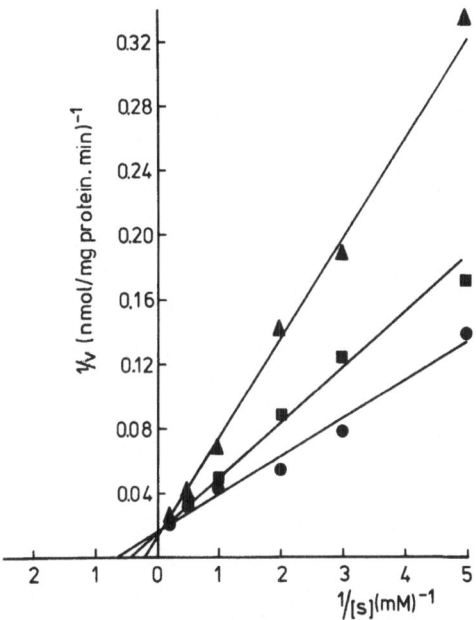

Fig. 2. Lineweaver-Burk plots of AIB transport across the plasma membrane of L929 cells after incubation with HPD (10 μg/ml, 1 h). ● - ●, without illumination. ■ - ■, illumination time 4 min. ▲ - ▲ illumination time 10 min.

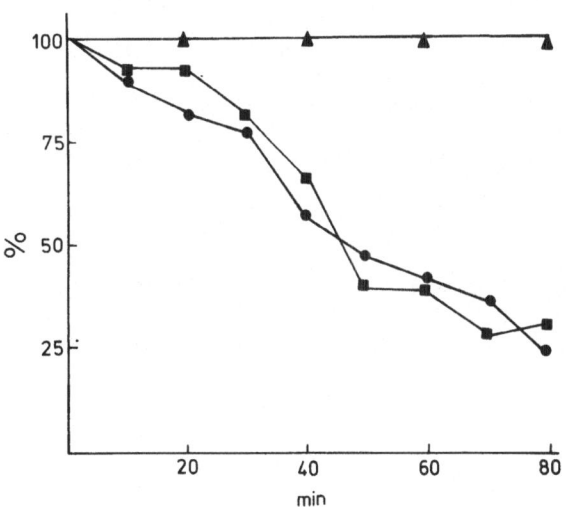

Fig. 3. Photodynamic crosslinking of the anion transporter. inhibition of sulfate transport and H_2-DIDS binding in intact red blood cells (10% suspension) at pH 7.9. Illumination in the presence of 0.8 mM DPS. ● - ●, sulfate transport, ▲ - ▲, amount of anion transport protein in an electropherogram of red blood cell membranes, ■ - ■, H_2-DIDS binding.

Porphyrin-induced inhibition of anion transport in erythrocytes

During illumination of intact red blood cells in the presence of deuteroporphyrin-2,4-disulfonic acid (DPS) the transmembrane sulfate transport velocity decreased dramatically (Dubbelman et al., 1981). Interpeptide crosslinking of the anion transport protein was absent (Fig. 3). Thus photodynamic inhibition of anion transport is caused either by direct photooxidation of an amino acid in the active center or by intrapeptide crosslinking.

Anion transport in erythrocytes is strongly inhibited by 4,4'-diisothiocyanodihydrostilbene-2,2'-disulfonate (H_2-DIDS). Binding of this compound is highly localized in the anion transporter (Ramjeesingh et al., 1981). The inhibition of both sulfate transport and H_2-DIDS binding by the photodynamic action of DPS closely parallel each other (Fig. 3). As H_2-DIDS binds to a site, presumably a lysine residue, that is apparently involved in anion translocation, it might well be that this NH_2 group is involved in intrapeptide crosslinking to a photooxidizable amino acid residue such as histidine.

REFERENCES

Boegheim, J.P.J., Scholte H., Dubbelman, T.M.A.R., Beems, E., Raap, A.K. and Van Steveninck, J., 1987, Photodynamic effects of hematoporphyrin-derivative on enzyme activities of murine L929 fibroblasts, J.Photochem.Photobiol., B: Biology, 1: 61.

Brown, P.A., Feinstein, M.B. and Sha'afi, R.I., 1975, Membrane proteins related to water transport in human erythrocytes, Nature, 254: 523.

Dubbelman, T.M.A.R., and Van Steveninck, J., 1984, Photodynamic effects of hematoporphyrin-derivative on transmembrane transport systems of murine L929 fibroblasts, Biochim.Biophys.Acta, 77: 201.

Dubbelman, T.M.A.R., De Goeij, A.F.P.M., Christianse, K., and Van Steveninck, J., 1981, Protoporphyrin-induced photodynamic effect on band 3 protein of human erythrocyte membranes. Biochim.Biophys.Acta, 649: 310.

Grollman, A.P., 1967, Inhibitors of protein biosynthesis, II, Mode of action of anisomycin, J.Biol.Chem., 242: 3226.

Ramjeesingh, M., Gaarn, A., and Rothstein, A., 1981, The amino acid conjungate formed by the interaction of the anion transport inhibitor DIDS with band 3 protein from human red blood cell membranes, Biochim.Biophys.Acta, 641: 173.

Van den Broek, P.J.A., and Van Steveninck, J., 1982, Kinetic analysis of H^+/methyl-β-D-thiogalactoside transport in Saccharmocyes fragilis, Biochim.Biophys.Acta, 693: 213.

Van Steveninck, J., Boegheim, J.P.J., and Dubbelman, T.M.A.R., 1985, The influence of cupric ions on porphyrin-induced photodynamic membrane damage in human red blood cells, Biochim.Biophys.Acta, 821:1.

SISTER CHROMATID EXCHANGES INDUCED BY PHOTODYNAMIC TREATMENT OF CELLS IN THE PRESENCE OF PHOTOFRIN II, ALUMINIUM PHTHALOCYANINE TETRASULFONATE AND TETRA(3-HYDROXYPHENYL)PORPHYRIN

Kristian Berg, Eivind Hovig and Johan Moan

Institute for Cancer Research
Montebello, 0310 Oslo 3, Norway

INTRODUCTION

The cellular effects and clinical applicability of photo-dynamic treatment (PDT) using hematoporphyrin derivative (HpD) or Photofrin II (PII) as photosensitizers have been extensively studied over the last years. The method is based on the partially selective retention of the sensitizers in tumor tissue and their photodynamic effects, which are mediated mainly through formation of singlet oxygen.

Several new phototoxic drugs have recently been found to improve the therapeutical usefulness of PDT of cancer. Major limiting factors for the usefulness of PDT have been the limited penetration of light into tissues and the low absorbance of the currently used sensitizers at wavelengths larger than those absorbed by hemoglobin, the main absorber in most tissues. The use of phthalocyanines as sensitizers improves this situation. The phthalocyanines have absorption peaks above 600 nm with relatively high extinction coefficients. Another class of new potent tumor photosensitizers are the meso-tetra(hydroxyphenyl)-porphyrins (THPPs), which have been shown to sensitize tumor tissue more selectively than HpD and PII. The 3-THPP was in spite of its low absorbance of red light shown to be 25-30 times more potent than HpD (Berenbaum et al., 1986).

When new treatment modalities are being developed for use in cancer therapy the mutagenic and carcinogenic potential of the treatments should be investigated before clinical use. Damage to DNA has been observed after PDT with hematoporphyrin and HpD: photodegradation of guanine residues (Gutter et al., 1977), alkali-labile sites (Moan et al., 1980), DNA-protein crosslinks (Dubbelman et al., 1982) and chromosome aberrations (Evensen and Moan, 1982). An increase in the frequency of sister chromatid exchanges (SCEs) has also been shown to occur (Christensen et al., 1985, Moan et al., 1980) while mutations have not been observed after HpD sensitized PDT of CHO cells

(Gomer et al., 1983). These different types of damage to DNA indicate that PDT may be potentially carcinogenic, which is the case for radiation- and chemotherapy. The work of Bungeler (1937) also indicated that Hp plus exposure to sunlight induced tumors in mice (see also Santamaria (1972)). The purpose of this study was to examine the capacity of aluminium phthalocyanine tetrasulfonate (AlPcTS), 3-THPP and PII sensitized photodynamic treatment to induce SCEs as a measure of damage to DNA and as an approach to the study of the carcinogenic potential of these drugs. These photosensitizers have very different abilities to incorporate into hydrophobic areas, according to their Triton X-114/H_2O partition coefficient (Moan et al., 1987). An interesting aspect to investigate is whether the widely different lipophilicity of these drugs plays any role for their ability to photosensitize DNA in cells.

MATERIAL AND METHODS

Cell Cultivation

V79 cells (Chinese hamster lung fibroblasts) were obtained from Dr.E. Pettersen, Dept. of Tissue Culture, Institute for Cancer Research, Oslo. The cell line was subcultured two times a week in MEM (Minimal Essential Medium) with Hank's salts, 100 U/ml penicillin and 100 µg/ml of streptomycin and 10 % fetal calf serum. The number of chromosomes per cell is 20.9 +0.15 (S.E.).

Chemicals

Aluminium phthalocyanine tetrasulfonate (AlPcTS), tetra(3-hydroxyphenyl)porphyrin (3-THPP) were provided by Porphyrin Products (Logan,UT) and Photofrin II (P II) was provided by Photofrin Medical Inc. (Raritan,N.J.). 5-bromodeoxyuridine (BrdU,Sigma) was dissolved in the cultivation medium at a concentration of 5 mg/ml shortly before use.

Synchronization, Sensitizer Labelling, Irradiation and Cell Survival

Mitotic cells were selected by shaking asynchronously growing cells with a reciprocal shaker for 10 sec.. Cells were seeded in 25 cm^2 flasks and incubated in a sensitizerfree medium for proper binding of the cells to the substratum. 1 1/2 h after the mitotic selection the cells were incubated with the sensitizers for 4.5 h in the cultivation medium, after which time the cells have reached the S-phase (fig.1). The following concentrations of sensitizers were used in all the experiments: 3-THPP:0.25 µg/ml, PII:2.0 µg/ml, AlPcTS:50 µg/ml. The cells were irradiated with lamps giving either red (AlPcTS) or blue (3-THPP and P II) light (fig.2) and subsequently given a sensitizerfree medium. All manipulations were performed at 37^{0}C. Cell survival was studied by measuring the colony-forming ability of the cells. 400 mitotic cells were inoculated in 25 cm^2 plastic flasks (Nunclon) and treated as described above. The cell survival experiments were done in parallel with the SCE measurements. The cells were incubated for 6 days at 37^{0}C before fixation, staining and counting.

SCE Assay

$1 \cdot 10^5$ mitotic cells were inoculated in 25 cm^2 plastic flasks and treated as described above. 5-bromodeoxyuridine (5 µg/ml) was supplemented to the medium after the mitotic shake off and till the preparation for the SCE assay. 0.1 µg/ml colchemid was added to the medium when the cells were seen to reach the second mitosis after synchronization. After 3 h in colchemid the cells were prepared for SCE measurements as described by Alves and Jonasson (1978). The fixed cells were dripped on cold, humidified cover slips before staining with the Giemsa solution. Coded slides were used for all experiments. 20-30 mitosis were scored at each dose.

Cell Cycle Measurements

The cell cycle distribution was examined on cells fixed in 70 % ethanol and stained for DNA with mithramycin (Mithracin,Pfizer Inc.,N.Y.). The measurements were performed with a Coulter Epics V flow cytometer. The 457.9 nm argon laser line was used to excite the mithramycin fluorescence. The fluorescence was registrated above 510 nm. For estimation of the fractions of cells in G_1, S and G_2+M a mathematical model developed by Dean and Jett (1974) and Lindmo and Aarnæs (1979) was used.

Cellular Uptake of the Sensitizers

Cellular uptake was measured on a Perkin Elmer LS 5 spectrofluorometer. $0.5 \cdot 10^5$ asynchronous cells were inoculated in 25 cm^2 Nunclon flasks overnight. The next day the cells were incubated with the same sensitizer concentrations as used for the other experiments. At different time intervals the cells were washed with icecold phosphate-buffered saline (PBS) 3 times and removed from the dish (in PBS) with a Costar cell scraper. The cell concentration were measured by light scattering at 600 nm with the Perkin Elmer LS 5 spectrofluorometer. The light scattering was found to increase linearly with the number of cells at this wavelength and was not influenced by fluorescence from the sensitizers. Afterwards 0.5 % sodium dodecyl sulfate (SDS, final concentration) was added and the mixture briefly sonicated (10 s) before the fluorescence measurements. The cell suspensions were diluted to approximately the same cell concentrations before sonication. The following wavelengths were used for the sensitizers: 3-THPP: λ_{ex}=419nm, λ_{em}=648nm; P II: λ_{ex}=405nm, λ_{em}=629nm; AlPcTS: λ_{ex}=360nm, λ_{em}=679nm. A cut-off filter (>345nm) was used on the excitation light and another on the emission side (>545nm).

RESULTS

All of the survival curves and the SCE measurements were performed on synchronized cells and illuminated in the beginning of the S-phase (fig.1) as described under Material and Methods. Fig.3 shows the survival curves after PDT treatment of the cells. The sensitizers had no cytotoxic effect in the absence of light and neither had light alone any effect. Table 1 lists the frequencies of SCEs obtained for all the treatments. Cells

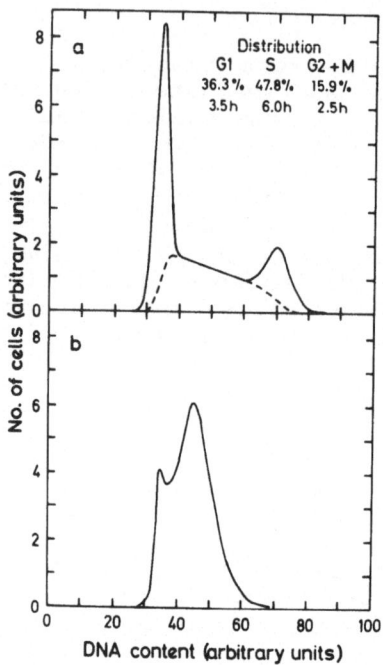

Fig.1. a)DNA-histogram of an exponentially growing population of V79 cells. The histogram has been analyzed by fitting a mathematical model (solid line) to the data. The broken line shows the model's estimate for cells in S phase within the histogram. Using these results and the observed generation time, the phase duration has been calculated assuming an age distribution as that of an exponentially growing population. b) DNA-histogram of a synchronized population of V79 cells 6 h's after mitotic shake off.

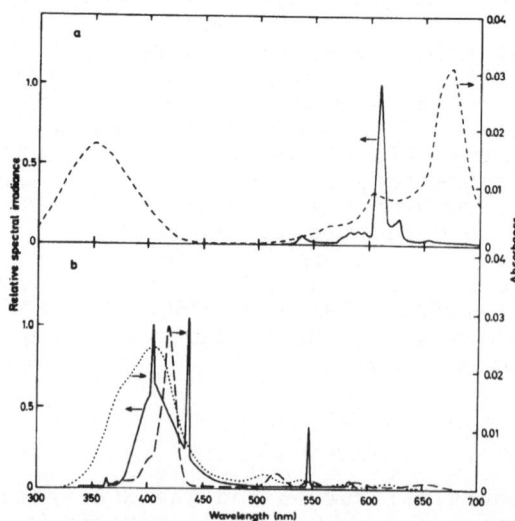

Fig.2. Absorption spectra of cells incubated with AlPcTs (a,- - -), 3-THPP (b, — —) and PII (b,····). The relative spectral irradiance of the Philips TLD/83 Fluorescent tubes filtered throught a Cinemoid 35 filter (40 W/m^2 at the position of the cells) (a,solid line) and the photophysics 3026 fluorescent tubes (36 W/m^2 at the position of the cells) (b, solid line) are also shown.

treated with sensitizers (in the absence of light) and cells
exposed to either red or blue light alone were not statistically
different from control levels. A 2-3 fold increase in SCE
frequency was obtained after PDT using either PII or 3-THPP as
sensitizers. The nonlinearity of the SCE frequency obtained
especially after PII sensitized PDT could be due to the technical
problem of measuring many SCEs closely related in space. PDT
with ALPcTS as sensitizer did not induce any increase in SCE
frequency.

To estimate the amount of sensitizer associated with the
outer parts of the cells uptake curves for the sensitizers
were performed with or without a 5 min. washout period at the
end of the incubation period. The cells were washed with sensitizerfree medium twice and subsequently incubated for 5 min.
in a sensitizerfree medium before measuring the uptake of sensitizer. The medium contained serum in all these treatments.
Fig.4 shows that the proportion of cellbound sensitizer that
was removed by washing increased with increasing watersolu-
bility of the sensitizer. Thus, ALPcTS was lost from the cell
to a much greater extent than 3-THPP and PII.

In view of the uptake curves (see Discussion) it was necessary to determine the survival curve and the induction of SCEs
on ALPcTS treated cells incubated for 5 min. in a sensitizerfree
medium before light exposure. Fig.3b shows the survival curve
after such a treatment. Neither in this case did the treatment
result in any increase in the frequency of SCEs (table 1).

DISCUSSION

It is reasonable to expect that induction of SCE over a
control level implicates a relatively close spatial relationship
between the photosensitizer and DNA. Increasing the hydrophobic
properties of a sensitizer may probably increase the amount of
sensitizer incorporated into the membrane relative to the amount
in the more aqueous phases of the cell, such as the nucleus.
If this is essensial for the induction of SCEs the use of different photosensitizers such as AlPcTS (very watersoluble) and
3-THPP (very hydrofobic) might elucidate the mechanisms behind
SCE induction after PDT.

Table 1 shows that PII sensitized PDT is an inducer of
SCEs. This correlates well with previously obtained results on
HpD- and Hp sensitized PDT which were also found to induce
SCE, but to a lesser extent than X-irradiation (Moan et al.,
1980, Christensen et al., 1985), which is known to be a weak
inducer of SCEs (Tofilon and Meyn, 1987). We conclude that
PII- and 3-THPP sensitized PDT induced SCEs to a similar level.

No increase in SCE frequency was observed after AlPcTS
sensitized PDT. When AlPcTS is present in the medium during
light exposure the cells will probably be killed by damage to
the outer parts of the cell. After a 4.5 h incubation period
the cells lost approximately 80 % of their content of AlPcTS
during a 5 min. incubation period in the absence of the sensitizer (fig.4) and the D_{10} dose increased by a factor of 2-3
(fig.3). However, this removal of weakly associated sensitizer
before light exposure did not result in any increase in the
SCE frequency at a D_5 survival level.

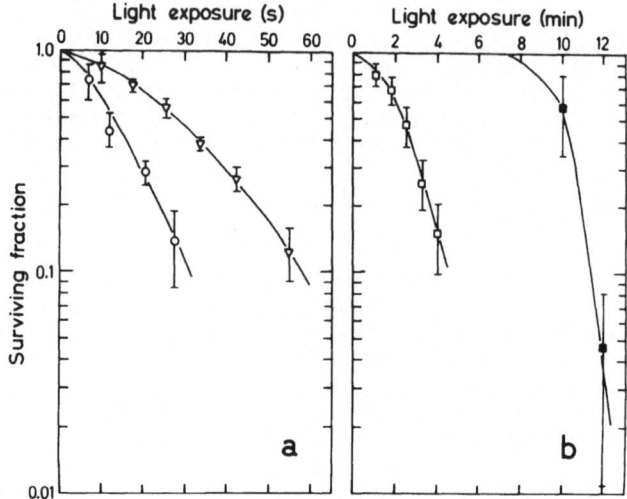

Fig.3. Survival curves for V 79 cells treated as described under Material and Methods. Symbols: O , 3-THPP; ∇ , PII; □ , ALPcTS; ■ , ALPcTS, exposed to light after 5 min. additional incubation in sensitizerfree medium. Bars, S.D. from 3 flasks.

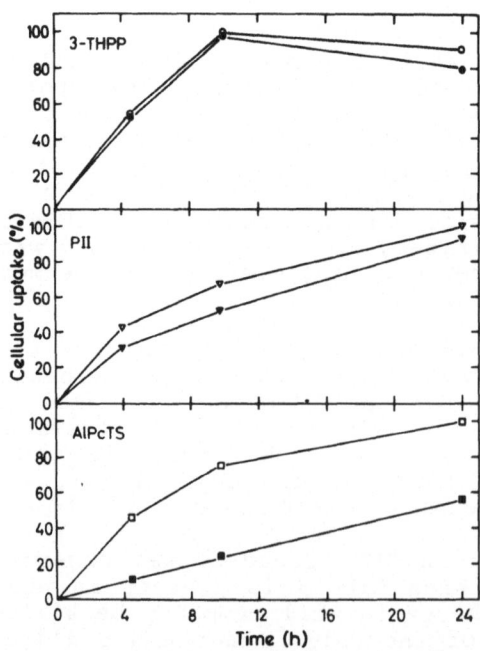

Fig.4. Time course of sensitizer uptake in asynchronously growing V 79 cells. The cells were incubated in the dark. The closed symbols correspond to samples washed for 5 min. in a sensitizer-free medium before the measurements were carried out, while the open symbols correspond to unwashed samples.

Table 1. SCE in V 79 cells after photoradiation.
a)(I),irradiated with blue light; (II),irradiated with red light (see fig.2). b)Irradiated after 5 min. in the absence of AlPcTS.
c)Mean + standard error
For relevant survival curves,see fig.3.

Treatment	Dose	BrdU labelling time (h)	SCE/cell
Control	0 s	25	6.80 ± 0.56[c]
	5 min (I)[a]	25	8.18 ± 0.57
	10 min (II)	25	6.71 ± 0.95
	20 min (II)	25	6.93 ± 0.79
AlPcTS	0 s	25	8.40 ± 0.75
	2 min (II)	27	8.33 ± 0.67
	3 min 15 s (II)	27.5	7.33 ± 0.43
	12 min (II)[b]	28	6.70 ± 0.76
PII	0 s	25	10.78 ± 0.88
	18 s (I)	28	19.70 ± 1.10
	43 s (I)	31	21.55 ± 1.58
3-THPP	0 s	25	8.15 ± 0.50
	7 s (I)	29	12.15 ± 0.77
	18 s (I)	30	19.70 ± 1.10

The lack of SCE induction by AlPcTS sensitized PDT cannot be explained by the data available on damage to DNA after AlPcTS sensitized PDT (Spikes, 1986,Hunting et al., 1987). However, some suggestions can be made:
1) The exposure rate used for AlPcTS was much lower than those used for 3-THPP- and PII sensitized PDT (fig.3). Thus, repair mechanisms may play a larger role during irradiation in the former case than in the latter cases. This has been considered to be the reason for the reduction in mutagenic frequency obtained by lowering the dose-rate of ionizing radiation (see Adams, 1987).
2) Differencies in the photodynamic effects on cells by AlPcTS sensitized PDT and porphyrin sensitized PDT must be considered. Damage to guanine residues in DNA, induction of alkali-labile sites in DNA and DNA-crosslinks have been observed after porphyrin sensitized PDT (see Introduction), while only alkali-labile sites have been described after phthalocyanine sensitized PDT (Hunting et al., 1987). DNA-crosslinking may for example probably induce SCEs after other types of treatments (see Tofilon and Meyn, 1987) and has not been demonstrated after AlPcTS sensitized PDT.
3) The uptake rate into the intracellular compartments of the cell is much lower for AlPcTS than for 3-THPP and PII, probably because of higher permeability barriers due to its polar nature. The amount of AlPcTS in the nucleus or at the nuclear membrane at the time of light exposure may be to low to induce any observable increase in the SCE frequency after treatment.

In this paper we have shown that PII- and 3-THPP sensitized PDT induce a significant increase in SCE frequencies. With the treatment regimes used here we were not able to detect any effects of AlPcTS sensitized PDT on the SCE frequency observed after treatment. This lack of SCE induction cannot be explained by the available data on DNA damage by AlPcTS sensitized PDT and has to be further studied.

ACKNOWLEDGEMENT

The present work was supported by The Norwegian Research Council for Science and The Humanities.

REFERENCES

Adams,G.E., 1987, Radiation and cancer:A two-edged sword.
Br.J.Cancer, 55,Suppl.VIII:11.

Alves, P., and Jonasson,J., 1978, New staining method for the detection of sister chromatid exchanges in BrdU-labeled chromosomes.
J.Cell Sci., 32:185.

Berenbaum,M.C.,Akande,S.L.,Bonnet,R.,Kaur,H.,Ioannou,S., White,R.D., and Winfield,U.-J., 1986, meso-Tetra(hydroxy-phenyl)porphyrins, a new class of potent tumour photo-sensitizers with favourable selectivity.
Br.J.Cancer, 54:717.

Bungeler,W., 1937, Uber den Einfluss photosensibilisierender Substanzen auf die Entstehung von Hautgeschwulsten.
Z.Krebsforsch., 46:130.

Christensen T.,Moan,J.,McGhie,J.B.,Waksvik,H., and Stigum,H., 1985, Studies of HpD: chemical composition and in vitro photosensitization, in "Porphyrin photosensitization," D.Kessel and T.J.Dougherty,eds., Plenum Press, New York. p.151.

Dean,P.N., and Jett,J.H., 1974, Mathematical analysis of DNA distributions derived from flow microfluorometry.
J.Cell.Biol., 60:523.

Dubbelman T.M.A.R.,Van Steveninck,A.L., and Van Steveninck,J., 1982, Hematoporphyrin-induced photo-oxidation and photo-dynamic cross-linking of nucleic acids and their consti-tuents.
Biochem.Biophys.Acta, 719:47.

Evensen,J.F., and Moan,J., 1982, Photodynamic action and chromo-somal damage: A comparison of haematoporphyrin derivative (HpD) and light with x-irradiation.
Br.J.Cancer, 45:456.

Gomer, C.J.,Rucker,N.,Banerjee,A., and Benedict,W.F., 1983, Comparison of mutagenicity and induction of sister chromatid exchange in chinese hamster cells exposed to hematoporphyrin derivative photoradiation, ionizing radiation, or ultraviolet radiation.
Cancer Res., 43:2622.

Gutter, B.,Speck,W.T., and Rosenkranz,H.S., 1977, The photo-dynamic modification of DNA by hematoporphyrin.
Biochem.Biophys.Acta, 475:307.

Hunting,D.J.,Gowand,B.J.,Brasseur,N., and van Lier,J.E., 1987, DNA damage and repair following treatment of V-79 cells with sulfonated phthalocyanines.
Photochem.Photobiol., 45:769.

Lindmo, T., and Aarnæs,E. 1979, Selection of optimal model for the DNA histogram by analysis of error of estimated parameters.
J.Histochem.Cytochem., 27(1):297.

Moan, J.,Waksvik,H., and Christensen,T. 1980, DNA single-strand breaks and sister chromatid exchanges induced by treat-ment with hematoporphyrin and light or by X-rays in human NHIK 3025 cells.
Cancer Res. 40:2915.

Moan, J.,Peng,Q.,Berg,K.,Western,A., and Rimington,C., 1987,
    Photosensitizing efficiencies, tumor- and cellular uptake
    of different photosensitizing drugs relevant for photo-
    dynamic therapy of cancer.
    Photochem.Photobiol., 46 :713.
Spikes,J.D., 1986, Phthalocyanines as photosensitizers in
    biological systems and for the photodynamic therapy of
    tumors.
    Photochem.Photobiol., 43:691.
Santamaria, L., 1972, Further consideration on photodynamic
    action and carcinogenicity, in: "Research progress in
    organic, biological and medical chemistry", U.Gallo
    and L.Santamaria,eds., Vol. 3, North-Holland Publ.
    Co.,Amsterdam, p.671.
Tofilon, P.J., and Meyn,R.E., 1987, Enhancement of X-ray-induced
    sister chromatid exchanges in hypoxic cells.
    Radiat.Res., 109:449.

THE ROLE OF THE ANATOMY, PHYSIOLOGY AND BIOCHEMISTRY OF TUMORS IN THE SELECTIVE RETENTION OF SENSITIZERS AND THE MECHANISMS OF PHOTOSENSITIZED TUMOR DESTRUCTION

John D. Spikes

Department of Biology
University of Utah
Salt Lake City, Utah 84112 U.S.A.

INTRODUCTION

During the last few years there has been a sharp increase in research on photomedicine, i.e., the applications of light (ultraviolet, visible, near infrared) as a tool in the prevention of disease, the preparation of pharmaceuticals, the elucidation of the etiology of disease, and in the diagnosis and treatment of disease (Regan and Parrish, 1982; Smith, 1984; Spikes, 1985). Light can be used in the treatment of disease in two ways, termed phototherapy and photochemotherapy. In phototherapy, the effective light is absorbed by endogenous chromophores, while in photochemotherapy, appropriate light absorbing drugs (termed "dyes" in this review) are administered to the patient prior to light treatment. Much of the recent interest in photochemotherapy has centered on the treatment of tumors. Such therapy offers the possibility of double selectivity for the target tumor by using a light absorbing drug that is taken up or retained specifically by the tumor, followed by illumination of the tumorous area. This review is concerned with those anatomical, physiological and biochemical properties of tumors and of malignant tumor cells that appear to be useful in the development of photochemotherapeutic modalities for tumor treatment.

Tumors differ in a number of ways from most normal tissues, and malignant tumor cells differ from normal cells (Creasey, 1981; Levine et al., 1984; Baldwin and Byers, 1985). Some of these differences have been exploited in commonly used tumor therapies, e.g., the use of ionizing radiation and antineoplastic drugs. Certain differences between tumors and normal tissues are also being vigorously exploited in the development of photochemotherapeutic approaches to cancer treatment. The principal phenomena being examined in this respect include the following:

a. Solid tumors somewhat selectively retain a variety of dyes at concentrations higher than are found in many kinds of normal tissues, perhaps as a result of differences in the properties of the microvasculature, stroma, etc. Illumination of tumors with retained photosensitizing dyes, such as certain porphyrins, can result in the destruction of the tumor with relatively little damage to the surrounding normal tissues.

b. The mitochondria of carcinoma cells differ from those in normal cells in that they accumulate very high concentrations of certain photosensitizing cationic dyes, such as some cyanine derivatives. Subsequent illumination results in the highly selective killing of the carcinoma cells with little effect on normal cells.

c. The membranes of leukemic lymphocytes have high affinity binding sites for the anionic photosensitizing dye, merocyanine 540. Thus the dye accumulates in high concentrations in the membranes of leukemic cells. Subsequent illumination in the presence of the dye efficiently kills leukemic, but not normal lymphocytes.

d. Malignant tumor cells have some cell surface antigens that are different from those of normal cells. Monoclonal antibodies directed toward tumor cell antigens can be covalently coupled to photosensitizing dyes. Such conjugates will bind specifically to tumor cells and sensitize their photokilling, without damaging normal cells.

The ways in which these four arrays of tumor/tumor cell properties are being exploited at present in the development of photochemotherapeutic modalities for the treatment of tumors are reviewed in the following sections.

DYE LOCALIZATION IN SOLID TUMORS

Since the time of Paul Ehrlich and the development of the concept of chemotherapy, investigators have searched for compounds that would localize in tumors in the hope that such substances might be used to control tumor growth. As part of this search for a "magic bullet" for tumors, large numbers of dyes of various chemical types have been injected into or fed to tumor-bearing laboratory animals, and their uptake, tissue localization, elimination, etc. followed. Dyes offer some convenience for this purpose, since their gross and microscopic localization can often be determined visually. A number of dyes have been found to somewhat selectively stain both spontaneous and implanted tumors of various types, as well as certain other tissues, depending on the dye. The pattern and intensity of staining depends on the dye, the route of introduction into the animal, the size of the tumor, the amount of necrotic material in the tumor, etc. For example, several azo dyes (Evans blue, Trypan blue) stain viable regions, but not necrotic areas of tumors, while others (Trypan red, Vital new red) stain both regions (Duran-Reynals, 1939). These azo dyes only stain components of the tumor stroma; they do not penetrate into the living malignant cells. In contrast, the oxazin dyes Nile blue and Cresyl violet, and the rhodamines (xanthene dyes) stain the entire tumor diffusely (Lewis et al., 1946). The phenyl methane dye, Lissamine green B, is retained only in a thin shell of tissue just outside of the necrotic region of the tumor; live tumor cells are not stained (Goldacre and Sylven, 1962). Other dyes that stain solid tumors to some extent include acridines, berberine sulfate, chlorins, a chlorin-porphyrin ester, fluorescein (a xanthene dye), phthalocyanines, and a number of different porphyrins (see refs. in Spikes, 1985; Spikes and Jori, 1987). The azo and phenyl methane dyes are not photochemical sensitizers; however, many representatives of the other categories are sensitizers. The porphyrins have been studied much more than the other types of dyes since they are being used as sensitizers for the photochemotherapy of tumors in a large number of patients (Dougherty, 1986, 1987).

Von Tappeiner and Jesionek in 1903 were the first to use photochemotherapy in the treatment of tumors. They applied a solution of eosin topically and by injection into skin tumors in three patients, and then exposed the tumorous areas to light from the sun and/or an arc lamp; improvement was noted in all three patients. This approach was not followed up to any great extent until Auler and Banzer (1942) found that carcinomas and sarcomas implanted into rats were selectively stained by injected hematoporphyrin. Irradiation of the stained tumors with a quartz arc lamp resulted in the partial regression of the tumors. Later, a series of studies of this type by a number of workers (see refs. in Spikes, 1985; Spikes and Jori, 1987) culminated in the initiation of full scale clinical trials on the photochemotherapy of tumors by Dougherty and coworkers in 1976. Over two thousand patients have now been treated using this modality, which has been termed photodynamic therapy (PDT); the results have been generally encouraging. Almost all of the human studies have been made using a porphyrin preparation termed hematoporphyrin derivative. A very large amount of work has been done in this area in recent years, as documented by many books and reviews (see Kessel, 1984, 1985, 1986; Jori and Perria, 1985; Dougherty, 1986, 1987; Moan, 1986; Spikes and Jori, 1987).

The mechanisms by which certain porphyrins and other dyes stain solid tumors are not well understood. It should be stressed that these porphyrins also localize in other rapidly proliferating tissues (healing wounds, embryonic structures) as well as in normal organs including liver, kidney and components of the reticulo-endothelial system (Dougherty, 1986). Some porphyrins are retained in the skin to some extent, resulting in cutaneous photosensitivity that can persist for a month or so; this is the only negative side effect that has been observed in photodynamic therapy.

A number of studies have been done on the sites of localization of porphyrins in tumors. With hematoporphyrin derivative and some other porphyrins, most of the retained sensitizer is found in the interstitial regions (stroma) of the tumor, in particular in the extracellular matrix. Significant amounts are also found in macrophages and fibroblasts, in necrotic areas, and associated with the endothelial cells and/or perivascular fibers of the microvasculature; little is found in the malignant cells of the tumor (see refs. listed in Straight and Spikes, 1985). Some lines of tumor and transformed cells in culture retain more porphyrin than normal cells; others do not. There does not yet seem to be any convincing evidence that the selective localization of porphyrins in tumors results from any characteristic differences between normal and malignant cells (see refs. in Bohmer and Morstyn, 1985).

The physical chemical properties of the sensitizer may play a role in localization. Very recently, based on studies with tetraphenylporphines sulfonated to different extents, Kessel (1987) has suggested that localization might occur by two pathways. One type could be mediated by more lipophilic sensitizers bound to lipoproteins, resulting in sensitizer incorporation into neoplastic cells, with the other involving binding of more hydrophilic sensitizers to serum albumin, resulting in sensitizer localization in the stroma of tumors (also see Jori et al., 1984). Those properties of tumors that form the basis for the selective retention of porphyrins are not known. There are, of course, some characteristic differences between solid tumors and most normal tissues. For example, the neovasculature of the proliferating tumor tissue is much "leakier" than the vasculature of most normal tissues (Duran-Reynals, 1939; Denekamp, 1984). Azo dyes, such as Evans blue, rapidly move out through tumor capillaries into the stroma, whereas they do not penetrate normal vasculature. Similarly, serum proteins penetrate out of tumor capillaries. However, the role that might be played by tumor microvasculature properties in dye retention is not known. The interstitial compartments of tumors also differ from those of normal tissues in that the large amounts of recently synthesized collagen present bind porphyrins more strongly than the crosslinked collagen of normal tissues (Musser et al., 1982).

We have some understanding of the mechanisms involved in the destruction of tumors during photodynamic therapy with porphyrins. It has been shown that vascular damage followed by decreased blood flow and blood stasis occurs almost immediately following illumination; hemorrhages are often observed (see Selman et al., 1984; Henderson et al., 1985; Star et al., 1986). The concentration of oxygen, a reactant necessary for photodynamic effects on cells, is decreased in the tumor (Freitas, 1985). There is little direct killing of malignant tumor cells during photodynamic therapy; clonogenic assays of cells in tumors in mice show that light doses giving essentially complete tumor destruction do not kill the malignant cells if the tumor is explanted shortly after illumination. However, cells in tumors left in place after illumination start dying progressively with time (Henderson et al., 1984, 1985). Thus, ischemia resulting from vascular destruction appears to be the main cause of tumor destruction in photodynamic therapy; direct photodynamic effects on malignant cells are probably of less importance with hematoporphyrin derivative as the sensitizer (see Musser and Datta-Gupta, 1984; Nelson et al., 1985; Shulock et al., 1986).

It should be remembered that the host animal, not the tumor, supplies the newly formed vascular and interstitial components of implanted tumors. We have used polyvinyl alcohol surgical sponges inserted subcutaneously as a model tumor system for studying the role of the neovasculature and neointerstitial components in the retention of sensitizing dyes in tumors (Straight and Spikes, 1985). Such implants in mice rapidly accumulate fluid and plasma proteins, including fibronectin, fibrinogen and fibrin. They then become vascularized, the plasma proteins are replaced by fibroblasts and newly synthesized collagen fibrils, and a connective tissue capsule is formed around the sponge. Vascularized sponge implants selectively retain hematoporphyrin derivative, tetraphenylporphine tetrasulfonate, zinc phthalocyanine sulfonate, and Evans blue to levels at least as high as in implanted tumors, even though no malignant cells are present. Illumination of vascularized sponges with retained hematoporphyrin derivative results in an immediate decrease in blood flow rate, clumping of red blood cells and circulatory stasis, i.e., the same pattern observed in tumors. Such treatment also markedly increases the permeability of the blood vessels in the sponge to Evans blue. Thus these implanted sponges appear to make a useful model for examining the mechanisms of dye retention in and photodynamic destruction of tumors, especially in terms of the possible roles of the tumor vasculature and stroma.

DYE ACCUMULATION BY MITOCHONDRIA OF CARCINOMA CELLS

It was reported a number of years ago that the xanthene dyes, rhodamines B, 5GN, 6G and 6GN, as administered orally, selectively stain implanted sarcomas in mice. The tumors are stained a bright salmon pink color throughout, while the surrounding tissues do not stain; there is also a slight coloring of the lymph nodes and thymus (Lewis et al., 1946). Tumor growth is somewhat inhibited by these dyes. More recently, Chen and his collaborators (see refs. in Davis et al., 1985) reported that the mitochondria in many carcinoma cell lines accumulate rhodamine 123 to very high concentrations and retain the dye much longer than do the mitochondria of normal cells; other lipophilic, cationic dyes behave in the same way. In theory, such dyes can be concentrated 10-fold across the plasma membrane and another 1000-fold within the mitochondria. This phenomenon may result from higher than normal transmembrane potentials in carcinoma cells. Rhodamine 123 shows some selective dark toxicity for carcinoma cells.

Very recently, Oseroff and his associates have exploited this phenomenon as a possible photochemotherapeutic modality for tumor treatment (Oseroff et al., 1986b, 1987abc). The most effective sensitizers reported are a substituted kryptocyanine dye (EDKC) and some of its derivatives.

EDKC is accumulated 300-fold into a human bladder carcinoma cell line (EJ) after a short incubation in the dye; subsequent illumination rapidly kills the cells. Normal cells are unaffected by this treatment, and the dye has a low dark toxicity for normal cells. EDKC phototoxicity is mediated largely by damage to Complex I in the electron transport chain of the mitochondria; this results in decreased ATP levels in the cells. If glycolytic energy-supplying processes are inhibited with 2-deoxy-D-glucose, EDKC-sensitized photokilling of carcinoma cells is enhanced 50-fold. Hyperthermia also increases cell killing. Cell killing is potentiated by the radiation sensitizer, metronidazole, and by hypoxia; this latter result suggests that mitochondrial damage with EDKC is not mediated by singlet oxygen. Human squamous cell and bladder carcinomas implanted subcutaneously into nude mice take up EDKC and one of its analogs efficiently, whereas skin retains very little dye. Illumination of squamous cell carcinoma nodules in mice after dye treatment essentially completely inhibits tumor growth, without damaging the overlying skin.

BINDING OF THE DYE MEROCYANINE 540 BY THE MEMBRANES OF LEUKEMIC CELLS

Circulating leukocytes purified from the blood of a number of leukemic patients are stained by the dye, merocyanine 540, as evaluated by fluorescence microscopy and by measurements using dye tagged with radioactive sulfur. In contrast, leukocytes from all normal individuals examined are not stained. Dye washes out of the leukemic cells only very slowly. Merocyanine 540 also stains excitable cells (nerve, muscle); the mechanism may be different with these cells, as indicated by different ionic requirements in the medium for staining. Illumination of leukocytes in the presence of merocyanine 540 destroys leukemic cells; however, this treatment has no effect on normal cells (Valinsky et al., 1978).

This phenomenon has been used in preclinical studies as a purging procedure to kill cancer cells in samples of bone marrow used for grafts. Destruction of bone marrow cells is typically the limiting factor for the maximum doses of ionizing radiation and of some cytotoxic drugs that can be used in the treatment of cancer patients. Doses can be increased significantly, however, if they are used in conjunction with bone marrow transplants. The most generally available material for transplants is the patient's own marrow (autologous transplant). However, bone marrow samples removed from an individual prior to antineoplastic therapy are often contaminated with viable tumor cells (Sieber, 1987). If mixtures of neuroblastoma plus normal marrow cells or leukemic plus normal cells are illuminated in the presence of merocyanine 540 and then injected into lethally irradiated mice, the mice survive without developing neuroblastomas or leukemia (Sieber and Sieber-Blum, 1986). This shows that the treatment destroys the tumor cells while sparing enough marrow cells to repopulate the mouse hematopoietic system. A similar treatment of mixtures of human leukemic and normal bone marrow cells results in the killing of the cancer cells; however, again, a significant number of normal cells survive and form colonies (Atzpodien et al., 1986). This modality has been tried on one patient with resistant Hodgkins disease (Sieber et al., 1986). Marrow was removed from the patient and illuminated in the presence of merocyanine 540. After receiving heavy drug therapy, the patient was injected with the treated marrow, which successfully reconstituted the hematopoietic system. Merocyanine 540 sensitizes the photooxidation of histidine, arachidonate, cholesterol and an unsaturated phospholipid; it also sensitizes the photoperoxidation of lipids in the membranes of erythrocyte ghosts. Mechanistic studies indicate that these various photosensitized reactions are mediated by singlet oxygen (Kalynaraman et al., 1987).

SPECIFIC BINDING OF ANTIBODY-DYE CONJUGATES BY TUMOR CELLS

A large amount of research has been done in recent years on the use of specific antibody-drug conjugates to target and destroy tumor cells; in particular, monoclonal antibodies directed toward cell surface antigens permit a highly specific reaction with a particular type of tumor cell (Goldberg, 1983; Baldwin and Byers, 1985). Some of this effort has centered on the photochemotherapy of tumors using tumor-directed antibodies coupled with photosensitizers. Such reagents might be expected to be much more specific for a given type of tumor than photosensitizers, such as hematoporphyrin derivative, used alone. For example, injection of hematoporphyrin conjugated directly with monoclonal antibodies specific for a myosarcoma into mice with implanted myosarcomas, followed by illumination with an incandescent lamp, suppresses tumor growth. The antibody-hematoporphyrin conjugate has no effect on the growth of a lymphoma in the mice, and conjugates prepared with a non-specific monoclonal antibody do not photosensitize the myosarcoma (Mew et al., 1983). Covalent conjugates of antibodies directed toward certain cell lines with hematoporphyrin react specifically with cells of those lines, as shown by fluorescence microscopy and by cell killing on illumination (Mew et al., 1985).

Direct binding of sensitizers to antibodies can interfere with antibody reactivity if attachment occurs near the antigen binding sites. To avoid this, Oseroff and coworkers (Oseroff et al., 1986a, 1987c) first coupled the sensitizer to a low molecular weight dextran; this complex was then coupled to carbohydrate moieties located well away from the antigen binding sites of the antibody. Chlorin e_6 was used as the sensitizer since it absorbs light more efficiently and at a longer wavelength than does hematoporphyrin; the use of longer wavelengths of light in photochemotherapy permits much better light penetration into mammalian tissues. Covalent conjugates of this type, prepared with monoclonal antibodies directed toward human T leukemia cells, efficiently and specifically sensitize the photokilling of T cells. Phototoxicity is enhanced in D_2O, suggesting that cell killing is mediated by singlet oxygen. In a variation on this photochemotherapeutic approach, monoclonal antibodies toward a T lymphocyte surface antigen were covalently coupled to liposomes containing the lipid-soluble photosensitizer, pyrene, incorporated into the lipid bilayer. Illumination of mixed T- and B-cell lines after treatment with the conjugate kills the T cells but has no effect on B cells (Yemul et al., 1987).

PHOTOCHEMICAL VERSUS PHOTOTHERMAL SENSITIZATION BY DYES

Light absorbing structures in cells can be selectively heated by using pulsed laser illumination. For effective heating, the duration of the laser pulse must be equal to or shorter than the thermal diffusion time leading to cooling of the organelle. For cell organelles, pulses must be in the nanosecond region for selective heating. Illumination of pigmented epidermal cells with a 351 nm excimer dye laser producing 20 ns pulses gives a selective destruction of melanosomes in the cells (Anderson and Parrish, 1983; Parrish et al., 1983). This suggests that carcinoma cells could be destroyed by a photothermal mechanism, since mitochondria with accumulated cationic dyes might be heated specifically on exposure to nanosecond pulsed illumination to temperatures that would interfere with their activity. Oxygen would not be required for such a mechanism, which could result in more efficient cell killing in the anaerobic regions of tumors. Sensitizing dyes for this technique should have large extinction coefficients, and should rapidly decay from the light-excited state to the ground state via non-radiative processes to maximize heat production. Such dyes might also sensitize via shock wave-damage (Anderson and Parrish, 1983; Parrish et al., 1983).

ACKNOWLEDGEMENTS

The preparation of this review, and the original work described, were supported in part by American Cancer Society Grant No. PDT-259A and USA Office of Naval Research Contract No. N00014-86-K-0258.

REFERENCES

Anderson, R.R., and Parish, J.A., 1983, Selective photothermolysis: precise microsurgery by selective absorption of pulsed radiation, Science, 220:524.

Atzpodien, J., Gulati, S.C., and Clarkson, B. D., 1986, Comparison of the cytotoxic effects of merocyanine 540 on leukemic and normal human bone marrow, Cancer Res. 46:4892.

Auler, H., and Banzer, G., 1942, Untersuchungen ueber die Rolle der Porphyrine bei geschwulstkranken Menschen und Tieren, Z. Krebsforsch., 53:65.

Baldwin, R.W., and Byers, V.S., eds., 1985, "Monoclonal Antibodies for Cancer Detection and Therapy", Academic, New York.

Bohmer, R.M., and Morstyn, G., 1985, Uptake of hematoporphyrin derivative by normal and malignant cells: effects of serum, pH, temperature, and cell size, Cancer Res., 45:5328.

Creasey, W.A., 1981, "Cancer. An Introduction", Oxford University Press, New York.

Davis, S., Weiss, M.J., Wong, J.R., Lampidis, T.J., and Chen, L.B., 1985, Mitochondrial and plasma membrane potentials cause unusual accumulation and retention of rhodamine 123 by human breast adenocarcinoma-derived MCF-7 cells, J. Biol. Chem., 260:13844.

Denekamp, J., 1984, Vasculature as a target for tumor therapy, Prog. Appl. Microcirc., 4:28.

Dougherty, T.J., 1986, Photosensitization of malignant tumors, Sem. Surg. Oncology, 2:24.

Dougherty, T.J., 1987, Photosensitizers: therapy and detection of malignant tumors, Photochem. Photobiol., 45:879.

Duran-Reynals, F., 1939, Studies on the localization of dyes and foreign proteins in normal and malignant tissues, Am. J. Cancer, 35:98.

Freitas, I., 1985, Role of hypoxia in photodynamic therapy of tumors, Tumori, 71:251.

Goldacre, R.J., and Sylven, B., 1962, On the access of blood-borne dyes to various tumor regions, Br. J. Cancer, 16:306.

Goldberg, E.P., 1983, "Targeted Drugs", Wiley, New York.

Henderson, B.W., Dougherty, T.J., and Malone, P.B., 1984, Studies on the mechanism of tumor destruction by photoradiaton therapy, p. 601, in: "Porphyrin Localization and Treatment of Tumors:, D.R. Doiron and C.J. Gomer, eds., Alan R. Liss, New York.

Henderson, B.W., Waldow, S.M., Mang, T.S., Potter, W.R., Malone, P.B., and Dougherty, T.J., 1985, Tumor destruction and kinetics of tumor cell death in two experimental mouse tumors following photodynamic therapy, Cancer Res., 45:572.

Jori, G., Beltramini, M., Reddi, E., Salvato, B., Pagvan, A., Ziron, L., Tomio, L., and Tsanov, T., 1984, Evidence for a major role of plasma lipoproteins as hematoporphyrin carriers in vivo, Cancer Lett., 24:291.

Jori, G., and Perria, C., eds., 1985, "Photodynamic Therapy of Tumors and Other Diseases", Edizioni Libreria Progetto, Padova.

Kalyanaraman, B., Felix, J.B., Sieber, F., Thomas, J.B., and Girotti, A.W., 1987, Photodynamic action of merocyanine 540 on artificial and natural cell membranes: involvement of molecular oxygen, Proc. Natl. Acad. Sci. USA, 84:2999.

Kessel, D., 1984, Porphyrin localization: a new modality for detection and therapy of tumors, Biochem. Pharmacol., 33:1389.

Kessel, D., ed., 1985, "Methods in Porphyrin Photosensitization", Plenum, New York.

Kessel, D., 1986, Photosensitization with derivatives of haematoporphyrin, Int. J. Radiat. Biol., 49:901.

Kessel, D., Thompson, P., Saatio, K., and Nantwi, K.D., 1987, Tumor localization and photosensitization by sulfonated derivatives of tetraphenyl-porphine, Photochem. Photobiol., 45:787.

Levine, A.J., Vande Woude, G.F., Topp, W.C., and Watson, J.D., 1984, "Cancer, Vol. 1", Cold Spring Harbor Laboratory, New York.

Lewis, M.R., Goland, P.P., and Sloviter, H.A., 1946, Selective action of certain dyestuffs on sarcomata and carcinomata, Anat. Record, 96:201.

Mew, D., Wat, C-K., Towers, G.H.N., and Levy, J.G., 1983, Photoimmunotherapy: treatment of animal tumors with tumor-specific monoclonal antibody-hematoporphyrin conjugates, J. Immunol, 130:1473.

Mew, D., Lum, V., Wat, C-K., Towers, G.H.N., Sun, C-H. C., Walter, R.J., Wright, W., Berns, M.W., and Levy, J.G., 1985, Ability of specific monoclonal antibodies and conventional antisera conjugated to hematoporphyrin to label and kill selected cell lines subsequent to light activation, Cancer Res., 45:4380.

Moan, J., 1986, Porphyrin photosensitization and phototherapy, Photochem. Photobiol., 43:681.

Musser, D.A., and Datta-Gupta, N., 1984, Inability to elicit rapid cytocidal effects on L1210 cells derived from porphyrin-injected mice following in vitro photoirradiation, J. Natl. Cancer Inst., 72:427.

Musser, D.A., Wagner, J.M., and Datta-Gupta, N., 1982, The interaction of tumor localizing porphyrins with collagen and elastin, Res. Commun. Chem. Pathol. Pharmacol., 36:251.

Nelson, S., Wright, W.H., and Berns, M.W., 1985, Histopathological comparisons of the effects of hematoporphyrin derivative on two different murine tumors using computer-enhanced digital video fluorescence microscopy, Cancer Res., 45:5781.

Oseroff, A.R., Ohuoha, D., Hasan, T., Bommer, J.C., and Yarmush, M.L., 1986a, Antibody-targeted photolysis: Selective photodestruction of human T-cell leukemia cells using monoclonal antibody-chlorin e_6 conjugates, Proc. Natl. Acad. Sci. USA, 83:8744.

Oseroff, A.R., Ohuoha, D., Ara, G., McAuliffe, D., Foley, J., and Cincotta, L., 1986b, Intramitochondrial dyes allow selective in vitro photolysis of carcinoma cells, Proc. Natl. Acad. Sci. USA, 83:9729.

Oseroff, A.R., Ohuoha, D., Ara, G., Foley, J., and Cincotta, L., 1987a, Selective carcinoma cell photolysis (SCCP), the preferential light-induced killing of carcinoma cells due to mitochondrial accumulation of cationic dyes, is potentiated by 2-deoxy-D-glucose (2-DG) and hyperthermia, Program, Clayton Foundation Conference on Photodynamic Therapy, February 1987 (Abstract).

Oseroff, A.R., Ohuoha, D., Ara, G., Kane, S.B., Foley, J., and Cincotta, L., 1987b, The carcinoma cell-selective phototoxicity of EDKC is potentiated by hypoxia and metronidazole, Photochem. Photobiol., 45:101S (Abstract).

Oseroff, A.R., Ara, G., Ohuoha, D., Aprille, J., Bommer, J.C., Yarmush, M.L., Foley, J., and Cincotta, L., 1987c, Strategies for selective cancer photochemotherapy: antibody-targeted and selective carcinoma cell photolysis, Photochem. Photobiol., 46:83.

Parrish, J.A., Anderson, R.R., Harrist, T., Paul, B., and Murphy, G.F., 1983, Selective thermal effects with pulsed irradiation from lasers: from organ to organelle, J. Invest. Dermatol., 80:75s.

Regan, J.D., and Parrish, J.A., eds., 1982, "The Science of Photomedicine", Plenum, New York.

Selman, S.H., Kreimer-Birnbaum, M., Klaunig, J.E., Goldblatt, P.J., Keck, R.W., and Britton, S.L., 1984, Blood flow in transplantable bladder tumors treated with hematoporphyrin and light, Cancer Res., 44:1924.

Shulok, J.R., Klaunig, J.E., Selman, S.H., Schafer, P.J., and Goldblatt, P1J., 1986, Cellular effects of hematoporphyrin derivative photodynamic therapy on normal and neoplastic rat bladder cells, Am. J. Pathol., 122:277.

Sieber, F., Craig, A., Krueger, G.J., Smith, R.E., and Ash, R.C., 1986, Autotransplantation of bone marrow after extracorporeal purging with merocyanine 540 and light, Blood, 68:292a (Abstract).

Sieber, F., and Sieber-Blum, M., 1986, Dye-mediated photosensitization of murine neuroblastoma cells, Cancer Res., 46:2072.

Sieber, F., 1987, Elimination of residual tumor cells from autologous bone marrow grafts by dye-mediated photolysis: preclinical data, Photochem. Photobiol., 46:71.

Smith, K.C., 1984, ed., "Topics in Photomedicine", Plenum, New York.

Spikes, J.D., 1985, The historical development of ideas on applications of photosensitized reactions in the health sciences, p. 209, in: "Primary Photo-Processes in Biology and Medicine", R.V. Bensasson, G. Jori, E.J. Land, and T.G. Truscott, eds., Plenum, New York.

Spikes, J.D., and Jori, G., 1987, Photodynamic therapy of tumours and other diseases using porphyrins, Lasers Med. Sci., 2:3.

Star, W.M., Marijnissen, P.A., van den Berg-Blok, A.E., Versteeg, J.A.C., Franken, K.A.P., and Reinhold, H.S., 1986, Destruction of rat mammary tumor and normal tissue microcirculation by hematoporphyrin derivative photoradiation observed in vivo in sandwich observation chambers, Cancer Res., 46:2532.

Straight, R.C., and Spikes, J.D., 1985, Preliminary studies with implanted polyvinyl alcohol sponges as a model for studying the role of neointerstitial and neovascular compartments of tumors in the localization, retention and photodynamic effects of photosensitizers, p. 77, in: "Methods in Porphyrin Photosensitization", D. Kessel, ed., Plenum, New York,

Tappeiner, H.v., and Jesionek, 1903, Therapeutische Versuche mit fluoreszierenden Stoffen, Muench. Med. Woch., 50:2042.

Valinsky, J.E., Easton, T.G., and Reich, E., 1978, Merocyanine 540 as a fluorescent probe of membranes: selective staining of leukemic and immature hematopoietic cells, Cell, 13:487.

Yemul, S., Berger, C., Estabrook, A., Suarez, S., Edelson, R., and Bayley, H., 1987, Selective killing of T lymphocytes by phototoxic liposomes, Proc. Natl. Acad. Sci., USA, 84:246.

NEW DEVELOPMENTS AND FUTURE PROSPECTS IN THE

CLINICAL APPLICATIONS OF PHOTODYNAMIC THERAPY

J.A.S. Carruth

Consultant Otolaryngologist and General Secretary of the
International Society of Laser Medicine and Surgery
Royal South Hants Hospital, Southampton, England, U.K.

INTRODUCTION

In 1900 Raab described the killing of paramoecium sensitized with an
acridine dye and exposed to light. The first photodynamic therapy of
tumours in man was carried out by von Tappenier and Jesionek who applied
eosin to skin tumours and then exposed the tumour area to sunlight or to
an arc lamp with some improvement in most cases. (12)

Since that time a large number of tumour sensitisers have been and
are being investigated, but a majority of the research work has been on the
porphyrins and on haematoporphyrin derivative (HPD) and its components in
particular.

For "ideal" photodynamic therapy a patient with a malignant tumour
should be given a photosensitive tumour-sensitiser which is non-toxic in
clinically useful doses and is either selectively taken up by or retained
by malignant tissues giving a high tumour/normal tissue ratio. The tumour
sensitiser should be photoactivated by a wavelength of light which
penetrates deeply into tissues and should have a high level of photochemical activity. A high power, monochromatic laser light source of the
appropriate wavelength should be available to photoactivate the sensitiser
with a wide range of delivery fibres to treat tumours in all sites, and
tumour destruction should be possible with minimal or no damage to adjacent
normal tissues. (5)

The ideal situation is some way off and it appears certain that
sensitisers superior to HPD and its components will be developed. However
it also appears certain that no sensitiser will be investigated to a point
at which clinical trials are appropriate for several years and during this
period HPD activated by red light at a wavelength of 630 n.m. will be the
only combination used in clinical practice. (13)

CLINICAL STUDIES

Following both in vitro and in vivo research work on photodynamic
therapy, much of which was carried out at Roswell Park Memorial Institute
by Dr. Tom Dougherty and his co-workers, clinical studies began approximately
a decade ago and estimates suggest that over 7,000 patients have been treated
in a number of centres to date.

Tumour Sensitiser

For all the clinical work carried out to date, either haematoporphyrin derivatives or its "active component" identified by Dougherty as dihaematoporphyrin ether (DHE) have been used.

Many of the early studies were carried out using an essentially standard product produced by Oncology Research and Development and subsequently Photomedica, but from the outset a small number of centres produced their own sensitiser and this number has increased dramatically recently and this will make it very difficult to compare accurately work carried out in different parts of the world.

Light Sources

For much of the early clinical work a wide range of both filtered and unfiltered light sources were used but it soon became apparent that to produce monochromatic light of the appropriate wavelength at high power levels which could be transmitted via a flexible fibre, a laser light source was necessary. At first the Argon pumped tunable dye laser was used to produce coherent light at a wavelength of 630 n.m. and recently the copper vapour pumped dye laser has been introduced. The gold vapour laser producing coherent light at 628.5 n.m. is being used in a number of clinical studies and although the wavelength of this laser is fixed and could not, therefore, be used to activate other sensitisers which may be developed in the future, it is a relatively easy matter to change the metal in the laser tube to copper and this laser could then be used to drive a tunable dye laser. (7)

Delivery Systems

A wide range of delivery fibres are now available enabling tumours in all sites to be treated. Many tumours have been treated by surface irradiation using a straight cut fibre or one with a micro lens tip and it has been estimated that light at 630 n.m. will penetrate up to 1 cm into tissue but this depth obviously varies widely depending on the type, vascularity and homogeneity of the tumour. Fibres with a diffusing cylinder tip are used to treat circumferential lesions or for interstitial treatment in which the tip is embedded into the tumour tissue. Fibres with diffusing bulb ends are available to treat the inside of hollow viscera such as the bladder and this can be either replaced or augmented by mechanical devices or by filling the whole of the inside of the bladder with a diffusing medium. (11)

Treatment Techniques

Allowing for the minor variations in the sensitiser described above essentially the same treatment technique has been used in all the clinical studies. On day one the sensitiser is injected intravenously in a dose of 3 mg/Kg body weight for HPD and 50-60% of this dose if the active component is used. No significant side effects from the injection have been described unless of course the injection is given outside the vein, but all patients develop severe skin photosensitisation which lasts for at least three to four weeks and in many cases for several months. This means the patients have to stay out of sunlight or even bright daylight and in effect the patients have to remain indoors during the day. This represents the only major side effect of this form of treatment and it is to be hoped that sensitisers which will be developed in the future will not have this problem.

After 72 hours the tumour is photo-irradiated using a laser with an appropriate delivery fibre or fibres. The delivery dose will obviously depend on the area of tumour to be treated and the power of the laser, and it has been suggested that a minimal delivery dose of 12 mW/cm^2 may be necessary although satisfactory clinical results have been obtained using a much lower dose. If very high power densities are used there may be a thermal effect and although there is evidence to suggest that hyperthermia may be synergistic with photodynamic therapy, it should be avoided as each modality must be investigated on its own.

The total energy dose remains somewhat empirical but it is suggested that for superficial skin lesions a total dose of 25-50 J/cm^2 may be appropriate, whereas for exophytic ulcerated lesions a total dose of 100-200 J/cm^2 is thought to be necessary.

All the treatment can be given without anaesthesia apart from any anaesthetic needed for associated surgical or endoscopic procedures but as the destruction of tissues by photodynamic therapy takes place, many patients will experience discomfort which on occasions may amount to severe pain.

Clinical Trials

A wide range of tumours have been treated by photodynamic therapy but many of the early trials were uncontrolled and the reported anecdotal. However with an increasing interest of specialist oncologists in the technique, the design of trials and the reporting has improved considerably and it is possible to identify areas in which this technique offers particular promise.

Carcinoma of the Bronchus. Control of early stage lesions in patients who were considered to be unsuitable for or who refused conventional resection with survival for more than three years has been reported by Konaka, Kato and Hayata and other workers are achieving similar results in this group of patients. However the problem remains that 65% of patients with carcinoma of the bronchus are untreatable at presentation and the five year survival figures are only 5%. Improved results depend on the identification of early stage lesions and this demands an adequate screening programme. The presence of an early lesion will be identified by the finding of increasingly abnormal cells in the sputum and the site of the lesions may then be found using a fluorescent bronchoscopic technique. A blue enhanced fibre-optic bronchoscope is used with a Krypton laser as a light source and an image intensifier to identify the areas of faint fluorescence and with this the areas of tumour can be identified much more easily than with standard white light bronchoscopy. Once the lesion has been identified it can then be treated by changing the light source to red at 630 n.m. with the potential for cure in early stage lesions. (1, 4, 6, 8)

Photodynamic therapy has also been used to achieve palliation in advanced lesions and studies are in progress to compare palliation achieved with this technique as opposed to tumour ablation using the Neodymium YAG laser.

Carcinoma of the Bladder. This is often multifocal and a fluorescent technique may be used to identify malignant foci with great accuracy. If the superficial lesions of carcinoma in situ of the bladder fail to respond to conventional therapy then treatment is almost always by cystectomy which is a very distressing operation for the patient.

A number of groups have reported exciting results in the treatment of this disease using whole bladder wall irradiation using a wide range of light delivery systems. (2)

Brain Tumours. The outlook for patients with malignant brain tumours is universally hopeless and treatment of these lesions by photo-irradiation of the tumour base following standard surgical resection appears to offer considerable promise. A number of feasibility studies have been carried out to show that this form of treatment can be carried out without significant morbidity and although it is too early to draw significant conclusions from the work carried out to date, it does appear that tumour control has been achieved in a number of patients with survival significantly prolonged over historical controls. (9, 10)

Head and Neck. The head and neck tumour is also very suitable for photodynamic therapy and some interesting preliminary results have already been reported. The head and neck tumour is accessible, remains localised with late metastasis, is relatively small and if conventional treatment by radiotherapy fails, then surgery is always mutilating either to the appearance of the patient or to his ability to talk and swallow. (3, 14)

After some interesting preliminary and pilot studies it is to be hoped that photodynamic therapy will be introduced as the first arm of treatment in multi-modality controlled trials followed by radiotherapy and/or surgery and in this way its role in the treatment of tumours at many sites in the head and neck region may be established.

Breast. Little work has been carried out on the treatment of primary breast tumours with this technique but in the treatment of multifocal superficial disease of the chest wall, Dougherty has established that control can be achieved in 60-80% of cases.

Skin. For many years it has been suggested that photodynamic therapy may represent the treatment of choice for multiple basal cell carcinomas which are so difficult to treat by other modalities. In the author's series this has been confirmed with local control in all cases and in single lesions unsuitable for other forms of treatment. In addition the multiple carcinoma in situ lesions of Bowen's disease appear to respond extremely well and the author has treated, successfully, more than 500 lesions in two patients.

Other Lesions. Some very interesting results are being obtained in the treatment of tumours of the upper and lower gastro-intestinal tracts, the eye and of gynaecological lesions. Some exciting results are also being obtained in the treatment of malignant ascites and there appear to be no problems related to the retention of the sensitiser by the liver, spleen and kidneys. Another suggested application for photodynamic therapy is in the sterilisation of the tumour bed after standard surgical resection, particularly in head and neck and abdominal surgery.

OTHER NON-MALIGNANT CONDITIONS

There is evidence to show that photodynamic therapy may be used to treat psoriasis with some encouraging early results.

A further possible application of photodynamic therapy is in the disobliteration of arteries obstructed by atheroma. Haematoporphyrin derivative appears to be selectively retained by the atheroma plaque and on exposure to light there is some retraction of the plaque with a widening of the vessel lumen but without any destruction of the surface of the plaque lining the vessel. In other "laser techniques" for the removal of atheroma there has to be some destruction of this surface with the possible risk of more distal obstruction of the vessel by detached fragments.

CONCLUSIONS

At its present state of development using haematoporphyrin derivative and red light, photodynamic therapy can play an important role in the management of many forms of superficial malignant disease. With the vast amount of research into this technique which is taking place at present, it appears certain that other drugs/light combinations will soon be developed with a marked improvement in results and in time photodynamic therapy will take its place as an important treatment modality for malignant tumours and other diseases.

REFERENCES

1. Balchum, O.J., Doiron, D.R., Profio, A.E. and Hutch, G.C. 1982. Fluorescence bronchoscopy for localizing early bronchial cancer and carcinoma in situ. In: Recent Results in Cancer Research. Vol. 82 (Berlin: Springer) p.98.

2. Benson, R.C. Jr., Farrow, G.M., Kinsey, J.H., Cortese, D.A., Zincke, H. and Utz, D.C. 1982. Detection and localization of in situ carcinoma of the bladder with haematoporphyrin derivative. Mayo Clin. Proc. 57, 548.

3. Carruth, J.A.S. and McKenzie, A.L. 1985. Pilot study on photoradiation therapy in the treatment of superficial tumours of the skin ane head and neck. Clin. Oncol. 11, 47-50.

4. Cortese, D.A., Kinsey, J.H., Woolner, L.B., Payne, W.S., Saunderson, D.R. and Fontana, R.S. 1979. Clinical application of new endoscopic technique for detection of in situ bronchial carcinoma. Mayo Clin. Proc. 54, 635-41.

5. El-Far, M.A. and Pimstone, N.R. 1983. Superiority of uroporphyrin I over other porphyrins in selective tumour localization. Proc. Clayton Found. Symp. on Porphyrin Localization and Treatment of Tumours, Santa Barbara, California. April 1983 (New York: Plenum) to be published.

6. Hayata, Y., Kato, H., Konaka, C., Ono, J., Matsushima, Y., Yoneyama, K. and Nishimiya, K. 1982. Fiberoptic bronchoscopic laser photoradiation for tumour localization in lung cancer. Chest, 82, 10.

7. Hisazumi, H., Naito, K., Misaki, T., Koshida, K. and Yamamoto, H. 1985. An experimental study of photodynamic therapy using a pulsed gold vapor laser. Proceedings of the VIth Conference of the International Society for Laser Surgery, Jerusalem, October 13-16, 1985.

8. Konaka, C., Kato, H. and Hayata, Y. 1987. Lung Cancer Treated by Photodynamic Therapy Alone: Survival for More than Three Years. Lasers in Medical Science, Vol. 2, No. 1, p.17-19.

9. Laws, E.R., Cortese, D.A., Kinsey, J.H., Eagan, R.T. and Anderson, R.C. 1981. Photoradiation therapy in the treatment of malignant brain tumours. A phase I (feasibility) study. Neurosurgery, 9, 672.

10. McCulloch, G.A.J., Forbes, I.J., Lee See, K., Cowled, P.A., Jacka, F.J. and Ward, A.D. Phototherapy in malignant brain tumours. Proc. Clayton Found. Symp. on Porphyrin Localization and Treatment of Tumours, Santa Barbara, California. April 1983 (New York: Plenum) to be published.

11. McKenzie, A.L. 1984. How to control beam profile during laser photoradiation therapy. Phys. Med. Biol. 29, 53-6.

12. Spikes, J.D. and Jori, G. 1987. Photodynamic therapy of tumours and other diseases using porphyrins. Lasers in Medical Science, Vol. 2, No. 1, p.3-15.

13. Weishaupt, K.R., Gomer, C.J. and Dougherty, T.J. 1976. Identification of singlet oxygen as the cytotoxic agent in photo-inactivation of a murine tumour. Cancer Res. 36, 2326.

14. Wile, A.G., Coffey, J., Nahabedian, M.Y., Baghdessarian, R., Mason, G.R. and Berns, M.W. 1984. Laser photoradiation therapy of cancer: an update of the experience at the University of California, Irvine. Lasers Surg. Med. 4, 5.

PDT OF BLADDER TUMOR

Dieter Jocham and E. Unsöld*

Urological Department of LM-University (Director:
Prof. Dr. Schmiedt), D-8000 Munich-70, FRG
* Central Laser Lab, GSF-Neuherberg, D-8042
Neuherberg, FRG

By now, the photodynamic laser therapy of bladder carcinoma has
proved to be a specially attractive application of photodynamic therapy(PDT).
Concerning the use of the photosensitizing substance hematoporphyrin-
derivative, the PDT of a bladder tumor was the first clinical application
ever of this procedure. Initial reports on the photodynamic therapy of
human bladder tumors go back to the year 1976. Kelly and Snell reported
the successful photodynamic destruction of bladder tissue by means of
endoscopically controlled exposure of the exophytically growing bladder
carcinoma to a mercury vapor lamp following intravenous administration
of the photosensitizing agent HpD (9).

After that spectacular result no further reports of success followed
from this research group - a circumstance which from today's urological
point of view is not surprising against the background of numerous investigational and increasing clinical experience gained with the PDT of
bladder carcinoma. In the mid-seventies, progress was hindered by technical problems relating to the transurethral transfer of sufficiently intense light to be photodynamically-therapeutical to the bladder. Moreover, in view of the special clinical picture of the common multifocally
growing carcinoma of the bladder of various developmental stages, direct
irradiation of endoscopically visible bladder tumors did not appear to
offer advantages over other forms of therapy.

In order to understand the principal advantages of PDT over other
local methods in the case of bladder carcinoma, the specific problem
of this tumor disease has to be visualized first.

The endoscopially visible exophytic tumors only represent the top
of the iceberg in the case of this tumor constellation. The established
endoscopic therapy procedures are at best suitable to eliminate these
endoscopically detectable tumor portions. Since, however, the multifocal
bladder carcinoma with large invisible tumor portions is present in 80 %
of the cases of bladder tumor diseases, it is easily understandable that
in most of the cases the disease cannot entirely be treated by mechanical or thermal elimination of the exophytic tumor alone. The mucosal
zones in which severe atypias or the carcinoma in situ are present - the
latter form being associated with an especially poor prognosis - often

give rise to new exophytic or infiltrating tumors. Especially the therapeutic detection of these tumor regions in their entirety raises the prospects for remarkably improving the therapy situation of multifocal bladder carcinoma.

A precondition for the useful application of PDT of bladder carcinoma was and still is, against the background of the above stated problems, the reliable marking of all tumorous regions of the bladder tumor by means of a photosensitizer. Fortunately, this precondition can be met. Investigations using experimentally induced tumors as well as the clinical data exhibit a good correlation between tumor foci and fluorescent component of HpD or DHE (2,3,7,9). Corresponding histologic analysis of degrees of atypia reveals that severe atypias and all tumor tissues of various differentiation from the superficial carcinoma in situ to the infiltrative carcinoma deposit the fluorescent component of the photosensitizer in substantially higher concentrations than normal tissues of the same origin (2).

Against this background, it seemed useful to use PDT for the treatment of bladder carcinoma. PDT has principally proven to be superior to other procedures. Fortunately, due to increasing clinical experience it becomes evident that PDT enables complete tumor-specific destruction of bladder carcinoma.

The urologist's main concern is the carcinoma in situ which is often not amenable to conventional methods of treatment and up to the present still indicates the surgical removal of the bladder, i.e. a cystectomy with all its consequences.

The application of suitable irradiation modalities is highly significant for the successful detection of all tumor portions of the multifocal bladder carcinoma. In our opinion, the direct PDT irradiation of endoscopically visible tumor foci, which is still applied by various urological research groups, is obsolete. At least in the hand of the experienced surgeon equipped with a laser, e.g. the Neodym-YAG-laser for thermal tumor destruction, no additional advantages are achieved as a result of limited tissue penetration of the photodynamically applied laser light. The situation is completely different if homogenous irradiation of all bladder wall sections results in selective destruction of endoscopically undetectable tumor parts. This type of PDT application is named "integral PDT" (8) or "whole bladder wall irradiation" (4,5).

Various physical-technical approaches (Tab. 1) which each have their own difficulties and require individual effort, provide homogenous irradiation of the whole bladder - an absolutely necessary precondition for achieving this aim.

CLINICAL RESULTS

Since the photosensitizer (HpD or DHE) is deposited in tumor tissue in higher concentrations by a factor 2-5 than in normal tissue, more damage is caused to tumor tissue than to normal tissue. The latter exhibits slight alterations under the chosen treatment conditions. These conditions are: Intralipid as light scattering medium 1:40, bladder volume 150 ccm, special bladder catheter, 2-2.5 mg HpD i.v. /kg body-weight, irradiation with 35 J/cm^2 48-65 h following the administration of the photosensitizer. Moreover, the healthy tissue appears to be equipped with much better repair-mechanisms than the tumor tissue.

Table 1.

Table 2. Literature review concerning therapeutic results of PDT of bladder carcinoma.

	N	stage of Tu	init. resp.	long-term response		
	86	$T_A - T_3$	91/117 (78 %)	60/86 (70 %) (3 years)	Zhang	1986
	15	CA i.s.	15/15 (100 %)	8/15 (53 %) (38 months)	Benson	1986
focal therapy	22	T_A/T_1	22/22 (100 %)	4/22 (18 %) (20 months)	Misaki	1986
	20	T_A/T_2	20/20 (100 %)	9/20 (45 %) (60 months)	Tsuchiya	1986
	24	T_A/T_2	24/24 (100 %)	6/24 (25 %) (6 months)	Misaki	1986
whole bladder irradiation	16	CA i.s.	16/16 (100 %)	13/16 (81 %) (12 months)	Benson	1986
	9/18	Ca i.s.	14/18 (78 %)	6/9 (67 %) (33 months)	Jocham	1987
	28	T_A/T_3	20/28 (71 %)	n.d. (30 months)	Nseyo	1987

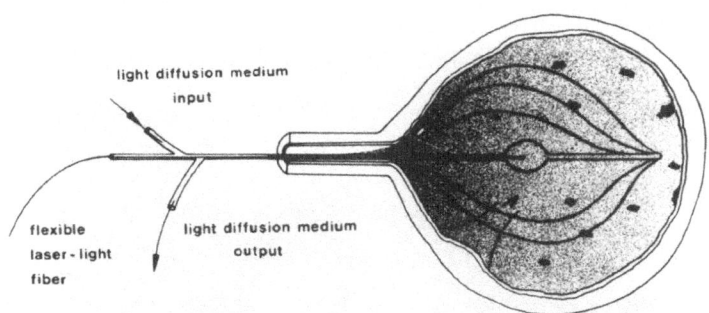

light diffusion medium input

flexible laser-light fiber

light diffusion medium output

Fig. 1. Scheme of integral PDT: The bladder is irrigated by a milky solution which allows diffuse light irradiation of all bladder regions. The fiber tip is positioned in the center of the bladder using a special catheter. Multifocal tumor tissue is destroyed photodynamically by the photodynamic process.

The integral PDT (Fig. 1) permits the detection of all superficial tumor foci nonwithstanding technical failure. Up to the present, the procedure has been confined to superficial tumors due to the tissue penetration depth of only 2-3 mm of red light of 630 nm at a photodynamically sufficient power density. As typical mucosal reaction after PDT, more or less circumscribed hemorrhagic zones are found endoscopically (Fig. 2). Such typical changes develop some hours following irradiation. In the following course, these hemorragic areas turn into yellowish necrotic material which partly disappear several months later. Even in cases of severe widespread mucosal damage a restitution of normal mucosal conditions of the bladder is observed.

The number of patients subjected to PDT is still extremely low. This is mainly due to the insufficient availability of the photosensitizer for clinical use. Furthermore, at least our group only accepts cases of recurrent bladder tumors where all other local therapy measures failed and cystectomy would be the only therapeutic alternative. In other words, this implies a particularly negative selection of patients.

Table 2 shows results obtained with direct irradiation (upper section) and homogenous irradiation (lower section). It should be noted that the results obtained by different groups are not exactly comparable, due to varying irradiation modalities and use of different photosensitizing substances.

Table 3 demonstrates the results obtained by our group. The radiation dose correlates with the therapeutic success but also with the rate of side effects. Follow-ups covering up to 34 months subsequent to PDT have shown a therapeutic success of 74 %.

Table 4 gives a summary of typical side effects of PDT. According to our experience all bladder tumor patients subjected to successful PDT complain about transient sensation of burning during miction as well as pollakisuria. If these symptoms do not occur, a therapeutic failure must be expected. The other complications have generally proven to be reversible.

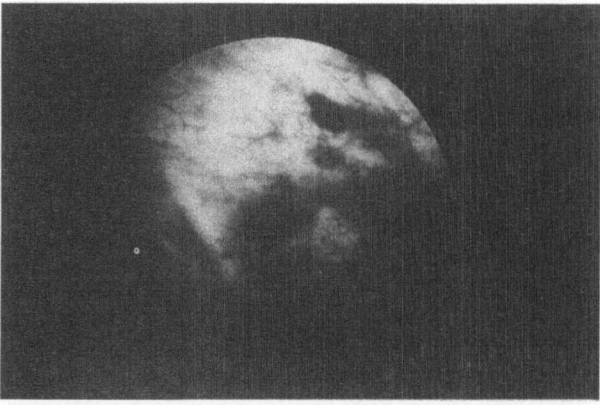

Fig. 2. Tumorselective hemorrhagic necrosis of CIS in a bladder region which looked normal before PDT.

Table 3. Results obtained by the Munich research group
regarding PDT of CIS of the bladder.

n	radiation (J/cm^2)	follow-up (months)	complete initial response (tumor-negative random biopsies/ cytology)	tumor recurrency post-PDT (months)	side effects (e.g. burning, pollakisuria)	bladder shrinkage (> 30 %)
14	35 - 70	7 - 32	11*	4	14	5**
2	25	14/16	2	1	2	-
2	15***	9	0	2*	1	-

* 2 more following a second treatment

** 1 cystectomy - no tumor material found

*** in 1 case following retreatment of "new" exophytic tumors with TUR and
Neodym-YAG-laser no tumor recurrency for 8 months

D. Jocham/5-87

Table 4.

Side effects of urological PDT

- pollakisuria

- burning

- long-term inflammation

- bladder shrinkage (dose dependent, transient)

- moderate urinary obstruction - upper tract

- photosensitization of skin and eyes

In one single case of slight dilatation of the ureter, it was necessary to dilate the contracted ostium of the ureter by slitting the
ostium roof, which is a very easy intervention. On the whole, PDT is a
promising procedure but there are also therapeutic failures.

All components involved in PDT, i.e. the irradiation modalities
as well as the photosensitizers require further optimization. Also, the
range of indications for PDT is still a matter of debate.

PDT will have to stand up to comparison with alternative procedures
in the future, and this process will take many years due to the complex
pattern of tumor diseases. In spite of these reservations, one can conclude that integral PDT represents a very consistent therapy concept
which is indubitably a very useful tool in the struggle for preservation
of a bladder affected by tumors. This prospect alone justifies the efforts
that have been made up to now and will have to be made in the future.

REFERENCES

1. R. Baghdassarian,M.W. Wright, S.A. Vaughn, M.W. Bern, D.C. Martin, A.G. Will: The use of lipid emulsion as an intravesical medium to disperse light in the potential treatment of bladder tumors. J. Urol. 133: 126 (1985)

2. R. Baumgartner, H. Stepp, L. Ruprecht, E. Unsöld, D. Jocham: Experimental study on fluorescence diagnosis of bladder cancer. In: Photodynamic therapy of tumors and other diseases, ed. by G. Jori, C. Perria, Liberia Progetto Editore Padova, p. 260 (1985)

3. R.C. Benson, Jr., G.M. Farrow, J.H. Kinsey, D.A. Cortese, H. Zincke, D.C. Utz: Detection and localization of in situ carcinoma of the bladder with hematoporphyrin derivative. Mayo Clin. Proc. 57:548 (1982)

4. R.C. Benson, Jr.: Treatment of diffuse transitional cell carcinoma in situ by whole bladder hematoporphyrin derivative photodynamic therapy, J. Urol. 134:675 (1985)

5. H. Hisazumi, N. Miyoshi, K. Naito, T. Misaki: Whole bladder wall photoradiation therapy for carcinoma in situ of the bladder: A preliminary report. J. Urol. 131:884 (1984)

6. D. Jocham, G. Staehler, Ch.Chaussy, C. Hammer, U. Löhrs: Laserbehandlung von Blasentumoren nach Photosensibilisierung mit Hämatoporphyrin-Derivat. Urologe A (Supp. Sept.) 20:340 (1981)

7. D. Jocham, G. Staehler, Ch. Chaussy, U. Löhrs, E. Unsöld; C.E. Alken Preis 1983: Integrale Photoradiotherapie des Blasenkarzinoms nach tumorselektiver Photosensibilisierung mit Hämatoporphyrin-Derivat (HpD). Akt. Urol. 15:109 (1984)

8. D. Jocham, G. Staehler, R. Baumgartner, E. Unsöld: Die integrale photodynamische Therapie beim multifokalen Blasenkarzinom - erste klinische Erfahrungen. Urologe A, 6:316-319 (1985)

9. J. F. Kelly, M.E. Snell: Hematoporphyrin derivatives: A possible aid in the diagnosis and therapy of carcinoma of the bladder. J. Urol. 115:150 (1976)

10. C.F. Rothauge, P. Röttger, J. Kraushaar, J. Kracht, H.D.Nöske: Die Phototherapie des Blasenkarzinoms im Licht der Histologie. Diagnostik & Intensivtherapie 8:17 (1983)

11. A. Tsuchiya, U. Obara, M. Miwa, T. Oi, H. Kato, Y. Hayata: Hematoporphyrin derivative and laser photoradiation in the diagnosis and treatment of bladder cancer. J. Urol. 130:79 (1983)

COMPARISON OF FLUORESCING AND PHOTOSENSITIZING PROPERTIES OF DIFFERENT
PORPHYRIN - DERIVATIVE - PREPARATIONS

Ronald Sroka [1,2], Christian Ell [1], Dieter Jocham [3],
Hans Mueller v.d. Haegen [4], Susanne Stocker [1,2], and
Eberhard Unsöld [2*]

[1] Medizinische Klinik, Universität Erlangen, D-8520 Erlangen
FRG

[2] Zentrales Laserlaboratorium, Gesellschaft für Strahlen- und
Umweltforschung, D-8042 Neuherberg, FRG

[3] Urologische Klinik und Poliklinik, Universität München
D-8000 München 70, FRG

[4] Institut für Chemie und chemische Technologie, Fachhoch-
schule Flensburg, D-2390 Flensburg, FRG

INTRODUCTION

Several substances are known to demonstrate a temporary difference in
concentration between normal and tumorous tissue after systemic application. This difference is due to the different metabolic rates in both tissues and enables the marking of tumors for fluorescence detection [1,2] and
for treatment by means of photodynamic methods [3,4].
The object of this study was:
- to examine five photosensitizing drugs for comparative judgement of
  their diagnostic and therapeutic effectiveness with respect to their
  efficiency in vivo
- to prove the correlation between fluorescence and photodynamic kinetics.

MATERIALS AND METHODS

The substances investigated were:

Photosan 1/S	Supplier: Seehof Laboratory, FRG [5]
Photosan 1/R	Seehof Laboratory, FRG [5]
Photosan 1/E	Seehof Laboratory, FRG [5]
HpD-Quentron	Quentron Optics, Australia [6]
Photofrin II	Photomedica, Inc., USA [7]

Photosan 1/R (Residue) and Photosan 1/E (Eluate) were separated by
single molecular membrane filtering of Photosan 1/S. All substances were
analyzed according to their polarity using high pressure liquid chromatography (HPLC) with a reverse phase column (detection wavelength: 402 nm). [5]

* author to whom correspondence should be addressed

The experiments were performed in a murine tumor model. The syngeneic fibrosarcoma SSK 2 was implanted subcutaneously into the flank of inbred C3H mice. This fibrosarcoma grows with a doubling time of approximately 1.5 days. The drugs were injected intraveneously into the tail vein of the mice at a dose of 9 mg/kg body weight[8],[9].

The fluorescence was excited by a Kr^+ laser operating at its UV wavelength of 407 nm with a power density of 25 mW/cm². This value is far below that at which the bleaching effect occurs [10]. The fluorescence emission between 600 and 700 nm was recorded via a SIT video system and a computer image processor [1,2,11].

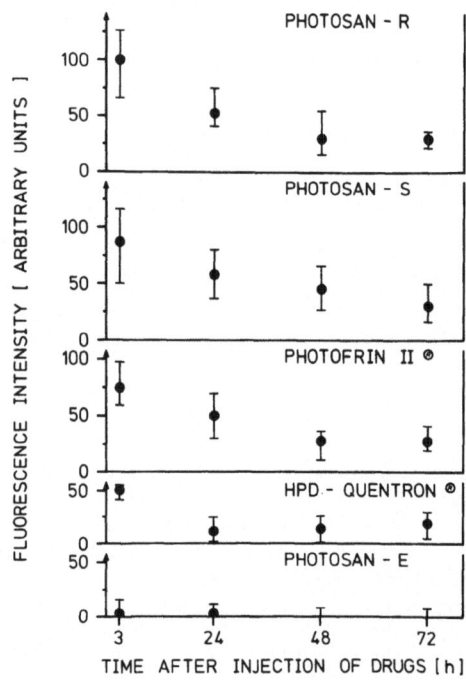

Fig. 1. Kinetics of fluorescence intensity (n=2)

The photodynamic therapy (PDT) was performed with an Ar^+ laser-pumped dye laser, continuous wave, tuned to 630 nm. The total energy density was 150 J/cm² at a power density of 400 mW/cm². A special tubus, covering the tumor, was fed by a 600 μm quartz fiber and guaranteed homogeneous irradiation. The growth curves of tumors with either HpD or light treatment (150 J/cm and 400 mW/cm²) alone did not differ from those of entirely untreated tumors. This implies that the regrowth delay measured is due to photodynamic effects and - dispite of the power density applied - is not thermally induced. In addition the tubus was cooled by N_2 gas flow. The decrease in body temperature, usually observed in anestesized mice, is not taken into account.

EXPERIMENTS AND RESULTS

The <u>fluorescence</u> of the tumor was measured 3, 24, 48, and 72 hours after injection of the drugs. Two mice were tested at each point for the fluorescence and a further two mice for autofluorescence. The images recorded from the fluorescing tissue were digitized by the image processor. The fluorescence intensity value is the average of the pixels intensity in the evaluation frame [11].

Fig. 1 shows that the absolute fluorescence emission intensities of the substances differ significantly and decrease with increasing time after injection.

Subcurative <u>treatment</u> of the tumors by PDT (n = 4-6) was performed at the same retention times. Five control mice experienced no irradiation and no substance application.

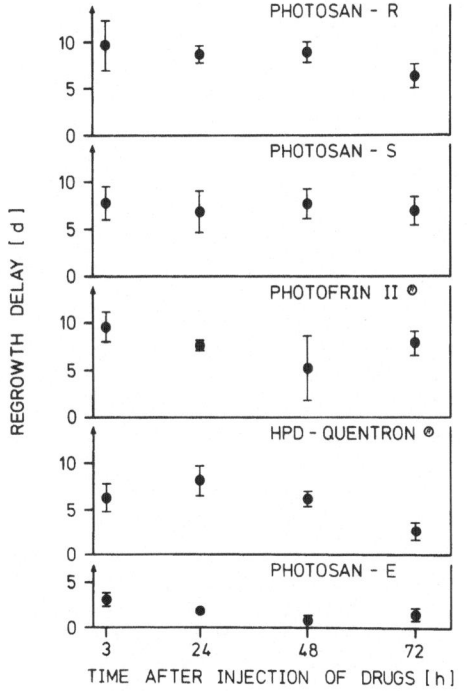

Fig. 2. Kinetics of regrowth delay (n = 4-6)

The photodynamic efficiency of the drugs was quantified by measuring the tumor regrowth delay, i.e. the additional time required for a tumor to grow to a defined size after subcurative treatment as compared with normal tumor development (control group). This size was chosen as four times the size at the time the tumor was treated [8,9].

Fig. 2 shows the tumor regrowth delay as a function of time after injection of the drugs. Photosan 1/E has nearly no therapeutic effect, whereas Photosan 1/R, Photosan 1/S, and Photofrin II, show similar pharmacokinetics in their subcurative efficiency. The treatment with these three substances also revealed curative effects at a retention time of 3 hours (not shown in fig. 2). The pharmacokinetic of HpD-Quentron differed entirely, peaked at 24 hours after injection, and could not yet be correlated with the temporal behaviour of the fluorescence signals.

The experimental data were summarized in a mathematical model in order to compare the pharmacokinetics of the fluorescence and the photo-dynamic effect. Their change with time t is approximated by an exponential function of the type

$$X(t) = (\exp A) \exp - Bt = \exp (A - Bt)$$

by aid of a least-square fit procedure. exp A represents the extrapolated value of $X(t=0)$, while B is the time constant of the decay observed. (In a semilogarithmic plot $\ln X(t) = A - Bt$ stands for a straight line). The extrapolated values of $X(t=0)$, the coefficient B, and the contrast ratio $F(m/t)$, are given in Table 1.

Table 1. Coefficients of the curve fit
Fluorescence: A,B; Regrowth delay: A^*, B^*

	Photosan 1/R	Photosan 1/S	Photofrin II	HpD-Quentron	Photosan 1/E	
exp A = Fluorescence (t=0)	10	9	9	5.5	0.5	
exp A^* = Regrowth delay (t=0)	10	8	9	6	3	
B = Time constant of fluorescence	2	2	2	2	2	$(\times 10^{-2})$
B^* = Time constant of regrowth delay	4	3	7	7	10	$(\times 10^{-3})$
F(m/t)= Ratio of fluorescence muscle/tumor	0.3	0.4	0.4	0.4	0.9	

10 = normalized value

The fluorescent and photodynamic properties, as derived from $X(t=0)$, were most efficient for Photosan 1/R. The corresponding values of Photo-san 1/S and Photofrin II were only slightly less, whereas those for Photo-san 1/E were considerably lower. The same numeric procedure was also applied to the unexpected kinetics of HpD-Quentron accepting a rather unsatisfying curve fit with large errors.

The time constant B of the kinetics indicates a slower decrease in the photodynamic properties than in the fluorescent properties. This might suggest that different molecular, cellular, or organic properties are re-sponsible for the kinetic behaviour. The fluorometric and growth-kinetic measurements were performed at the laser wavelengths 407 nm and 630 nm, respectively. The penetration depth into the tumor at these wavelengths may differ significantly, affecting the volumes irradiated[13]. Its influence on the kinetics observed has not yet been taken into account.

When the contrast ratio F(m/t) is also taken into account, Photosan 1/R is seen to be a substance very similar to the well established Photofrin II and slightly exceeding it in contrast, fluorescent and photodynamic properties. Photosan 1/S should provide similar results to Photosan 1/R whereas Photosan 1/E is substantially less efficient. The kinetics of HpD-Quentron differ entirely from those of the other substances and should be investigated further. These results indicate that the timing of application of HpD-Quentron in clinical treatment should be considered carefully.

Although the test population must be enlarged for proper statistical evaluation our investigations indicate that:
- the method is useful for a semiquantitative evaluation of the pharmacokinetics of the photosensitizers
- there is no obvious correlation between fluorescence and tumor regrowth delay
- it is possible to make a simple qualitative judgement of sensitizers using the derived coefficients A and B.

ACKNOWLEDGEMENT

The authors thank R. Oswald and P. Zandl for their assistance, the Bundesministerium für Forschung und Technologie (BMFT), Bonn, (grant no. MMT 51), and the Wilhelm Sander-Stiftung, Neustadt an der Donau, (grant no. 85.001.1), for supporting the project.

REFERENCES

1 R. Baumgartner, J. Feyh, A. Götz, D. Jocham, H. Schneckenburger, H. Stepp, E. Unsöld, Experimental Study on Laser-induced Fluorescence of Hematoporphyrin Derivative (HpD) in Tumor Cells and Animal Tissue. Laser in medicine and surgery 2 (1): 4-9 (1986).

2 R. Baumgartner, E. Unsöld, High Contrast Fluorescence Imaging Using two Wavelength Laser Excitation and Image Processing. J. Photochem. Photobiol.,B: Biology 1(1): 130-132 (1987).

3 S.G. Bown, Photodynamic Therapy, in: "Laser in Gastroenterology", J.F. Riemann, ed., Georg Thieme, Stuttgart (in press) (1987).

4 T.J. Dougherty, Photosensitization of Malignant Tumors. Seminars in Surgical Oncology 2: 24-37 (1986).

5 H. Mueller von der Haegen, Hochleistungsflüssigkeits-Chromatographie (HPLC) von HpD's. (unpubl. results) (1987).

6 A. Stanco, HpD for Phototherapy. (priv. comm.) (1986)

7 T.J. Dougherty, W.R. Potter, K.R. Weishaupt, The Structure of the Active Component of Hematoporphyrin Derivative, in: "Porphyrin Localization and Treatment of Tumors", D.R. Doiron, C.J. Gomer, eds., Alan R Liss, New York, pp 301-314 (1984).

8 J. Kummermehr, K.R. Trott, Rate of Repopulation in a Slow and a Fast Growing Mouse Tumour. in: "Progress in Radio-Oncology", K.H. Kärcher et al., eds., Vol II. Raven Press, New York, pp 299-307 (1982).

9 S. Stocker, Photosensibilisierende Substanzen für die Tumordiagnose und
   -therapie unter besonderer Berücksichtigung gastrointestinaler Tumoren.
   Inaugural dissertation, University of Munich, Med. Dept., Munich (in
   preparation)

10 H. Schneckenburger, E. Unsöld, W. Weinsheimer, D. Jocham, Time-resolved Laser Fluorescence and Photobleaching of Single Cells after Photosensitization with Hematoporphyrin Derivative (HpD), in: "Porphyrins
   in Tumor Phototherapy", A. Andreoni, R. Cubeddu, eds., Plenum Press,
   New York London, pp 137-141 (1984).

11 R. Baumgartner, H. Fisslinger, D. Jocham, H. Lenz, L. Ruprecht, H.
   Stepp, E. Unsöld, A Fluorescence Imaging Device for Endoscopic Detec-
   tion of Early Stage Cancer - Instrumental and Experimental Studies,
   Photochem. Photobiol. 46(5): 759-764 (1987).

12 R. Sroka, Einfluß physikalischer und biologischer Parameter auf die
   Laserbestrahlung photosensibilisierter Gewebe, Inaugural dissertation,
   University of Munich, Med. Dept., Munich (in preparation)

13 J. Moan (priv. comm.) (09.09.1987)

NEW SENSITIZERS FOR PHOTODYNAMIC THERAPY OF CANCER

Johan E. van Lier

MRC Group in the Radiation Sciences
Faculty of Medicine, University of Sherbrooke
Sherbrooke, QC, Canada   J1H 5N4

INTRODUCTION

The sensitizer preparations currently used in clinical trials of photodynamic therapy (PDT) of cancer (Dougherty, 1987) consist of mixtures of hematoporphyrin derivatives (HPD) obtained via alkaline hydrolysis of hematoporphyrin acetates (Lipson et al., 1961; Bonnett et al., 1981). Although the different components of HPD all exhibit good in vitro photodynamic properties, only the aggregated fraction of HPD is sufficiently retained by neoplasms to exhibit in vivo photodynamic action. This active component of HPD has been labeled DHE and is characterized as a mixture of dihematoporphyrins and oligomers containing 2-5 hematoporphyrin (HP) units (Kessel, 1987) linked via ether and ester bonds (Dougherty et al., 1984; Kessel, 1986a). It has recently been shown by Kessel (1987) that both ester and ether linked material localizes in tumors and furthermore, that upon storage hydrolysis of the ester bonds occurs resulting in enrichment of ether linked hematoporphyrins in the DHE fraction. In view of the increased understanding of the chemical nature of the active components of HPD several attempts have been made to rationalize the preparation of DHE. The development of photosensitizers with improved photochemical properties is also actively

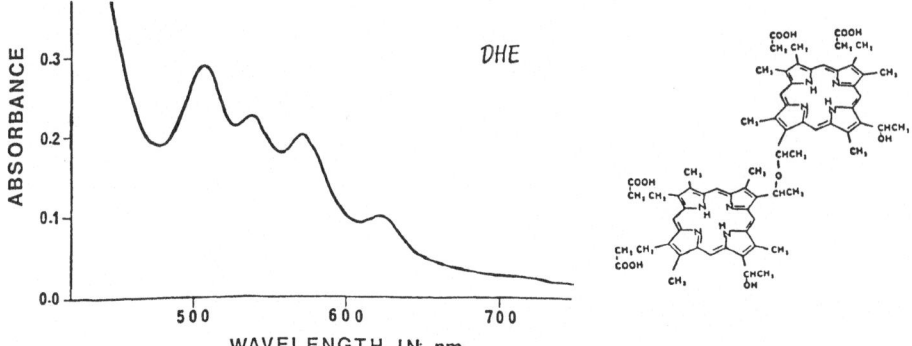

Figure 1.   Structure and absorption spectrum of dihematoporphyrin ether. DHE consists of a mixture of 2-5 ether and ester linked hematoporphyrin units which, like the parent molecule, absorb weakly above 600 nm.

pursued. Hematoporphyrins absorb only weakly above 600 nm (Fig. 1), where light exhibits optimal penetration through tissue, and several porphin analogs with red-shifted absorption maxima as compared to DHE are at present under evaluation as alternative photosensitizers for PDT.

HEMATOPORPHYRIN ETHERS AND ESTERS

The conventional method to prepare HPD involves alkaline hydrolysis of HP acetates and the formation of tumor-localizing dimers and oligomers during this procedure is rather coincidental (Lipson et al., 1961, Bonnett et al., 1981). The unambiguous synthesis of HP dimers and oligomers covalently linked via ether bonds (HE) has recently been reported by Scourides et al. (1987). HE exhibited identical chromatographic and spectroscopic properties as HPD and based on fluorescence data, the authors show that the synthetic HE samples are as efficient tumor localizers as HPD. Preparation of the HE derivatives involves the reaction of the HP dimethyl ester with its dibromide adduct via the in situ formation of the former during the controlled hydrolysis of the dibromide (Fig. 2A). This mild procedure yields mixtures of ether linked dimers and oligomers of HP methyl esters from which free HE is obtained upon hydrolysis of the methyl ester substituents (Scourides, 1987). In another approach Rimington et al. (1987) prepared a series of highly purified HP di-ether derivatives substituted with linear carbon chains consisting of two to five carbon atoms. The derivatives were tested on experimental mouse tumors and the most hydrophobic analog was shown to approach DHE in terms of efficiency as a sensitizer for PDT (Evensen et al., 1987).

A simple method to prepare ester linked HP oligomers consisting of 3-7 porphin units has also recently been reported (Lai et al., 1987). The procedure involves a reaction between the carboxyl and hydroxyl groups of HP (Fig. 2B) in aqueous phosphate buffer at elevated temperatures with the degree of polymerization being controlled by the reaction time and temperature. The poly-HP esters thus obtained again exhibited similar in vivo photodynamic properties as DHE (Lai et al., 1987). It is apparent that the availability of chemically well characterized HP ether and ester derivatives should foster rapid advances towards the elucidation of possible mechanisms of tumor uptake and retention.

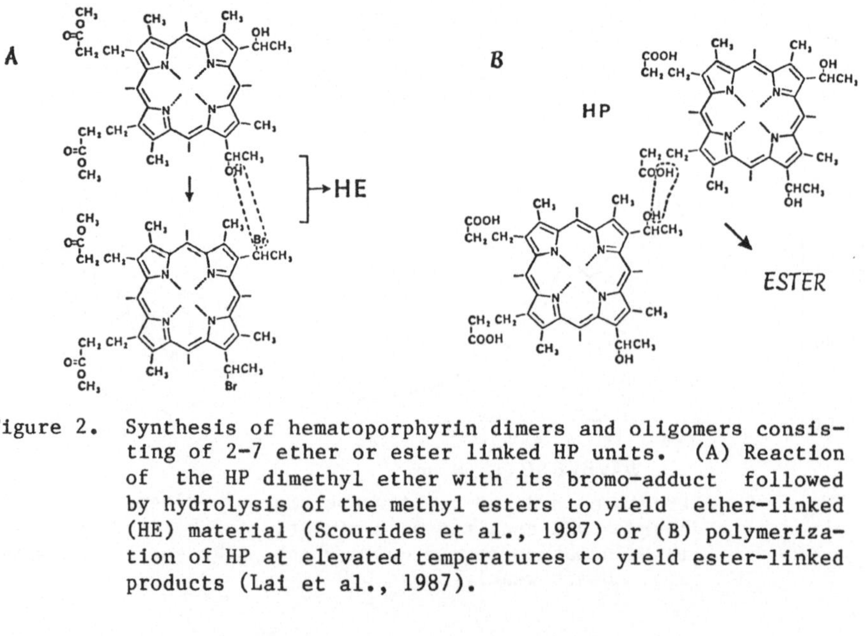

Figure 2. Synthesis of hematoporphyrin dimers and oligomers consisting of 2-7 ether or ester linked HP units. (A) Reaction of the HP dimethyl ether with its bromo-adduct followed by hydrolysis of the methyl esters to yield ether-linked (HE) material (Scourides et al., 1987) or (B) polymerization of HP at elevated temperatures to yield ester-linked products (Lai et al., 1987).

The effectiveness of DHE in PDT results from two important properties: (i) selective retention by neoplasm, and (ii) induction of cellular damage upon light excitation. There are however a number of features
of DHE that could be improved upon in order to increase the efficiency of
PDT. They include, (i) chemical purity and stability, (ii) tumor specificity, and (iii) photochemical and photophysical properties of the dye.

Approaches to improve tumor localizing properties of photosensitizers include coupling of existing dyes with site-selective carriers.
Monoclonal antibodies against tumor antigens have been used as carriers
of porphyrin analogs with the aim to combine the phototoxic effect of the
photosensitizer with the target-seeking ability of antibodies (Daphne et
al., 1983; Oseroff et al., 1986). Both direct hematoporphyrin coupling
onto the antibody via the carbodiimide method (Daphne et al., 1983) as
well as preloading of chlorin onto dextran followed by coupling of the
modified dextran molecules to monoclonal antibodies (Oseroff et al., 1986
and 1987) gave conjugates which have been used successfully in PDT of
tumor bearing mice. Such an approach may allow for lower dose levels of
the porphin sensitizers to induce the required tumor response resulting
in diminished damage to skin and normal tissue.

An alternative approach towards site-specificity of photosensitizers
involves the evaluation of known antitumor agents for possible photodynamic properties. Many antitumor drugs act on the rapidly growing cancer
cells through interaction with DNA and several of such agents contain
heterocyclic and aromatic chromophores. Positively charged porphin analogs are also capable to intercalate with DNA and some examples are
discussed below.

Porphin Analogs

Most novel chromophores currently advanced as photosensitizers for
PDT involve chemically well characterized porphin analogs. Their absorption maxima are selected between 650-800 nm in order to capitalize on the
good tissue penetrating properties of light at the red end of the spectrum. A number of modifications of the basic porphyrin structure resulting in improved red light absorption properties are depicted in Figure
3. Addition of phenyl rings on the meso positions of porphin (Fig. 3A)
does increase absorption above 600 nm only marginally but this symmetric
and stable structure is of interest because of its ready availability by
chemical syntheses. A number of derivatives are known (Fig. 3E) and
several have been tested for their photodynamic effect, cell uptake and
tissue distribution properties (Berenbaum et al., 1986). The tetraphenyl

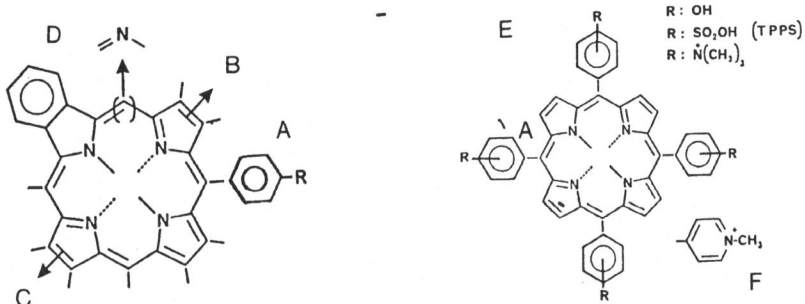

Figure 3. Modifications of the basic porphyrin structure resulting
in improved red light absorption properties (A-D). Derivatives of tetraphenyl porphin (E,F).

porphin tetrasulfonate (TPPS$_4$) has been studied by Winkelman (1962, 1985) and this synthetic dye was found to have similar or better tumor-localizing properties than HPD. More recently, Kessel (1987) compared the mono- to tetrasulfonated TPPS, including two constitutional isomers of the disulfonated dye with adjacent and opposite substituents. The product with 2 adjacent sulfonates was the most potent photosensitizer in vitro but all derivatives were found to localize equally well in vivo. Sites of localization varied considerably, however, the mono- and adjacent disubstituted derivative showed preferential binding to HDL and LDL while the more polar analogs bound progressively more to albumin. The author concludes that tumor localization proceeds through two routes: (i) via lipoprotein binding leading to dye accumulation in neoplastic cells, and (ii) in the case of the more polar dyes, via albumin binding resulting in dye accumulation in stromal elements of neoplastic tissues. It is likely that similar pathway selection, based on dye hydrophobicity, occurs with most porphin analogs lacking affinity for more specific receptor sites. Tetrahydroxyphenyl porphin (Fig. 3E) has also been tested in animals and particularly the para- and metahydroxy derivatives were found to exhibit superior PDT efficiencies as compared to HPD (Berenbaum et al., 1986). Trimethylaniline (R = $N^+(CH_3)_3$; Fig. 3E) and methylpyridine (Fig. 3F) meso-substituted porphins are cationic species with 1 to 4 positive charges. They are capable to intercalate with DNA (Sari et al., 1986). and accordingly may provide a complementary model for studies on photodynamic action mechanisms.

Hydrogenation of a pyrrole double bond to yield dihydroporphin gives the parent molecule chlorin (Figs. 3B and 4). Depending on the nature of the substituents, chlorin analogs, including the sub-class of purpurins (Fig. 4), exhibit red-shifted absorption maxima from 640-700 nm with substantially enhanced extinction coefficients. Chlorin e$_6$ and its aspartate esters were shown to induce photodynamic response in vitro and in vivo (Kessel and Dutton, 1984; Bommer and Burnham, 1985) while hematoporphin-chlorin esters, prepared in an analogous manner as HPD, showed better photosensitization efficiency in vivo, all in accordance with their stronger absorption above 600 nm as compared to HPD (Kessel, 1986b). Purpurin derivatives (Fig. 4) absorb further into the red end of the spectrum (λ_{max} 695 nm) and have recently been advanced as photosensi-

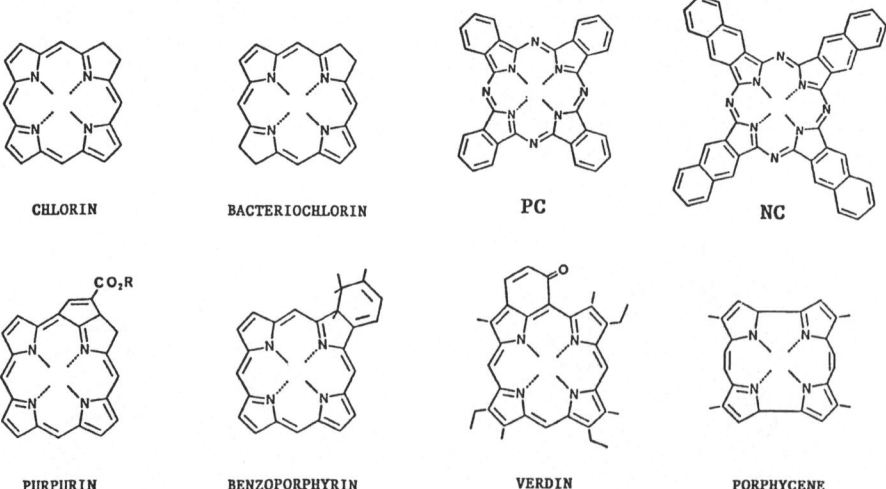

Figure 4. Porphin based parent molecules of porphyrin derivatives currently under evaluation as new photosensitizers for PDT.

tizers for PDT by Morgan et al. (1987a). They can be synthesized from naturally occurring porphyrin substrates (Morgan and Tertel, 1987). Other chlorin derivatives with attractive red-shifted absorption properties include the benzoporphyrin derivatives (Figs. 3D and 4) obtained via a Diels-Alder addition to protoporphyrin IX (Pangka et al., 1986). In order to evaluate the role of the reduced pyrrole ring of chlorin derivatives in possible tumor retention phenomena of porphyrin analogs Morgan et al. (1987b) prepared a verdin (Fig. 4), a porphyrin derivative with an isocyclic ring, like in purpurins, but lacking the reduced pyrrole ring. The in vivo cytotoxic effect of verdins appeared to be comparable to that observed with HPD or purpurins.

Hydration of a second porphin pyrrole ring (Fig. 3B and C) gives tetrahydroporphin, known as the parent molecule bacteriochlorin (Fig. 4). Bacteriochlorin derivatives derived from chlorophyl have further red-shifted major absorption bands as compared to chlorin with λ_{max} about 765 nm. Although less stable than hematoporphyrin their strong red absorption make them interesting candidates for PDT (Beems et al., 1987; Borland et al., 1987), particularly in view of the availability of compatible, simple diode laser systems.

Finally, porphycene derivatives (Fig. 4) have recently been suggested by Aramendia et al. (1986) as potential sensitizers for PDT based on their photophysical properties. Although their $\lambda_{max} > 600$ nm is not red-shifted as compared to hematoporphyrin, the extinction coefficient is markedly increased.

A fourth modification of the basic porphin structure involves substitution of the methine bridges for nitrogen atoms to yield azaporphin derivatives (Fig. 3D). Tetraazaporphin, combined with the addition of benzo rings to the 4 pyrrole units, gives the highly stable phthalocyanine macroring system (PC, Fig. 4). Phthalocyanines (PC) and their sulfonated water-soluble metallo derivatives (M-SPC) have received increasing attention as sensitizers for PDT. Some of the earlier work on PC derivatives as well as current results on their photodynamic properties has recently been reviewed (Spikes, 1986; Ben Hur, 1987; van Lier et al., 1987). Like porphyrins, they have an intense Soret absorption maximum between 300 and 400 nm, but in addition they also exhibit strong absorption bands between 650 and 700 nm. PC form stable complexes with many metal ions or metal oxides and depending on the size of the ion M-PC can be prepared by direct chelation or by condensation at elevated temperature of the appropriate phthalic acid or anhydrid precursors in the presence of metal salt (Weber and Busch, 1965). Sulfonated M-PC may be obtained by the condensation procedure using mixtures of sulfophthalic and phthalic acid, or by direct sulfonation of the M-PC in oleum (Linstead and Weiss, 1950). The differently sulfonated products and their constitutional isomers can be separated by reverse phase HPLC (Ali et al., 1987). Although the phthalocyanines are more stable towards acid and heat as compared to most other porphin analogs, the aza bridges render the molecule vulnerable towards oxidizing agents. Oxidation of M-SPC and M-PC in HNO_3 results in quantitative degradation of the macroring system to give the corresponding sulfophthalimide and phthalimide fragments and this procedure has been adapted to determine the average number of sulfonates per PC molecule of SPC preparations (Ali et al., 1987). The first evidence for the potential of PC in PDT was provided through observations of their efficiency as photosensitizers of in vitro cell killing (Ben-Hur and Rosenthal, 1985a,b; Brasseur et al., 1985; Chan et al., 1986). The metal free H_2-SPC and several metal chelates including AlCl-, GaCl- and Zn-SPC were found to be photoactive. This correlates well with the high triplet yield and long lifetimes of such complexes and is in conformity with their capacity to generate $^1O_2$ (Darwent et al., 1982; Langlois et al., 1986; Rosenthal et al., 1986). The effect of the degree of sulfonation and hydrophobicity of several M-SPC on their photochemical and photodynamic properties has also been studied in detail. In spite of the reduced photochemical efficiency of the lower sulfonated dyes to generate $^1O_2$ due to their higher tendency to form photoinactive aggregates (Wagner et al., 1987), the lower sulfonated M-SPC exhibited the highest photodynamic activity (Brasseur et al., 1987a,b). The disulfonated M-SPC with sulfonate substituents located on adjacent benzyl groups of the macroring system showed particularly strong in vitro photoactivity whereas the constitutional isomer with substituents on opposite benzyl groups was inactive. The photoactivities were shown to parallel the in vitro cell uptake of the M-SPC and the facility of M-SPC with two adjacent sulfonate substituents to cross the cell membrane was attributed to their amphiphilic nature (Paquette et al., 1987). Several M-SPC have been shown to induce necrosis and tumor cure in conjunction with red light during in vivo PDT (Selman et al., 1986; Bown et al., 1986; Tralau et al., 1987; Brasseur et al., 1987b). Among the differently sulfonated metal chelates the disulfonated Zn-SPC showed the highest activity whereas the tetrasulfonated dye did not induce any tumor response (Brasseur et al., unpublished results). Most monosulfonated M-SPC are poorly water soluble and like the nonsulfonated M-PC, are more likely to show in vivo PDT activity upon incorporation in liposomes or lipoproteins (Valduga et al., 1987).

The absorption spectrum of naphthalocyanines (NC, Fig. 4) resembles that of PC, but the extended conjugation results in a red-shift of the Q band maximum of almost 100 nm to well above 750 nm. The Soret band remains in the same position as for the PC and the widely separated intense maxima render the Al- and Zn-NC complexes colourless in solution (McCubben and Phillips, 1986). The photophysical properties of the sulfonated Al- and Zn-NC (McCubben and Phillips, 1986) as well as of a bis (tri-n-hexyl siloxy) Si-NC (Firey and Rodgers, 1987) suggests that the NC could be suitable for photodynamic applications. The increase in ligand size from PC to NC results however in a 1000 fold increase in sensitivity of the macrocycle towards photooxidation (McCubben and Phillips, 1986) and this could be a drawback for biological applications.

ACKNOWLEDGEMENTS

The author is grateful to the Medical Research Council of Canada and the National Cancer Institute of Canada for generous financial support.

REFERENCES

Ali, H., Langlois, R., Wagner, J.R., Brasseur, N., Paquette, B., and van Lier, J.E., 1988, Biological activities of phthalocyanines X. Syntheses and analyses of sulfonated phthalocyanines, to be published.

Aramandia, P.F., Redmond, R.W., Nonell, S., Schuster, W., Braslavsky, S.E., Schaffner, K., and Vogel, E., 1986, The photophysical properties of porphycenes: potential photodynamic therapy agents, Photochem. Photobiol., 44: 555.

Beems, E.M., Dubbelman, T.M.A.R., Truscott, J.G., van Delft, J.L., van Best, E.M., and Oosterhuis, J.A., 1987, Bacteriochlorin, a new photosensitizer for photodynamic therapy, Photochem. Photobiol., 46: in press.

Ben-Hur, E., 1987, Photochemistry and photobiology of phthalocyanines: new sensitizers for photodynamic therapy of cancer, in: "From photophysics to photobiology", A. Favre, R. Tyrrell and J. Cadet, eds, Elsevier, New York.

Ben-Hur, E., Rosenthal, L., 1985a, The phthalocyanines: a new class of mammalian cell photosensitizers with a potential for cancer photo-therapy, Int. J. Radiat. Biol., 47: 145.

Ben-Hur, E., Rosenthal, I, 1985b, Photosensitized inactivation of Chinese hamster cells by phthalocyanines, Photochem. Photobiol., 42: 129.

Berenbaum, M.D., Akande, S.L., Bonnett, R., Kaur, H., Ivannow, S., White, R.D., and Winfield, J., 1986, Meso-tetra-(hydroxyphenyl)porphyrins, a new class of potent tumour photosensitizers with favorable selectivity, Br. J. Cancer, 54: 717.

Bommer, J.C., and Burnham, B.F., 1985, Tetrapyrrole compounds and process for the production of same and pharmaceutical composition containing same, US Patent, 72874.

Bonnett, R., Ridge, R.J., Scourides, P.A., Berenbaum, M.C., 1981, On the nature of Haematoporphyrin Derivative, JCS Perkin, 1: 3135.

Borland, C.F., McGarvey, D.J., Truscott, T.G., Codgell, R.J., and Land, E.J., 1987, Photophysical studies of bacteriochlorophyll a and bacteriopheophytin a - Singlet oxygen generation, J. Photochem. Photobiol., 1: 93.

Bown, S.G., Tralau, C.J., Coleridge-Smith, P.D., Akdemir, D., Wieman, T.J., 1986, Photodynamic therapy with porphyrin and phthalocyanine sensitization: quantitative studies in normal rat liver, Br. J. Cancer, 54: 43.

Brasseur, N., Ali, H., Autenrieth, D., Langlois, R., van Lier, J.E., 1985, Biological activities of phthalocyanines - III. Photoinactivation of V-79 Chinese hamster cells by tetrasulfophthalocyanines, Photochem. Photobiol., 42: 515.

Brasseur, N., Ali, H., Langlois, R., van Lier, J.E., 1987a, Biological activities of phthalocyanines - VII. Photoinactivation of V-79 Chinese hamster cells by selectively sulfonated gallium phthalocyanines, Photochem. Photobiol., 46: in press.

Brasseur, N., Ali, H., Langlois, R., Wagner, J.R., Rousseau, J., van Lier, J.E., 1987b, Biological activities of phthalocyanines - V. Photodynamic therapy of EMT-6 mammary tumors in mice with sulfonated phthalocyanines, Photochem. Photobiol., 45: 581-587.

Daphne, M., Wat, C-K., Towers, G.H.N., and Levy, J.G., 1983, Photoimmuno-therapy: Treatment of animal tumors with tumor-specific monoclo-nal antibody-hematoporphyrin conjugates, J. Immunol., 130: 1473.

Darwent, J.R., Douglas, P., Harriman, A., Porter, G., Richoux, M.-C., 1982, Metal phthalocyanines and porphyrins as photosensitizers for reduction of water to hydrogen, Coord. Chem. Rev., 44: 83.

Dougherty, T.J., 1987, Photosensitizers: Therapy and detection of malignant tumors, Photochem. Photobiol., 45: 879.

Dougherty, T.J., Potter, W.R., Weishaupt, K.R., 1984, The structure of the active component of hematoporphyrin derivative, in: "Porphyrin localization and treatment of tumors", D.R. Doiron and C.J. Gomer, eds., Alan R. Liss, New York.

Duff, G.A., Yeager, S.A., Linghal, A.K., Pestel, B.C., Ressner, J.M., and Foster, N., 1987, Separation of metalloporphyrins from metallation reactions by liquid chromatography and electrophoresis, J. Chromatog., 416: 71.

Evensen, J.F., Sommer, S., Remington, C., and Moan, J., 1987, Photodynamic therapy of C$_3$H mouse mammary carcinoma with hematoporphyrin diethers as sensitizers, Br. J. Cancer, 55: 483.

Firey, P.A., and Rodgers, M.A.J, 1987, Photo-properties of a silicon naphthalocyanine: a potential photosensitizer for photodynamic therapy. Photochem. Photobiol., 45:535.

Jackson, A.H., 1978, Phthalocyanines, in: "The porphyrins", D. Dolphin, ed., Academic Press, New York.

Kessel, D., 1986a, Proposed structure of the tumor-localizing fraction of HPD (hematoporphyrin derivative), Photochem. Photobiol., 44: 193.

Kessel, D., 1986b, Localization and photosensitization of murine tumors in vivo and in vitro by a chlorin-porphyrin ester, Cancer Res., 46: 2248.

Kessel, D., 1987, HPD: Chemical and biophysical studies, in: "Photosensitization: Molecular and Medical Aspects," G. Moreno, R.H. Pottier and T.G. Truscott, eds. Springer-Verlag, Berlin, in press.

Kessel, D., and Dutton, C., 1984, Photodynamic effects: porphyrin vs chlorin, Photochem. Photobiol., 40: 403.

Lai, J.-J., McCaul, B., Smith, K., and Straight, R.C., 1987, Polyhematoporphyrin esters: preparation, properties and biological activity of a new stable derivative of hematoporphyrin for photodynamic therapy, Photochem. Photobiol., 46: in press.

Langlois, R., Ali, H., Brasseur, N., Wagner, R., van Lier, J.E., 1986, Biological activities of phthalocyanines IV. Type II sensitized photooxidation of L-tryptophan and cholesterol, Photochem. Photobiol., 44: 117.

Linstead, R.P., Weiss, F.T., 1950, Phthalocyanines and related compounds - Part XX. Further investigations on tetrabenzoporphin and allied substances, J. Chem. Soc., 2975.

Lipson, R.L., Baldes, E.J., and Olsen, A.M., 1961, The use of a derivative of haematoporphyrin in tumor detection, J. Natl. Cancer Inst., 226: 1.

McCubben, I., and Phillips, D., 1986, The photophysics and photostability of zinc (II) and alumimium (III) sulfonated naphthalocyanines, J. Photochem., 34: 187.

Morgan, A.R., Garbo, G.M., Kreimer-Birnbaum, M., Keck, R.W., Chandhuri, K., and Selman, S.H., 1987a, Morphological study of the combined effect of purpurin derivatives and light on transplantable rat bladder tumors, Cancer Res., 47: 496.

Morgan, A.R., Rampersaud, A., Keck, R.W., and Selman, S.H., 1987b, Verdins: new photosensitizers for photodynamic therapy, Photochem. Photobiol., 46: 441.

Morgan, A.R., and Tertel, N.C., 1986, Observations on the synthesis and spectroscopic characteristics of purpurins, J. Org. Chem., 51: 1348.

Oseroff, A.R., Ohuoha, D., Hasan, T., Bommer, J.C., and Yarmush, M.L., 1986, Antibody-targeted photolysis: Selective photodestruction of human T-cell leukemia cells using monoclonal antibody-chlorin e_6 conjugates, Proc. Natl. Acad. Sci. USA, 83: 8744.

Oseroff, A.R., Ara, G., Ohuoha, D., Aprille, J., Bommer, J.C., Yarmush, M.L., Foley, J., and Cincotta, L., 1987, Strategies for selective cancer photochemotherapy: Antibody-targeted and selective carcinoma cell photolysis, Photochem. Photobiol., 46: 83.

Pangka, V.S., Morgan, A.R., and Dolphin, D., 1986, Diels-Alder reactions of protoporphyrin IX dimethyl ester with electron-deficient alkynes, J. Org. Chem., 51: 1094.

Paquette, B., Ali, H., Langlois, R., van Lier, J.E., 1987, Biological activities of phthalocyanines - VIII. Cellular distribution in V-79 Chinese hamster cells and phototoxicity of selectively sulfonated aluminum phthalocyanines, Photochem. Photobiol., in press.

Rimington, C., Sommer, S., and Moan, J., 1987, Hematoporphyrin ethers I. Generalized synthesis and chemical properties, Int. J. Biochem., 19: 315.

Rosenthal, I., Krishna, C.M., Riesz, P., Ben-Hur, E., 1986, The role of molecular oxygen in the photodynamic effect of phthalocyanines, Radiat. Res., 107: 136.

Sari, M.A., Battioni, J.P., Mansuy, D., and Le Pecq, J.B., 1986, Mode of interaction and apparent binding constants of meso-tetraaryl porphyrins bearing between one and four positive charges with DNA, Biochem. Biophys. Res. Comm., 141: 643.

Scourides, P.A., Böhmer, R.M., Kaye, A.H., and Morstyn, G., 1987, Nature of the tumor-localizing components of hematoporphyrin derivative, Cancer Res., 47: 3439.

Selman, S.H., Kreimer-Birnbaum, M., Chaudhuri, K., Garbo, G.H., Seaman, D.A., Keck, R.W., Ben-Hur, E., Rosenthal I, 1986, Photodynamic treatment of transplantable bladder tumors in rodents after pretreatment with chloroaluminium tetrasulfophthalocyanine, J. Urol., 136: 141.

Spikes, J.E., 1986, Yearly review: phthalocyanines as photosensitizers in biological systems and for the photodynamic therapy of tumors, Photochem. Photobiol., 43: 691.

Tralau, C.J., MacRobert, A.J., Coleridge-Smith, P.D., Barr, H., Bown, S.G., 1987, Photodynamic therapy with phthalocyanine sensitization: quantitative studies in a transplantable rat fibrosarcoma, Br. J. Cancer, 55: 389.

Valduga, G., Reddi, E., Jori, G., 1987, Spectroscopic studies on Zn(II)-phthalocyanine in homogeneous and microheterogeneous systems, J. Inorg. Biochem., 29: 59.

van Lier, J.E., Brasseur, N., Paquette, B., Wagner, J.R., Ali, H., Langlois, R., and Rousseau, J., 1987, Phthalocyanines as sensitizers for photodynamic therapy of cancer, in: "Photosensitization: Molecular and Medical Aspects," G. Moreno, R.H. Pottier and T.G. Truscott, eds., Springer-Verlag, Berlin, in press.

Wagner, J.R., Ali, H., Langlois, R., Brasseur, N., van Lier, J.E., 1987, Biological activities of phthalocyanines - VI. Photooxidation of L-tryptophan by selectively sulfonated gallium phthalocyanines: singlet oxygen yields and effect of aggregation, Photochem. Photobiol., 45: 587.

Weber, H.J., Busch, D.H., 1965, Complexes derived from strong field ligands - XIX. Magnetic properties of transition metal derivatives of 4,4´,4´´,4´´´-tetrasulfophthalocyanines, Inorg. Chem., 4: 469.

Winkelman, J., 1962, The distribution of tetraphenylporphinesulfonate in the tumor-bearing rat, Cancer Res., 22: 589.

Winkelman, J., 1985, Quantitative studies of tetraphenylporphinesulfonate and hematoporphyrine derivative distribution in animal tumor systems, in: "Methods in Porphyrin Photosensitization", D. Kessel, ed., pp. 91-96, Plenum Press, New York.

CLINICAL AND IN VITRO PHOTOCHEMISTRY OF BILIRUBIN

John F. Ennever

Department of Pediatrics
Case Western Reserve University
Cleveland, Ohio

INTRODUCTION

The use of light in treatment of neonatal jaundice began nearly 30 years ago following the empirical observations of a nurse, J. Ward, and a biochemist, P.W. Perryman, on the clinical and in vitro photochemistry of bilirubin (Dobbs and Cremer, 1975). In this presentation I shall discuss more recent observations on how phototherapy appears to work in jaundiced infants and ways in which we may be able to improve the efficiency of the treatment. Jaundice occurs in the newborn period because the normal hepatic mechanism for solubilization of bilirubin, esterification with glucuronic acid, is insufficient for the load of pigment formed in the first few days of life. This chemical modification of bilirubin is required for excretion of the pigment in bile or urine. Phototherapy works by changing bilirubin photochemically into molecules or fragments which can be excreted by the liver or kidney without the need for chemical modification. Three types of photochemical reactions of bilirubin occur in jaundiced infants treated with phototherapy: configurational isomerization, structural isomerization and photooxidation. A brief summary of the different photoreactions of bilirubin is given below. For a more complete discussion, see Lightner and McDonagh, 1984.

PHOTOCHEMICAL REACTIONS OF BILIRUBIN

The native configuration of bilirubin is 4Z,15Z (Figure 1). In this native configuration, bilirubin is able to adopt a conformation in which all of the polar functional groups are internally hydrogen bonded. This preferred conformation explains the lipophilic properties of the pigment and why addition of a polar group (e.g., glucuronic acid) is required for excretion. When exposed to visible light falling within the main absorption band, the fastest reaction of bilirubin is a Z to E isomerization about one of the exocyclic double bonds. While this isomerization can occur at either the 4 or 15 position, for bilirubin which is bound to human albumin, the reaction is highly regiospecific for the 15 position (McDonagh et al., 1982). The quantum yield for formation of 4Z,15E-bilirubin from bilirubin bound to human albumin at 37 °C is approximately 0.2 (Lamola et al., 1982). The 4Z,15E-isomer is more water soluble than the native molecule because not all of the polar functions can be involved in internal hydrogen bonding (Figure 1). The 4Z,15E-isomer is thermally unstable, and in the absence of

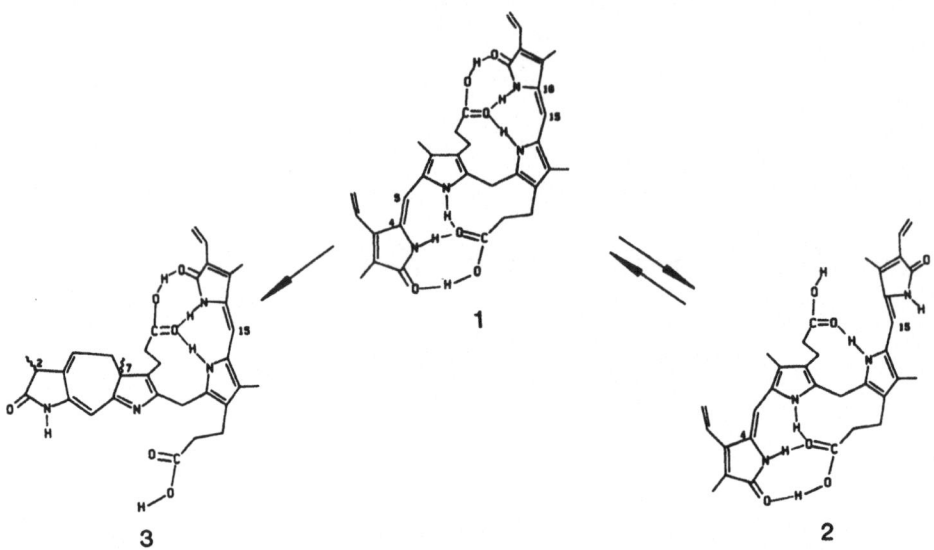

Fig. 1. Photoisomerization reactions of bilirubin. **1**, 4$\underline{Z}$,15$\underline{Z}$-bilirubin; **2**, 4$\underline{Z}$,15$\underline{E}$-bilirubin; **3**, $\underline{Z}$-lumirubin.

the stabilizing effect of albumin binding, rapidly reverts to native bili-rubin. The $\underline{Z}$ to $\underline{E}$ reaction is photochemically reversible and because the absorption spectrum of the 4$\underline{Z}$,15$\underline{E}$-isomer overlaps but is not identical to that of the native molecule, a photoequilibrium is quickly established when bilirubin bound to albumin is irradiated in a closed system. The composi-tion of the photoequilibrium depends upon the wavelength of light driving the reaction: light at the blue end of the absorption band drives the reac-tion toward the 4$\underline{Z}$,15$\underline{E}$-isomer; light at the long-wavelength end of absorp-tion band drives the reaction toward the native 4$\underline{Z}$,15$\underline{Z}$ form (Ennever et al., 1984).

A second type of photoisomerization reaction of bilirubin is an intra-molecular cyclization involving the endo vinyl group at C-3 (Figure 1). The name given to this structural isomer of bilirubin by McDonagh et al. (1982) is $\underline{Z}$-lumirubin. [Because this molecule has chiral centers at C-2 and C-7 there are four possible diastereoisomers. McDonagh et al. (1986) have shown that $\underline{Z}$-lumirubin formed from bilirubin bound to human albumin is optically active due to a specific configuration at C-7.] The quantum yield for lumi-rubin formation is approximately 0.0015 (Greenberg et al., 1987), far lower than that for the configurational isomerization reaction. As with the 4$\underline{Z}$,15$\underline{E}$-isomer, $\underline{Z}$-lumirubin is more water soluble than native bilirubin because of a disruption of internal hydrogen bonding. In contrast to con-figurational isomerization, however, the structural isomerization reaction is not readily reversible either photochemically or thermally.

In the presence of oxygen, bilirubin can also undergo photooxidation to a variety of colorless mono and bicyclic compounds. Lightner et al. (1984) have characterized the photooxidative products and identified them in the urine of jaundiced infants undergoing phototherapy. While it clear that photooxidation does contribute to bilirubin elimination during phototherapy, current evidence suggest that it is not as quantitatively important as the two types of photoisomerization reactions.

In summary, the two photochemical reactions of bilirubin that are important during phototherapy produce isomers which are more water soluble than native bilirubin. One is a very rapid, reversible Z to E configurational isomerization, occurring predominately at the 15 position; the other is a much slower and irreversible structural isomerization. Based upon the large difference in quantum yields, one would predict that the configurational isomerization would be the quantitatively more important process. Although it is clear that the 4Z,15E-isomer is the most rapidly-formed photoproduct in babies treated with phototherapy, a variety of evidence, albeit indirect, suggest that lumirubin is more important because of more rapid clearance.

CLINICAL PHOTOCHEMISTRY OF BILIRUBIN

Within hours of beginning phototherapy, the concentration of configurational and structural isomers achieve steady-state levels in the serum. A chromatograph of serum obtained from a jaundiced newborn infant receiving phototherapy is shown in Figure 2. The configurational isomers, 4Z,15E and 4E,15Z, are the most abundant photoproducts found in the serum. The 4Z,15E-isomer is the principal configurational isomer in the serum, presumably as the result of the regioselectivity of the reaction for bilirubin bound to human albumin. The structural isomer, Z-lumirubin, is found in far lower concentrations in the serum than the configurational isomers. [It should be noted that the relative chromatographic peak areas do not reflect the relative molar amounts of each bilirubin species present because the various isomers do not have the same detector response. Under conditions used for the chromatogram in Figure 2, the relative response ratios are: 1.0:0.7:0.43 for native bilirubin:4Z,15E-isomer:Z-lumirubin (Malhotra and Ennever, 1986).] In a group of nine infants who had been under phototherapy for at least 24 hours we found that the structural isomer Z-lumirubin was 3.2 ± 1.4% (±SD) and the configurational isomers, principally 4Z,15E, were 17.0 ± 2.1% of the total pigment in the serum (Ennever et al., 1987).

The high steady-state levels of the bilirubin isomers in the serum of babies treated with phototherapy indicates that the rate of excretion of

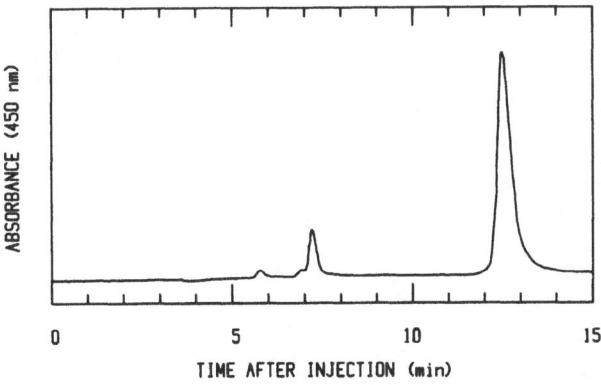

Fig. 2. HPLC of serum from jaundiced infant receiving phototherapy. The principal peaks are: 4Z,15Z-bilirubin eluting at 12.7 min, 4Z,15E-bilirubin at 7.2 min (the 4E,15Z-isomer is the unresolved component just before the 4Z,15E peak), and Z-lumirubin at 5.8 min. Conditions for separation were: ODS column, 0.1 M di-n-octylamine acetate in methanol, pH 7.6, flow rate 0.7 ml/min.

these photoproducts is not rapid compared with their rate of formation. This is quite different from what is found in the phototherapy-treated Gunn rat, where the photoisomers are nearly undetectable in the serum unless bile flow is mechanically interrupted (McDonagh, 1984). In the Gunn rat excretion of both configurational and structural isomers in bile is so rapid that the relative contributions of the two types of photoisomerization reactions to bilirubin elimination is simply a reflection of the relative rates of the two photochemical reactions. As a result, configurational isomerization is the far more important process in the rat. In the human infant, the relative rates of excretion are an important determinant of the relative importance of configurational and structural isomerization to the elimination of bilirubin during phototherapy.

We have measured the rate of disappearance of the configurational isomers and Z-lumirubin from the serum of jaundiced infants after stopping phototherapy. The decline in both types of isomers from the serum could be approximated by first-order rate equations. For the configurational isomers, the half-life ranged from 12 to 21 hours with a mean of 15 hours (Ennever et al., 1985). For Z-lumirubin, the half-life ranged from 80 to 158 minutes, with a mean of 112 minutes (Ennever et al., 1987). It should be cautioned that these measurements are apparent serum half-lives, and not true pharmacokinetic half-lives. They were determined by simply measuring concentrations of the isomers in the serum after stopping phototherapy. Since a large portion of the total pool of bilirubin isomers is probably extravascular, even after stopping phototherapy photoisomers must continue to migrate from the site of formation in light-exposed tissues into the vasculature. This complication probably has its largest effect on the half-life of Z-lumirubin, causing the measured value to be greater than the true half-life. Despite this limitation, these data indicate that in human newborn infants the excretion of Z-lumirubin is much faster than the excretion of the configurational isomers. It would be hazardous to extrapolate from these results alone to the conclusion that formation and excretion of Z-lumirubin is the quantitatively most important pathway for bilirubin elimination during phototherapy. However, the results of other clinical studies are consistent with this conclusion.

In one such study we compared the concentration of configurational and structural isomers in the serum of two groups of infants treated with two different intensities of light of the same color. There was no difference in the steady-state concentration of the configurational isomers between those infants receiving the lower dose of light and those receiving the higher dose. In contrast, however, the serum concentration of lumirubin was significantly higher in those infants receiving the higher dose of light (Costarino et al., 1985). These results can be understood in terms of the difference between configurational and structural isomerization of bilirubin. The former is a readily-reversible reaction, and in a closed system reaches a photostationary state. In this study, the lower dose of light was of sufficient intensity to produce the maximal amount of the configurational isomer and greater intensity could increase it no further. A corollary of these observations is that in terms of the configurational isomer, the human infant acts like a closed system. Structural isomerization is an essentially irreversible reaction, thus more light produces more Z-lumirubin. Since a number of clinical studies have demonstrated that higher dose phototherapy results in more rapid decline in bilirubin (e.g., Tan, 1982), these results suggest that the improved efficacy of higher dose phototherapy is not due to increased formation of configurational isomers, but rather to increased formation of Z-lumirubin.

In 1983, Vecchi and co-workers reported that green fluorescent lights were more effective in reducing serum bilirubin in jaundiced infants than the standard daylight (white) fluorescent light. These findings were quite surprising because the emission spectrum of the green lights overlap only with the long-wavelength end of the bilirubin absorption band. Moreover, the percentage of configurational isomers at photoequilibrium produced by irradiation of bilirubin bound to human albumin with green fluorescent light in vitro is less than one-half of that produced by daylight fluorescent light (Ennever et al., 1984). We have compared the steady-state concentrations of configurational and structural isomers in the serum of jaundiced infants under green and daylight fluorescent phototherapy (Ennever et al., 1986). As expected from the in vitro results, the configurational isomer concentration was less (12.6% vs. 17.0% of total pigment) in those infants treated with green light. Unexpectedly, however, the concentration of Z-lumirubin was higher in those infants treated with green light (3.8% vs. 2.2%). Thus, the lights that which produce more Z-lumirubin and less configurational isomers are more effective in lowering serum bilirubin. These results are further support for the hypothesis that formation and excretion of Z-lumirubin is quantitatively most important during phototherapy.

Quantitation of the relative importance of the two classes of bilirubin photoproducts to pigment elimination during phototherapy requires direct measurement of the two photoproducts in excreta. We have shown that Z-lumirubin is the only isomer of bilirubin in the urine of infants treated with phototherapy (Knox et al., 1985). Although urinary excretion of Z-lumirubin and photooxidation products does contribute to bilirubin elimination, most

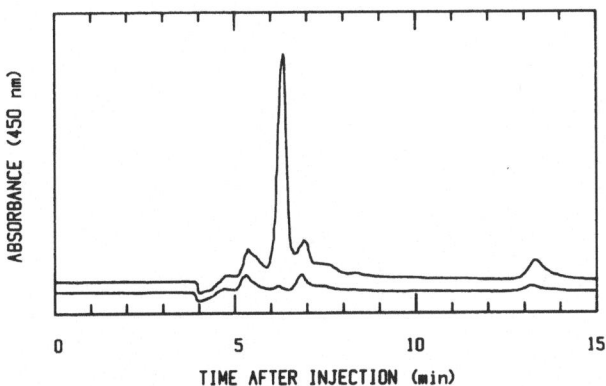

Fig. 3. HPLC of bile obtained during phototherapy (upper tracing) and 24 hours after stopping phototherapy (lower tracing). The principal peaks are: Z-lumirubin eluting at 6.2 min, 4Z,15Z-bilirubin at 13 min, bilirubin di-glucuronide at 5.3 min and bilirubin mono-glucuronide at 6.9 min. Conditions for separation were as described in legend to Fig. 2.

evidence points to biliary excretion as the primary pathway. We have found that Z-lumirubin is the major bilirubin species in aspirates of duodenal bile from premature infants under phototherapy and that it is not found in bile before beginning, or after stopping, phototherapy (Figure 3). Onishi et al. (1986) have reported similar results. Although it is not possible to duplicate in human infants the elegant experiments performed by McDonagh on Gunn rats to establish definitely the relative contributions of configurational and structural isomerization of bilirubin to the elimination of bilirubin during phototherapy, the available clinical evidence supports the hypothesis that Z-lumirubin excretion is most important.

If the therapeutic effect of phototherapy is dependent upon formation and excretion of Z-1umirubin, then strategies to improving the therapy should be directed toward enhancing either the formation or excretion of this photoproduct. (Alternatively, one could try to enhance the excretion of the much more rapidly-formed configurational isomers.) One way in which formation of Z-1umirubin has been shown to be enhanced is by use of green light. We have measured the wavelength-dependence of the quantum yield for the formation of Z-1umirubin from bilirubin bound to human albumin (Greenberg et al., 1987). The results are shown in Figure 4. There is a more than twofold increase in the quantum yield at the long wavelength edge of the bilirubin absorption band. The rate of rise in the quantum yield is of sufficient magnitude to compensate for the fall in extinction coefficient, so that the relative photochemical yield, normalized to equal photon fluence, is greatest at 500 nm. This phenomenon must, in part, explain the clinical observation of greater production of Z-1umirubin by green fluorescent light.

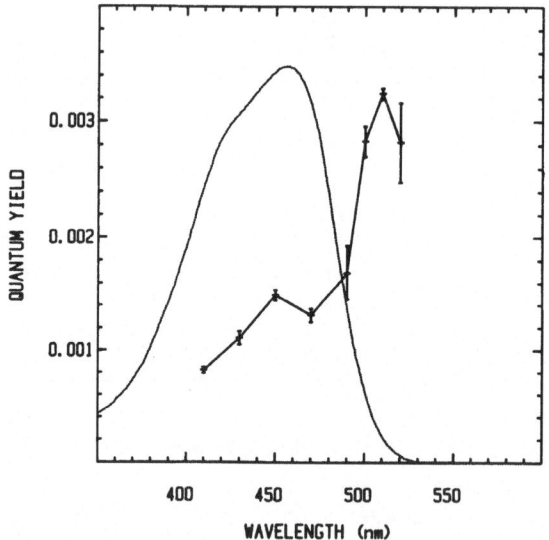

Fig. 4. Quantum yield for formation of Z-1umirubin from bilirubin bound to human albumin at 8 wavelengths (5 nm half-bandwidth). Data points are mean and error bars one standard deviation of triplicate determinations. Bilirubin absorption band is shown as dotted line.

McDonagh and Lightner (1985) have presented evidence that there is rapid intramolecular transfer between vibronic energy levels that formally correspond to excitation of the first excited state of one or the other dipyrromethenone chromophores of bilirubin. Because of this exciton coupling between the two halves of bilirubin and the rapid relaxational pathway available on the half of the molecule with the exo vinyl group (Greene et al., 1981), most of the photochemistry occurs on this half of the molecule. Because the absorption bands for the two chromophores overlap but do not coincide there is the possibility for anomalous behavior with absorption of light at the extreme long-wavelength edge of the absorption band (Lamola, 1985). This will occur if the difference in the energy of the zero-zero transition of the two chromophores is significantly greater than the thermal kinetic energy. In this case, following the absorption of long-wavelength light the excitational energy would not have as much opportunity to "jump" to the other chromophore as it would have following excitation with higher energy light. Our quantum yield results are consistent with an "edge effect" phenomenon if the absorption spectrum for the dipyrromethenone with the endo vinyl group is shifted lower in energy relative to the dipyrromethenone with the exo vinyl group. Although this ordering of the absorption bands of the dipyrromethenone chromophores of bilirubin is opposite of that reported for the symmetrical isomers of bilirubin in various organic solvents (McDonagh, 1979), the effects of the microenvironment surrounding the chromophores when bound to human albumin are not known.

We have also found that the binding of fatty acids of carbon chain length greater than 10 to albumin markedly enhances the rate of formation of Z-1umirubin from bilirubin bound to albumin (Malhotra et al., 1987). The results for representative fatty acids are shown in Figure 5. The binding of these same fatty acids produced bathochromic and hyperchromic shifts in the bilirubin absorption band. Although some of the increased rate of formation of lumirubin could be attributed to these changes in the absorption band, the major contributor was an increase in the quantum yield for Z-lumirubin formation. In addition to changes in the absorption spectrum, the binding of these long-chain fatty acids also produced changes in the circular dichroism spectrum of bilirubin, which have been interpreted to indicate an increase in the dihedral angle between the two dipyrromethenone chromophores (Blauer and Wagniere, 1975). Since for bilirubin bound to human albumin in the absence of fatty acids this angle is ~109° (Lightner et al., 1986), only slightly greater than the optimal angle calculated for exciton coupling (Blauer and Wagniere, 1975), the increase in the dihedral angle produced by fatty acid binding would tend to decrease the coupling of

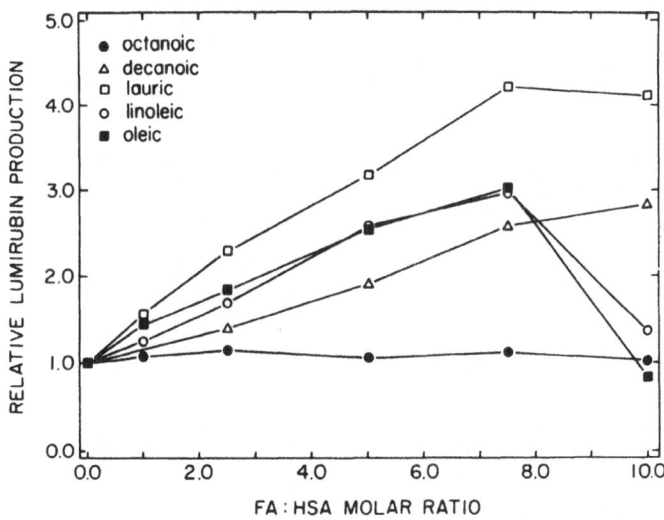

Fig. 5. Effect of fatty acid binding on the rate of Z-1umirubin formation. Amount of Z-1umirubin formed with fatty acid bound to human albumin relative to amount formed in a reference solution without fatty acid is plotted as a function of the number of equivalents of fatty acid per mole of albumin.

the excited states of the two dipyrromethenone chromophores. If, as described above, exciton coupling between the two halves of bilirubin and the rapid relaxational pathway available on the half with the exo vinyl group limits the rate of formation of $\underline{Z}$-lumirubin on the other half, then this conformational change produced by fatty acid binding would be expected to enhance the rate of structural isomerization. Our data cannot exclude other mechanisms for the increased formation of $\underline{Z}$-lumirubin produced by fatty acid binding, such as a conformational change within the bilirubin molecule which in some way simply facilitates the cyclization reaction. There is, however, evidence that energy transfer processes are important in the photochemistry of bilirubin (McDonagh and Lightner, 1985), and our results suggest that biologically-important ligands may affect these processes.

CONCLUSIONS

Nearly thirty years have passed since the original observations were made that light could affect bilirubin metabolism. Although we now know far more about the photochemistry of bilirubin and what happens to the bilirubin products in jaundiced infants, the treatment today is no different than it was thirty years ago. This new knowledge has not yet led to improved therapy. The empirical observation that green fluorescent light was more effective than conventional white light was inconsistent with the widely-held belief in the importance of configurational isomerization, and was the clue which led to the conclusion that formation and excretion of $\underline{Z}$-lumirubin is important during phototherapy. More recent work suggests that we may be able to manipulate our therapy, for example through use of selected wavelengths of light or use of other agents such as fatty acids, to enhance the formation of $\underline{Z}$-lumirubin. Much work remains to be done to apply our knowledge of the photochemistry of bilirubin to future improvements in the therapeutic use of light.

ACKNOWLEDGMENTS

This work was supported by grants DK-36517 and DK-38575 from the National Institutes of Health. The collaboration of Drs. L.L Dunn, J.W. Greenberg, I. Knox, V. Malhotra, A.F. McDonagh, R.A. Polin, and W.T. Speck is greatly appreciated.

REFERENCES

Blauer, G., and Wagniere, G., 1975, Conformation of bilirubin and biliverdin in their complexes with albumin, J. Am. Chem. Soc., 97:1949.

Costarino, A.T., Ennever, J.F., Baumgart, S., Speck, W.T., Paul, M., and Polin, R.A., 1985, Bilirubin photoisomerization in premature neonates under low- and high-dose phototherapy, Pediatrics, 75:519.

Dobbs, R.H., and Cremer, R.J., 1975, Phototherapy (Looking back), Arch. Dis. Child., 50:833.

Ennever, J.F., Sobel, M., McDonagh, A.F., and Speck, W.T., 1984, Phototherapy for neonatal jaundice: In vitro comparison of light sources, Pediatr. Res., 18:667.

Ennever, J.F., Knox, I., Denne, S.C., and Speck, W.T., 1985, Phototherapy for neonatal jaundice: In vivo clearance of bilirubin photoproducts, Pediatr. Res., 19:205.

Ennever, J.F., Knox, I., and Speck, W.T., 1986, Differences in bilirubin isomer composition in infants treated with green and white light phototherapy, J. Pediatr., 109:119.

Ennever, J.F., Costarino, A.T., Polin, R.A., and Speck, W.T., 1987, Rapid clearance of a structural isomer of bilirubin during phototherapy, J. Clin. Invest., 79:1674.

Greenberg, J.M., Malhotra, V., and Ennever, J.F., 1987, Wavelength dependence of the quantum yield for the structural isomerization of bilirubin, Photochem. Photobiol., 46:453.

Greene, B.I., Lamola, A.A, and Shank, C.V., 1981, Picosecond primary photoprocesses of bilirubin bound to human serum albumin, Proc. Natl. Acad. Sci. USA, 78:2008.

Knox, I., Ennever, J.F., and Speck, W.T., 1985, Urinary excretion of an isomer of bilirubin during phototherapy, Pediatr. Res., 19:198.

Lamola, A.A., 1985, Effects of environment on photophysical processes of bilirubin, In "Optical properties and structure of tetrapyrroles," G. Blauer, H. Sund eds. Walter de Gruyter, Berlin.

Lamola, A.A., Flores, J., and Doleiden, F.H., 1982, Quantum yield and equilibrium position of the configurational photoisomerization of bilirubin bound to human serum albumin, Photochem. Photobiol., 35:649.

Lightner, D.A., and McDonagh, A.F., 1984, Molecular mechanisms of phototherapy for neonatal jaundice, Acc. Chem. Res., 17:417.

Lightner, D.A., Linnane, W.P. III, and Ahlfors, C.E., 1984, Bilirubin photooxidation products in the urine of jaundiced infants receiving phototherapy, Pediatr. Res., 18:696.

Lightner, D.A., Reisinger, M., and Landen, G.L., 1986, On the structure of albumin-bound bilirubin. Selective binding of intramolecularly hydrogen-bonded conformational enantiomers, J. Biol. Chem., 261:6034.

Malhotra, V., and Ennever, J.F., 1986, Determination of the relative detector response for unstable bilirubin photoproducts without isolation, J. Chromatog., 383:153.

Malhotra, V., Greenberg, J.W., Dunn, L.L., and Ennever, J.F., 1987, Fatty acid enhancement of the quantum yield for the formation of lumirubin from bilirubin bound to human albumin, Pediatr. Res., 21:530.

McDonagh, A.F., 1979, Bile pigments: Bilatrienes and 5,15-biladienes, In "The Porphyrins, Vol. VI," D. Dolphin ed. Academic Press, New York.

McDonagh, A.F., 1984, Molecular mechanisms of phototherapy for neonatal jaundice, In "Neonatal Jaundice," F.F. Rubaltelli, G. Jori eds. Plenum Press, New York.

McDonagh, A.F., and Lightner, D.A., 1985, Intramolecular energy transfer in bilirubins, In "NATO advanced study institute on primary photoprocesses in biology and medicine," R.V. Bensasson, G. Jori, E.J. Land, T.G. Truscott eds. Plenum Press, New York.

McDonagh, A.F., Palma, L.A., and Lightner, D.A., 1982, Phototherapy for neonatal jaundice: Stereospecific and regioselective photoisomerization of bilirubin bound to human serum albumin and NMR characterization of intramolecular cyclized photoproducts, J. Am. Chem. Soc., 104:6867.

McDonagh, A.F., Lightner, D.A., Reisinger, M., and Palma, L.A., 1986, Human serum albumin as a chiral template. Stereoselective photocyclization of bilirubin, J. Chem. Soc. Chem. Commun.,249.

Onishi, S., Isobe, K., Itoh, S., 1986, Metabolism of bilirubin and its photoisomers in newborn infants during phototherapy, J. Biochem., 100:789.

Tan, K.L., 1982, The pattern of bilirubin response to phototherapy for neonatal hyperbilirubinemia, Pediatr. Res., 16:670.

Vecchi, C., Donzelli, G.P., Migliorini, M.G., and Sbrana, G., 1983, Green light in phototherapy, Pediatr. Res., 17:461.

UNWANTED SIDE EFFECTS AND OPTIMIZATION OF PHOTOTHERAPY

Terje Christensen, Anne Støttum
Gunnar Brunborg[*] and Jon B. Reitan

National Institute of Radiation Hygiene
P.O. Box 55, N-1345 Østeras, Norway
[*]National Institute of Public Health
Geitmyrsveien 75, N-0462 Oslo 4, Norway

INTRODUCTION

Phototherapy of hyperbilirubinemia in newborns is a relatively new modality which is mainly based on empirical observations of the beneficial effects of light upon icteric children (Cremer et. al., 1958). Bilirubin is poorly water soluble, and the development of icterus in newborns is caused by a defect in the conjugation mechanism that converts bilirubin to a more water-soluble form that can be excreted. Light irradiation converts bilirubin to other types of water soluble products. The nature of these products and their excretion route have been a matter of discussion during the last years (McDonagh, 1971; McDonagh et al., 1980; Pratesi, 1983). The current view is that light converts bilirubin to photoisomers that can be excreted through the bile or the urine. A number of photoproducts are formed simultaneously. The photoreactions may be complicated and sensitive to a number of factors as light intensity and wavelength (Ennever, 1986).

It cannot be excluded that certain long term side effects of phototherapy may still not be recognized. Induction of cancer is of particular interest, since the expression time in the treated individuals is maximal.

It can be estimated that between one and ten per cent of all newborns are currently given phototherapy. The number of treated children may vary between countries, and often the exact number is not known although it is high in most countries Therefore any unwanted side effects are potentially harmful to a high number of patients even if they occur at a low frequency.

On this background we initiated our study which is divided into two separate parts: One study of the clinical use of phototherapy and one _in vitro_ study of the toxic effects of irradiated bilirubin.

At the outset of this study the use of phototherapy in Norway was poorly evaluated. There was a need for information that could be used for the planning of a project aimed at optimization of phototherapy. Therefore three Norwegian hospitals were visited. Two of the hospitals were university hospitals. They all had a department of pediatrics and phototherapy was used routinely. No systematic, statistical analysis of the practice was attempted, but a list of questions was prepared and used as a guide when interviewing the medical staff in each hospital.

The list contained questions about:

1. The indications for phototherapy
2. The methods routinely used
3. The clinical effects of the phototherapy
4. Side effects

The results may be summed up as follows:

In all hospitals the <u>indications</u> for phototherapy were based on routine blood bilirubin measurements together with data on weight, gestation age etc. The application of such criteria was relatively similar in the different hospitals. The criteria used were all based on internationally accepted values and the physicians seemed to be open to change them if new knowledge indicates that a change should be made. The frequency of exchange transfusions has decreased markedly after the introduction of phototherapy and transfusions are now used very seldom.

With respect to the therapeutic <u>method</u>, only the irradiation set-up will be mentioned here. The light sources consisted of mobile phototherapy units, usually equipped with eight 20 W fluorescent tubes, either blue or white. In front of the tubes a plastic shield was mounted. The light was kept on continuously except during observation of and care for the child. Irradiation was applied until the desired result was obtained, normally until the bilirubin concentration had fallen somewhat below the values used as an indication for phototherapy in each case. The light intensity was not measured routinely, and equipment for determination of the light dose was usually not available. Furthermore, no distinction was made between the application of white or blue lamps or a mixture of lamps with the two colours mounted in the same phototherapy unit.

The <u>clinical effectiveness</u> of treating hyperbilirubinemia with phototherapy has been excellent in all three hospitals. No cases of clinically manifest kernicterus were reported.

Only less serious <u>side effects</u> of phototherapy were observed. All of these were reversible upon discontinuing light treatment. Among the most frequently mentioned side effects were: Diarrhea, rash, dehydration and hyperthermia. Most of these side effects could be counteracted by relatively simple means.

OPTIMIZATION OF PHOTOTHERAPY

The International Commission on Radiological Protection (ICRP) has given recommendations for the use of ionizing radiation (ICRP, 1977). Some of the recommendations may be useful for evaluating the use of non-ionizing radiation as well. For example the use of radiation should be _justified_. The use of light in phototherapy is clearly justified since it aims at, e.g., avoiding kernicterus.

Another important recommendation from the ICRP is that the use of radiation should be _optimised_, meaning that procedures involving radiation should be designed to give a maximal benefit with as few side effects as possible. If the current use of phototherapy is evaluated according to these criteria, a number of questions arise.

The only dose-related parameter under some form of control, was the total treatment time. Functional photometers or radiometers are normally not available. Thus, the main basis for a sound assay of total dose is missing. Similarly, the emission spectra of the fluorescent tubes were not under adequate control. In some cases the difference between white and blue light was not taken into account properly. Technical service of the phototherapy units was normally performed by the technical staff at each hospital. It seemed as if the contact between them and the medical staff was not systematically used for the purpose of optimization of technical parameters, like the choice of lamp types or replacement of tubes at appropriate intervals.

Application of the recommendations given by the ICRP also for non-ionizing radiation would imply that the dose should be limited to the dose that gives the intended effect. It has been shown that fractionation of the light dose may limit the total dose needed to obtain a certain lowering of the bilirubin concentration (Jährig et al., 1985). No attempts on fractionation of the dose had been made in any of the hospitals.

If one wants to perform optimization of the use of phototherapy, it is important that the parameters used and the results of treatment of previous patients are registered. Such registration is normally not carried out. Therefore an important tool for the future planning of treatment regimens and epidemiology seems to be lost in many cases.

EXPERIMENTAL

Cell culture. Cultured V79 Chinese hamster cells were used. The V79 cells were grown in Eagle's MEM with supplements as described previously (Holme et al., 1987). The cells were subcultured twice a week in numbers allowing almost continuous exponential growth. When seeded in 6 cm tissue culture dishes (Costar) or 75 cm^2 tissue culture flasks (Costar) at a density of 8 cells cm^{-2}, 70 - 100% of the cells formed visible colonies within one week of incubation. The relative number of colonies formed by treated cells was scored as surviving fraction in the experiments.

Irradiation. The irradiation equipment consisted of an Air-Shields phototherapy unit containing 8 Philips 20W/52 tubes. At a distance of 50 cm the incident light fluence was 20 Wm^{-2}, and the emission spectrum of the lamps peaked at about 450 nm (measured with a UDT-detector and an EG & G, Princeton Applied Research, model 1460 spectrophotometer). The solutions to be irradiated were contained in tissue culture vessels at room temperature or at about 4°C on an ice bath during irradiation.

Bilirubin solutions. Bilirubin (Sigma) was dissolved in 0.01 M NaOH at 10 times the concentration to be used. This solution was diluted in serum-containing saline. The pH of the final solutions containing serum was about 8.5. Freshly prepared solutions were used.

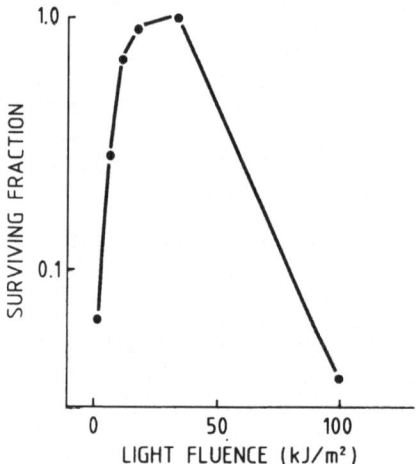

Fig. 1.   Survival of cells treated with bilirubin solutions previously irradiated at 20°C with the light fluences indicated. A light fluence of 72 kJ/m^2 corresponds to an irradiation time of 1 h. The bilirubin was dissolved at 100 µg/ml in physiological saline containing 10% fetal calf serum.

Toxicity testing. The toxicity of bilirubin solutions or the phototoxicity of blue light combined with bilirubin was tested by scoring the surviving fractions of treated cells. The cells were inoculated and left for about 2 h to attach. The solutions were either irradiated before being added to the cells, or with the cells present in the plastic vessels during irradiation. In the former case the irradiated solutions

156

were immediately added to culture dishes containing 200 cells per dish and incubated at 37°C for 30 min followed by a change to growth medium. Thereafter the cells were incubated for 6-7 days to form macroscopic colonies.

Genotoxicity testing. Cells treated as above were processed for the scoring of sister chromatid exchanges or mutations at the HGPRT locus as described previously (Holme et al., 1987). Strand breaks in the DNA were assayed by the alkaline elution technique in a semi-automated setup (Brunborg et al., in preparation). DNA measurements involved the use of the fluorochrome HOECHST 33258.

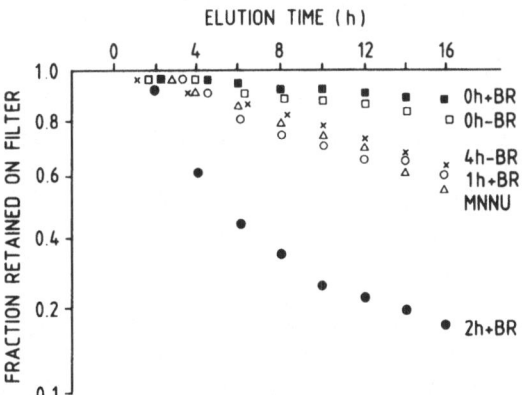

Fig. 2.   Alkaline elution of DNA from V79 hamster cells irradiated at 4°C with blue light, in the presence or absence of 100 μg/ml bilirubin (BR) dissolved in physiological saline containing 10% fetal calf serum. N-nitroso-N-methylurea (MNNU) is used as a positive control at a concentration of 10 μg/ml for 1 h at 37°C. More than 50% of the cells survived the treatment under all the conditions used, except for 0 h light + bilirubin, where less than 10% of the cells survived.

RESULTS

Fig. 1 shows that the unirradiated bilirubin solution has a certain toxicity to hamster cells. Under similar conditions human cells seemed to be inactivated to a lower degree (Christensen, 1986). Upon irradiation, the toxicity of the solutions decreased to a minimum level which was obtained at different fluences under a variety of conditions (data not shown). When the irradiation was continued, a dose dependent decrease in cell survival was observed. These effects have

Table I. Mutations at the HGPRT locus in cells treated as described in the text of Fig. 1. Nitroquinoline oxide (NQO) was used at a concentration of 1 μg/ml for 1 h as a positive control.

CONDITIONS	MUTATION FREQUENCY 10^{-6}	
	Exp. 1	Exp. 2
Control	13	6
1 μg/ml NQO	77	38
100 μg/ml bilirubin: No light	30*	0*
- " - 20 min light	22	8
- " - 30 min light	46	8
- " - 45 min light	52	14

* Uncertain data due to low cell survival.

been observed in solutions containing calf serum, but not in the presence of human serum or human serum albumin (data not shown). When the cells were present during irradiation, almost similar dose response curves were obtained (data not shown).

A genotoxic response is demonstrated in Fig. 2 and Table I. Strand breaks in the DNA were induced at 4°C; at 20°C there was less damage (data not shown). The mutation frequency was elevated when the cells were treated in the presence of calf serum (Table I). On the other hand no increase in the frequency of sister chromatid exchanges could be found (data not shown).

DISCUSSION

Bilirubin solutions were toxic to the hamster cells under the conditions used in the present study (Fig. 1). The toxicity decreased upon irradiation of the solutions with a low light dose. It may be assumed that this effect is due to the conversion of bilirubin to more water soluble and less cytotoxic photoproducts, e.g. photoisomers. After irradiation with higher doses of light, photoproducts with a higher toxicity were formed.

Previously an increased cytotoxicity after high doses of light has been indicated in solutions containing newborn calf serum and in the absence of serum (Christensen, 1986; Rosenstein et al., 1983). In the latter study only effects on the DNA was measured. Thus, the light doses may have been supralethal. The previous studies indicate that the toxic products may be peroxides. In the present study the toxic effect in calf serum has been reproduced, while bilirubin solutions made up in human serum have been found not to exert the same light induced toxic effect on the cells.

The fact that more DNA damage is measured in cells treated at 4°C than at 20°C is most probably explained by DNA strand break repair at room temperature. With respect to mutation induction, Table I shows somewhat variable results. However, there is a consistent increase in the mutation frequency in both experiments.

Further studies are needed to decide whether the toxic and genotoxic effects observed have some relevance to the treatment of human patients.

CONCLUSIONS

This study has shown that phototherapy has genotoxic effects under certain _in vitro_ conditions. It is possible that similar effects may occur _in vivo_ during clinical phototherapy of newborns. This is an argument for further experimental studies. Measures should be taken to administer phototherapy only when indicated and under conditions optimised with regard to physical and biological parameters. The present study calls for a careful registration of the use of photo-therapy to facilitate future follow-up studies.

REFERENCES

Christensen, T., 1986, Cytotoxicity of bilirubin photoproducts. _Photobiochem., Photobiophys._, 10: 253.

Cremer, R.J., Perryman, P.W. & Richards, D.H., 1958, Influence of light on the hyperbilirubinemia of infants. _Lancet_, 24 may: 1094.

Ennever, J.F., 1986, Phototherapy in a new light. _Pediatr. Clin. North Am._, 33: 603.

Holme, J.A., Hongslo, J.K., Søderlund, E., Brunborg, G., Christensen, T., Alexander, J. & Dybing E., 1987, Comparative genotoxic effects of IQ and MeIQ in _Salmonella typhimurium_ and cultured mammalian cells, _Mutation Res._ 187: 181.

ICRP, 1977, Publication no 26. Recommendations of the International Commission on Radiological Protection. _Annals of the ICRP_, 1.

Jährig, D., Berck, A., Meisel, P & Jährig, K., 1985, Die Kalkulation des Phototherapieeffektes unter Berücksichtigung verschiedener Bestrahlungsvarianten, I. Der Einfluss des Therapiemodus auf den Photoeffect. _Kinderärztl. Praxis_, 53: 171.

McDonagh, A.F., 1971, The role of oxygen in bilirubin photo-oxidation. _Biochem. Biophys. Res. Comm._ 44: 1306.

McDonagh, A.F., Palma, L.A. & Lightner, D.A., 1980, Blue light and bilirubin excretion. _Science_, 208: 145.

Pratesi, R., 1983, Two lights for phototherapy. _Lancet_ October, 8: 859.

Rosenstein, B.S., Ducore, J.M. & Cummings, S.W., 1983, The mechanism of bilirubin-photosensitized DNA strand breakage in human cells exposed to phototherapy light. _Mutation Res._, 112: 397.

LONG-WAVELENGTH PHOTOTHERAPY

Gian Paolo Donzelli

Department of Pediatrics
University of Florence, Italy

INTRODUCTION

Long-wavelength phototherapy (PT) using narrow-spectrum fluorescent green lamps has been introduced in 1981[1,2]. The positive results of green lamp PT came unexpected: they were accepted with skepticism as they were in conflict with the current dogma on blue light mechanism of PT action in man[3,4]. Today, green lamp PT is a well documented procedure[5], and several "in vitro" and animal experiments have been reported showing the peculiar characteristics of green light photochemistry of bilirubin (BR)[6-10].

As it is well known, the therapeutical action of light is based on the modification induced by light on the target molecules of BR. The efficiency of the various spectral bands are expressed by the action spectrum of the process. In general, the action spectrum follows closely the absorption spectrum of the chromophore determined "in vitro". However, in complex systems as babies the effective absorption spectrum of the chromophore in the various binding configurations that may result are only partially known, and additional wavelength-dependent mechanisms may modify to some extent the "in vitro" action spectrum. A complete action spectrum of PT in babies is still lacking. Its determination would have required the use of unefficient UV and far green light, with obvious ethical problems; moreover, since PT has always worked nicely, no pressing need for the PT action spectrum was present in the past, and only a few coloured lamps were tested clinically. In addition, a not cautious transposal to man of the action spectrum of PT in Gunn rats established the superiority of spectral bands matching the absorption spectrum of BR "in vitro" for clinical PT.

To day we know, however, that although the basic photochemistry of BR occurs in Gunn rats as well as in babies, the metabolism of the high-quantum yield configurational isomers (ZE-BR, EZ-BR) is much more efficient in Gunn rats than in babies[11]. The knowledge of the action spectrum of PT of jaundiced newborns is a necessary step in the progress of PT. Only detailed information on spectral dependence of the various photoproducts of BR and the kinetics of their metabolism will make it possible to establish an optimal protocol for PT in man.

TOWARDS GREEN LIGHT PT

In 1981 we started a multidisciplinary work with the aim of improving the protocol and the phototechnology of PT[12,13]. A shift to longer wavelength light appeared desirable in view of the reduction of the potential side effects of PT expected with UV-violet-blue radiation[14].

A support for green lamp PT can be found in the old literature on PT. In the 1970 paper by Sisson and coworkers[15] we find that fluorescent green and white lamps have almost equal efficacy in modifying the optical density of a BR solution. Sisson and coworkers again in 1970[16] reported that also green lamps were able to reduce BR concentration in Gunn rats. However, the action of green lamps was attributed to the portion of the spectrum with wavelength shorter than 500 nm.

In a more detailed comparison of coloured lamps published by Ballowitz and coworkers in 1977[17] it was found that green lamps are only 23% less efficient than white lamps to produce the final reduction of serum BR in Gunn rats.

As already mentioned, similar investigations in babies were lacking. However, some indications on the relative inefficiency of short wavelength PT can be found in the literature.

Ebbesen in 1975[18] compared the BR decline produced by violet-blue lamps with the emission peak at about 420 nm with that produced by white lamps: the drop of BR was sensibly larger with white than with violet-blue lamps.

A more accurate indication of the low clinical efficiency of short-wavelength light can be found in the work published by Tan in 1977[19]. He was very surprised[20] to get only 10% of increment of BR decline per day when the 20 W white lamps were replaced by a similar set of more powerful 40 W violet-blue lamps (Philips TLAK40W/03) delivering to the infant skin almost two times the white lamp power. The spectrum of these lamps has a sufficiently good overlap with the BR/HSA absorption spectrum "in vitro". These lamps, the 03 type, were developed by Philips specifically for PT. They are still in use in some neonatal units: their employment should be discouraged due to their lower efficacy and higher risks of phototoxicity.

As discussed later, the reasons for the limited efficiency of the violet-blue lamps are: i) these lamps are expected to produce larger amount of the ZE-BR isomer, which is inefficiently excreted in babies; ii) light scattering by dermal tissue is much higher for violet than for blue or green light; iii) hemoglobin absorbs very strongly near 420 nm and may drastically reduce the intensity of the radiation in the skin.

GREEN LAMP OR GREEN LIGHT PT?

Recent clinical trials performed by our group showed the PT equivalence of narrow-spectrum green lamps and narrow-spectrum blue lamps matching the BR absorption "in vitro"[5].

In order to understand whether the efficacy of green lamp PT was due or not to the small fraction of blue light emitted by the lamp, we performed another experiment in which a yellow filter was inserted in the PT unit to cut-off radiation with wavelength shorter than 490 nm. As can be seen in Fig.1 the high intensity violet and blue Hg lines are suppressed, and only a very small fluorescent intensity is present below 500 nm.

A group of 23 jaundiced newborns, with a mean gestational age of 34 weeks, mean birthweight of 1980 grams, with age at onset of PT of 87 hours and an initial serum bilirubin concentration of 13.5 mg/dl, underwent green filtered lamps PT. A second group of 22 infants, with comparable characteristics, underwent PT with green lamps without filter. No statistically significant differences were found between the declines of serum BR level in the two groups[21].

This "in vivo" investigation with green filtered lamps shows that a satisfactory clinical effect can be achieved even in absence of the blue

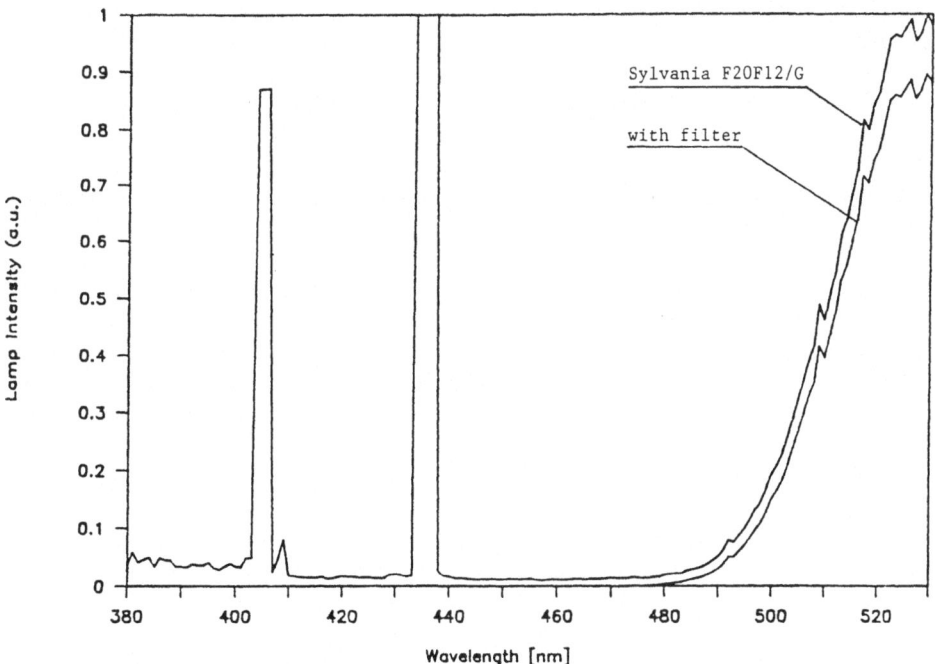

Fig. 1. Emission spectra of Sylvania F20F12/G lamp without and with cut-off filter.

component in the emission spectrum, and proves the intrinsic efficacy of green light PT.

POSSIBLE MECHANISM OF GREEN LIGHT PT

The peculiar behavior of BR photochemistry in the longwavelength part of its absorption spectrum has been recently pointed out by several groups[6,7,9,10,22,23]. At the present state of our knowledge three main effects may concurrently make long wavelength PT efficient:

a) the increase of the quantum yield of lumirubin in the green[10,22]

b) the modification of the "in vitro" absorption spectrum of BR when bound to skin constituents and to serum albumin "in vivo"[23-25]

c) the decrease of skin attenuation at longer wavelength[26].

With reference to item a) our group has performed a detailed investigation of BR photochemistry "in vitro". A relevant result is that the quantum yield of lumirubin formation depends on the excitation wavelength, and its value in the green is sensibly larger than in the blue[10,27], as independently reported by Ennever and cowoekers[22]. Moreover, recent measurements of the quantum yield $\phi_{ZZ/ZE}$ of the reaction leading to the configurational ZE isomer show a strong decrease of $\phi_{ZZ/ZE}$ from blue to green[28]. These two results together, if they occur "in vivo", may contribute to the observed efficacy of green light in decreasing serum BR in babies.

In regard to the "in vivo" absorption spectrum of BR Malhotra and coworkers[25] have shown that the extinction coefficient of BR in the long-wavelength side of the spectrum increases in the presence of fatty acids.

Finally, in regard to item c) it is known since long time that skin may influence the results of BR photochemistry established "in vitro"[29]. Due to the wavelength dependence of the light scattering processes and of the absorption coefficients of the various skin pigments a polychromatic light changes its spectrum as it penetrates into the skin.

Figure 2 shows the computed spectra of various colored lamps used currently for PT after transmission through a skin layer constituted by an epidermal layer 0.1 mm thick and a dermal layer of various thicknesses (0.4, 0.8, 1.2, 1.6 mm). The optical model of skin introduced by Parrish and coworkers[30] has been used for computation[31]. The spectra of the lamps have been measured by using a calibrated spectrometer equipped with an OMA-II detection system.

In Figure 2 the effect of increasing skin thickness on spectral shape and light intensity is shown for each of the four fluorescent lamps. The peaks of the fluorescent part of the spectra have been normalized to the same value so to have 100% transmission for zero-thickness of skin.

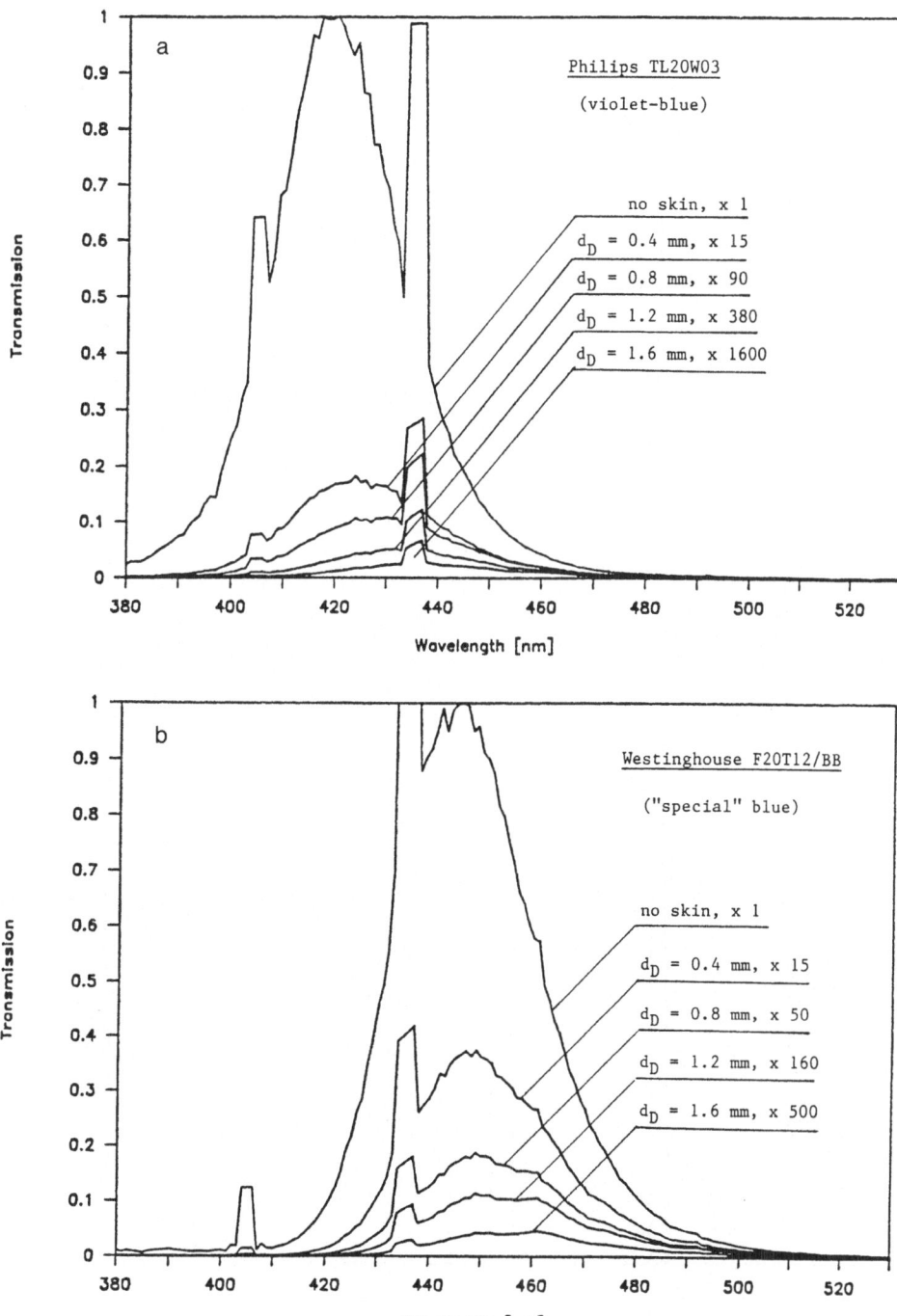

Fig 2a, 2b.   Transmission of epidermal (ED) and dermal (D) layers of
              different thickness: d_{ED} = 0.1 mm; d_D = 0.4, 0.8, 1.2, 1.6
              mm. The intensity of the fluorescent peak at the skin
              surface has been assumed equal to 1. The various amplifi-
              cation factors are shown in correspondence of the dermal
              layer thickness. a) Philips TL20W03T (violet-blue); b)
              Westinghouse F20T12/BB (special-blue).

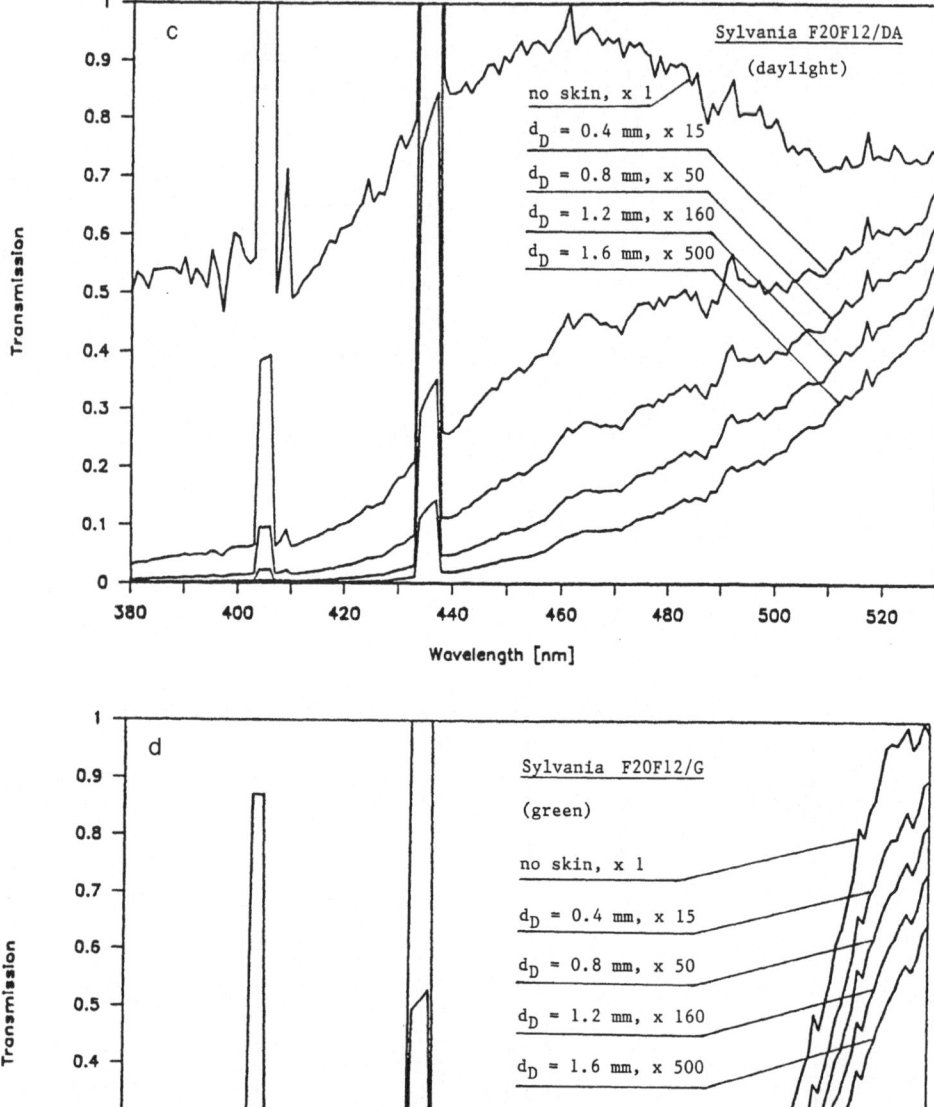

Fig 2c, 2d. Transmission of epidermal (ED) and dermal (D) layers of different thickness: $d_{ED} = 0.1$ mm; $d_D = 0.4, 0.8, 1.2, 1.6$ mm. The intensity of the fluorescent peak at the skin surface has been assumed equal to 1. The various amplification factors are shown in correspondence of the dermal layer thickness. c) Sylvania F20F12/DA (daylight); d) Sylvania F20F12/G (green).

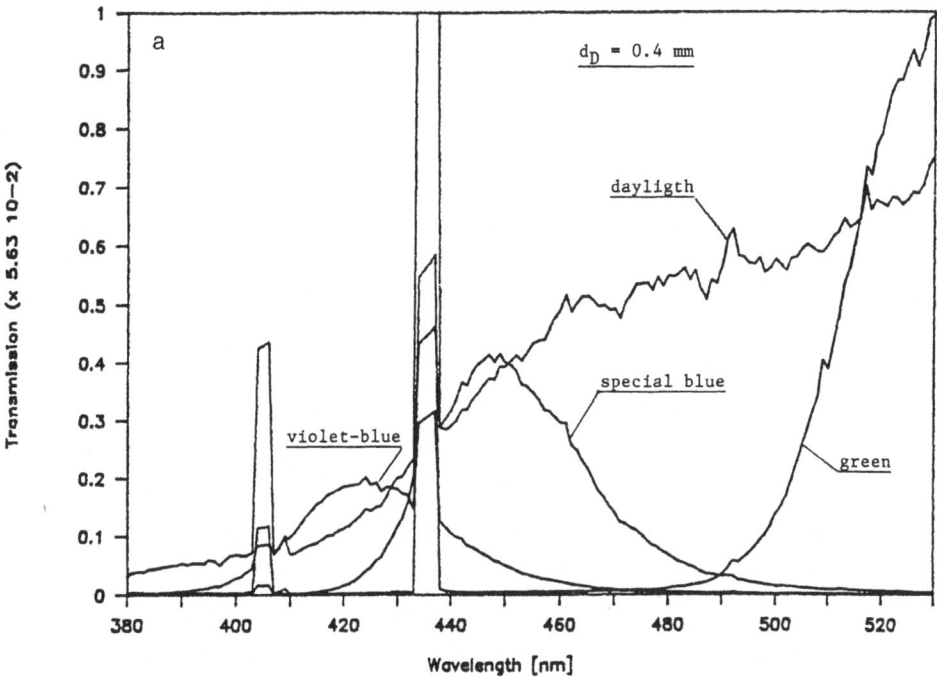

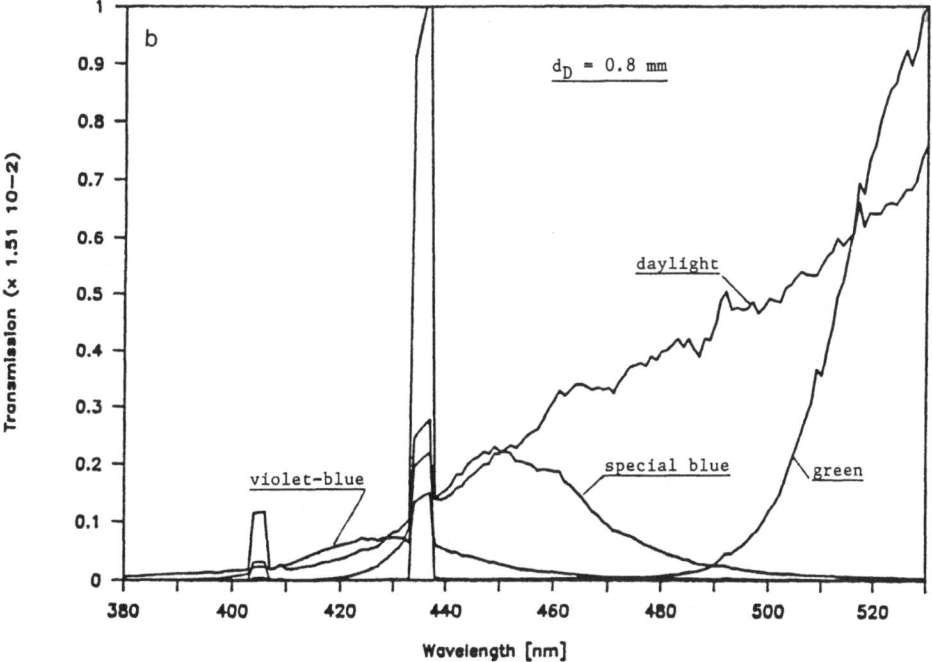

Fig 3a,3b.   Transmission spectra of PT coloured fluorescent lamps for
              different dermal layer thicknesses (d_{ED} = 0.1 mm). The
              intensity of the fluorescent peak of the lamps at the sur-
              face of the skin has been assumed equal to 1 (arbitrary
              units). a) d_D = 0.4; b) 0.8 mm.

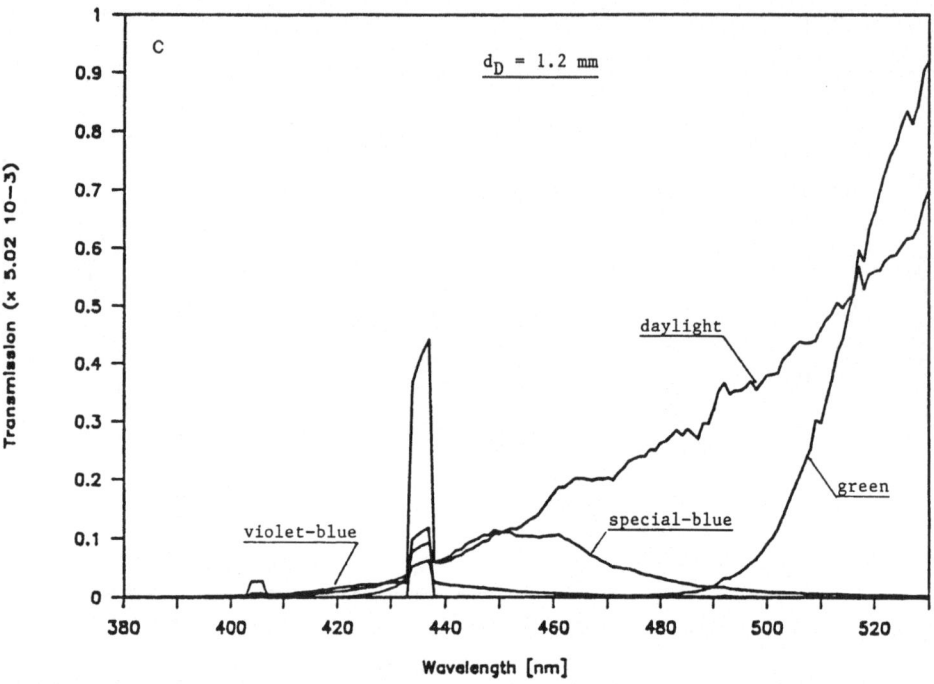

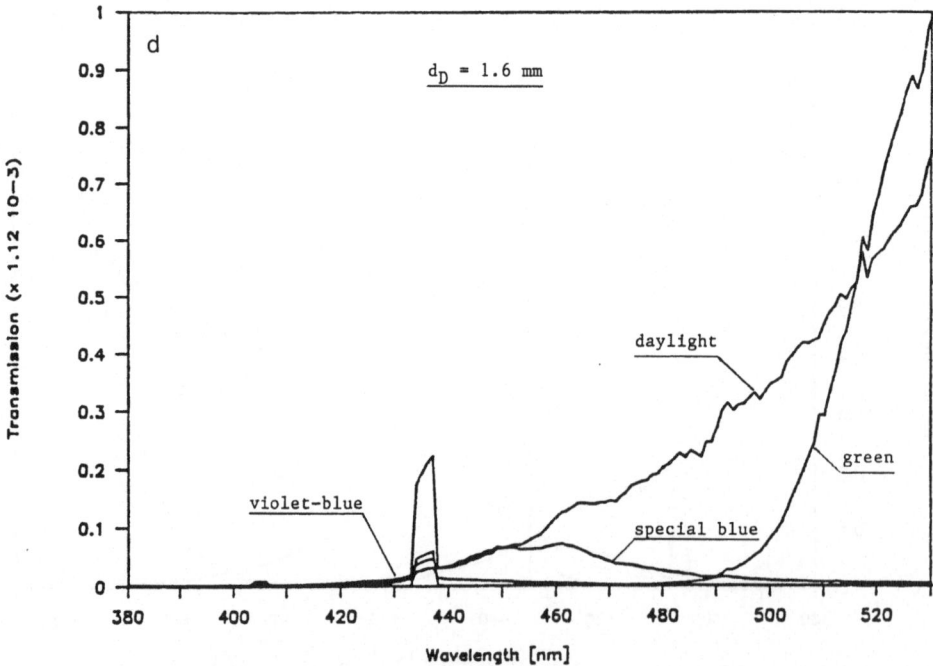

Fig 3c,3d.   Transmission spectra of PT coloured fluorescent lamps for different dermal layer thicknesses (d_{ED} = 0.1 mm). The intensity of the fluorescent peak of the lamps at the surface of the skin has been assumed equal to 1 (arbitrary units). c) d_D = 1.2; d) 1.6 mm.

In Fig.3 we compare the spectra of violet-blue, "special" blue, daylight, and green lamps for several values of the thickness of the dermal layer, d_D (the thickness of the epidermal layer is 0.1 mm).

As can be seen, the violet-blue lamp has the strongest attenuation, followed by the "special-blue" lamp. The attenuation is particularly strong in the short-wavelength side of the absorption spectrum of bilirubin, i.e. between 400 and 450 nm. This attenuation is expected to be still larger in the infant due to blood absorption, not considered in this computation. Daylight and green lamps have the highest intensity transmitted in the spectral band from 480 to 520 nm, and for wavelengths greater than 520 nm, respectively.

ACKNOWLEDGMENTS

The author thanks Dr.G.Agati and Dr.F.Fusi for computation of transmission spectra. This work was supported by the CNR Special Project "Medicina Preventiva e Riabilitativa", contract n.840256356.

REFERENCES

1. C. Vecchi, G. P. Donzelli, M. G. Migliorini, G. Sbrana, and R. Pratesi, New light for phototherapy, The Lancet ii:390 (1982).
2. C. Vecchi, G. P. Donzelli, M. G. Migliorini, and G. Sbrana. Green light phototherapy, Pediatr. Res. 17:461 (1983).
3. J. Ennever, A. F. Mc Donagh, and W. T. Speck, Phototherapy for neonatal jaundice: optimal wavelengths of light, J. Pediatr. 103:295 (1984).
4. L. Ballowitz, F. Hanefeld, and G. Wiese, The Gunn rat: a model for phototherapy, in: "Phototherapy for Neonatal Jaundice", F. Rubaltelli, and G. Jori, eds., Plenum Press, New York (1984).
5. C. Vecchi, G. P. Donzelli, G. Sbrana, and R. Pratesi, Phototherapy for neonatal jaundice: clinical equivalence of fluorescent green and "special" blue lamps, J. Pediatr. 108:452 (1986).
6. J. F. Ennever, M. Sobel, A. F. Mc Donagh, and W. T. Speck, Phototherapy for neonatal jaundice: "in vitro" comparison of light sources, Pediatr. Res. 18:667 (1984).
7. J. F. Ennever, Preferential formation of lumirubin by green light, Pediatr. Res. (abstract) 19:218A (1985).
8. J. F. Ennever, I. Knox, and T. W. Speck, Differences in bilirubin isomer composition in infants treated with green and white light phototherapy, J. Pediatr. 109:119 (1986).
9. A. F. Mc Donagh, and L. A. Palma, Bilirubin photoisomer excretion in Gunn rats: effects of green and blue light, Photochem. Photobiol. 45:96S (1987).
10. G. Agati, A. F. Mc Donagh, R. Pratesi, and F. Fusi, Bilirubin photoisomerization: wavelength effects, Photochem. Photobiol. 45:52S (1987).

11. J. F. Ennever, J. Knox, and S. C. Denne, Phototherapy for neonatal jaundice: "in vivo" clearance of bilirubin photoproducts, Pediatr. Res. 19:205 (1985).

12. G. Sbrana, M. G. Migliorini, C. Vecchi, and G. P. Donzelli, Laser photolysis of bilirubin, Pediatr. Res. 15:1517 (1981).

13. G. P. Donzelli, M. G. Migliorini, R. Pratesi, G. Sbrana, and C. Vecchi, Laser-oriented search of the optimum light for phototherapy, in: "Phototherapy for Neonatal Jaundice", F. Rubaltelli, and G. Jori, eds., Plenum Press, New York (1984).

14. B. S. Rosenstein, and J. M. Ducore, Introduction of DNA strand breaks in normal human fibroblasts exposed to monochromatic UV and visible wavelengths in the 240-546 nm range, Photochem. Photobiol. 38:51 (1983).

15. T. R. C. Sisson, N. Kendall, R. E. Davies, and D. Berger, Factors influencing the effectiveness of phototherapy in neonatal hyperbilirubinemia, Birth Defects, Original Article Series 6:100 (1970).

16. S. M. Goldberg, S. Kendall, and T. R. C. Sisson, Photodecomposition of bilirubin "in vivo", Clinical Res. 18:692 (1970).

17. L. Ballowitz, G. Gentler, J. Krochmann, R. Pannitschka, G. Roemer, and I. Roemer, Phototherapy in Gunn rats, Biol. Neonate 31:229 (1977).

18. F. Ebbesen, Phototherapy with "daylight" and blue light, Danish Med. Bull. 22:207 (1975).

19. K. L. Tan, The nature of the dose-response relationship of phototherapy for neonatal bilirubinemia, J. Pediatr. 90:448 (1977).

20. K. L. Tan, personal communication (1983).

21. G. Sbrana, G. P. Donzelli, and C. Vecchi, Efficacy of phototherapy in the management of neonatal hyperbilirubinemia with light sources emitting above 500 nm, Pediatrics 80:395 (1987).

22. J. W. Greenberg, J. F. Ennever, and V. Malhotra, Wavelength dependence of the quantum yield for the structural isomerization of bilirubin, Photochem. Photobiol. 46:453 (1987).

23. J. F. Ennever, Clinical and "in vitro" photochemistry of bilirubin, in this volume.

24. C. L. Kapoor, C. R. K. Murti, and P, C. Bajpaj, Uptake and release of bilirubin by skin, Biochem. J. 136:35 (1973).

25. V. Malhotra, J. W. Greenberg, L. L. Dum, and J. F. Ennever, Fatty acid enhancement of the quantum yield for the formation of lumirubin from bilirubin bound to human albumin, Pediatr. Res. 21:530 (1987).

26. R. R. Anderson, and J. A. Parrish, The optics of human skin, J. Invest. Dermatol. 77:13 (1981).

27. G. Agati, F. Fusi, P. Galvan, A. F. M. Donagh, and R. Pratesi, Quantum yields for laser photocyclization of bilirubin in the presence of human serum albumin. Dependence of quantum yield on excitation wavelength (submitted for publication).

28. G. Agati, F. Fusi, R. Pratesi, A. F. Mc Donagh (unpublished data).

29. J. D. Hardy, H. T. Hammell, and D. Murgatroyd, Spectral transmittance and reflectance of excised human skin, J. Appl. Physiol. 9:257 (1956).

30. S. Wan, R. R. Anderson, J. A. Parrish, Analytical modeling for the optical properties of the skin with "in vitro" and "in vivo" applications, Photochem. Photobiol. 34:493 (1981).

31. R. Pratesi, L. Ronchi, G. Cecchi, G. Sbrana, M. G. Migliorini, C. Vecchi, and G. P. Donzelli, Skin optics and phototherapy, Photochem. Photobiol. 40:77 (1984).

ULTRAVIOLET RADIATION PHOTOTHERAPY FOR PSORIASIS

THE USE OF A NEW NARROW BAND UVB FLUORESCENT LAMP

Brian E. Johnson, Cathy Green, Thiruvellor Lakshmipathi
and James Ferguson

Department of Dermatology
University of Dundee
Dundee, Scotland, U.K.

INTRODUCTION

Phototherapy with ultraviolet (UV) radiation, alone, or as part of a
multi-faceted therapy, has been used for Psoriasis with varying degrees of
success depending mainly on the radiation source, the treatment regimen used
and the type of Psoriasis treated (Goeckermann, 1931; Ingram, 1953; Petrozzi
et al, 1978; Levine and Parrish, 1980; Larkö, 1982; Morison, 1983; Eells et
al, 1984).

Where UV radiation alone has been used, studies using artificial UV-
sources have shown that the UVC, wavelengths below around 280nm, are relatively ineffective, producing burning rather than any therapeutic effect
(van Weelden et al, 1980; Parrish, 1982). The UVA, wavelengths longer than
320nm, requires impractically long exposures to be effective (Young and van
der Leun, 1975; Parrish, 1977) and it is clear that the major therapeutic
wavelengths are in the UVB, wavelengths between 280 and 315nm (van Weelden
et al, 1980; Parrish, 1982). Even within this wavelength range, there is
some indication that the longer wavelengths are more effective (Tronnier
and Heidbuchel, 1976; Pullman et al, 1978; van der Leun and van Weelden,
1984; Diffey and Farr, 1987). Fischer et al (1984) combining the results
of a number of studies concluded that 313nm radiation produced clearing of
Psoriasis with a dose schedule of $\frac{1}{2}$ that required to produce erythema while
with wavelengths around 300nm, erythema doses were required. Action spectrum
studies have indicated that the peak therapeutic effect is around 295-310nm
but with a rapid fall off in activity below 295nm (fig. 1) as opposed to the
erythema action spectrum (Parrish and Jaenicke, 1981) and probably also that
for UV-carcinogenesis (Cole et al, 1986; Sterenborg and van der Leun, 1987).

The value of exposure to sunlight in any successful seaside therapy for
Psoriasis at the unlikely geographical locations of the original 19th century
treatment centres on the North East coast of Kent in England and the islands
off the north coast of Germany (Hartung, 1969) as well as those of the unique
Dead Sea resorts at present (Abels and Kattan-Byron, 1985) must surely also
reside in the longer wavelengths of the UVB.

Van Weelden and van der Leun (1984) concluded that it should be possible
to improve the effectiveness of phototherapy for Psoriasis by producing
controlled irradiation conditions in which the radiation was restricted to
these longer UVB wavelengths with the exclusion of the shorter wavelength
more erythemogenic radiatiom.

For whole body exposures, an irradiation cabinet equipped with fluor-
escenttube sources appears to be the most convenient and cost effective form

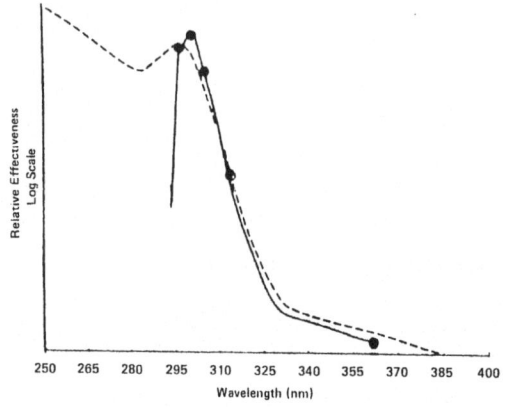

Fig. 1. Action spectra for erythema ------
and clearing of psoriasis ————

of equipment. A typical source of UVB is the Philps TL12 lamp with a broad
spectrum from around 280 - 350nm, peaking around 315nm (fig. 2). One method
for producing mainly longer wavelength UVB would be to use cabinets fitted
with these lamps and filters designed to cut off at about 300nm. However,
van der Leun and van Weelden (1986) were unable to find a suitable filter
which was stable on exposure to UV radiation. Nor were they able to confirm the superiority of SUP therapy in which mercury arc lamps with amplification of the longer wavelength UVB are used (Tronnier and Heidbuchel,
1976). They therefore adopted a new approach to the problem with the help
of Meule,anns of the Philips company and examined the effects of an experimental fluorescent lamp with an emission mainly restricted to a narrow
band around 311nm with a $\frac{1}{2}$ band-width value of 3nm (fig. 2). At a meeting
of the Netherlands Dermatological Society in February, 1984, (Van Weelden
and van der Leun, 1984) they reported that, in 9 patients with wide spread,
symmetrically distributed Psoriasis, when half of the body was treated with
TL12 lamps while the other half was treated with the new '311' lamps using
similar, erythema producing dose schedules, better results were obtained
with the '311' lamps (p 0.05). Similarly, a comparison of whole body
treatments in a further 8 patients showed best results with the '311' lamps.
A comparative study of the carcinogenic effects of exposure to the emission
of these two lamps showed, not surprisingly, that skin cancers were produced
in mice by both. However, the '311' lamps were significantly (p 0.01) less
carcinogenic. These results were allowed to be introduced to a wider aud-
ience in a Plenary Lecture on "Phototreatments in Human Subjects" at the

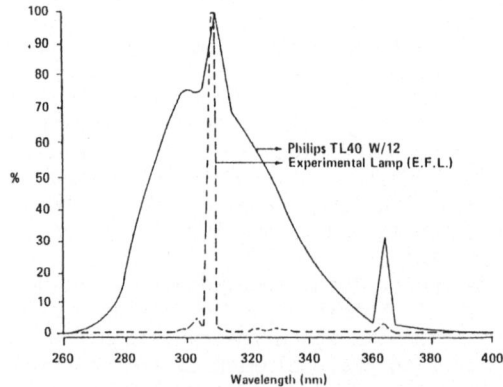

Fig. 2. Relative spectral emission from
TL12 and TL01 fluorescent lamps.

Ninth International Photobiology Congress in Philadelphia in July, 1984 (Johnson, 1985)

As indicated by van der Leun and van Weelden (1986) this preliminary study has been succeeded by further trials in different institutions, the regimen being determined by each institution according to the conditions prevailing there rather than following a specific pattern. We report here the results obtained in our own initial studies.

PATIENTS AND METHODS OF TREATMENT

Psoriasis patients

In all, 52 patients were treated. In an attempt to obtain some differentiation of the efficacy of the therapy, the patients were sub-divided according to the clinical presentation of their psoriasis (Table 1).

Table 1. Clinical types of Psoriasis patients treated.

Plaque	n = 17
Guttate	n = 20
Plaque/Guttate	n = 15

Phototherapy cabinet

The phototherapy cabinet, 2m high x 1m x1m, originally contained 32 40W fluorescent tubes of the '311' type, now designated as Philips TL40-01, arranged to provide whole body exposure with an average irradiance of 1.07 mW/cm^2, measured with a calibrated UVB meter (Jagger, 1967). To reduce exposure times, the irradiance was increased, first to 1.42mW/cm^2 by placing reflecting foil behind the lamps, and then to 2.09mW/cm^2 by increasing the number of lamps to 49. From November 1986 - May 1987, over a period of use of 318 hours, the output of the lamps has fallen by approximately 15% to give an irradiance of 1.77mW/cm^2.

Phototherapy regimen

Exposures were given three times per week, a schedule which appears to produce best therapeutic results with maximal patient compliance. The regimen used (Table 2) is typical of many described in the literature.

Table 2. Regimen for phototherapy with TL01 lamps.

1. Determination of minimal erythema dose (MED)
2. First treatment   70% of MED
3. Second treatment: if no erythema   increase by 40%
   if slight erythema   increase by 20%
   if more than slight erythema   same dose
   mild burn   decrease by 50%
   severe burn   omit dose
4. Aim for slight erythema response with each subsequent exposure.

Three exposures per week.

The determination of the MED was carried out using an independent array of the lamps according to the exposure schedule in Table 3.

Table 3. Exposure chedule for MED determination with TL01 lamps.

Distance 20cm   Irradiance 2.46mW/cm^2

Exposure dose

mJ/cm^2	minutes' seconds"
50	20"
100	41"
250	1' 42"
500	3' 24"
750	5' 06"
1000	6' 48"
1250	8' 30"
1500	10' 12"

Skin type II subjects typically had MEDs of 500mJ/cm^2 (Table 4).

Table 4. MEDs obtained with the TL01 lamps prior to phototherapy.

Skin type	Mean MED (mJ/cm^2)	Range	Number
I/II	480	50 - 1250	38
III/IV	750	250 - 1500	14
Total	536	50 - 1500	52

An assessment of improvement or otherwise, and of side effects such as burning, was made weekly, a satisfactory response being defined as the clearing of lesions with or without minimal residual activity. As a rule, the duration of treatment and total exposure dose required to achieve a satisfactory response were recorded and the treatment was then continued to provide for maximal remission according to clinical judgement.

RESULTS

A satisfactory response was obtained in 92% of the patients with plaque Psoriasis while in those classified as gutatte and plaque/guttate, the rate was 100%.

The duration of treatment, range 3 - 12 weeks, and total exposure dose, range 4 - 42J/cm^2, were very similar for the different types (Table 5).

Table 5. Duration and exposure dose for different Psoriasis types.

Type of Psoriasis	Mean duration of therapy (weeks)	Mean total exposure dose (J/cm^2)
Plaque	6.7	18
Guttate	6.6	18
Plaque/Guttate	6.6	19

We are able to compare these results with those from a previous trial (Kenicer et al, 1981) in which the broad spectrum UVB lamps were used (Table 6). Overall, with 52 subjects treated with TL01 lamps, a satisfactory response was obtained in 96% while in a group of 25, treated with conventional broad band UVB, a satisfactory response was obtained in 92%. For both the first achievement of a satisfactory response and the total course of treatment, the duration of treatment with the TL01 lamps was significantly less than that for the TL12 broad band lamps although the total exposure dose in

J/cm^2 was similar in each case. A typical skin type II MED for the TL12 lamps is 100mJ/cm^2 so that with equal exposure doses of the two lamps, the effects of the TL01 are obviously less damaging.

Table 6. Comparative results with TL01 and conventional TL12 phototherapy.

	Satisfactory Result		Total Treatment Course	
	Mean duration weeks (range)	Mean Exposure J/cm^2 (range)	Mean duration weeks (range)	Mean Exposure J/cm^2 (range)
TL01	4.3 (2 - 9)	10 (3 - 26)	6.6 (3 - 12)	18 (4 - 42)
TL12	10.6 (3 - 26)	9 (0.7 - 16)	22 (8 - 43)	19 (0.5 - 41)

However, in our use of the TL12 lamps, with the aggressive treatment shedule used, it was found that patient compliance was lost in too many cases because of burning if exposures were given three times per week. Therefore, for successful therapy, exposures were given only 2 times per week. Even with this schedule, there was a 28% incidence of burning episodes. With the TL01 lamps, the problem of burning did not occur to such an extent (10%) so that the intended treatment regimen was possible. It is clear that the comparison here is one of what is clinically possible rather than of a carefully designed comparative trial.

We have used oral Psoralen and UVA (PUVA) treatment in large numbers of psoriasis patients, especially those which are more difficult to treat with other, more conventional therapies. When a comparison between conventional UVB, TL01 and PUVA therapies is made (Table 7) the inclusion of these difficult subjects in the PUVA group is obvious. However, the data are included so that a reasonable comparison of remission time following the different treatments might also be illustrated (fig. 3).

Table 7. Comparison of TL12, TL01 and PUVA therapy.

Form of Therapy	Number of patients	Satisfactory Response %	Number of Treatments	Total Exposure dose (J/cm^2)
TL01	52	96	19	18
TL12	25	92	40	19
PUVA	329	81	25	127

With our standard PUVA therapy, more than 50% of the subjects who achieved a satisfactory response remained clear for more than 12 months. The results with conventional UVB therapy are disappointing in this respect, the majority requiring further treatment within 10 months. The group treated with '311' UVB however were much closer to the PUVA subjects, nearly 40% not requiring treatment for 12 months at least.

In conclusion, in our population of Psoriatics, UVB phototherapy with the new, narrow band '311' lamp has proved as effective as phototherapy with conventional broad band lamps and provides faster clearance, less burning and a longer period of remission. To pursue an aggressive therapeutic regimen, the exposure doses required with the new lamps are some 4 - 8 times those with the broad band lamps and this is a potential disadvantage, especially when more obsessional patient compliance allows more frequent, less aggressive but, apparently equally successful treatments with the conventional UVB sources (Halprin et al, 1982; Larkö and Swanbeck, 1982; Eells et al, 1984). However, the exposure time required for the treatment doses may be reduced by increasing the irradiance at the skin surface and this, up to a certain point, may be done by increasing the numbers of lamps in the cabinet. Moreover, further controlled trials may well show that satisfactory results might be obtained with less aggressive therapy using the narrow band lamps if the results reported by Fischer et al

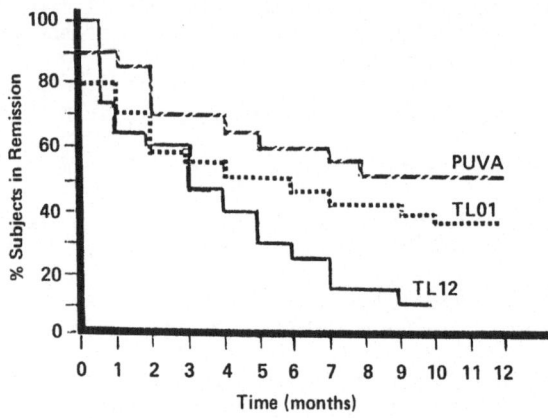

Fig. 3. Periods of remission after treatment
with TL12 and TL01 phototherapy and PUVA.

(1984) can be duplicated and the exposure dose schedule for clearing with
this wavelength region is indeed only ½ the MED rather than close to the
MED required for the shorter wavelengths in the broad band lamps. In addition,it seems that simultaneous or subsequent exposure to the shorter
wavelength radiation not only lessens the therapeutic effect of the longer
wavelength UVB but may even obviate it completely (Parrish, 1982) and the
use of the new lamps with the narrow band around 311nm alone avoids these
problems.

  We have confirmed that phototherapy with the new narrow band lamps
has clinical advantages over that with the broad band UVB lamps. The
experimental finding that the use of these lamps has less risk in terms of
UV-carcinogenesis is also encouraging. For routine treatment of the less
severe forms of Psoriasis, this new lamp treatment appears to be particularly useful. Nonetheless, at present, PUVA remains the mainstay of therapy
for chronic Psoriasis in our Department.

REFERENCES

Abels, D.J. and Kattan-Byron, J., 1985, Psoriasis treatment at the Dead Sea:
    a natural selective ultraviolet phototherapy. J. Am. Acad. Dermatol.,
    12: 639.
Diffey, B.L. and Farr, P.M., 1987, An appraisal of ultraviolet lamps used
    for phototherapy of psoriasis. Brit. J. Dermatol., 117:49.
Eells, L.D., Wolff, J.M., Garloff, J. and Eaglstein, W.H., 1984, Comparison
    of suberythemogenic and maximally aggressive ultraviolet B therapy for
    psoriasis. J. Am. Acad. Dermatol., 11: 105.
Fischer, T., Alsins, J. and Berne, B., 1984, Ultraviolet-action spectrum
    and evaluation of ultraviolet lamps for psoriasis healing. Int. J.
    Dermatol., 23: 633.
Cole, C.A., Forbes, P.D. and Davies, R.E., 1986, An action spectrum for
    photocarcinogenesis. Photochem. Photobiol., 43: 275.
Goeckermann, W.H., 1931, Treatment of psoriasis. Continued observations of
    the use of coal tar and ultraviolet light. Arch. Dermatol., 24: 446.
Halprin, K.M., Comerford, M. and Taylor, J.R., 1982, Constant low-dose
    ultraviolet light therapy for psoriasis. J. Am. Acad. Dermatol., 7:
    614.
Hartung, J., 1969, Ultraviolet therapy at the North Sea Coast. In "The
    Biologic Effects of Ultraviolet Radiation (with emphasis on the skin)."
    F. Urbach, ed., Pergamon Press, p. 657.
Ingram, J.T., 1953, The approach to psoriasis. Brit. Med. J., ii: 592.
Johnson, B.E., 1985, Phototreatments of human subjects. Am. Soc. Photobiol≠

Kenicer, K.J.A., Lakshmipathi, T., Addo, H.A., Johnson, B.E. and Frain-Bell, W., 1981, An assessment of the effect of photochemotherapy (PUVA) and UV-B phototherapy in the treatment of psoriasis. Brit. J. Dermatol., 105: 629.

Larkö, O., 1982, Phototherapy of psoriasis - clinical aspects and risk evaluation. Acta Derm. Venereol., Suppl. 103.

Larkö, O. and Swanbeck, G., 1982, Psoriasis treatment and day-care centre: clinical aspects and an attempt at a cost-benefit analysis. Acta Derm. Venereol., 62: 413.

LeVine, J. and Parrish, J.A., 1980, Outpatient phototherapy of psoriasis. Arch. Dermatol., 116: 552.

Morison, W.L., 1983, "Phototherapy and Photochemotherapy of Skin Diseases." Praeger.

Parrish, J.A., 1977, The treatment of psoriasis with longwave ultraviolet light (UV-A). Arch Dermatol., 113: 1525.

Parrish, J.A., 1982, Phototherapy of psoriasis and other skin diseases. In, "The Science of Photomedicine." J.D. Regan and J.A. Parrish, eds., Plenum Press, New York and London, p. 511.

Parrish, J.A. and Jaenicke, K.F., 1981, Action spectrum for phototherapy of psoriasis. J. invest. Dermatol., 76: 359.

Petrozzi, J.W., Barton, J.O., Kaidbey, K.K. and Kligman, A.M., 1978, Updating the Goekermann regimen for psoriasis. Brit. J. Dermatol., 98: 437.

Pullman, H., Wichmann, A.C. and Steigleder, G.K., 1978, Praktische Erfahrungen mit verschiedenen Phototherapieformen der Psoriasis - PUVA, SUP -, Teer - UV - Therapie. Z. Hautkr., 53: 641.

Sterenborg, H.J.C.M. and van der Leun, J.C., 1987, Action spectra for tumorigenesis by ultraviolet radiation. In "Human Exposure to Ultraviolet Radiation; Risks and Regulations." W.F. Passchier and B.F.M. Bosnjakovic, eds., Excerpta Medica, Int. Congr. Ser. 744, p 173.

Tronnier, H. and Heidbüchel, H., 1976, Zur Therapie der Psoriasis vulgaris mit ultravioletten Strahlen. Z. Hautkr., 51: 405.

van der Leun, J.C. and van Weelden, H., 1986, UVB phototherapy; principles, radiation sources, regimens. Curr. Probl. Derm., 15: 39.

van Weelden, H., Young, E. and van der Leun, J.C., 1980, Therapy of psoriasis: comparison of photochemotherapy and several variants of phototherapy. Brit. J. Dermatol., 103:1.

van Weelden, H. and van der Leun, J.C., 1984, Improving the effectiveness of phototherapy for psoriasis. Brit. J. Dermatol., 111: 484. (Abs.)

Young, E. and van der Leun, J.C., 1975, Treatment of psoriasis with longwave ultraviolet light. Dermatologica, 150: 352

---

*Video.

HOME UV TREATMENT OF PSORIASIS - AN UPDATE

Olle Larkö

Department of Dermatology
Sahlgren's Hospital
Göteborg, Sweden

INTRODUCTION

In 1977 we started a trial in Sweden where we used home UV treatment of psoriasis. The introduction of part-time day care centers has increased the number of patients that can get this treatment and the cost of treatment has been considerably reduced compared to hospital treatment. An even more economic way of giving treatment is to do this in the patient's home after education of the patient. An important factor is also the safety connected with the therapy and this has also been investigated.

In 1979 we (Larkö and Swanbeck) reported the results of treating 28 patients with plaque psoriasis with a home UVB unit. The dose was increased until a slight erythema developed.

As topical treatment only emollients were allowed. After clearing maintenance treatment was commenced. The results revealed that 20/28 cleared. Only 1/28 got worse. The median time to clearing was 6.5 weeks. Nineteen patients were put on a maintenance program and of these 11 relapsed on treatment twice a week. In another study Jordan et al. 1981 described their findings among 56 patients with more than 15% body involvement of plaque type psoriasis. They used a modified Goeckerman treatment with daily treatments and after clearing, maintenance treatment was given 1-3 times/week. 55/66 completed. All 55 cleared. Among patients on treatment for more than a year 80% were doing well.

In a recent study (Swanbeck, unpublished data, 1987) a follow up was made. A questionnaire was sent to 500 patients. 8% did not use the lamp at all. The level of satisfaction with this mode of treatment was high. The median number of treatments per year were well within the acceptable limits.

Gothenburg guidelines

In Gothenburg we also use day care centers for training and education of patients. Patients can come in the evening and learn how to treat themselves. Home UV equipment is then borrowed by the patient or the patients buy their own equipment. The indications or home UV treatment in an urban area is patients with special kinds of jobs or difficulties in using day care center facilities. In sparsely populated areas home UV equipment could be the only way of giving UV treatments. Patients are recommended to register all treatments. In Gothenburg we let the patients borrow the unit for 2 months and control the patients every 2 months. General maintenance treatments are not given but if there is a rapid recurrence this could be the case. We often combine the treatment with other modalities such as dithranol or retinoids.

The problems connected with home UV treatment is often patient cooperation. It is important to select the right type of patient. Another problem is that other persons occasionally use the equipment for tanning purposes. There has been some instances of burning as unexperienced family members have used the UVB solaria as UVA solaria and consequently got a burn. Bad desquamation is also often seen.

Cost of treatment

The annual cost of treatment for a model patient with psoriasis for Swedish conditions is 13000 US dollars in a hospital ward and 650 US dollars when treated in a day care center. If the patient is treated in his home with a home UVB equipment the annual cost is less than 80 US dollars.

In conclusion home UV therapy has a good clinical effect at a low cost. However, it is important to educate the patients and select the patient suitable for this type of therapy. If used correctly this is an effective and safe therapy.

References

Jordan WP, Clarke AM, Hale RK, 1981. Long-term modified Goeckerman regimen for psoriasis using an ultraviolet B light source in the home. J Am Acad Dermatol 4: 584.

Larkö O, Swanbeck G, 1979. Home solarium treatment of psoriasis. Br J Dermatol 101: 13.

Larkö O, Swanbeck G, 1982. Psoriasis treatment at a day care center: Clinical aspects and an attempt at a cost-benefit analysis. Acta Derm Venereol (Stockh) 62: 413.

5-METHOXYPSORALEN AND OTHER NEW

FUROCOUMARINS IN THE TREATMENT OF PSORIASIS

A. Tanew, B. Ortel, and H. Hönigsmann

Division of Photobiology, Department of
Dermatology I, University of Vienna
Vienna, Austria

INTRODUCTION

Two psoralens are presently established as standard photosensitizers in photochemotherapy of psoriasis. These are 8-methoxypsoralen (8-MOP), administered mainly orally for systemic treatment and trimethylpsoralen (TMP), which is used topically in the form of PUVA-baths as whole body treatment.

Both modalities are very effective in the management of psoriasis. However, there are restrictions to their use. Not infrequently, 8-MOP is only poorly tolerated by the patients because of nausea and vomiting. The practicability and acceptance of the TMP-bath treatment, on the other hand, is hampered in many centres by the logistics of the bathing procedure. With both treatment forms dosimetry may become a problem, in particular, when dealing with light-sensitive patients, and overdosage reactions such as severe phototoxic erythema, blistering or severe pruritus may occur.

Besides the occurrence of acute side effects there exist always concerns about long-term risks, in particular, the risk of carcinogenicity. Consequently, in the recent past new furocoumarins have been developed and tested for their potential clinical use with the aim of expanding the therapeutical range of the photosensitizer, and of reducing either short-term side effects and/or long-term hazards.

This paper summarizes our preliminary clinical experience with three investigational furocoumarins, namely 5-methoxypsoralen (5-MOP), 7-methylpyridopsoralen (7-MPP), and 4,6,4'-trimethylangelicin (4,6,4',-TMA).

5-METHOXYPSORALEN

5-MOP is a naturally occurring linear psoralen. Already in 1979 our group performed a first pilot study in which the photosensitizing and phototherapeutical properties of a microcrystalline form of 5-MOP was investigated (1). Given either in a dose of 0.6 or 1.2 mg/kg body weight 5-MOP seemed to have a therapeutical efficacy comparable to that of 8-MOP. On the other hand, 5-MOP appeared to be barely erythematogenic and practically devoid of acute side effects.

In the recent past a new liquid preparation of 5-MOP has been developed, which we used to extend our earlier study and investigate the following issues: (1) the kinetics and the range of the serum levels; (2) the phototoxic (erythemogenic) properties of the new preparation; (3) the therapeutic efficacy in psoriasis as compared to 8-MOP and (4) the spectrum and incidence of short-term side effects.

169 patients were included in this study and assigned randomly to treatment with either 0.6 or 1.2 mg/kg 5-MOP or 0.6 mg/kg 8-MOP respectively. The distribution of skin types and types of psoriasis was almost identical in all treatment groups. In each group approximately one third of the patients had acute guttate or seborrheic type of psoriasis and two thirds presented with chronic plaque type psoriasis.

Both psoralens were used as a liquid preparation in soft gelatine capsules (Gerot Pharmazeutika, Vienna) and given orally one hour (8-MOP) or two hours (5-MOP) prior to irradiation. Minimal phototoxicity dose (MPD) testing and treatment were performed according to the standard European protocol (2). The MPD served as the initial UVA treatment dose. In patients with a negative MPD test the initial exposure dose was determined arbitrarily according to skin typing (3). Treatment consisted of four PUVA exposures per week until complete clearing of psoriasis.

Pharmacokinetic results

The evaluation of 17 24-h-serum profiles confirmed the previous clinical experience that treatment is most effective if the UVA irradiation is performed 2 hours after drug administration, because a definite peak of the mean serum concentration was found between 90 and 120 minutes after 5-MOP ingestion. Thereafter the serum level curves exhibited a steep decline, and 5-MOP was almost completely eliminated from the serum after 12 hours.
Under therapeutic conditions, i.e. at the time of irradiation, the mean serum level of patients with the low dose 5-MOP regimen was 81 ng/ml. This indicates an absorption rate of only 25 % as compared to that of liquid 8-MOP, which already peaks 1 hour after ingestion with a mean serum level of 305 ng/ml (4). In high dose 5-MOP treated patients a mean serum concentration of 164 ng/ml was determined 2 hours after intake.

Phototoxic properties

As expected from the serum level determinations, skin photosensitivity as assessed by the MPD reaction was clearly dependent on the 5-MOP dosage. A mean MPD of 6.4 J/cm² was determined in the low dose 5-MOP cohort as opposed to 3.5 J/cm² in the high dose group. The latter value came close to the mean MPD seen one hour after ingestion of 0.6 mg/kg 8-MOP which was 2.7 J/cm².

These findings indicate, that the absence of phototoxic reactions reported previously is not due to absent or lower phototoxic properties but presumably related to the lower serum levels of 5-MOP as compared to 8-MOP.

Therapeutic results

Treatment results were assessed as a function of total UVA dose, number of exposures, and time needed for inducing complete remission of psoriasis (table 1). For all three parameters there was a significant difference between the low dose 5-MOP group and the two other groups.
Although the high dose 5-MOP regimen also required significantly more exposures and a longer treatment period for clearing patients as compared to 8-MOP, the UVA requirements were not substantially different.
These data demonstrate, that 5-MOP, when given in the same dosage as 8-MOP, is not comparably effective. However, by doubling the dose of 5-MOP, the therapeutic results come close to those obtained with 8-MOP. In particular, there is no significant difference in the most important parameter, namely the cumulative UVA dose required for clearing.

Table 1 Treatment results

regimen	number of patients	total UVA dose (J/cm^2)	number of exposures	duration of treatment (days)
5-MOP (0.6mg/kg)	58 (3)[+]	132 ± 87	18 ± 6	36 ± 14
8-MOP (0.6mg/kg)	48	45 ± 32	12 ± 4	24 ± 9
5-MOP (1.2mg/kg)	63	53 ± 33	15 ± 7	31 ± 14

[+] Figure in brackets indicates number of treatment failures

Short term side effects

Severe erythema reaction, nausea, and pruritus were seen rather frequently in the 8-MOP treated patient cohort (table 2). None of these were associated with the low dose 5-MOP regimen.
In the high dose 5-MOP group two patients (3 %) had a severe phototoxic reaction, another two experienced pruritus. Nausea was not reported by any patient.

Table 2 Side effects

	Frequency of side effects (%)		
	8-MOP (0.6mg/kg)	5-MOP (0.6mg/kg)	5-MOP (1.2mg/kg)
	48 patients	58 patients	63 patients
Severe erythema reaction	19%	0	3%
Pruritus	17%	0	3%
Nausea/Vomiting	10%	0	0

CONCLUSION

With regard to the overall high tolerability and good therapeutic efficacy high dose 5-MOP can be considered as a valuable alternative drug for oral photochemotherapy of psoriasis. Due to its increased benefit-side effect ratio 5-MOP seems to be particularly advantageous for the treatment of light sensitive subjects and patients with 8-MOP intolerance.

7-METHYLPYRIDOPSORALEN

7-MPP is a recently synthesized monofunctional furocoumarin (5). In vitro studies have demonstrated that it has a high binding affinity for DNA and, upon irradiation with UVA, only forms monoadducts. 7-MPP is less mutagenic and 3 to 4 times less carcinogenic in the albino mouse than the bifunctional psoralens 8-MOP and 5-MOP (6).

In our first attempt to evaluate its therapeutic properties we used an 0.5 % alcoholic solution of 7-MPP and compared the antipsoriatic activity with that of 0.15 % 8-MOP in glyceroformal.
Topical treatment was carried out four times per week in 6 patients with chronic plaque type psoriasis. After three weeks 4 patients had cleared completely or improved markedly with 8-MOP while showing only a moderate to poor response to 7-MPP. In the other 2 patients both compounds induced only a moderate response of the psoriatic plaques.
Since these unsatisfactory results with 7-MPP were thought to be due to its poor solubility in alcohol, a second trial was carried out in 4 patients with a 10^{-2}M concentration of 7-MPP and 8-MOP in the lipophilic hydrocerine ROC^R excipient (kindly provided by Drs.Dubertret & Bisagni). UVA irradiations were given 20 minutes after topical application.
Both psoralens in this preparation induced complete remission in 3 patients and a marked improvement in the fourth patient after a mean of 15 exposures. To achieve this result, howe-

ver, a much higher mean cumulative UVA dose was needed with 7-MPP than with 8-MOP. In contrast to 8-MOP, 7-MPP did not cause phototoxic reactions even with high irradiation doses and exhibited only moderate melanogenic activity.

These preliminary data indicate, that 7-MPP in a lipophilic cream base is a highly efficient compound for the topical photochemotherapy of psoriasis. However, while lacking acute side effects, it seems to require considerable higher irradiation doses than 8-MOP.

4,6,4'-TRIMETHYLANGELICIN

4,6,4'-TMA is another monofunctional furocoumarin, which has been developed recently as a potential photochemotherapeutic agent for the treatment of psoriasis (7). It has been shown in vitro and in preliminary clinical studies, that 4,6,4'-TMA has a strong antiproliferative activity and apparently lacks phototoxicity (8).

To assess the therapeutic efficacy in psoriasis we followed a protocol similar to that used with 7-MPP. 12 patients with chronic plaque type psoriasis were treated with an 0.1 % ethanolic solution of 4,6,4'-TMA (kindly provided by Dr.Dall' Acqua) and an 0.15 % solution of 8-MOP in glyceroformal, respectively. Irradiation was performed 20 minutes after application of the psoralens four times weekly.
In 6 out of the 12 patients both compounds induced complete remission or marked improvement of psoriasis. 5 patients responded better to 8-MOP, whereas one patient cleared completely with 4,6,4'-TMA while exhibiting only moderate improvement in the 8-MOP-treated skin area.
Similar to the observation with 7-MPP, also with 4,6,4'-TMA the UVA irradiation dose requirements were substantially higher than those needed with 8-MOP. With regard to phototoxicity only 2 patients showed a slight erythema reaction to 4,6,4'-TMA treatment as opposed to 5 patients with a moderate or marked phototoxic response to the 8-MOP treatment. The melanogenic activity of both compounds was not substantially different.

In summary, a definite antipsoriatic activity could be demonstrated for 4,6,4'-TMA. However, the therapeutic efficacy is lower than that of 8-MOP and UVA dose requirements are higher.
Whether in analogy to 7-MPP, the phototherapeutic potential of this compound could be enhanced by modification of the vehicle remains to be clarified.

REFERENCES
1. Hönigsmann H, Jaschke E, Gschnait F, Brenner W, Fritsch P, Wolff K. 5-methoxypsoralen (Bergapten) in photochemotherapy of psoriasis. Brit J Dermatol 1979; 101:369-378.
2. Henseler T, Hönigsmann H, Wolff K, Christophers E. Oral 8-methoxypsoralen photochemotherapy of psoriasis. The European PUVA study: A cooperative study among 18 European centres. Lancet 1981; 1:853-857.
3. Current status of oral PUVA therapy for psoriasis. J Am Acad Dermatol 1979; 1:106-117.

4. Hönigsmann H, Jaschke E, Nitsche V, Brenner W, Rausch-
    meier W, Wolff K. Serum levels of 8-methoxypsoralen in
    two different drug preparations: correlation with
    photosensitivity and UV-A dose requirements for photo-
    chemotherapy. J Invest Dermatol 1982; 79:233-236.
5. Moron J, Chi Hung N, Bisagni E. Synthesis of 5H-Furo
    (3',2': 6,7)benzopyrano(3,4-c)pyridin-5-ones and 8H-
    Pyrano(3', 2':5,6)benzofuro(3,2-c)pyridin-8-ones(Pyrido-
    psoralens). J Chem Soc 1983; 225-229.
6. Dubertret L, Averbeck D, Bisagni E, et al. Photochemo-
    therapy using pyridopsoralens. Biochimie 1985; 67:417-
    422.
7. Guiotto A, Rodighiero P, Manzini P, et al. 6-methylange-
    licins: a new series of potential photochemotherapeutic
    agents for the treatment of psoriasis. J Medicin Chem
    1984; 27:959-967.
8. Cristofolini M, Guiotto A, Rodighiero P, et al. Synthe-
    sis of new 6-methylangelicins as potential agents for
    the photochemotherapy of psoriasis. Acta Derm Venereol
    (Stockh) 1984; Suppl. 113:170-172.

PHOTOTHERAPY OF PRURITUS

Erhard Hölzle, Renate von Kries
and Angela Höveler

Department of Dermatology
University of Dusseldorf
Dusseldorf

INTRODUCTION AND HISTORICAL REVIEW

Pruritus is a very frequent symptom accompanying skin diseases, but also occurs in connection with systemic disorders, e.g., uremia, cholestasis, diabetes, and myelo-proliferative disorders. Phototherapy was introduced first in uremia and cholestasis to relieve itching.

The idea to treat cholestatic pruritus with light originated from the observation that phototherapy decreased hyperbilirubinemia in infants.(1)

Knodell et al(2) applied this approach to patients with alcohol-induced cirrhosis. Although these authors were using visible light (400-800nm) and were focussing on parameters of liver function, they found that three out of eight patients experienced significant relief from pruritus. UVB phototherapy was carried out by Hanid and Levi(3) in patients with primary biliary cirrhosis. Five from six patients responded favourably. These investigators were led to their study by the observation that one of their patients' pruritus disappeared while she was sunbathing in Italy.

In the following there were anecdotal reports on the efficiacy of UVB (4,5) and UVA(6). Cerio et al(5) found bile acids to decrease first in skin blister fluid and subsequently in serum indicating that the cutaneous pool of bile acids is depleted by phototherapy.

Saltzer(7) was the first to provide experimental proof for the effect of UV phototherapy in uremic patients, but this study was uncontrolled. Gilchrest(8) and her group compared UVB phototherapy with UVA therapy, which was regarded as a placebo treatment. It was shown that UVA was less effective; two out of eight patients responded to UVA versus nine out of ten to UVB. When half-body-exposures were performed, one side receiving UVA, the other UVB, all patients noticed a generalized improvement and the authors suggested a systemic effect of UVB therapy.(9)

The usefulness of UVB therapy in uremic pruritus was confirmed by other investigators.(10,11) Hindson et al(12) found some relief in all of their nine patients treated with UVA.

In recent studies, however, the effect of phototherapy was questioned. Simpson and Davison(13) found merely a placebo effect of either UVA or UVB in a double-blind cross-over study in 12 patients. Taylor et al(14) compared UVA with blue light as a placebo treatment and observed 50% improvement in all patients. They attributed the effect to either a placebo response or a biological effect of blue light. Finally, blue light versus red light was tested by the same group and no effect of either treatment could be shown.(15) The authors concluded that the earlier observed UVA response must have been a placebo effect.

This conflicting data stimulated us to conduct some studies to investigate the efficacy of phototherapy on pruritus using UVB, UVA, and UVA plus UVB.

EXPERIMENTAL STUDIES

Patients and Methods

A total of 36 patients were enrolled in the study. They suffered from uremic pruritus (9), cholestatic pruritus (6), diabetic pruritus (2), pruritus sine materia (7), and pruritus associated with a topic dermatitis (10).

Patients were evaluated before treatment and weekly under phototherapy which was carried out for three to four weeks. Patients were carefully interviewed and asked to express intensity of pruritus on a six-point scale (no pruritus, slight, moderate, severe, very severe, incapacitating). Evaluations were made for each day of the previous week and each hour of the last 24 hours. The use of topical steroids and systemic antipruritic therapy, mainly anti-histamines, was also monitored. Before and after treatment laboratory data were obtained (IgE, RBC, WBC, ESR, SGOT, SGPT, creatinine) and skin tests performed. These included intracutaneous injection and prick test of histamine (0.01%) and codeine phosphate (0.06%) as well as allergens, e.g., pollen, dust, mite, if the patient was sensitized.

Phototherapy was carried out three to five times weekly using a convenient light cabin (8001K, Waldmann, Villingen-Schwenningen, FRG) equipped with fluorescent bulbs emitting UVA (Sylvania F85 100W/PUVA) (315-400nm) or mainly UVB (Sylvania F75 85W/UV6) (285-360 nm). Single doses of UVA were increased from 4J/cm² to 7, 10 and 15 J/cm² and then kept constant. UVB doses started at 0.05-0.1 J/cm². Then, doses were adjusted to approximately one MED (minimal erythema dose) according to skin type of patients. 16 patients were treated with UVA, 12 with UVB, and 9 with UVA plus UVB.

Results

Intensity of Pruritus. Patients treated either with UVA or UVB showed improvement of pruritus in 62.5% and 66.7% respectively. The remaining patients experienced no change or their condition worsened. All patients treated with UVA plus UVB reported at least some relief. In 14 patients pruritus was evalutated quantitatively as is shown in Fig. 1. Although there was a reduction of itching in the majority of patients, the response was only a gradual one.

When itching was monitored hourly, intensive pruritus was recorded in the morning and late at night. During daytime the patient was distracted by his duties and itching was perceived less disturbing. Fig. 2 depicts a typical course with gradual improvement of pruritus during phototherapy.

Topical and drug therapy. Evaluation of use of topical corticosteroids or systemic antihistamines revealed a decrease in consumption of

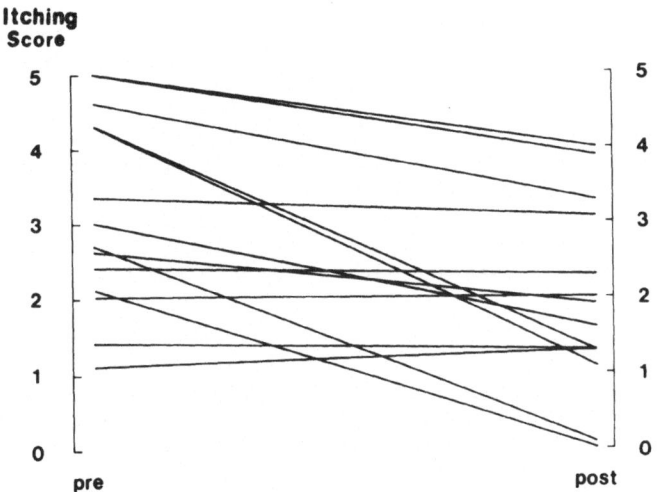

Fig.1　Itching score before and after phototherapy.
About two thirds of the patients responded to
treatment, however, only gradually.

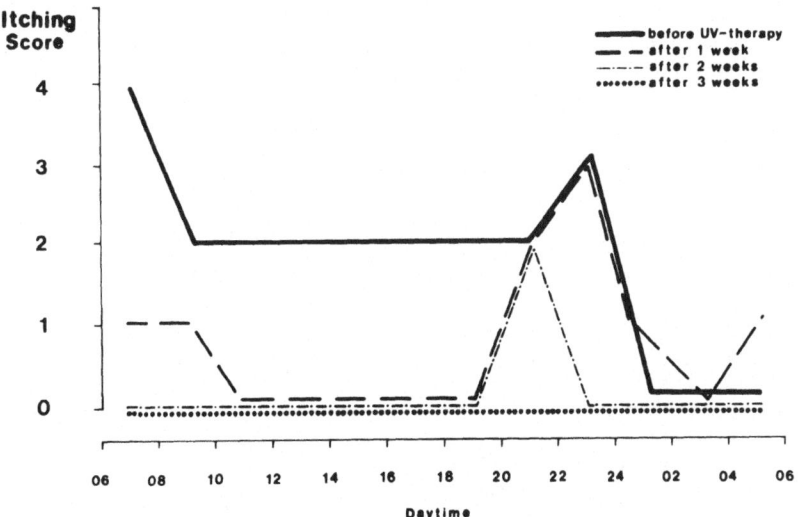

Fig.2　Pruritus recorded over 24 hours shows peak
intensities in the morning and late in the
evening.　It improves gradually and finally ceases
during phototherapy with UVA plus UVB

medicaments in ten out of eleven patients. This holds true, even if pruritus was not significantly relieved. This indicates that phototherapy can aid in the management of pruritus and exerts a therapeutic effect in addition to topical and systemic treatment.

Laboratory data. Except for IgE no change was observed. In atopic patients with significantly elevated IgE levels, IgE was reduced strikingly. Levels fell from an average of 650 kU/l before treatment to 420 kU/l after phototherapy. Only a slight decrease was found in patients with normal IgE (100kU/l) at the beginning of therapy.

Skin tests. Neither size of weal nor area of erythematous flare was found different before and after phototherapy. This applies to intracutaneous testing as well as to prick tests. No difference was seen between histamine, codein phosphate or allergens. This indicates that neither susceptibility of histamine receptors nor immunologic or non-immunologic mechanisms of histamine release were altered by phototherapy.

CONCLUSION

UV-phototherapy proved helpful in the management of pruritic disorders. Even though the therapeutic response is far from being complete in most patients, the treatment is to be regarded as a useful adjunctive therapy. It relieves pruritus to some degree and can save additional medication. In this regard our results are consistent with a critical review of the findings reported in the literature.

Since phototherapy is inexpensive, convenient, easily available and safe, it can be recommended as a routine treatment of pruritic disorders. The combination of UVA plus UVB seems superior. The mechanism of action, however, is still elusive.

A striking finding in our study is the decrease of serum IgE levels in atopic patients. There is, as yet, no explanation for this effect.

Histamine release by pharmacological and immunological mechanisms remains unaltered by phototherapy as well as response to intracutaneous injection of histamine itself. This indicates that histamine may not be the major mediator in the pruritic disorders investigated in the present study.

REFERENCES

1. R. J. Cremer, P. W. Perryman, and D. H. Richards, Influence of light on the hyperbilirubinaemia of infants. Lancet I:1094-1097 (1958)
2. R. G. Knodell, H. Cheney, and D. J.Ostrow, Effects of phototherapy on hepatic function in human alcoholic cirrhosis. Gastroenterology 70: 1112-1116 (1976)
3. M. A. Hanid, and A. J. Levi, Phototherapy for pruritus in primary biliary cirrhosis. Lancet II: 530 (1980)
4. S. M. Perlstein SM, Phototherpy for primary biliary cirrhosis. Arch Dermatol 117:608 (1981)
5. R. Cerio, G. M. Murphy, G. E. Sladen, and D. M. Mac Donald, A combination of phototherapy and cholestyramine for the relief of pruritus in primary biliary cirrhosis. Br J Dermatol 116: 265-267 (1987)
6. J. R. Person JR, Ultraviolet A (UV-A) and cholestatic pruritus. Arch Dermatol 117:684 (1981)
7. E. I. Saltzer, Relief from uremic pruritus: A therapeutic approach. Cutis 16:298-299 (1975)

8. B. A. Gilchrest, J. W. Rowe, R. S. Brown, T. I. Steinman and K. A. Arndt KA, Relief of uremic pruritus with ultraviolet phototherapy. N Engl J Med 8:136-138 (1977)

9. B. A. Gilchrest, J. W. Rowe, R. S. Brown, T. I. Steinman, 0and K. A. Arndt, Ultraviolet phototherapy of uremic pruritus. Long-term results and possible mechanism of action. Ann Intern Med 91:17-21 (1979)

10. B. C. Shultz, and H. H. Roenigk HH JR, Uremic pruritus treated with ultraviolet light. JAMA 243:1836-1837 (1980)

11. D. Lubach, and B. W. Kock, Behandlung des schweren Pruritus bei Dialyse-Patienten mit der selektiven UV-Phototherapie (SUP). Fortschr Med 101:2125-2120 (1983)

12. C. Hindson, A. Taylor, A. Martin, and A. Downey, UVA light for relief of uraemic pruritus. Lancet I:215 (1981)

13. N. B. Simpson, and A. M. Davison, Ultraviolet pruritus for uraemic pruritus. Lancet I:781 (1981)

14. R. Taylor, A. E. M. Taylor, B. l. Diffey, and T. C. Hindson, A placebo-controlled trial of UV-A phototherapy for the treatment of uraemic pruritus. Nephron 33:14-16 (1983)

15. J. G. Spiro, S. Scott, J. MacMillan, B. l. Diffey, T. C. Hindson, R. Taylor, E. A. M. Taylor, and A. Downey, Treatment of uremic pruritus with blue light. Photodermatol 2:319-321 (1985)

THE CURRENT STATUS OF LASER ANGIOPLASTY

Jonathan A Michaels, Frank W Cross, Timothy J Bowker, and
Stephen G Bown

National Medical Laser Centre, Dept. of Surgery
The Rayne Institute, University College London
5, University Street, London WC1E 6JJ

Laser angioplasty is the reopening of a blocked or narrowed vessel by
the intraluminal application of laser energy.

Narrowing or blockage of arteries by atherosclerosis is the commonest
cause of death in Europe and the United States. It is responsible for angina
and myocardial infarction, cerebro-vascular disease and peripheral vascular
disease and it is thus the cause of considerable morbidity as well as
mortality. Until recently symptoms due to an occluded or stenosed artery
could only be relieved by surgery to remove or bypass the occlusion.

BALLOON ANGIOPLASTY

In 1964 Dotter and Judkins suggested dilating a stenosed vessel from
within via a percutaneous approach. Since the description of a balloon for
such a dilatation by Grüntzig and Hopff (1974), 'percutaneous transluminal
balloon angioplasty' (PTBA) has gained rapidly in popularity. It is now a
commonly used method for the treatment of stenoses or short occlusions in
the coronary and peripheral circulations.

Briefly, the method of PTBA is to introduce a guidewire through an
arterial puncture, usually of the femoral artery. By the use of standard
radiological techniques a guidewire can be advanced into most areas of the
arterial tree. Once it is placed across the narrowed segment a balloon
catheter is advanced over the wire and inflated to dilate the vessel.

Unfortunately, balloon angioplasty is only applicable to a limited
number of lesions. The main problem is that a guidewire must be through the
lesion before it can be dilated. In many cases it is not possible to pass
the obstruction with a guidewire and even in those that can be passed it has
been shown that balloon dilatation of long occluded segments result in an
unacceptable level of early reocclusion (Murray et al., 1987).

LASER ANGIOPLASTY

The idea of laser angioplasty, or laser assisted balloon angioplasty is
very attractive. An optical fibre is used in place of the conventional
guidewire and when an obstruction is met the fibre is advanced whilst

cutting a channel through the atheromatous material by the application of laser energy. If this does not create a sufficiently large channel it should be possible to enlarge it by conventional balloon angioplasty.

The main advantage of such a technique would be a much wider application to lesions presently considered unsuitable for balloon angioplasty. Lesions may be unsuitable because there is a long segment of complete occlusion, previous attempts to pass a guidewire have failed or because radiology shows the proximal end of the block to be a difficult one to approach with standard techniques. Another potential advantage is that the post-angioplasty patency may be improved by the fact that the occluding material is vaporised rather than compressed.

The idea of laser as a source for energy to ablate atheroma was first suggested in 1963 by McGuff et al. (1963), but experimental work is not reported until Daniel Choy filed a patent for a fibre-optic delivery device in 1979. Since that time it has been shown that the procedure is feasible and much research over the past few years has been centred upon determining the optimum system and parameters for successful laser angioplasty.

The system used for laser angioplasty consists of two main components, the laser itself and the device for delivering laser energy to the desired site within the artery.

LASER

Atheromatous tissue has been ablated in the experimental situation by a large range of lasers including Argon, Neodymium-YAG, CO_2, Excimer and Dye. Clinical work has been carried out with Argon and Neodymium-YAG lasers. All of these are capable of destroying atheromatous material and therefore the main thrust of research has been in trying to find a system which is of practical clinical use with a minimum of complications.

Three possible complications have caused the most concern. These are perforation of the vessel wall, embolisation of material dislodged by the laser action or of clot formed on the tip of the delivery system and diffuse damage to the vessel wall sufficient to cause aneurysm formation or reocclusion. Of these only perforation has been found to be a major problem. Several studies have failed to show production of particulate material by laser angioplasty (Isner et al., 1985) and although one early animal study did show aneurysm formation in rabbit aorta after laser treatment (Lee et al., 1984a) this has not been confirmed in other studies or clinical work.

Thus the majority of recent research has been aimed at finding the best devices and the optimal parameters for successful recanalisation of vessels without perforation of the arterial wall.

Wavelength

Work with lasers producing wavelengths from ultraviolet (UV), such as excimer lasers (193nm) (Bowker et al., 1986) to the infra-red CO2 (10,600nm) (Eldar et al., 1986) has shown that all are capable of destroying atheromatous material. There are, however, several considerations which may determine the optimum wavelength for laser angioplasty.

The absorption and scattering coefficients for normal and diseased vessel wall have been investigated and at some wavelengths, particularly around 400 to 500nm there is greater energy absorption by atheroma compared to normal vessel wall (Prince et al., 1986). Experimental work has confirmed that there is a selective effect in that the crater size produced by a given exposure is greater in atheromatous than in normal tissue. This includes exposures at longer wavelengths, for example 1064nm (Cross et al., 1987a).

The figures for coefficients of scattering and absorption are obtained for tissue exposed to low energy levels and may be altered in the clinical situation where tissue destruction is taking place. The selectivity is not great, craters are generally no more than 50% larger in diseased tissue (Cross et al., 1987a) and thus the difference is unlikely to give significant protection to the arterial wall in situations where sufficient energy is being delivered to destroy atheromatous tissue.

The mechanism of tissue ablation by laser energy is not fully understood but it may be that the process depends upon wavelength (Boulnois, 1986). With UV wavelengths the individual photon energy exceeds that of the chemical bonds linking organic polymers. For example the C-C bond in organic polymers has a bond energy of approximately 3.6eV which is exceeded by photons with wavelength of less than 340nm. Thus, a 'photochemical' break-down may occur which would be unlikely at longer wavelengths as it would require multiple photon events. In the case of lower photon energy it may be that tissue destruction is mainly by a thermal mechanism. Some of the conclusions drawn from this model of tissue ablation have been confirmed in work with organic polymers (Garrison and Srinivasan, 1985) but there is no evidence at present that the same mechanisms occur in heterogeneous animal tissues. Experimental lesions produced with excimer lasers have sharp edges with little surrounding damage (Bowker et al., 1986), but this may be more closely related to the pulse length than the wavelength as it is also seen with short pulses of longer wavelength as discussed below.

Another consideration regarding wavelength is the practicality of transmitting energy along an optical fibre. At present this rules out the use of the CO_2 and excimer lasers as suitable devices are not yet available (Wolbarsht, 1984). Other points to be considered are the possible mutagenic effect of UV light (Parrish and Deutsch, 1984), and reliability.

Continuous Wave or Pulsed Lasers

Exposure of tissue to laser energy produces a lesion with three zones. There is a zone of ablation where the tissue is removed. This may be surrounded by a zone of charred tissue and around this is an area of damaged tissue which may show necrosis and vacuolation (Abela et al., 1985).

Continuous wave lasers were the first to be used for laser angioplasty. They create a saucer shaped crater usually with a charred base and a large area of surrounding tissue damage. In contrast to this pulsed lasers have been shown to produce an almost cylindrical crater with no charring and a much smaller area of damage (Deckelbaum et al., 1985; Cross et al., 1987b).

Theoretically the laser energy should be delivered in a short time compared to the thermal relaxation time of the exposed tissue. The thermal relaxation time for different systems have been calculated and a typical figure for 100μm radius spot size is about 25ms and is proportional to the square of the radius (Selzer et al., 1985). If the pulse is short compared to this then there is little dissipation of heat to the surrounding tissue during the exposure which minimises unwanted thermal effects on surrounding tissue. Also, a break between pulses which is long compared to this figure should allow sufficient cooling between pulses to prevent heat build-up.

A pulsed Neodymium-YAG laser is now being used in our centre for laser angioplasty and with a pulse length of 100μs and repetition rate of 10Hz, in keeping with the theoretical optimum, there is little thermal damage surrounding the zone of ablation. In theory it may be possible to use even shorter pulse length by Q-switching but this is not possible with the present delivery devices due to damage of the fibres and interfaces by the very high peak powers produced. Since the pulses used are already short compared to the thermal relaxation time there may be little advantage in shortening them further. With the ultrashort pulses produced by Q-switched lasers there is also a possibility of surrounding photoacoustic damage due to the production of shockwaves (Reichel et al., 1987).

Power parameters

Apart from whether the laser is pulsed or not delivery may be varied in other ways. Continuous wave lasers may be switched on and off for short periods, or 'chopped' and this produces greater effect than one exposure of the same total energy (Kramer et al., 1987). With pulsed lasers, repetition rate and number of pulses can be altered. At present these parameters are mainly limited by the capabilities of the delivery system in use.

The total power output, or average output in the case of pulsed lasers may also be altered. The important figure seems to be the power density and this is a function of both the laser output and the area over which it is distributed by the delivery system. Thus the power density at the distal end of a 200μm fibre is 9 times that of a 600μm fibre carrying the same power.

With power density below a threshold there is no ablation of tissue. Above the threshold with continuous wave lasers there seems to be a short period of rapid ablation followed by an almost constant slower rate (van Gemert et al.,1985). This may partly explain the greater effects of chopped delivery, and pulsed lasers. With pulsed lasers and repeated short exposures from a CW laser the dose response is almost linear (Kramer et al., 1987).

DELIVERY SYSTEM

Bare Fibre

The simplest delivery system is a bare optical fibre which can take the place of the guidewire in the usual set-up for PTBA. Unfortunately, early work showed that the bare fibre caused a high rate of perforation. One suggested way of overcoming this is to use the fibre within an angioplasty balloon which can be inflated proximal to the obstruction in the vessel to keep the fibre in a coaxial position. This method has been used with some success in the superficial femoral artery which remains fairly straight throughout its course (Geschwind et al., 1984) but the method fails in situations where the artery is curved or bifurcates (Lee et al., 1984b).

Metal Tip

The largest clinical experience has been accumulated with a device in which a metal cap is used to cover the distal end of the optical fibre. The effect of this is to convert all the light energy into heat at the fibre tip, which has been shown to reach temperatures above 400°C (Sanborn et al., 1985; Verdaasdonk et al., 1987a). Thus the laser energy is not used directly to perform angioplasty. Clinical results in the leg have shown that this device is capable of recanalising occluded femoral vessels with a low risk of complications and fair early results when combined with balloon angioplasty (Cumberland et al., 1986a). In theory the use of a hot tip to which charred tissue adheres would be expected to increase the risk of surrounding thermal damage and embolisation. In practice this does not seem to cause clinical problems in the peripheral circulation but in the few cases of coronary angioplasty using this technique there has been a high rate of subsequent myocardial infarction (Cumberland et al., 1986b).

Sapphire Tip

The sapphire tipped device has a catheter within which the fibre is abutted against a hemispherical sapphire of about 2mm diameter. There is some heating of the sapphire, but about 60% of the energy is transmitted as light (Verdaasdonk et al., 1987b), and by constant infusion and using repeated short exposures heating can be minimised and the accumulation of charred tissue on the tip is avoided (Douville et al., 1987). This device has been used in experiments with an artificial circulation (Cross et al., 1986), and in some early clinical trials with promising results.

Ball-ended

Another possible approach is to make a ball of the same material as the fibre. This can be fabricated separately and welded onto the fibre tip, or a bulbous end can be produced by using energy passed down the fibre to produce a 'microfurnace' which will melt the end of the fibre (Russo et al., 1984). In this way a fibre with a less traumatic end can be produced with very low energy losses at the interface so that heating of the tip is minimal (White et al., 1986). This is a new system and is still undergoing evaluation.

OTHER TECHNIQUES

Angioscopy and Guidance

An alternative approach to the problem of damage to the arterial wall is the use of a device which can be steered by the use of a guidewire. Anderson et al. (1986), of 42 exposures in rabbit aortas, had a perforation rate of 12% with such a device, an improvement on the unguided fibre. An extension of this idea is the simultaneous visualization and laser delivery by use of an angioscope. Experiment with such a device has been described (Lee et al., 1983), but at present there are still problems to be overcome concerning the size of the instrumentation before it can be applied clinically.

Enhancement of laser effects

It may be possible, by using drugs, to enhance the effects of the laser on the atheroma and increase the selectivity between normal and diseased tissue. Tetracycline has been shown to bind to atheroma and absorbs UV light at 355nm (Murphy-Chutorian et al., 1985). At present this is not of clinical use as it is limited to the UV but other substances such as haematoporphyrin derivative may prove useful at other wavelengths (Spears et al., 1983).

CLINICAL RESULTS

As mentioned above the device with which there is the most clinical experience is the metal capped fibre which is in clinical use by Cumberland et al. (1986a). They report 56 patients who underwent laser assisted balloon angioplasty mostly for lesions of the femoral or popliteal artery. Of these 53% were considered difficult or impossible for conventional balloon angioplasty. 89% of the lesions were recanalised by the laser device. There was one perforation thought to be partly due to mechanical pressure. There was no evidence of embolism or arterial spasm. Two patients had acute reocclusion, one requiring further surgery. Of 26 patients with 3 to 10 months follow up there were two further reocclusions, both within three months. These results compare favourably with the results of balloon angioplasty alone and are on a group of patients with more advanced disease.

Other devices have also been used in clinical work. Geschwind et al. (1984), and Ginsburg et al. (1985), have reported results using a bare fibre within a balloon catheter with limited success. The larger number of cases was reported by Ginsburg, of 16 attempts the lumen was improved in 8 but with several complications. There were three perforations and 2 dissections and 2 stenoses became completely occluded.

Early clinical experience has been obtained in our centre using a sapphire tipped device with a pulsed Neodymium-YAG laser. It was initially applied in patients requiring amputation due to disease affecting the superficial femoral artery. Of the first five patients, limb salvage was obtained in two and in one the level of amputation was lower than it would otherwise have been. There were no complications and after these promising results the technique is now undergoing more extensive clinical evaluation.

CONCLUSIONS

Certain characteristics are desirable in the ideal system for laser angioplasty.

1. Ablation should take place with the minimum of surrounding damage. This can be achieved by using short pulses of laser energy applied to small areas to attain high peak powers and a high power density. Further development of delivery devices is still required to allow transmission of these high powers down fine fibres.
2. Any angioplasty device should be able to accurately apply the energy to the occluding material rather than the vessel wall. There are several possible ways of achieving this. The fibre may have properties that make it likely to remain directed coaxially in the vessel or it may be directed by a guidewire or inflatable balloon. It may be possible to image the device during exposure with biplanar arteriography or duplex ultrasound scanning or to visualise the plaque and fibre with an angioscope. This is probably the area in which there is most scope for continuing improvements in devices.
3. It would be useful if there could be selection such that atheroma is ablated in preference to normal tissue. This would help to minimise damage if accurate aiming could not be achieved. There is some degree of selectivity with present devices but it would be necessary to enhance this effects with dyes or drugs before it would be clinically useful.
4. Any device which is to be of clinical use must be safe and reliable.

THE FUTURE

In the case of peripheral vascular disease, particularly of the more distal vessels, the results of surgery and the fact that the general health of the patients is often poor, makes any potential for non-operative treatment an attractive proposition. It has already been shown that it is possible to use the laser as an adjunct to conventional balloon angioplasty, and thus greatly increase the number of patients in whom the technique may be applicable. In an unfit patient who is at risk of limb loss, any new technique which has been proved safe can be justified, so there may already be a place for laser angioplasty in the clinical setting.

Further clinical work needs to be performed to establish the longer term patency following laser recanalisation before the procedure can be generally used to treat intermittent claudication, but if early results are confirmed then it would be reasonable to attempt laser angioplasty in selected cases before submitting patients to surgery.

In the UK alone there are about 5000 legs amputated each year, the majority for vascular disease, with a considerably larger number of patients seeking treatment for disabling symptoms. There is thus considerable potential for laser angioplasty in peripheral vascular disease.

In the case of coronary laser angioplasty however, there is still considerable work to be done before clinical trials can be started in order to reduce the risks of perforation or embolism in a situation where either of these may be fatal.

REFERENCES

Abela,G.S., Crea,F.,Seeger,J.M.,Franzini,D., Fenech,A., Normann,S.J., Feldman,R.L., Pepine,C.J. and Conti,C.R., 1985, The healing process in normal canine arteries and in atherosclerotic monkey arteries after transluminal laser irradiation, Am J Cardiol 56: 983

Anderson,H.V., Zaatari,G.S., Roubin,G.S.,Leimgruber,P.P. and Gruentzig,A.R., 1986, Steerable fibreoptic catheter delivery of laser energy in atherosclerotic rabbits, Am Heart J 111: 1065

Boulnois,J-L., 1986, Photophysical processes in recent medical laser developments: a review, Lasers Med Sci 1: 47

Bowker,T.J. , Cross,F.W., Rumsby,P.T. , Gower,M.C., Poole-Wilson,P.A., Fox,K.M. and Bown,S.G., 1986, Excimer laser angioplasty: quantitative comparison in vitro of three ultraviolet wavelengths on tissue ablation and haemolysis, Lasers Med Sci 1: 91

Choy,D.S.J., 1979, US Patent 4 207 874

Cross,F.W., Bowker,T.J., Marston,A., Adiseshiah,M. and Bown,S.G., 1986, Artificial circulation assessment of sapphire fibretips for use in laser angioplasty, Vascular Surgical Society, Nov. 1986, Br J Surg, 74: 329

Cross,F.W., Mills,T.N. and Bown,S.G., 1987a, Pulsed Neodymium-YAG laser effects on normal and atheromatous aorta in vitro, Lasers in the Life Sciences 1: 193

Cross,F.W., Wright,J.K., Bowker,T.J. and Bown,S.G., 1987b, The role of pulse length in limiting distant damage to vascular tissue caused by the Nd-YAG laser, Lasers Med Sci., In press.

Cumberland,D.C., Tayler,D.I. and Procter,A.E., 1986a, Laser-assisted percutaneous angioplasty: initial clinical experience in peripheral arteries, Clin Radiol 37: 423

Cumberland,D. C., Starkey,I. R., Oakley,G. D., Fleming,J. S., Smith,G. H., Goiti,J. J., Tayler,D. I. and Davis, J.,1986b, Percutaneous laser-assisted coronary angioplasty, Lancet 2: 214

Deckelbaum,L. I., Isner,J. M., Donaldson,R. F., Clarke,R. H., Laliberte,S. M., Ahron,A. S. and Bernstein,J. S., 1985, Reduction of laser induced pathologic tissue injury using pulsed energy delivery, Am J Cardiol 56: 662

Dotter,C. T. and Judkins,M. P., 1964, Transluminal treatment of arteriosclerotic obstruction. Description of a new technique and a preliminary report of its application, Circulation 30: 654

Douville,E. C., Kempczinski,R. F., Doerger,P. T., van der Bel-Kahn,J., Sankar,M. Y. and Joffe, S. N.,1987, Effects of Nd-YAG laser energy on the arterial wall: evaluation of a new contact delivery system, J Surg Res 42: 185

Eldar,M., Battler,A., Gal,D., Rath,S., Rotstein,Z., Neufeld,H. N., Akselrod,S., Katzir,A., Gaton,E. and Wolman, M.,1986, The effects of varying lengths and powers of CO2 laser pulses transmitted through an optical fiber on atherosclerotic plaques, Clin Cardiol 9: 89

Garrison,B. J. and Srinivasan,R., 1985, Laser ablation of organic polymers: microscopic models for photochemical and thermal processes, J Appl Phys 57: 2909

Geschwind,H.J., Boussignac,G., Teisseire,B., Benhaiem,N., Bittoun,R. and Laurent,D., 1984, Conditions for effective Nd-YAG laser angioplasty, Br Heart J 52: 484

Ginsburg,R., Wexler,L., Mitchell,R. S. and Profitt D, 1985, Percutaneous
    transluminal laser angioplasty for treatment of peripheral vascular
    disease., Radiology 156: 619
Grüntzig,A. R. and Hopff,H., 1974, Perkutane Rekanalisation chronischer
    arterieller Verschlusse mit einem neuen Dilationskatheter. Modifikation
    der Dotter-Technik., Dtsch Med Wochenschr 99:2502
Isner,J. M.,Clarke,R. H., Donaldson,R. F. and Aharon,A., 1985,
    Identification of photoproducts liberated by in vitro argon laser
    irradiation of atherosclerotic plaque, calcified cardiac valves and
    myocardium, Amer J Cardiol 55: 1192
Kramer,J. R., Bott-Silverman,C., Ratliff,N. B., Strikwerda,S., Loop,F. D.,
    Shearin,A., Cothren,R. M., Kittrell,C. and Feld, M. S., 1987, Removal of
    atherosclerotic plaque using multiple short exposures of argon ion laser
    light, Am Heart J. 113:1038
Lee,G., Ikeda,RM., Stobbe,D., Ogata,C., Theis,J. H., Hussein,H. and Mason,
    D. T ., 1983, Laser irradiation of human atherosclerotic obstructive
    disease: simultaneous visualization and vaporization achieved by dual
    fiberoptic catheter, Am Heart J., 105:163
Lee,G., Ikeda,RM., Theis,J. H., Chan,M. C., Stobbe,D., Ogata,C., Kumagai,A.
    and Mason,D. T ., 1984a, Acute and chronic complications of laser
    angioplasty: vascular wall damage and formation of aneurysms in the
    atherosclerotic rabbit, Amer J Cardiol 53: 290
Lee,G., Seckinger,D., Chan,M. C., Embi,A., Stobbe,D., Thomson,R. V.,
    Sanchez,N. A., Ikeda,R. M., Reis,R. M. and Mason,D. T., 1984b, Potential
    complications of coronary laser angioplasty., Amer Heart J 108: 1577
McGuff,P. E., Bushnell,D., Soroff,H. S. and Deterling, R. A.,1963, Studies
    of the surgical applications of laser (Light Amplification by Stimulated
    Emission of Radiation), Surg Forum 14: 143
Murphy-Chutorian,D., Kosek,J., Mok,W., Quay,S., Huestis,W., Mehigan,J.,
    Profitt,D. and Ginsburg, R.,1985, Selective absorption of ultraviolet
    laser energy by human atherosclerotic plaque treated with tetracycline,
    Amer J Cardiol 55: 1293
Murray,R. R., Hewes,R. C., White,R. I., Mitchell,S. E., Auster,M., Chang,R.,
    Kadir,S., Kinnison,M. L. and Kaufman,S. L.,1987, Long-segment femoro-
    popliteal stenoses: is angioplasty a boon or a bust, Radiology., 162:473
Parrish,J.A. and Deutsch,T.F., 1984, Laser photomedicine, IEEE J Quantum
    Electron QE20:1386
Prince,M.R., Deutsch,T.F., Mathews-Roth,M.M., Margolis,R., Parrish,J.A. and
    Oseroff,A.R., 1986, Preferential light absorption in atheromas in vitro.
    Implications for laser angioplasty, J Clin Invest 78: 295
Sanborn,T.A., Faxon,D.P., Haudenschild,C.C. and Ryan,T.J., 1985,
    Experimental angioplasty: circumferential distribution of laser thermal
    energy with a laser probe, J Am Coll Cardiol 5: 934
Selzer,P.M., Murphy-Chutorian,D., Ginsburg,R. and Wexler,L., 1985,
    Optimizing strategies for laser angioplasty, Invest Radiol 20: 860
Spears,J.R., Serur,J., Shropshire,D. and Paulin,S., 1983, Fluorescence of
    experimental atheromatous plaques with haematoporphyrin derivative, J
    Clin Invest 71: 395
Reichel,E., Schmidt-Kloiber,H., Schoffmann,H., Dohr,G. and Eherer,A., 1987,
    Interaction of short laser pulses with biological structures, Opt and
    Laser Technol., 19:40
Russo,V., Righini,G.C., Sottini,S. and Trigari,S., 1984, Lens-ended fibers
    for medical applications: a new fabrication technique, Appl Opt.,
    23:3277
van Gemert,M.J.C., Schets,G.A.C.M., Stassen,E.G. and Bonnier,J.J., 1985,
    Modeling of (Coronary) laser-angioplasty, Lasers Surg Med 5: 219
Verdaasdonk,R.M., Borst,C., Boulanger,L.H.M.A. and van Gemert,M.J.C., 1987a,
    Laser angioplasty with a metal laser probe ("hot tip"): probe
    temperature in blood, Lasers Med Sci. In press.
Verdaasdonk,R.M., Cross,F.W. and Borst,C., 1987b, Physical properties of
    sapphire fibretips for laser angioplasty, Lasers Med Sci. In press.
White,C.,Ramee,S.,Aita,M.,Abrahams,L.,Card,H.,Aiwa,G.,Samson,G.,Wade, C. and
    Virmani,R., 1986, Enhanced efficacy and safety of silica ball tip laser
    fibers for in-vivo laser angioplasty (abstr). Lasers Med Sci., 1:313
Wolbarsht,M.L.,1984,Laser surgery:CO or HF,IEEE J Quantum Electron QE20:1427

202

PHOTOBIOCHEMISTRY WITHOUT LIGHT:

INTRACELLULAR GENERATION AND TRANSFER OF ELECTRONIC ENERGY

Giuseppe Cilento

Department of Biochemistry, Instituto de Química
Universidade de São Paulo, C.P. 20780
01498 - São Paulo, Brazil

INTRODUCTION

Photobiochemistry without light (Cilento, 1973, 1984)-the biochemical counterpart of photochemistry without light pioneered by E.H. White and his associates (White et al., 1974) - encompasses the biochemical/biological generation of electronically excited species, followed by reaction of the excited species or induced photochemistry resulting from energy transfer. Electronically excited triplet species are much better candidates than excited singlet species for photobiochemistry without light, essentially because of the much longer intrinsic triplet lifetime, a fundamental factor given the low concentrations that may prevail in biological systems. Indeed, photochemical reactions accompanying bioluminescence (where an excited singlet species is formed) apparently have never been described. The potential importance of photobiochemistry without light stems from the fact that it might provide a rationale for the occurrence of several typically photochemical processes in vivo in the complete absence of light (Cilento, 1973,1984; White et al., 1974).

Although a possible biological formation of excited triplet species has often been mentioned, prior to our work a systematic study had never been undertaken. The clue we pursued in the search for triplet species was the observation that thermolysis of dioxetanes with simple (e.g., alkyl, alcoxy) substituents generates one of the carbonyl compound in the triplet state (Kopecky and Mumford, 1969; Adam, 1982):

$$R_1 - \underset{\underset{O}{|}}{\overset{\overset{R_2}{|}}{C}} - \underset{\underset{O}{|}}{\overset{\overset{R_4}{|}}{C}} - R_3 \longrightarrow {}^3\underset{\underset{O}{\|}}{\overset{R_1 \diagdown \diagup R_2^*}{C}} + \underset{\underset{O}{\|}}{\overset{R_3 \diagdown \diagup R_4}{C}}$$

ENZYMATIC SYSTEMS

We soon realized that there are many enzymatic systems that generate products of the type expected from the cleavage of a hypothetical dioxetane/dioxetanone intermediate and therefore qualify for investigation. Among these several appeared to be of special interst, either because the expected triplet species was well-known photochemically and in the closely related field of dioxetane chemistry, or because of the natural

importance of the enzymatic system. It was fully realized at the   outset
that, since triplet species are readily quenched by oxygen and the
enzymatic systems selected require oxygen, difficulties might be
encountered in detecting triplet reaction products.

<u>Peroxidase-catalyzed aerobic oxidation of isobutanal</u>. This reaction –which
was studied by Kenten (1953)- generates acetone and formic acid, the
expected products from a dioxetane intermediate:

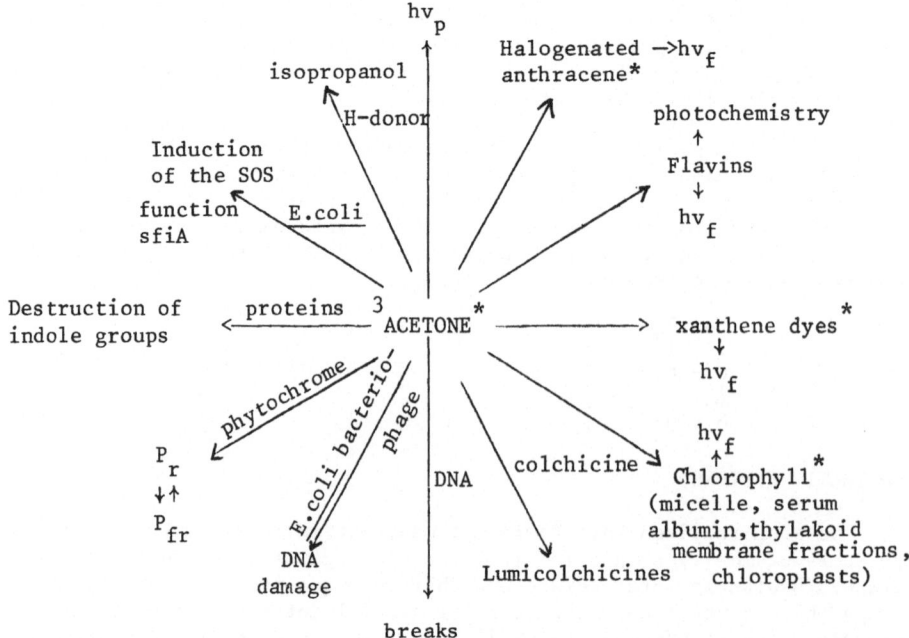

The reaction was found to be accompanied by emission. That acetone is
formed in the triplet state was nicely indicated by the dramatic promotion
of emission from 9,10-dibromoanthracene (or its sulfonate) and partial
quenching of the emission by the non-halogenated analogos.   Several other
emissive acceptors could be excited as a result of energy transfer.Rather
surprisingly, this acetone triplet is generated within the enzyme;largely
protected from O_2 collisions, it thus phosphoresces (Rivas-Suarez and
Cilento, 1981; Adam et al., 1986). Indeed, the emission spectrum is
identical to that of acetone phosphorescence. The kinetics of the system
have recently been fully elucidated (Baader et al., 1985).   The true
reactive substrate is the enol form of the aldehyde; under varying
conditions, the reaction may depend upon the concentration of the enol,the
initiator (the peracid formed in the autoxidation of isobutanal) or enzyme.

Clearly, since sensitized emission can be elicited, sensitized photochemistry should also be possible. Scheme 1 presents the main processes
observed to date with horseradish peroxidase generated triplet acetone;it
encompasses recent results obtained in this (Rivas et al.,1984; Guillo et
al., 1986) and other laboratories (Menck et al., 1986; Nassi et al; Rojas
and Silva):

Scheme 1

204

What has yet to be fully elucidated is the detailed mechanism of energy transfer. The transfer presumably occurs in the vicinity of the enzyme, where triplet acetone is located. Thus, the behavior of the halogenated aromatics and xanthene dyes as acceptors indicates the operation of a heavy atom effect (Cilento, 1984); this implies a reverse $T_n \leadsto S_1$ intersystem crossing in which the prior step must be a T_1-T_2 energy transfer requiring contact. In some cases, transfer may occur by the Förster mechanism when spectral overlap criterium is satisfied, e.g., as in flavin sensitized emission. However, this may not necessarily be the dominant mechanism since reverse $T_n \leadsto S_1$ ISC in flavin has recently been reported (Richter et al., 1987). That the interaction occurs on the enzyme is also indicated by the different rates of reaction of epimeric steroidal dienes. Of course, when the T_1 level is favourably located energetically, the transfer may occur directly to this state. This may well be one reason why enzyme-generated triplet acetone can be more efficient than irradiation, e.g., in the conversion of colchicine into lumicolchicines (Brunetti et al., 1982; Cilento, 1984).

Peroxidase-catalyzed aerobic oxidation of phenylacetaldehyde. This reaction was also described by Kenten (1953), the products being benzaldehyde and formic acid. The formation of an excited species is attested to by xanthene dye (Nascimento and Cilento, 1985) and chlorophyll (Brunetti et al., 1983) sensitized emission. By analogy, the excited species must be triplet benzaldehyde. This species is able to induce lipid peroxidation in the fungus Blastocladiella emersonii (Nascimento and Cilento, 1985). As we shall see, phenylacetaldehyde displays a most interesting behavior with organelles and cells rich in peroxidase.

Peroxidase-catalyzed aerobic oxidation of indole-3-acetic acid. This reaction is of particular importance as indole-3-acetic acid (IAA) is a plant hormone (auxin) and the above reaction occurs in plants. Excited species are indeed formed, as indicated by sensitized emission (Cilento, 1984). At low pHs, the excited species appears to be the expected triplet indole-3-aldehyde as suggested by trapping (uridine) experiments (De Mello et al., 1980). At neutral pHs, a different excited species appears to be formed. The mechanism of chemiexcitation has not yet been fully elucidated. Nevertheless, as a model this system has already proved very valuable. It is capable of activating the light activable enzyme urocanase (Venema and Hug, 1985). Important results obtained with this system are summarized in Scheme 2 (Cilento, 1984; Nassi and Cilento, 1985; Durán et al., 1986).

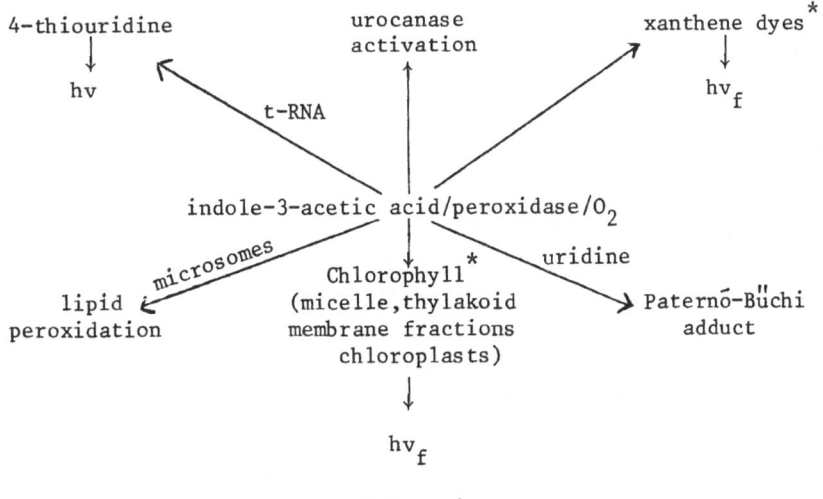

Scheme 2

The α-oxidase system of higher plants. This system converts a long chain fatty acid to the lower aldehyde and CO_2. This reaction is of the type expected for peroxidase activity:

$$R-(CH_2)_n-COOH + O_2 \longrightarrow \left[R-(CH_2)_{n-1} - \overset{\overset{\displaystyle H}{|}}{\underset{\underset{\displaystyle O-O}{|}}{C}} - \overset{\displaystyle O}{\overset{\displaystyle \diagup\!\diagup}{C}} \right] \longrightarrow R-(CH_2)_{n-1} - \overset{\displaystyle O}{\overset{\displaystyle \diagup\!\diagup}{\underset{\underset{\displaystyle H}{\diagdown}}{C}}} + CO_2$$

When prepared from peas (Pisum sativum) this system contains broken chloroplasts; addition of the fatty acid results in red emission and Hill activity (Salim-Hanna et al., 1987). The excitation of the chloroplasts appears to be brought about by a CIEEL (Schmidt and Schuster, 1980) process between the α-peroxylactone and exposed chlorophylls.

Using $1-^{14}C$ labelled myristic acid, a correlation between release of radioactivity (as CO_2) and emission is observed (Salim-Hanna et al.,1987). It may be significant that the excitation of the chloroplasts is observed only with preparations from very young leaves.

GENERATION OF EXCITED STATES IN ORGANIZED BIOLOGICAL STRUCTURES

An important step would be to switch from purely enzymatic systems to organized structures. One of the first attempts was to investigate the true extent of excited state formation in microsomal lipid peroxidation. For this purpose, microsomes were labelled with chlorophyll, an excellent detector of excited states (Bohne et al., 1986). When lipid peroxidation was induced by addition of t-BuOOH the emission -under optimized conditions- increased by two orders of magnitude and shifted to the red spectral region (Cadenas et al., 1984). Considering the low ϕ_F of chlorophyll, the poor sensitivity of the equipment in the red and efficiency of energy transfer, it can be inferred that the formation of excited species is several orders of magnitude greater than one might have inferred from the weak spontaneous emission.

It is not surprising that chloroplasts, an organelle in which the excellent enhancer chlorophyll is built in, have given interesting results. The promoter here is phenylacetaldehyde, which, upon addition to chloroplasts, results in malondialdehyde formation, sustained red emission and Hill activity, indicating excitation of the organelle (Nassi and Cilento, 1982; De Mello et al.). Since chloroplasts have peroxidase activity, it is likely that triplet benzaldehyde is formed and, in turn, promotes lipid peroxidation. Triplet benzaldehyde and/or the excited species formed in lipid peroxidation then transfer energy to the chlorophylls. Essentially similar results were obtained by adding phenylacetaldehyde to Euglena gracilis, a protozoan that contains chloroplasts (De Mello et al.).

The unique effects of phenylacetaldehyde were also strikingly observed with polymorphonuclear leukocytes (Nascimento and Cilento,1987) (PMNL), cells which are very rich in myeloperoxidase (MPO). Here too, lipid peroxidation occurred as attested to by oxygen consumption, malondi- aldehyde formation and biphasic light emission. By analogy to the other systems, the first step must be triplet benzaldehyde formation which then promotes lipid peroxidation. Energy transfer occurs in situ. Thus, neutrophils take up chlorophyll, the pelleted cells appearing red under the fluorescence microscope. Following addition of phenylacetaldehyde to these cells, biphasic red emission could be observed with a long wawlength (< 630 nm) cut-off filter. This result clearly indicated that the excited species generated in lipid peroxidation are able to transfer energy to chlorophylls. All the evidence indicates that the process occurs at the membrane level. The transfer step appears to be specific because no chlorophyll sensitization was observed with other inducers of lipid per-oxidation or with other excited species generated in situ.

Chronologically, the first indication that an excited triplet species may be generated within the cell was obtained by adding a precursor of triplet acetone, i.e., the trimethylsilylenolether of isobutanal, to PMNL (Nascimento et al., 1986). The enol underwent the expected MPO catalyzed oxidation to triplet acetone and formic acid. The involvment of MPO was indicated by N_3^- and N_3^-/H_2O_2 inhibition of the oxygen consumption and emission which acompanies the process. Rigorous controls have shown that the MPO-catalyzed chemiexcitation step does not occur in the bulk solution. The emitter was identified as triplet acetone on the basis of sensitization of the fluorescence of the 9,10-dibromoanthracene-2-sulfonate ion and quenching by the analogous non-halogenated ion. When the leukocytes were exposed to high concentrations of the enol, considerable damage occurred; lymphocytes, which have no peroxidase, were not damaged.

CONCLUSIONS AND PERSPECTIVES

The fact that electronically excited states may be generated enzymatically in high yields and are able to undergo or/and promote photo-chemical processes provides a rationale for explaining the occurrence, in vivo, of typical photochemical processes in the absence of light. Photoreceptors and macromolecules with informational value can be excited in the absence of light. Some of these macromolecules may be altered by energy transfer and/or by reaction with the initially excited species. Of the several photochemical-like processes that have been promoted, the following stand out: (i) the conversion of colchicine into lumicolchicines, which occurs in Colchicum autmnale, even in parts of the plant not exposed to light; (ii) the activation of the photoactivable enzyme urocanase from Pseudomona putida; (iii) phytochrome interconversions, thus possibly accounting for the P_r — P_{fr} dark transformation in vivo. Unambigous proof for the occurrence in vivo of an endogenous photobiology may, however, prove to be extremely difficult because disruption or perturbation of the system in which coupled chemiexcitation/induced photochemical processes occur is required (Steele, 1963).

Endogenously generated excited species might also exert detrimental effects, a possible example being certain mutations in bacteria (Smith and Sargentini, 1985). The detrimental effects might also be exerted indirectly through energy transfer. This latter inference stems from the possibility that the triplet state of polycyclic hydrocarbons may play a role in carcinogenesis; is has been suggested that the excitation is provided by enzymatically generated triplet carbonyl species (Leška et al., 1986).

Under special circumstances, a new strategy for chemotherapy may be possible via the use of excited state precursors as substrate directed towards target microorganisms containing enzyme(s) capable of generating excited species therefrom.

This potential of electronically excited species may be especially valuable in a "dark" photomedicine, particularly since high degrees of specificity are now being achieved in areas of photomedicine such as generation of $^1\Delta_g$ $^1O_2$. A recent example is the use of a pyrene-labelled, liposome-bounded monoclonal antibody against an antigen on the surface of T lymphocytes (Yemul et al., 1987). The corresponding possibilities in the case of a "dark" photomedicine would include the use of antibody-dioxetane ($\longrightarrow$ triplet carbonyls) and antibody-endoperoxide ($\longrightarrow$ $^1\Delta_g$ $^1O_2$) conjugates. Thus appropriate endoperoxides may release $^1O_2$ at a significant rate even at relatively low temperatures (Wilson et al., 1986), which, incidentally, rises the question as to whether the antitumor properties of the endoperoxide of provitamin D (Matsueda and Katsukura, 1985) and the well-known antihelmintic properties of the α-terpinene-derived ascaridole may be due to release of $^1O_2$.

Conversely the possibility exists that the toxic properties of certain agents are exerted through excited state formation. As a possible example, it has been suggested that peroxidase is involved in the toxicity of the pollutant 2-nitropropane; interestingly, the peroxidase catalyzed oxidation of 2-nitropropane does lead to excited acetone (Indig and Cilento, 1987).

The following scheme, in which P^* represents the excited species summarizes these potentialities of biologically generated species.

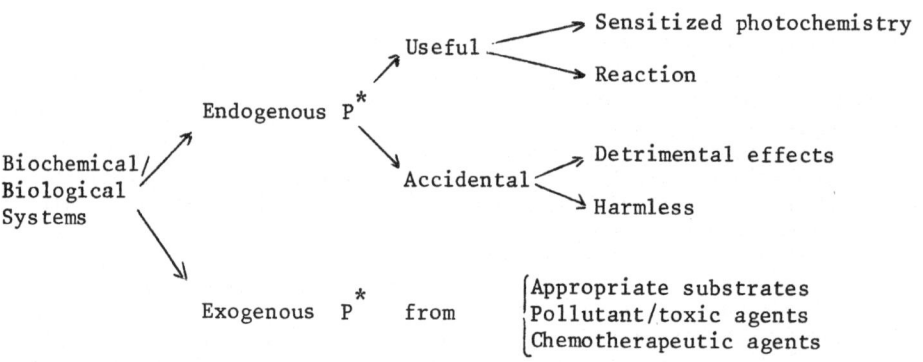

ACKNOWLEDGEMENTS

I wish to acknowledge the fundamental contribution of my colleagues. The author is deeply indebted to Prof. Frank H. Quina for unfailing and generous theoretical assistance.

The financial supports by FINEP (Rio de Janeiro), FAPESP (São Paulo), CNPq (Brasília) and the Volkswagenwerk Stiftung is gratefully acknowledged.

REFERENCES

Adam, W., Determination of chemiexcitation yields in the thermal generation of electronic excitation from 1,2-dioxetanes, in "Chemical and Biological Generation of Excited States", W. Adam and G. Cilento, eds., Academic Press, New York (1982).
Adam, W., Baader, W.J. and Cilento, G., 1986, Enol of aldehydes in the peroxidase/oxidase-promoted generation of excited triplet species, Biochim.Biophys.Acta 88, 330.
Baader, W.J., Bohne, C., Cilento, G. and Dunford, H.B., 1985, Peroxidase-catalyzed formation of triplet acetone and chemiluminescence from isobutyraldehyde and molecular oxygen, J.Biol.Chem. 260, 10217.
Bohne, C., Campa, A., Cilento, G., Nassi, L. and Villablanca, M., 1986, Chlorophyll: an efficient detector of electronically excited species in biochemical systems, Anal.Biochem. 155, 1.

Brunetti, I.L., Bechara, E.J.H., Cilento, G. and White, E.H., 1982, Possible in vivo formation of lumicolchicines from colchicine by endogenously generated triplet species, Photochem.Photobiol. 36: 245.

Brunetti, I.L., Cilento, G. and Nassi, L., 1983, Energy transfer from enzyme-generated triplet species to acceptors in micelles, Photochem. Photobiol. 38: 511.

Cadenas, E., Sies, H., Campa, A. and Cilento, G., 1984, Electronically excited states in microsomal membranes: use of chlorophyll-a as an indicator of triplet carbonyls, Photochem.Photobiol. 40: 661.

Cilento, G., 1973, Excited electronic states in dark biological processes, Quart.Rev.Biophys. 6: 485.

Cilento, G., 1984, Generation of electronically excited triplet species in biochemical systems, Pure Appl.Chem. 56: 1179.

De Mello, M.P., De Toledo, S.M., Haun, M., Cilento, G. and Durán, N., 1980, Excited indole-3-aldehyde from the peroxidase-catalyzed aerobic oxidation of indole-3-acetic acid. Reaction with and energy transfer to transfer ribonucleic acid, Biochemistry 19: 5270.

De Mello, M.P., Nascimento, A.L.T.O., Bohne, C. and Cilento, G., Excitation of chloroplasts in Euglena gracilis in the absence of light, Photochem.Photobiol., in press.

Durán, N., Campa, A., Leite, L.C.C., Cilento, G. and Cadenas, E., 1986, Microsomal lipid peroxidation concomitant to peroxidase-catalyzed aerobic oxidation of indole-3-acetate, Photobiochem.Photobiophys. 11: 281.

Guillo, L.A., Faljoni-Alário, A. and Cilento, G., 1986, Interaction between enzyme-generated triplet carbonyls and molecules intercalated into DNA, Biochim.Biophys.Acta 884: 39.

Indig, G.L. and Cilento, G., 1987, Peroxidase-promoted aerobic oxidation of 2-nitropropane: mechanism of excited state formation, Biochim. Biophys.Acta 923: 347.

Kenten, R.H., 1953, The oxidation of phenylacetaldehyde by plant saps, Biochem.J. 55: 350.

Kopecky, K.R. and Mumford, C., 1969, Luminescence in the thermal decomposition of 3,3,4-trimethyl-1,2-dioxetane, Can.J.Chem. 47: 709.

Leška, J., Krčulová, G. and Loos, D., 1986, Contribution to the study of carcinogenecity of polycyclic aromatic hydrocarbons and their azaanalogues. Carcinogenesis in the triplet state, Chem.Papers 40: 257.

Matsueda, S. and Katsukura, Y., 1985, Antitumour-active photochemical oxidation products of provitamin D., Chemistry and Industry (London) 12: 411.

Menck, C.F., Cabral-Neto, J.B., Faljoni-Alário, A. and Alcantara-Gomes, R., 1986, Damages induced in λ-phage DNA by enzyme-generated triplet acetone, Mutation Res. 165: 9.

Nascimento, A.L.T.O. and Cilento, G., 1985, Induction of chemiluminescent processes in the fungus Blastocladiella emersonii by exposure to enzyme-generated triplet benzaldehyde, Biochim.Biophys.Acta 843: 254.

Nascimento, A.L.T.O. and Cilento, G., 1987, Generation of Electronically excited states in situ. Polymorphonuclear leukocytes treated with phenylacetaldehyde, Photochem.Photobiol., 46: 137-141.

Nascimento, A.L.T.O., Da Fonseca, L.M., Brunetti, I.L. and Cilento, G., 1986, Intracellular generation of electronically excited states. Polymorphonuclear leukocytes challenged with a precursor of triplet acetone, Biochim.Biophys.Acta 88: 337.

Nassi, L. and Cilento, G., 1982, Excitation of chloroplasts induced by phenylacetaldehyde, Photochem.Photobiol. 36: 121.

Nassi, L. and Cilento, G., 1985, Energy transfer from enzyme-generated triplet carbonyls to thylakoid membrane fractions enriched in Photosystems I and II, Photochem.Photobiol. 41: 195.

Nassi, L., Schiffmann, D., Favre, A., Adam, W. and Fuchs, R., Induction of the SOS function sfiA in E.coli by systems which generate triplet ketones, Mutation Res., in press.

Richter, C., Hub, W., Traber, R. and Schneider, S., 1987, Excited triplet-singlet intersystem crossing in isoalloxazines, Photochem. Photobiol. 45: 671.

Rivas-Suárez, E. and Cilento, G., 1981, Quenching of enzyme-generated acetone phosphorescence by indole compounds. Stereo-specific effects of D- and L-tryptophan. Photochemical-like effects, Biochemistry 20: 7329.

Rivas, E., Paladini Jr., A. and Cilento, G., 1984, Photochemical-like destruction of tryptophan in serum albumins induced by enzyme--generated triplet species, Photochem.Photobiol. 40: 565.

Rojas, J. and Silva, E., Photochemical-like adduct formation between tryptophan and riboflavin induced by enzymically generated triplet acetone, Photochem.Photobiol., in press.

Salim-Hanna, M., Campa, A. and Cilento, G., 1987, The α-oxidase system of young pea leaves (Pisum sativum) as generator of electronically excited states. Excitation in the dark under natural conditions, Photochem.Photobiol. 45: 695.

Schmidt, S.P. and Schuster, G.B., 1980, Anomalous metalloporphyrin and chlorophyll-a activated chemiluminescence of dimethyldioxetanone Chemically initiated electron exchange chemiluminescence, J.Am.Chem. Soc. 102: 7100.

Smith, K.-C. and Sargentini, N.J., 1985, Metabolically-produced "UV-like" DNA damage and its role in spontaneous mutagenesis, Photochem.Photobiol. 42: 801.

Steele, R.H., 1963, A photoinduced chemiluminescence of riboflavin in water containing hydrogen peroxide. I. The primary photochemical phase, Biochemistry 2: 529.

Venema, R.C. and Hug, D.H., 1985, Activation of urocanase from Pseudomonas putida by electronically excited triplet species, J.Biol. Chem. 260: 12190.

White, E.H., Miano, D.J., Watkins, C.J. and Breaux, E.J., 1974, Chemically produced excited states, Ang.Chem. (Int.Ed.Engl.) 13: 229.

Wilson, T., Khan, A.U. and Mehrotra, M.M., 1986, Spectral observation of singlet molecular oxygen from aromatic endoperoxides in solution, Photochem.Photobiol. 43: 661.

Yemul, S., Berger, C., Estabrook, A., Suarez, S., Edelson, R. and Bayley, H., 1987, Selective killing of T lymphocytes by phototoxic liposomes, Proc.Natl.Acad.Sci. USA 84: 246.

HISTIDINE : A CLASTOGENIC FACTOR

Pierre Tachon and Paolo U. Giacomoni

Laboratoires de Recherche Fondamentale de L'Oreal
1 avenue Eugène Schueller
93600 Aulnay sous bois, France

INTRODUCTION

H_2O_2 induces breaks into purified DNA (1,2) but we
wanted to have a deeper insight into this phenomenon, and we
decided to investigate the effect of H_2O_2 on double-stranded
covalently closed circular DNA.
One nick in such a molecule promotes its conversion from a
supercoiled to a relaxed circle (3). The electrophoretic
mobility of a relaxed circle in an agarose gel is nearly
half that of a supercoil having more than fifteen superhelical turns (4,5). The DNA in the gel can be stained with
ethidium bromide and photographed. Conditions can be found,
in which the densitometric scanning of the negatives allows
the measurement of band areas which are proportional to the
amount of DNA present in the gel. It is thus possible to
analyze the rate of formation of single strand's breaks into
supercoiled DNA by measuring the decrease of the
fluorescence intensity of the stain associated with the high
mobility material.

RESULTS

We have incubated for different times supercoiled DNA
with 0.5 M H_2O_2 in the presence of MEM or of PBS (both from
Gibco). The results are reported in figure 1. From these
results it appears that the kinetics of degradation of the
supercoiled DNA is much slower in PBS than in MEM. We
concluded that some components of MEM play an important role
in catalizing the degradation of DNA induced by H2O2 and we
decided to look for them.
It is well known that unirradiated cells die, when incubated
in UV-irradiated culture media, and that tryptophan and
riboflavin are responsible for the acquired cytotoxicity of
irradiated media (6). Moreover it is well known that
irradiated tryptophan promotes the formation of H_2O_2 (7).

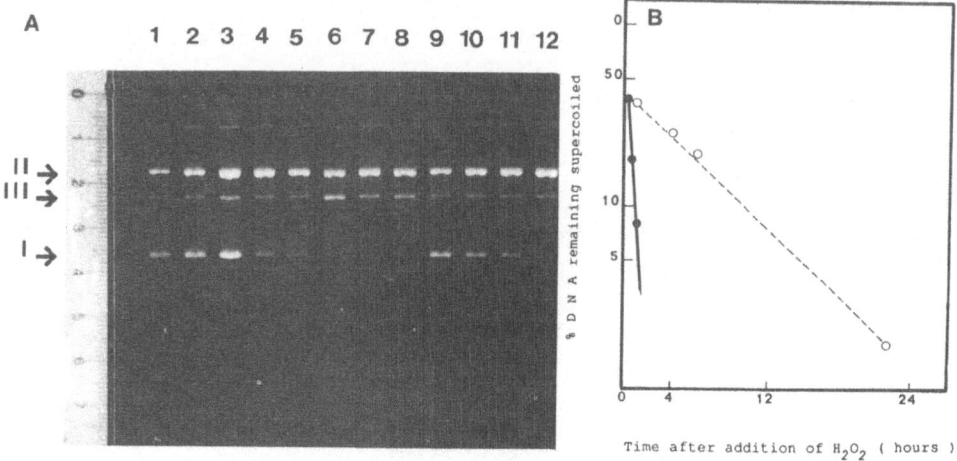

FIGURE 1 : Kinetics of degradation of supercoiled DNA by
0.5 M H$_2$O$_2$ in 7.5 µM EDTA at 37°C.

Panel A : Lane 1, 2 and 3 : 0.1 ; 0.2 and 0.3 µg of DNA
respectively, no H$_2$O$_2$. Lane 4 to 12 : 0.15 µg of DNA after
incubation with H$_2$O$_2$: lane 4 : 30 min., lane 5 : 1 hour,
lane 6 : 2 hours, lane 7 : 3 hours, lane 8 : 4 hours (in
MEM). Lane 9 : 1 hour, lane 10 : 4 hours, lane 11 : 6 hours,
lane 12 : 22 hours (in PBS).

The DNA used was the replicative form of phage fd DNA (fd RF
DNA) purified as described (9), ethanol precipitated several
times and dissolved in 10 mM Tris-Cl pH 7.5, 10 mM NaCl, 0.1
mM EDTA at a concentration of 100 µg/ml. This DNA was
diluted to 30 µg/ml in Calcium and Magnesium free PBS
(solution A). The samples in lane 4 to 8 consisted of 5 µl
of solution A, 10 µl of MEM and were added with 5 µl of 2 M
H$_2$O$_2$ (Prolabo, Paris) in PBS. The samples in lanes 9 to 12
consisted of 5 µl of solution A, 10 µl of PBS and were added
with 5 µl of 2M H$_2$O$_2$ in PBS. The times of addition of H$_2$O$_2$
were such that all the samples arrived simultaneously to the
end-time of the reaction.
The reaction was stopped by the addition of 10 µl of
electrophoresis sample buffer (solution B : 4 M Urea, 50 %
sucrose, 50 mM EDTA, 0.1 % bromophenol blue) and the samples
were loaded in the slots of a horizontal, 1 % slab agarose
gel. The buffer for the gel and for running the electrophoresis was 40 mM Tris Acetate pH 8.4, 10 mM EDTA. After
electrophoresis, the gel was stained in running buffer
containing 0.5 µg/ml ethidium bromide, destained in distilled water, laid on a UV-transilluminator and photographied
with a Polaroid camera equipped with a yellow filter. The
negatives were scanned with a Joyce-Loebl densitometer.
The arrows show the position of supercoiled (I), relaxed
(II) and linear (III) molecules.

Panel B : Densitometric analysis of the lanes of the gel :
    —●— MEM   —○— PBS

The ordinates are on a logarithmic scale.

212

Therefore we incubated for a fixed time supercoiled DNA with H_2O_2 in the presence of several amino acids, riboflavin and phenol red, individually or in several kinds of combinations, and subjected the reaction mixtures to agarose gel electrophoresis. The results are reported in figure 2. From these results it appears that among the chemicals tested the only one able to promote a striking increase of the rate of degradation of supercoiled DNA is histidine.

The kinetics of degradation of supercoiled DNA by H_2O_2 in the presence and in the absence of histidine, as well as the dose-response effect at a single time are reported in figure 3. From these data it appears that the degradation of DNA is exponential with time, with different rate constants in the

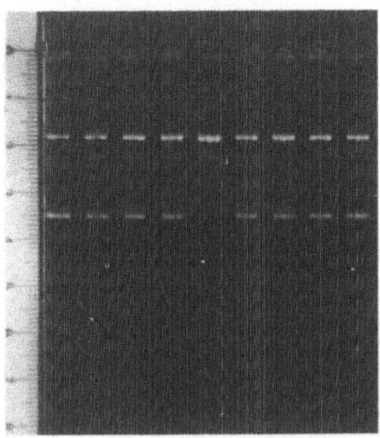

FIGURE 2 : Degradation of supercoiled DNA by H_2O_2 in the presence of various components of MEM

Every sample consisted of 5 μl solution A (0.15 μg of fd RF DNA) (see legend to figure 1), 5 μl of PBS and 5 μl PBS containing the component to be tested.
At time t = 0, 5 μl of 2 M H_2O_2 in PBS were added to the sample.
Every component was at a concentration equal to its concentration in the samples in figure 1, Lanes 4 to 7.(i.e. half its molar concentration in MEM).
After two hours and a half at 37°C, the sample was subjected to electrophoresis.

Lane 1 : PBS, no H2O2, Lane 2 : no chemical, Lane 3 : Riboflavin (0.12 μM), Lane 4 : Cystein (0.1 mM), Lane 5 : Histidine (0.1 mM), Lane 6 : Phenylalanine (0.1 mM), Lane 7 : Tryptophane (0.025 mM), Lane 8 : Tyrosin (0.1 mM), Lane 9 : Methionin (0.05 mM)

In another serie of gels we did observe that Mg^{++}, Ca^{++}, phenol red and cystin do not increase the rate of degradation of cccDNA by H2O2.

presence and in the absence of histidine. On the other hand, the fraction of DNA remaining supercoiled after 2 1/2 hours decreases linearly with increasing the logarithm of the concentration of histidine, thus suggesting that histidine plays a catalytic role in the degradation of DNA.

We have observed that 1 mM sodium azide or 95 % D_2O do not affect the rate of DNA degradation by H_2O_2 and histidine, thus ruling out a posible role for singlet oxygen in this reaction. On the other hand 20 mM mannitol reduces noticeably the rate of this reaction, thus suggesting a role for OH·. Metal chelators such as ethylen diamino tetra acetic acid or diethyl pentaacetic acid (100 μM) completely block the reaction.

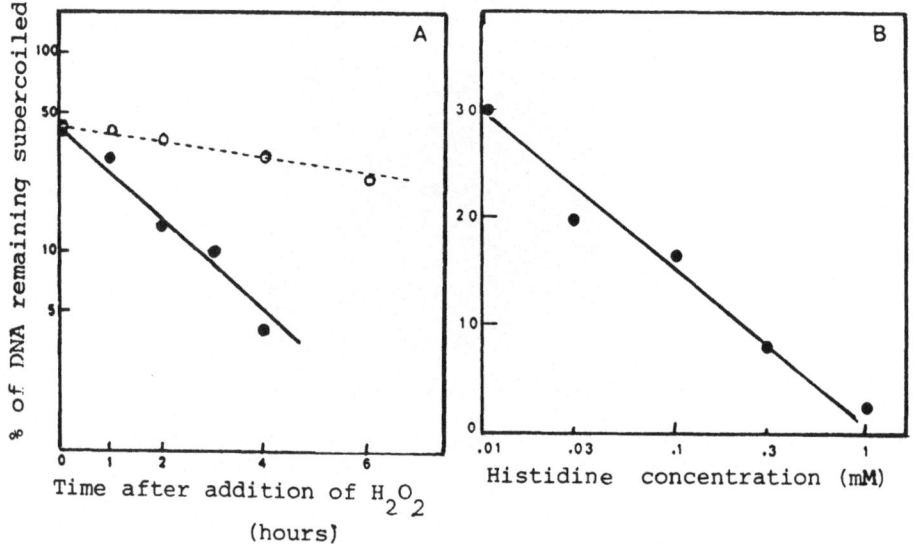

FIGURE 3 : Kinetic and dose-response effects of histidine

PANEL A: Kinetics of nicking of supercoiled DNA by H_2O_2 with or without 0.1 mM histidine
The ordinates are on a logarithmic scale.
—O— no histidine added —●— histine added

PANEL B: Dose-response effect of histidine on the nicking of supercoiled DNA by H2O2. The percent of DNA remaining supercoiled after 150 minutes incubation with 0.5 M H2O2 and histidine is plotted versus the concentration of histidine. The abscissae are on a logarithmic scale.

One might then suspect that DNA is degraded by OH· radicals, liberated in a Fenton-type reaction, from H_2O_2 and iron (our commercial preparation of PBS contains 1 uM iron, as checked by atomic absorption spectroscopy).
Histidine and H_2O_2 might be able to re-cycle ferric iron to ferrous iron, and the role of the free electron doublet on the imidazole ring of histidine appears of importance in this reaction (at pH 7.25) because when the imidazole is protonated (at pH 5.2), histidine loses the capability of increasing the rate of degradation of DNA by H_2O_2.

In order to make sure that these in vitro findings are of biological relevance, we have analyzed the effect of histidine on the cytotoxicity of H_2O_2 on V79 cells.
The results are shown in figure 4 A, where it appears that the survival of these cells treated with up to 20 µM H_2O_2 is dramatically reduced in the presence of histidine. These results point out the role of histidine in increasing the toxicity of H_2O_2 at physiological concentrations, but also seem to indicate the extreme complexity of the phenomenon, since in the explored range of concentrations H_2O_2 is non toxic if the incubation is performed in the presence of MEM. H_2O_2 resumes toxicity in MEM only at 50 µM (data not shown).

It was thus just natural to ask the question concerning the clastogenicity of H_2O_2 in MEM or in PBS with or without histidine and we have measured the number of sister chromatid exchanges (SCE) per metaphase. The data are reported in figure 4 B. The average number of SCE per metaphase, in the absence of H_2O_2, is about six. This number increases with increasing H_2O_2 concentration, up to seventeen in 20 µM H_2O_2, and up to twenty one when histidine is added. These data might lead one to conclude that histidine is a clastogenic factor.
It was suggested that a clastogenic factor is more likely to induce chromosomal aberrations than sister chromatid exchanges (8), we have therefore determined the effect of H_2O_2 and histidine on the formation of micronuclei since it is known that the induction of micronuclei as a criterion for chromosomal damage has been extensively developped (10). The results are reported in figure 4 C.

A striking increase in the number of micronuclei per thousand cells is induced by H_2O_2 and histidine : 15 to 20 times the control without H_2O_2, and histidine increases fourfold the efficiency of H_2O_2 in inducing micronuclei.These results are in keeping with the findings of Shibuya and coworkers (12), who reported that histidine increases the frequency of chromosomal aberrations induced by the hypoxanthine - xanthine oxydase system in V79 cells.

Histidine thus appears to fulfill the criteria for being recognized as a clastogenic factor.

These findings might be of physiological importance since it is well known that in the course of the differentiation of epidermal cells, the appearence of histidine rich proteins in the stratum granulosum is concomitant with the beginning of the degradation of the nucleus in the keratinocytes of that same epidermal layer.

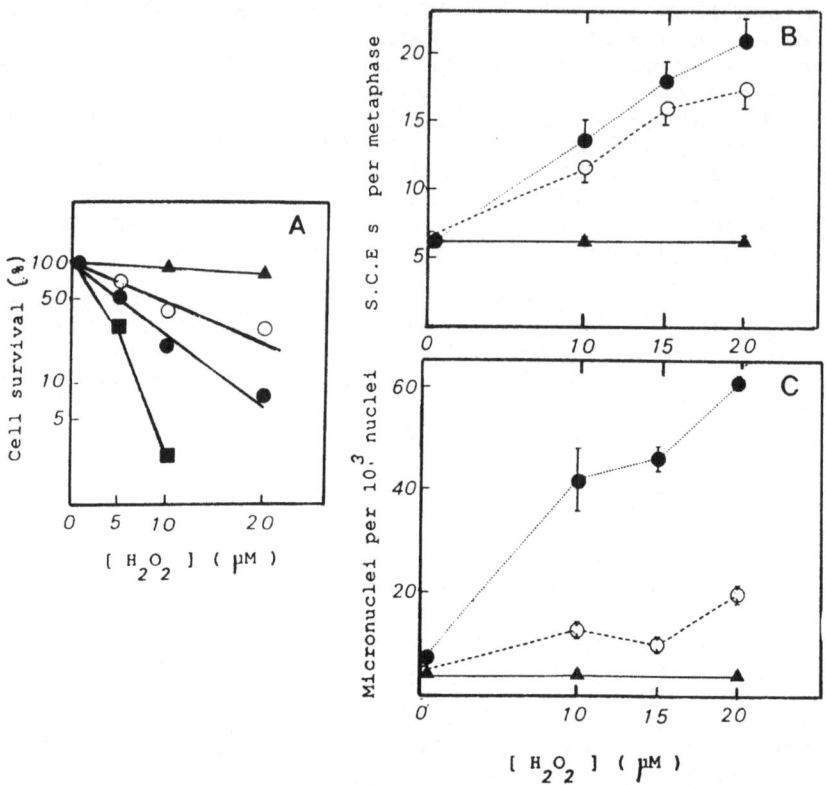

FIGURE 4 : Toxic and clastogenic effects of histidine

PANEL A : Cell survival in H_2O_2
Exponentially growing V 79 cells were rinced in prewarmed
PBS, added with 5 ml MEM or PBS or PBS + histidine in
quadruplicate. At time t=0, H_2O_2 was added. After one hour,
the medium was removed, the cells were rinced, added with
complete medium and allowed one week in the incubator. The
cells were then fixed with 80 % ethanol and stained with
10 % Giemsa. The colonies containing more than fifty cells
were counted and the results are expressed in percent of the
control versus the concentration of H_2O_2.

—▲—MEM; —O—PBS; —●—0.1 mM Histidine; —■—1 mM Histidine.

PANEL B: Sister chromatid exchanges per metaphase versus
the concentration of H2O2
—▲— MEM; —O—PBS; —●—PBS + 0.2 mM histidine

PANEL C: Number of micronuclei per thousand nuclei versus
the concentration of H2O2

Micronuclei and SCE were counted according to the methods of
Lasne and coworkers (10), and of Perry and Wolf (11).
—▲—MEM; —O—PBS; —●—PBS + 0.2 mM histidine

The bars express the standard error of the mean.

REFERENCES

1. Rhaese, H.J. and Freese, E. (1968).
   Biochem. Biophys. Acta. 155, 476-490

2. Ananthaswami, H.N. and Eisenstark, A. (1976).
   Photochem. Photobiol. 24, 439-442

3. Vinograd, J., Lebowitz, J., Radloff, R., Watson, R. and
   Laipis, P. (1965).
   Proc. Nat. Acad. Sci. (USA) 53, 1104-1111

4. Dugaiczyck, A., Boyer, H. and Goodman, H.M. (1975).
   J. Mol. Biol. 96, 171-184

5. Giacomoni, P.U. (1982).
   Biopolymers 21, 117-129

6. Stoien, J.D. and Wang, R.J. (1974).
   Proc. Nat. Acad. Sci. (USA) 71, 3961-3965

7. Mc Cormick, J.P., Fischer, J.R., Patchlatko, J.P. and
   Eisenstark, A. (1976).
   Science 191, 468-469

8. Emerit, I. (1984)
   in Sister Chromatid Exchanges, Part A : The nature of
   Sister Chromatid Exchanges. Tice, R.R. and Hollaender,
   A. Editors. Plenum Press, New-York, London, pp. 127-140

9. Giacomoni, P.U., Delain, E. and Le Pecq, J.B (1977).
   Eur. J. Biochem. 78, 205-213

10. Lasne,C.,Gu,Z.W.,Venegas,W. and Chouroulinkov, I.(1984).
    Mut. Res. 130, 273-282

11. Perry, P. and Wolf, S. (1974).
    Nature 251, 156-159

12. Shibuya, H., Iwata, K., Ohkawa, Y. and Inui, N. (1985).
    Toxicol. Lett. 28, 117-123

ALTERATION OF DNA STRUCTURE: THE COMMON RESULT OF THE INTERACTION BETWEEN DNA AND PHYSICAL AND CHEMICAL AGENTS

G. Ciarrocchi, A. Montecucco, G. Pedrali-Noy and S. Spadari
Istituto di Genetica Biochimica ed Evoluzionistica del C.N.R.
Pavia, Italy

INTRODUCTION

It is becoming evident that the majority, if not all, of mutagens and carcinogens bind, either covalently or not, to nucleic acids or induce chemical modifications in these target molecules. This interaction seems critical in both mutagenic and carcinogenic processes. Since the alterations in nucleic acid structure are likely to have functional consequences, it is important to gain a better understanding of the modifications in the target structure.

There are no a priori reasons to exclude an active role for DNA in the regulation of different biological functions. The constraints of secondary and tertiary structure interactions may in fact induce enough variations which could constitute the bases for such active role. For example the affinity between DNA and protein in sequence specific interaction varies depending on the superhelical density of DNA (Maniatis and Ptashne, 1973). Therefore variations in one or more of parameters such as helical conformation (Dickerson et al., 1983), twist angle between adjacent base pairs (Kabsch et al., 1982), angle between planal base pairs (Trifonov and Sussman, 1980), local denaturations (Lan and Gray, 1979) and hairpin extrusion (Lilley, 1980), may play an important role in the regulation of biological functions of proteins. In addition the interaction between a specific protein and the DNA might generate other alterations of the DNA structure. For example when the RecA protein binds to douplex DNA, the fiber length increases by a factor of 1.6 and this deformation is matched to an unwinding of 10.5° per covered base pair (Register et al., 1983). In the case of the lac repressor-operator interaction, the result is an unwinding of the DNA by about 54.5° (Kim and Kim, 1983). The effect of the altered conformation can be transmitted along the DNA molecule to a second site, in the same way as the interaction of RNA polymerase at one site can be influenced by a regulatory protein molecule bound at another site, without any direct protein-protein interaction (Dickson et al., 1975).

It is known that DNA damaging agents interact with various bases and positions in DNA and one can predict appreciable changes in DNA conformation resulting from alterations in conformation at the glycosyl bond, rotations of the backbone residues and possible changes in sugar puckering. It is reasonable to suppose that such structural alterations (including those deriving from non-covalent binding) might interfere with the normal action of proteins usually acting on DNA, in addition to being a recognition signal for DNA repair mechanisms operating on a broad variety of damages.

METHODOLOGY

Conformational changes deriving from the interaction of various modifying agents with DNA can be studied by different kinds of spectroscopy, by electron microscopy, velocity sedimentation, dye titration, electric linear dichroism, gel electrophoresis, sensitivity to specific endonucleases and even by using antinucleoside antibodies.

In our lab we have mainly applied the electrophoresis on agarose gel, exploiting the property of this technique to separate the topological isomers of an identical circular double-stranded covalently closed sequence of DNA. The choice of this simple technique is due to its great potential deriving from the possibility to detect, after the necessary specific modifications of the assay, all kinds of interaction spanning from covalent to non-covalent modifications of the DNA molecule. In fact the agarose gel can give informations on length and shape of a molecule within an acceptable approximation but it is extremely sensitive in the detection of twist angles modifications of DNA molecules (Keller and Wendell, 1975).

UV radiation seemed useful as an experimental model for the basic mechanism of action of several mutagens and carcinogens since many of them depend upon the same type of DNA repair system (Regan and Setlow, 1974). In analogy with chemical agents, UV radiation produces varions lesions in DNA. The chemical composition of the majority of such photodamages is known (Wang, 1976) but their influence on the secondary and tertiary structure of DNA is still quite unclear.

EFFECTS OF UV RADIATION ON DNA STRUCTURE

Major alterations of physical properties of double helix

The major class of photodamages, the cyclobutyl dipyrimidine, has been studied by several authors utilizing different techniques. The conclusions on the structural alterations imposed by the presence of a dimer were different. Those utilizing conditions generally favoring denaturation (Hayes et al., 1971; Kahn, 1974; Shafranowskaya et al., 1973) concluded that a region of denaturation sorrounds the dimer. Those utilizing non destructive methods excluded denaturation as a major conformational change (Vorlickova and Palecek, 1978; Triebel et al., 1979).

Denhart and Kato (1973) detected an increase in the sedimentation coefficient of oX174 RFI DNA after irradiation with UV light and this increase was accompanied by a decrease in the number of superhelical turns as detected by electron microscopy. These observations were confirmed by the results of titration of irradiated and unirradiated DNA with ethidium bromide, but since at that time the unwinding angle of ethidium bromide was erroneously supposed to be -12° only, they estimated an unwinding for pyrimidine dimer equivalent to 5.5°. However, by correcting their data utilizing a more realistic value of unwinding for ethidium bromide, the indirectly determined unwinding angle would result to be -11.9° per pyrimidine dimer. On the other hand values of -14.3° and -13.7° were obtained by measuring the alteration of electrophoretic mobility of single topoisomers (Ciarrocchi and Pedrini, 1982; Ciarrocchi et al., 1982). These values of unwinding were based on the assumption that the major photoproduct, the pyrimidine dimer, was responsible for the great majority of unwinding. Since the denaturation of a single base pair would correspond to an unwinding of an average of -34.6° (360°/10.4 base pairs per helical turn) in B DNA under torsional stress, one can conclude that in the absence of base pairing as at the level of pyrimidine dimer, the DNA structure is still fairly well conserved. In fact in the case of an ample denaturation, as the one averaging 4.3 base pairs suggested by other authors (Hayes et al., 1971) we whould expect an unwinding of about 150°, a value 10 times greater than the one observed by direct measurement. Then this good conservation of DNA structure at the dimer level suggests some considerations: the pyrimidine rings in the dimers are saturated and therefore they should not effectively contribute to the stacking interactions; however in the opposite strand stacking is still possible. It seems reasonable to suppose that those residual stacking forces are great enough to maintain the DNA structure.

Deformation of the axis of the douplex helix

UV-induced lesions in DNA have been reported to act as "kinks" in the molecule, increasing the flexibility and decreasing the statistical segment length of DNA (Triebel et al., 1979). UV irradiated douplex DNA, no matter if under linear, nicked or covalently closed, fully supercoiled circular form, shows a little but definitely detectable increase in electrophoretic mobility that appears as a minor effect when compared to the reduction of electrophoretic mobility of single topoisomers (Ciarrocchi and Pedrini, 1982). Since the electrophoretic mobility of those forms of DNA is inversely proportional to the molecular weight, an increase in electrophoretic mobility would suggest a reduction of the effective cross-section of the molecule. Therefore a more compact structure derives from the previously mentioned "kinks" in the douplex DNA molecule. This minor effect could originate either from transient denaturation at photodamage level, where the photodamage could act as nucleation site, or from change in the interplanar angle as determined in the case of psoralen adducts (Kim et al., 1983) and suggested from theoretical studies (Pearlman et al., 1985). To explain the removal of supercoiling upon repressor binding by the lac operator, Kim and Kim (1983) suggested three possibilities: a) DNA douplex unwinding; b) DNA bending; c) any combination of the above two modes to give the total net removal of supercoiling. Can this model be utilized to explain the unwinding by UV photoproducts? We have just seen that a bent or kinked irradiated DNA molecule is affected in its electrophoretic mobility independently from its topological state. Since a phenomenon detectable under our conditions of analysis in topologically unconstrained DNA does not influence the unwinding but interfere with its measurement, we have overcome the problem by simply expressing the mobility of irradiated topoisomers as mobility relative to irradiated RFI, taken as relative front (Ciarrocchi and Pedrini, 1982).

DNA unwinding by thymine dimers

Camerman and Camerman (1968) have resolved the structure of the cis 5,5:6,6 isomer of thymine dimers, the same obtained by ultraviolet irradiation of DNA, and found that the angle of rotation between the two saturated thymine residues is 28°. Since the angular displacement between non dimerized thymines in DNA is 35,62° (Kabsch et al., 1982) it follows that the contribution to unwinding attributable to the simple angular displacement between bases should be of about -7,6°. Utilizing the agarose gel technique it has been possible to more precisely determine the unwinding angle for photoproducts obtained by sensitization of DNA topoisomers in the presence of silver ions and UV radiation at 313 nm, conditions known to excite thymines only. In addition, by evaluating the photoreactivation of unwinding it has been possible to attribute to the thymine dimer the more specific unwinding angle of -8.7° (Ciarrocchi and Sutherland, 1983). Therefore, almost 90% of the total unwinding seems to depend solely on the angular displacement between dimerized thymines. These data suggest that if phenomena other than the angular displacement must be taken into account, they should contribute for just about 10% to the total unwinding. In conclusion, if only the stacking forces on the strand opposite to the dimer are in charge of mantaining the DNA structure almost intact, it follows that this damage might easily become a nucleation site for local denaturation even under mild denaturating conditions, as also suggested by theoretical calculations (Benham, 1981).

Unwinding by photodamages other than pyrimidine dimers

But the agarose gel technique has also provided evidence about the contribution by other non-photoreactivable photoproducts to the total unwinding by UV. In fact, after full photoreactivation in the presence of E. coli photolyase 30 and 23% of the unwinding are still present in DNAs irradiated with 254 nm or sensitized with 313 nm radiation in the presence of silver ions, respectively. These results clearly indicate that the tertiary structure of DNA is altered by UV photodamages other than pyrimidine dimers (Ciarrocchi and Sutherland, 1983).

Unwinding by chemical agents

It is evident that the unwinding of douplex DNA can be easily detected by this technique also in the case of chemical modifing agents. In addition, we have recently developed a new modification of the technique to study the effects of those agents which reversibly alter the DNA structure. In this case the assay utilizes the DNA topoisomers generated by ligation of an appropriate substrate in the presence of the agent under examination. The population of ligated molecules, after removal of the DNA binding agent, is analyzed on the usual agarose gel. This assay already proved to be very useful for studying both intercalating and externally binding agents. Obviously, the unwinding angle per modified site can be determined only when the precise number of modified sites per molecule can be measured.

The unwinding angles of a number of douplex-DNA modifying agents is listed in Table I. This table also includes values of unwinding by agents binding non-covalently to DNA. In all but two cases reported in Table I, the effect is the introduction of positive supertwists. Only netropsin seems to

Table I. Average unwinding angle per modified site in douplex DNA

Agents	$\emptyset$	References
Steroidal diamines	-5°	Waring and Chisholm, 1972
Ethidium bromide	-26°	Wang, 1974
Netropsin	+7.5°	Malcolin and Snounou, 1983
N-acetoxy-2-acetylaminofluorene	-22°	Drinkwater et al., 1978
Benzo(a)pyrene-diol-epoxide	-13/-330°	Meehan et al., 1982; Gamper et al., 1980
4,5',8-trimethylpsoralen	-28°	Wiesehahn end Hearst, 1982
cis-diamminedichloroplatinum(II)	-12/-60°	Scovell and Collart, 1985
Trans- " "	-6/-45°	" " "
Methyl methanesulfonate	-3°	Ciomei et al., 1984
Pyrimidine dimer (UV 254 nm)	-10.1°	Ciarrocchi and Sutherland, 1983
Thymine dimer	-8.7°	Ciarrocchi and Sutherland, 1983
Apurinic site	-12°	Ciomei et al., 1984
C-5 methyl cytosine	undetectable	Ciomei et al., 1984
lac repressor	-55°	Kim and Kim, 1983
RNA polymerase	-240/-580°	Wang et al., 1977; Gamper and Hearst, 1982

increase the number of negative supercoils. On the other hand the methylation of the C5 position of cytosine, which occurs under the regulation of cellular systems (physiological modification) is totally ineffective in altering the DNA tertiary structure (Ciomei et al., 1984).

INFLUENCE OF DNA ALTERED STRUCTURE ON CELLULAR PROCESSES

If we exclude the DNA-protein interactions, only the pathological modifications of DNA generated by physical and chemical agents seem to locally alter the tertiary structure of DNA. Then these observations allow to speculate about the involvement of DNA tertiary structure in some cellular processes. Phathological modifications are known to inhibit the action of enzymes controlling the topological state of DNA. The inhibition of bacterial topoisomerase I is already detectable at UV doses generating less than one pyrimidine dimer per circular molecule. Interestingly UV damages also alter the mode of action of the enzyme which then acts more processively (Pedrini and Ciarrocchi, 1983). Also a mammalian DNA methylase has been found to be inhibited in its action by UV damages, depurination and alkylation (Wilson and Jones, 1983). On the contrary HpaII methylase is neigther inhibited by depurination or alkylation and is only partially inhibited by UV photodamages (Sangalli et al. 1984). However the mode of action of Hpa II methylase has been found to depend on the topological state of the DNA substrate (Spadari et al. 1984). We have found that intercalating and externally binding drugs such as antracyclines are good inhibitors of the human topoisomerase II (Spadari et al., 1986) and of the human DNA ligase.

In conclusion the results of this kind of studies suggest that also other proteins involved in DNA repair and cellular functions might be influenced by the presence of local alterations in DNA structure, just as the binding of cis diammine dichloroplatinum(II) is influenced by the presence of ethidium bromide (Merkel and Lippard, 1983). DNA modified sites might therefore influence cellular processes in at least two ways: by driving the interactions of DNA repair enzymes (specific interaction), or by either inhibiting or changing the mode of action of proteins which normally interact with unmodified DNA.

ACKNOWLEDGMENTS

This work was supported by the Progetto Finalizzato "Oncologia" and by the Progetto Strategico "Aspetti chimici e fisici dei sistemi biologici" del Consiglio Nazionale delle Ricerche.

REFERENCES

Benham, C.J., 1981, Theoretical analysis of competitive conformational transitions in torsionally stressed DNA, J. Mol. Biol. 150: 43-68.

Camerman, N., and Camerman, A., 1968, Photodimer of thymine in ultraviolet-irradiated DNA: proof of structure by X-ray diffraction, Science, 160: 1451-1452.

Ciarrocchi, G. and Pedrini, A.M., 1982, Determination of pyrimidine dimer unwinding angle by measurement of DNA electrophoretic mobility, J. Mol. Biol. 155: 177-183.

Ciarrocchi, G. and Southerland, B.M., 1983, Irradiation of circular DNA with 254 nm radiation or sensitization in the presence of Ag^+: evidence for unwinding by photoproducts other than pyrimidine dimers, Photochem. Photobiol. 38: 259-263.

Ciarrocchi, G., Sutherland, B.M. and Pedrini, A.M., 1982, Photoreversal of DNA unwinding caused by pyrimidine dimers. Biochimie 64: 665-668.

Ciomei, M., Spadari, S., Pedrali-Noy, G. and Ciarrocchi, G., 1984, Structural alterations of pathologically or physiologically modified DNA. Nucleic Acid Res. 12: 1977-1989.

Denhart, D.J. and Kato, A.C., 1973, Comparison of the effect of ultraviolet radiation and ethidium bromide intercalation on the conformation of superhelical oX174 replicative form DNA. J. Mol. Biol. 77: 479-494.

Dickerson, R.E., Drew, H.R., Concer B.N., Kopka M.L. and Pjuva, P.E., 1983, Helix geometry and hidration in A-DNA, B-DNA and Z-DNA. Cold Spring Harbor Symp. Quant. Biol. 47: 13-24.

Dickson, R.C., Abelson, J., Barnes, W.M. and Reznikoff, W.S., 1975, Genetic regulation: the lac control region. Science 187: 27-35.

Drinkwater, N.R., Miller, J.A., Miller, E.C. and Yang, N.-C., 1978, Covalent intercalative binding to DNA in relation to the mutagenicity of hydro carbon epoxides and N-acetoxy-2-acetylaminofluorene. Cancer Res. 38: 3247-3255.

Gamper, H.B. and Hearst, J.E., 1982, A topological model for transcription based on unwinding angle analysis of E. coli RNA polymerase binary, initiation and ternary complexe. Cell 29: 81-90.

Gamper, H.B., Straub, K., Calvin, M. and Bartholomew, J.C., 1980, DNA alkylation and unwinding induced by benzo(a)pyrene diolepoxide: modulation by ionic strength and superhelicity. Proc. Natl. Acad. Sci. USA 77: 2000-2004.

Kabsch, W., Sander, C., and Trifonov, E.N., 1982, The 10 helical twist angles of B-DNA. Nucleic Acids Res. 10: 1097-1104.

Kahn, M., 1974, The effect of thymine dimers on DNA:DNA hybridization. Biopolymers 13: 669.

Keller, W. and Wendell, I., 1974, Stepwise relaxation of supercoiled SV40 DNA 39:199-208

Kim, R. and Kim, S.-H, 1983, Direct measurement of DNA unwinding angle in specific interaction between lac operator and repressor. Cold Spring Harbor Symp. Quant. Biol. 47: 451-454.

Kim, S-H, Peckler, S., Graves, B., Kanne, D., Rapoport, H. and Hearst, J.E., 1983, Sharp kink of DNA at psoralen-cross-link site deduced from crystal structure of psoralen-thymine monoadduct. Cold Spring Harbor Symp. Quant. Biol. 47: 361-365.

Hayes, F.N., Williams, R.L., Ratliff, R.L., Varghese, A.J., and Rupert, C.S., 1971, Effect of a single thymine photodimer on the oligo-deoxythymidilate-polydeoxyadenilate interaction. J. Am. Chem. Soc. 93: 4940-4942.

Lan, P., and Gray, H., 1979 Extracellular nucleases of Alteromonas espejiana BAL 31. IV. The single strand-specific deoxiriboendo nuclease activity as a probe for regions of altered secondary structure in negatively and positively supercoiled closed circular DNA. Nucleic Acids Res. 6: 331-357.

Lilley, D., 1980, The inverted repeat as a recognizable structural feature in supercoiled DNA molecules. Proc. Natl. Acad. Sci. U.S.A. 77: 6468-6472.

Malcolm, A.D.B., and Shnounou, G., 1983, Netropsin increases the linking number of DNA. Cold Spring Harbor Symp. Quant. Biol. 47, 323-326.

Maniatis, T. and Ptashne, M., 1973, Multiple repressor binding at the operators in bacteriophage lambda. Proc. Natl. Acad. Sci. U.S.A. 70: 1535-1531.

Meehan, T., Gamper, H. and Becker J.F., 1982, Characterization of reversible, physical binding of benzo(a)pyrene derivatives to DNA. J. Biol. Chem. 257: 10479-10485.

Merkel, C.M. and Lippard, S.J. (1983) Ethidium bromide alters the binding mode of cis-diammine dichloroplatinum(II) to pBR322 DNA. Cold Spring Harbor Symp. Quant. Biol. 47: 355-360.

Pearlman, D.A., Hoibrook, S.R., Pirkle, D.H. and Kim, S-H, 1985, Molecular models for DNA damaged by photoreaction. Science 227: 1304-1308.

Pedrini, A.M. and Ciarrocchi, G., 1983, Inhibition of Micrococcus luteus DNA topoisomerase I by UV photoproducts. Proc. Natl. Acad. Sci. U.S.A., 80: 1787-1791.

Regan, J.D., and Setlow, R.B., 1974, Two forms of repair in the DNA of human cells damaged by chemical carcinogens and mutagens. Cancer Res., 34: 3318-3325.

Register, J.C., Sperrazza, J.M. and Griffith, J., 1983, RecA protein unwinds douplex DNA by 180 degrees for every 17 base pairs in the fiber formed with ATP S. UCLA Symposia on Molecular and Cellular Biology, 10: 731-738.

Sangalli, S., Rebuzzini, A., Spadari, S., Pedrali-Noy, G., Focher, F., and Ciarrocchi, G., 1984, Methylation of UV-irradiated and chemically modified DNAs. Medecine Biologie Environnement, 12: 527-530.

Scowell, W.M. and Collart, F., 1985, Unwinding of supercoiled DNA by cis- and trans-diammineichloroplatinum(II): influence of the torsional strain on DNA unwinding. Nucleic Acids Res., 13: 2881-2895.

Shafranovskaya, N.N., Trifonov, E.N., Lazurkin, Yu.S. and Franck--Kamenetskii, M.D., 1973, Clustering of the thymine dimers in ultraviolet irradiated DNA and the long-range transfer of electronic excitation along the molecule. Nature 241: 58-60.

Spadari, S., Pedrali-Noy, G., Ciomei, M., Rebuzzini, A., Hubscher, U., and Ciarrocchi, G., 1984, DNA methylation and DNA structure, pp. 551-556. In: Proteins involved in DNA replication (ed. Ubscher and S. Spadari) Plenum Publishing Corp.

Spadari, S., Pedrali-Noy, G., Focher, F., Montecucco, A., Bordoni, T., Geroni, C., Giuliani, F.C., Ventrella, G., Arcamone, F. and Ciarrocchi, G., 1986, DNA polymerases and topoisomerases as targets for the development of anticancer drugs. Anticancer Res., 6: 935-940.

Triebel, H., Reinert, K.-E., Bar, H. and Lang, H., 1979, Structural changes of ultraviolet-irradiated DNA derived from hydrodynamic measurements. Biochim. Biophys. Acta, 561: 59-68.

Trifonov, E.N. and Sussman, J.L., 1980, The pitch of chromatin DNA is reflected in its nucleotide sequence, Proc. Natl. Acad.Sci, U.S.A., 77: 3816-3820.

Vinograd, J., Lebowitz, J. and Watson, R., 1968, Early and late helix-coil transitions in closed circular DNA. The number of superhelical turns in polyoma DNA. J. Mol. Biol. 33: 173-197.

Vorlickova, M. and Palecek, E., 1978, Changes in properties of DNA caused by gamma and ultraviolet radiation. Dependence of conformational changes on the chemical nature of the damage. Biochim. Biophys. Acta, 517: 308-318.

Wang, J.C., 1974, The degree of unwinding of the DNA helix by ethydium I. titration of twisted PM2 DNA molecules in alkaline cesium chloride density gradients. J. Mol. Biol., 89: 783-801.

Wang, J.C., Jacobsen J.H., and Saucier, J.M., 1977, Physicochemical studies on interactions between DNA and RNA polymerase. Unwinding of the DNA helix by Escherichia coli RNA polymerase. Nucleic Acids Res., 4: 1225-1241.

Wang, S.Y., 1976, Photochemistry and photobiology of Nucleic Acids, Vol. II, Academic Press, N.Y.

Waring, M.J. and Chisholm, J.W., 1972, Uncoiling of bacteriophage PM2 DNA by binding of steroidal diamines, Biochim. Biophys. Acta, 262: 18-23.

Wiesehahn, G. and Hearst, J.E., 1978, DNA unwinding induced by photoaddition of psoralen derivatives and determination of dark binding equilibrium constants by gel electrophoresis. Proc. Natl. Acad. Sci. U.S.A. 75: 2703-2707.

UV-ENHANCED REACTIVATION OF UV-IRRADIATED SV40 IS DUE TO

FACILITATED TRANSCRIPTION OF THE VIRAL EARLY GENE

Thomas C. Brown and Peter A. Cerutti

Swiss Institute for Experimental Cancer Research (ISREC)
Chemin des Boveresses
1066 Epalinges, Switzerland

SUMMARY

This chapter will present evidence that UV-enhanced reactivation (ER) of SV40 is due to the restoration of viral early gene function. The observation supporting this conclusion is, simply, that ER acts only on damage in the viral early gene region; lesions elsewhere in the SV40 genome are not subject to ER. The implication of this finding is that cells respond to UV-induced damage by inducing the ability to synthesize RNA from a damaged DNA template.

Damage in different regions of the SV40 genome inactivate the DNA by disrupting different sets of genetic functions. We will first describe results indicating that different genomic regions of SV40 differ considerably in their sensitivity to UV-induced damage. We will then show how these different sensitivities allow sectors of lethality to be assigned to the disruption of each viral genetic function. Finally, we will describe experiments linking the UV-enhanced reactivation of UV-irradiated SV40 DNA to the disappearance of the lethality sector associated with early gene transcription.

LETHAL EFFECT OF DAMAGE IN DIFFERENT REGIONS OF SV40 DNA

Damage in DNA disrupts DNA replication (Rasmussen and Painter, 1964; Sarasin and Hanawalt, 1980), RNA synthesis (Mayne, (1984) and specific protein-DNA interactions (Siebenlist et al., 1980; Ptashne et al., 1980). We examined the relative sensitivities of these genetic functions to UV-induced damage. For these experiments we introduced defined amounts of damage into specific regions of the genome of SV40 and determined the effect of the damage on the survival of transfected viral DNA. We expected that the disruption of especially sensitive functions within a particular region would render the DNA hypersensitive to lesions.

SV40 was chosen for this study because its genome can be divided into functional domains (Tooze, 1980). A 302-bp noncoding sequence between the BglI and KpnI restriction sites contains the promotor and enhancer elements that regulate the transcription of viral genes (Takahashi et al., 1986; Hartzell et al., 1984) and also contains part of the origin of viral DNA replication (Li et al., 1986). We expected that the survival of trans-fected DNA would be hypersensitive to lesions in this regulatory region if the disruption of its control functions played a role in the toxicity of DNA damage. A 1969-bp region between the TaqI and BclI restriction sites contains most of the transcribed sequences of the viral A gene, which codes for the T antigen. Expression of this gene is essential for the early events of viral infection, including the initiation of viral DNA replication (Tooze, 1980). Accordingly, hypersensitivity to damage within the A gene region would be expected if lesions inactivated the gene by blocking RNA synthesis. In contrast, a 1488-bp region between the EcoRI and KpnI restriction sites was expected to be relatively insensitive to damage because this part of the viral genome plays no known role in the early events of viral infection. Although this region contains the coding sequences for viral coat proteins these genes are necessary only late in infection and would therefore be transcribed from undamaged progeny viral genomes, provided DNA replication occurs. Lesions in this late genes region should thus have little effect other than to inhibit DNA replication.

Our strategy for constructing damaged SV40 molecules exploited the occurrence of several unique restriction sites in the viral genome. Cleavage of viral DNA with combinations of two such enzymes yielded pairs of restriction fragments. Ligation of an undamaged fragment to a complementing fragment damaged by exposure to UV radiation yielded a viral genome with a defined amount of damage in a defined region. Transfection of such viral genomes revealed the amount of damage required in each affected region to inactivate viral DNA.

We used a chromatographic procedure (Niggli and Cerutti, 1982) to directly determine the amount of damage introduced by UV radiation in each region of SV40 relevant to this study. Results demonstrated that for all regions the same fraction of thymine was converted to TT dimers per dose of UV radiation (Brown and Cerutti, 1986). Different sensitivities could not, therefore, be attributed to differences in the amounts of damage formed in each region.

We found that different genomic sequences of SV40 varied considerably in their sensitivity to damage. The regulatory region was the most sensitive; lesions in this region were more than three-fold more effective in inactivating SV40 DNA than was the same amount of damage distributed randomly in the viral genome. As a control, we showed that the same amounts of damage in another small fragment that does not encode essential viral functions did not inactivate viral DNA. Lesions in the region coding for the viral early genes were about as effective (1.1-fold) in inactivating DNA as was randomly distributed damage. In contrast, the 1488-bp late genes region was remarkably insensitive to the presence of damage. Lesions in this area of the genome inactivated viral DNA, but only 45% as effectively as randomly distributed damage. The sum of sensitivities for all regions of the SV40 genome was equal to the overall sensitivity of viral DNA to randomly distributed damage (Brown and Cerutti, 1986).

The different sensitivities of each region of the SV40 genome allowed us to estimate the proportion of the lethal effect of damage attributable to the disruption of individual viral functions. We assumed that lesions throughout the viral genome would inhibit DNA replication to an equal extent. Accordingly, sensitivity due to the inhibition of DNA replication was set at 45% (relative to 100% for randomly distributed damage), which is the sensitivity of the least affected, late genes region. This value, when subtracted from the higher sensitivities of the early genes and regulatory regions, yielded residual sensitivities attributable to the disruption of genetic functions peculiar to these regions. Thus, 45% of the overall lethality of randomly distributed damage was assigned to the inhibition of viral DNA replication; 30% was attributed to the inhibition of RNA synthesis from the viral early gene (residual sensitivity of the entire early gene transcription unit from the BglI to the BclI restriction sites)., and 15% was attributed to the disruption of regulatory functions encoded by sequences between the BglI and KpnI restriction sites.

COMPLEMENTATION OF INACTIVATED EARLY GENES IN COS-1 CELLS

We confirmed the dual-component character of lethal damage in the SV40 early genes and regulatory regions by showing that lethality ascribed to the inactivation of viral early functions is fully reversed if damaged virus is transfected into host cells modified to constitutively express SV40 early gene functions. Data presented in figure 1 compare the survival of damaged SV40 DNA in CV-1 and cos-1 African green monkey cells. CV-1 cells are the normal hosts for SV40 in tissue culture and were used for the survival experiments described above. cos-1 are CV-1 cells transformed with a derivative of SV40 lacking a functional origin of viral DNA replication. SV40 early gene sequences integrated into the genomic DNA of cos-1 cells constitutively produce viral early gene products and complement the growth of mutant viruses whose own early genes are inactive (Glutzman, 1981). We reasoned that cos-1 cells would also complement the inability of irradiated virus to express UV-inactivated early genes. Accordingly, viral sensitivity to lesions in the early genes and regulatory regions should be reduced in cos-1 cells relative to CV-1 cells whereas sensitivity in the late genes region should remain unchanged.

Our results (Brown and Cerutti, 1987) indicated that the lethal effect of damage in the SV40 early genes and regulatory regions was reduced in cos-1 cells to the level of sensitivity observed for the viral late genes region (figure 1). In contrast, the lethality of damage in the late genes region was not reduced in cos-1 cells, a result that confirmed that these cells complement only early gene functions of SV40, and also indicates that these cells are not merely more proficient in host-cell reactivation. The reduced sensitivity of SV40 DNA to randomly distributed damage in cos-1 cells reflects the complementation of lethality in the early genes and regulatory regions, and confirms that half of the lethal effect of random damage in the viral genome is due to the inactivation of viral gene function.

The overall lethal effect of UV-induced damage in SV40 DNA can thus be partitioned into three sectors, each assigned to the disruption of a particular genetic function. The three sectors correspond to (i) the lethal disruption of gene regulation, comprising almost all of the sensitivity of the viral regulatory region and 15% of the overall lethality of randomly distributed damage; (ii) the disruption of mRNA transcription, corresponding to the cos-1 cell-reversible sensitivity of the early genes region and 30-40% of the lethality of randomly distributed damage; and (iii) the inhibition of DNA replication, corresponding to the sensitivity of the late genes region in CV-1 and cos-1 cells, and ~50% of the lethality of randomly distributed damage. These sectors distinguish between lethality due to the disruption of DNA function and DNA integrity. Sectors 1 and 2 affect DNA function, or the ability of DNA to encode essential genes and regulatory activities. Sector 3 affects DNA integrity, or the ability of DNA to maintain an intact structure and to propigate as a macromolecule.

CORRELATION OF ER WITH THE REACTIVATION OF SV40 EARLY GENE EXPRESSION.

Mammalian cells respond to UV radiation by inducing an increased ability to reactivate UV-damaged virus (Cornelius et al., 1980; Sarasin and Benoit, 1980). Enhanced viral survival is most evident if cells are infected 2 days after they are irradiated. We used the three lethality sectors described for SV40 above to investigate which genetic function is altered to produce ER. We reasoned that ER would be most pronounced for viral DNA damaged in the region interacting with the major enhanced function.

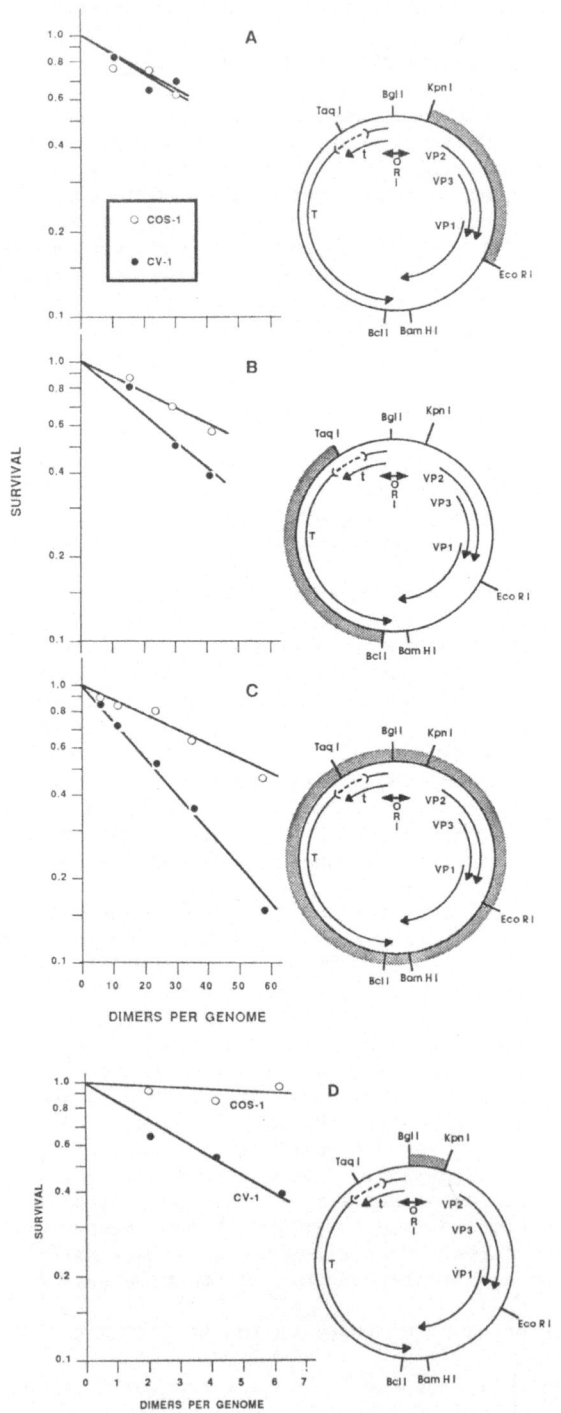

Figure 1. Survival in CV-1 (●) and cos-1 (O) of SV40
DNA containing the indicated number of pyrimidine
dimers in the shaded region shown on the
map to the right. From Brown and Cerutti (1987).

We induced ER by exposing CV-1 cells to 7.5 J/m^2 of UV radiation. Induced and control cells were transfected 48 hours later with UV-damaged SV40. Lesions were either randomly distributed or confined to specific regions. The results of five experiments can be summarized as follows: (i) the lethal effect of randomly distributed damage is reduced about 20% in ER-induced cells relative to unirradiated controls. This value of 20% is a dose reduction factor (DRF) obtained by dividing the difference between the survival slopes for induced and control cells by the slope for control cells. A DRF of 20% corresponds to a 2-fold increase in viral viability at 10% survival, equal to ER values previously reported. (ii) A larger DRF of about 40% is apparent for SV40 specifically damaged in the early genes region. (iii) No enhanced survival is observed for SV40 damaged in either the regulatory region or in the late genes region.

These results indicate that ER of SV40 counteracts lethal damage in the viral early genes region but not elsewhere in the viral genome. Thus, ER restores a UV-sensitive function peculiar to the early genes region. This function is almost certainly transcription.

Enhanced transcription of UV-damaged SV40 DNA could be explained in principle either by induced preferential excision repair of the early genes region or by the induced ability to transcribe despite the presence of UV-induced damage that would terminate mRNA synthesis in uninduced cells (Protic-Sabljic and Kraemer, 1985). Induced excision is improbable because ER is observed in excision-deficient Xeroderma pigmentosum cells (Lytle et al., 1976; Abrahams et al., 1984). We therefore believe that ER reflects a UV-induced cellular process allowing mRNA to be synthesized using a damaged DNA template.

Inducible processes leading to ER are thought to also mediate enhanced cell survival (Tyrrell, 1984). Our interpretation of ER therefore implies that the inhibition of transcription plays an important role in the UV-inactivation of cells, and that the induction of enhanced transcription using a damaged template could lead to enhanced cellular resistance to UV radiation. This view is supported by experiments showing that UV-induced damage in transcribed sequences is especially lethal (Mayne, 1984) and is preferentially repaired by constitutive cellular excision mechanisms (Bohr et al., 1985; Mellon et al., 1986). As discussed by Konze-Thomas et al. (1982), this view does not contradict earlier work that identified DNA replication as a major target for the lethal effect of DNA damage.

Enhanced viral reactivation is often accompanied by elevated viral mutation rates (Sarasin and Benoit, 1980), prompting the suggestion that reactivation is due to the induction of an inherently mutagenic process. ER is not always accompanied by elevated mutagenesis, however, (Cornelius et al., 1981; Taylor et al., 1981), indicating that ER is not invariably error-prone. Our results and interpretation imply that ER is not mutagenic. Transcription should not, in itself, introduce sequence changes in DNA. Enhanced reactivation and elevated mutagenesis may therefore be due to different processes that can be coordinately induced under some circumstances.

ACKNOWLEDGEMENTS

Work described here was supported by grants from the Swiss National Science Foundation. T.C.B. was supported by an EMBO postdoctoral fellowship.

231

REFERENCES

Abrahams, P.J., Huitema, B.A. and Van der Eb, 1984, <u>Mol</u>. <u>Cell</u>. <u>Biol</u>.
 4:2341.

Bohr, V.A., Smith, C.A., Okumoto, D.S. and Hanawalt, P.C., 1985, <u>Cell</u>
 40:359.

Brown, T.C. and Cerutti, P.A., 1986, <u>EMBO</u> J. 5:197.

Brown, T.C. and Cerutti, P.A., 1987, <u>Carcinogenesis</u> 8:1133.

Cornelius, J.J., Lupker, J.H. and Van der Eb, A.J., 1980, <u>Mutat</u>. <u>Res</u>.
 71:139.

Cornelius, J.J., Lupker, J.H., Klein, B. and Van der Eb, A.J., 1981,
 <u>Mutat</u>. <u>Res</u>. 82:1.

Glutzman, Y., 1981, <u>Cell</u> 23:175.

Hartzell, S.W., Byrne, B.J. and Subramanian, K.N., 1984, <u>Proc</u>. <u>Natl</u>. <u>Acad</u>.
 <u>Sci</u>. (<u>USA</u>) 81:23.

Konze-Thomas, B., Hazard, R.M., Maher, V.M. and McCormic, J.J., 1982,
 <u>Mutat</u>. <u>Res</u>. 94:421.

Li, J.J., Peden, K.W.C., Dixon, R.A.F. and Kelly, T., 1986, <u>Mol</u>. <u>Cell</u>.
 <u>Biol</u>. 6:1117.

Lytle, C.D., Day, R.S., Hellman, K.B. and Bockstahler, L.E., 1976, <u>Mutat</u>.
 <u>Res</u>. 36:257.

Mayne, L.V., 1984, <u>Mutat</u>. <u>Res</u>. 131:187.

Mellon, I., Bohr, V.A., Smith, C.A. and Hanawalt, P.C., 1986, <u>Proc</u>. <u>Natl</u>.
 <u>Acad</u>. <u>Sci</u>. (<u>USA</u>) 83:8878.

Niggli, H.J. and Cerutti, P.A., 1982, <u>Biochem</u> <u>Biophys</u> <u>Res</u>. <u>Commun</u>. 105:1251

Protic-Sabljic, M. and Kraemer, K.H., 1985, <u>Proc</u>. <u>Natl</u>. <u>Acad</u>. <u>Sci</u>. (<u>USA</u>)
 82:6622.

Ptashne, M., Jeffrey, A., Johnson, A.D., Maurer, R., Meyer, B.J., Pabo, C.O.,
 Roberts, T.M. and Sauer, R.T., 1980, <u>Cell</u>, 19:1.

Rasmussen, R.E. and Painter, R.B., 1964, <u>Nature</u>, 203:1360.

Sarasin, A. and Benoit, A., 1980, <u>Mutat</u>. <u>Res</u>., 70:71.

Sarasin, A.R. and Hanawalt, P.C., 1978, <u>Proc</u>. <u>Natl</u>. <u>Acad</u>. <u>Sci</u>. (<u>USA</u>) 75:346.

Siebenlist, U., Simpson, R.B. and Gilbert, W., 1980, <u>Cell</u> 20:269.

Takahashi, K., Vigneron, M., Matthes, H., Wildeman, A., Kenke, M. and
 Chambon, P., 1986, <u>Nature</u> 319:121.

Taylor, W.D., Bockstahler, L.E., Montes, J., Babich, M.A. and Lytle, C.D.,
 1981, Mutat. Res. 105:291.

Tooze, J., ed., 1980 <u>DNA</u> <u>Tumor</u> <u>Viruses</u>, 2nd edition, Cold Spring Harbor
 Laboratory Press, NY.

Tyrrell, R.M., 1984, <u>Proc</u>. <u>Natl</u>. <u>Acad</u>. <u>Sci</u>. (<u>USA</u>) 81:781.

DIFFERENTIAL REACTIVATION AND MUTAGENESIS OF SINGLE- AND DOUBLE-STRANDED DNA VIRUSES IN IRRADIATED CELLS FROM ATAXIA TELANGIECTASIA PATIENTS

Geneviève Hilgers[1,2], Jan Cornelis[1,3], Peter Abrahams[2], Ron Schouten[2], Alex van der Eb[2], and Jean Rommelaere[1,3]

[1]Department of Molecular Biology
Université Libre de Bruxelles
B-1640 Rhode St Genèse, Belgium

[2]Department of Medical Biochemistry
State University of Leiden
P.O.Box 9503, 2300 RA Leiden, The Netherlands

[3]Molecular Oncology Unit, INSERM U186 and CNRS UA 041160
Institut Pasteur de Lille
F-59019 Lille Cédex, France

INTRODUCTION

Still much has to be learned about the cellular and molecular mechanisms involved in the induction of gene mutations in mammalian cells. The induction of mutations by UV-light and chemicals producing bulky DNA lesions in E.coli depends to a great extent on an inducible pathway belonging to the so-called SOS regulatory network. Thanks to numerous bacterial mutants, the SOS system has been well studied at the molecular, phenomenological and genetic levels (Walker, 1984). Bacteriophages proved particularly useful as probes to unravel repair and mutagenic processes in SOS-induced E.coli (Defais et al., 1983). The increase in the survival of damaged phages in bacteria treated with UV-light, ionizing radiation or various chemical mutagens prior to infection, a phenomenon denoted Weigle Reactivation, was among the first evidence for the inducibility of a repair component of the SOS system. Similarly, SOS mutagenesis was first revealed as an enhanced induction of phage mutations under Weigle Reactivation conditions, a response known as Weigle Mutagenesis.

No conclusive evidence for the existence of a SOS type of regulation of repair and mutagenesis in mammalian cells has been reported so far (Radman, 1980 ; Rossman and Klein, 1985 ; Sarasin, 1985). However, conditional recovery responses could be convincingly demonstrated in eucaryotic cells, using various animal viruses as probes (Lytle, 1978 ; Defais et al., 1983 ; Cornelis and Rommelaere, 1988). Thus, "induced" mammalian cells achieve an Enhanced Reactivation (ER) of damaged single- and double-stranded DNA viruses, i.e. virus survival is greater in pretreated cell cultures than in mock-treated ones. ER was found to be accompanied with

Enhanced viral Mutagenesis (EM) in some but not all systems (Sarasin and Benoit, 1980 ; Taylor et al., 1982 ; Defais et al., 1983 ; Cornelis and Rommelaere, 1988). Like Weigle Reactivation and Mutagenesis in SOS-induced bacteria, ER and EM are often activated coordinatedly and their expression is both transient and dependent on de novo protein synthesis (Das Gupta and Summers, 1978 ; Su et al., 1981 ; Su et al., 1985).

Mammalian cells deficient for ER or EM would be of primordial value to characterize SOS-like responses in higher eucaryotes. Cells from patients with the cancer-prone hereditary syndrome ataxia telangiectasia (AT) may be interesting in this respect. AT cells are hypersensitive to the killing effect of ionizing radiation and some chemical clastogens (Shiloh et al., 1985, and references therein). Moreover, mutation induction by X-rays is reduced in AT cells compared with those from normal individuals (Arlett, 1980). Hence, AT cells might constitute natural mutants deficient in the control of repair and error-prone processes. In order to test this possibility, ER and EM of two nuclear-replicating viruses, parvovirus H-1 and herpes virus HSV-1,were studied in irradiated AT cells. H-1 and HSV-1 viruses contain a single- and double-stranded DNA genome, respectively, and are thus susceptible of revealing different cellular repair mechanisms. Preliminary results presented in this report do not support the view that AT cells are impaired for the overall induction of a SOS-like phenotype upon irradiation. The comparison of AT and normal cells on the one hand, and of single- and double-stranded DNA viruses on the other hand, suggests that radiation-induced ER can be subdivided into a least two components, only one of which is deficient in AT cells and may be related to EM.

MATERIALS AND METHODS

Cells and Viruses

SV40-transformed human cells of normal (NB-E and GM0637) or AT AT5BIVA) origin were maintained as described (Chen et al., 1986 ; Hilgers et al., 1987). Parvovirus H-1 and herpes simplex virus type 1 (HSV-1) were propagated, purified and titered as reported by Hilgers et al. (1987) and Abrahams et al. (1984), respectively.

Irradiation

H-1 and HSV-1 were UV-irradiated according to Hilgers et al. (1987) and Abrahams et al. (1984), respectively. For Fig.1a, cell X-irradiation conditions are given by Hilgers et al. (1987). For Fig. 1b and Table 1, cell pellets in plastic tubes were X-irradiated at an incident dose rate of 2.3-2.5 Gy/min (150 KV, 3mA) at room temperature.

Virus Survival

Irradiated or mock-treated growing cells were infected with irradiated and unirradiated virus. Virus surviving fractions were determined by infectious center assays as described previously by Hilgers et al. (1987) and Abrahams et al. (1984) for H-1 and HSV-1, respectively.

HSV-1 Mutation Assay

A forward mutation assay was used, in which mutations in the HSV-1 thymidine kinase (TK) gene were scored by determining the fraction of the virus population able to produce plaques in the presence of 10 μ g/ml 5-iododeoxycytidine (Icdr) (Abrahams et al., 1984).

Definitions

The multiplicity of infection (MOI) is the number of infectious viral units per cell for an equivalent inoculum of unirradiated virus. The level of ER expression is given by the reactivation factor which is the ratio of virus survival in UV (UVER) or X-ray (XRER)-treated cells over that in control cultures.

RESULTS AND DISCUSSION

Differential Capacity of AT Cells for Radiation-Induced Enhanced Reactivation of Single- and Double-Stranded DNA Viruses

Upon mild UV- or X-irradiation, normal human cells acquire an enhanced ability to reactivate UV-damaged H-1, a single-stranded DNA parvovirus (Cornelis et al., 1982 ; Hilgers et al., 1987). As illustrated by Fig. 1a for X-irradiated cells, the expression of this conditional recovery phenotype, indicated by reactivation factors greater than 1.0, is both delayed and transient. In contrast, a cell line derived from a patient with ataxia telangiectasia failed to achieve a detectable ER of damaged H-1 at any time post-irradiation (Fig. 1a), irrespective of the type and dose of radiation given to the cells (Hilgers et al., 1987). However, AT cells did not appear deficient in ER when UV-damaged HSV-1, a double-stranded DNA virus, was used as a probe (Fig. 1b). The magnitude of ER of HSV-1 in preirradiated AT cells was similar to that reported previously for normal human cells (Abrahams et al., 1984). UV-light was as efficient as X-rays for triggering ER in AT cells (Fig. 1b). Yet, both types of radiations may elicit the ER response by at least partly independent mechanisms since UV- and XRER were maximally expressed after different induction periods, i.e. 24 h and 72 h, respectively.

These observations suggest that ER involves several processes whose relative contributions to virus survival depend on the nature of the probe. One possibility would be that AT cells are deficient for a component of ER which is of little importance for the rescue of damaged double-stranded DNA but participates in the replication of single-stranded DNA containing lesions. Such a case would be reminiscent of the different DNA processing mechanisms which are responsible for most of the Weigle Reactivation of single- versus double-stranded DNA phages in SOS-induced bacteria (Defais et al., 1983 ; Silber and Achey, 1984). Alternatively, the differential susceptibility of H-1 and HSV-1 to ER in AT cells may be related to the fact that the former virus relies on the host DNA replicating machinery while the latter encodes its own DNA polymerase. One should consider at this point that the molecular basis of ER of animal viruses remains speculative, although reactivation may take place at the level of parvoviral single-stranded DNA conversion to double-stranded forms (Rommelaere and Ward, 1982) and SV40 gene expression (Brown and Cerutti, 1987 ; and this volume). In any case, our results suggest that

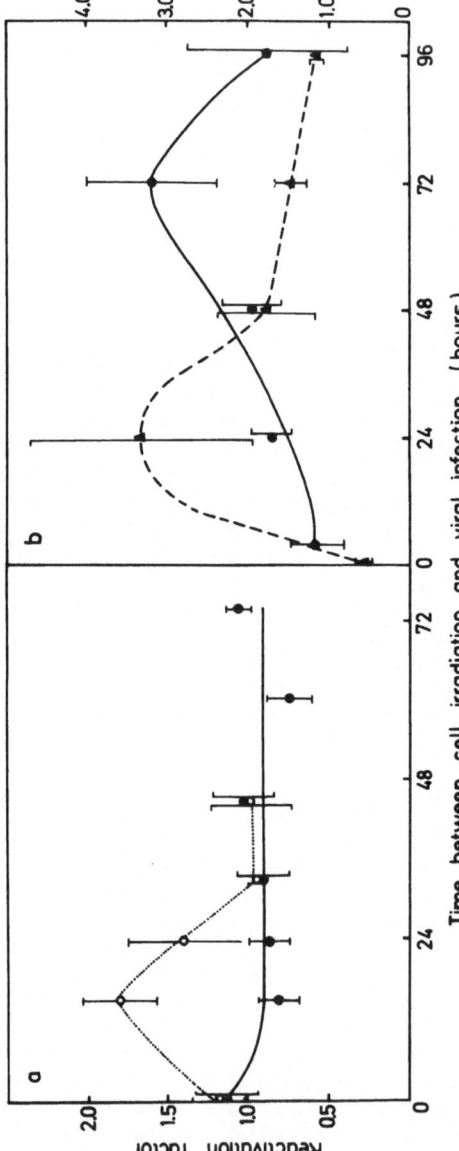

Fig.1. Time course of Enhanced Reactivation of damaged H-1 and HSV-1
viruses in irradiated normal and AT cells. Parallel cultures
were UV-, X- or mock-irradiated, incubated for various times at
37°C and infected with UV-damaged or intact virus. Average
values and standard deviation bars were calculated from 2-3
experiments. (a) ER of H-1 in X-irradiated normal and AT cells.
Data reproduced, with permission, from Hilgers et al. (1987).
O, NB-E cells ; ●, AT5BIVA cells. (b) ER of HSV-1 in UV- and X-
irradiated AT5BIVA cells. MOIs for mock-treated and UV-irradiated
(300 J m^{-2}) HSV-1 were 8 x 10^{-4} and 8 x 10^{-2},
respectively. ▲, UV-irradiated (4.5 J m^{-2}) cells ; ●, X-irradiated
(3.85 Gy) cells.

AT cells are not equivalent to induction-minus mutants. Although ER of
H-1 infectivity (Hilgers et al., 1987) and adenovirus Vag antigen
synthesis (Jeeves and Rainbow, 1986) is impaired in AT cells, the
occurrence of ER of HSV-1 in these cells indicates that they are not
intrinsically deficient in the signalling of conditional responses.

Differential Mutagenesis of HSV-1 in Normal and AT Cells

Contradictory evidence has been obtained as to whether ER of HSV-1
is accompanied by the Enhanced Mutagenesis (EM) of surviving viruses (Das
Gupta and Summers, 1978 ; Lytle and Knott, 1982 ; Takimoto, 1983). This
uncertainty, together with the aberrant ER phenotype of AT cells, prompted
us to compare normal and AT cells for HSV-1 mutagenesis under constitutive
and induced conditions. A forward mutation assay was used to measure the
fraction of mutants among the progeny of intact or UV-damaged HSV-1
produced by mock- or X-preirradiated cells. Observed viral mutation
frequencies are given in Table 1 ; mutation induction rates per UV unit
dose were calculated assuming linear dose-responses (Lytle et al., 1982).

UV-irradiation of HSV-1 gave rise to an increase in viral mutagenesis
taking place in untreated cells of both normal and AT origin. Together
with previous reports concerning HSV-1 (Das Gupta and Summers, 1978 ;
Lytle and Knott, 1982) and other viruses (Day and Ziolkowski, 1978 ;
Cornelis et al., 1981 ; Cornelis et al., 1982), this result indicates that,
contrary to E.coli, mammalian cells achieve a significant level of
constitutive mutagenesis of damaged viruses.

In cells of normal origin, HSV-1 mutations were induced at an increased rate by virus UV-irradiation if cultures were exposed to a mild
X-ray treatment prior to infection. A similar enhancement of the mutation
rate of UV-damaged HSV-1 was reported by Lytle and Knott (1982) for UV-
preirradiated normal cells. In contrast, X-irradiation of AT cells under
conditions triggering ER of UV-damaged HSV-1, resulted in a decrease of
the rate of UV-induced viral mutations. These data argue against the
error-proneness of the ER process present in AT cells, since its activation coincided with a lower incidence of viral mutations. However, the
fact that an EM response was detected in preirradiated normal human fibroblasts, raises the possibility that an additional error-prone process can
be triggered by radiation but is deficient in AT cells. Such a putative
conditional mutator does not appear to play a major role in ER of HSV-1
but may contribute to ER of parvovirus H-1 since the latter response is
also impaired in AT cells. One may speculate that the failure of AT cells
to achieve ER and EM of irradiated viruses with single- and double-
stranded DNA genomes, respectively, reflects a common defect at the level
of DNA synthesis opposite damaged templates.

CONCLUSION

Altogether, our data emphasize the complexity of ER, a conditional
recovery phenotype activated by radiation in human and animal cells. ER
appears to involve at least two processes which could be distinguished on
the basis of their respective contributions to the reactivation of HSV-1,
a double-stranded DNA virus and H-1, a single-stranded DNA virus. Enhanced
Mutagenesis may only accompany one ER component which is apparently
deficient in cells from AT patients. The relevance of the latter defect to
the aberrant reactions of AT cells to radiation remains to be elucidated.

Table 1. Radiation-Induced Mutagenesis of HSV-1 in Normal and AT Cells.[a]

UV Dose to Virus (J m^{-2})	X-Ray Dose to Cells (Gy)	Virus Mutation Frequency (x10^5)		Virus Mutation UV-Induction Rate (x10^5 m^2/J)	
		GMO637	AT5BIVA	GMO637	AT5BIVA
0	0	55	120		
40	0	71	167	0.4$\pm$0.3	1.2$\pm$0.6
0	8	40	167		
40	8	80	170	1.0$\pm$0.3	0.1$\pm$0.6

[a] Growing cells were X- or mock-irradiated and infected 72 h later with UV (surviving fraction = 0.2) or mock-treated HSV-1 (MOI = 2.5 x 10^{-1}). Average values and standard deviations from 3-4 experiments.

ACKNOWLEDGEMENTS

We thank R. Filon, Ch. Kumps and H.J. Knaken for their excellent technical assistance. We are also grateful to Drs. C.F. Arlett and A.R. Lehmann (Brighton) for providing AT5BIVA cells. This work was supported by grants from the Commission of the European Communities (contracts B16-78-B and B16-169-NL) and an Action de Recherche Concertée from the Services Belges de Programmation de la Politique Scientifique. GH was supported by I.R.S.I.A.

REFERENCES

Abrahams, P.J., Huitema, B.A., and van der Eb, A.J., 1984, Enhanced reactivation and enhanced mutagenesis of herpes simplex virus in normal human and xeroderma pigmentosum cells, Mol.Cell.Biol., 4:2341.

Arlett, C.F., 1980, Mutagenesis in repair-deficient human cell strains,in: "Progress in environmental mutagenesis", M. Alacevic, ed., Elsevier, North Holland.

Brown, T.C., and Cerutti, P., 1987, Gene expression rather than the initiation of DNA replication is the principal target of lethal UV-induced damage in a regulatory region of SV40 DNA, Carcinogenesis, 8:1133.

Chen, Y.Q., de Foresta, F., Hertoghs, J., Avalosse, B.L., Cornelis, J.J., and Rommelaere, J., 1986, Selective killing of simian virus 40-transformed human fibroblasts by parvovirus H-1, Cancer Res., 46:3574.

Cornelis, J.J., Lupker, J.H., Klein, B., and van der Eb, A.J., 1981, The effect of cell irradiation on mutation in ultraviolet irradiated and intact simian virus 40, Mut.Res., 82:1.

Cornelis, J.J., Su, Z.Z., and Rommelaere, J., 1982, Direct and indirect effects of ultraviolet light on the mutagenesis of parvovirus H-1 in human cells, EMBO J., 1:693.

Cornelis, J.J., and Rommelaere, J., 1988, Parvoviral probes for cellular responses to DNA damage, in : "Parvovirus handbook", P. Tijssen, ed., CRC Press, Boca Raton (in press).

Das Gupta, U.B., and Summers, W.C., 1978, Ultraviolet reactivation of herpes simplex virus is mutagenic and inducible in mammalian cells, Proc. Natl. Acad. Sci. USA, 75:2378.

Day III, R.S., and Ziolkowski, C.H.J., 1981, UV-induced reversion of adenovirus 5ts2 infecting human cells, Photochem.Photobiol., 34;403.

Defais, M.J., Hanawalt, P.C., and Sarasin, A.R., 1983, Viral probes for DNA repair, Adv.Radiat.Biol., 10:1.

Hilgers, G., Chen, Y.Q., Cornelis, J.J., and Rommelaere, J., 1987, Deficient expression of enhanced reactivation of parvovirus H-1 in ataxia telangiectasia cells irradiated with X-rays or UV-light, Carcinogenesis, 8:315

Jeeves, W.P., and Rainbow, A.J., 1986, An aberration in gamma-ray-enhanced reactivation of irradiated adenovirus in ataxia telangiectasia fibroblasts, Carcinogenesis, 7:381.

Lytle, C.D., 1978, Radiation-enhanced virus reactivation in mammalian cells, Natl.Cancer Inst.Monogr., 50:145.

Lytle, C.D., and Knott, D.C., 1982, Enhanced mutagenesis parallels enhanced reactivation of herpes virus in a human cell line, EMBO J., 1:701.

Lytle, C.D., Nikaido, O., Hitchins, V.M., and Jacobson, E.D., 1982, Host cell reactivation by excision repair is error-free in human cells, Mut.Res., 94:405.

Radman, M., 1980, Is there SOS induction in mammalian cells ?, Photochem.Photobiol., 32:823.

Rommelaere, J., and Ward, D.C., 1982, Effect of UV-irradiation on DNA replication of the parvovirus minute-virus-of-mice in mouse fibroblasts, Nucleic Acids Res., 10:2577.

Rossman, T.G., and Klein, C.B., 1985, Mammalian SOS system : a case of misplaced analogies, Cancer Invest., 3:175.

Sarasin, A., and Benoit, A., 1980, Induction of an error-prone mode of DNA repair in UV-irradiated monkey kidney cells, Mut.Res., 70:71.

Sarasin, A., 1985, SOS response in mammalian cells, Cancer Invest., 3:163.

Shiloh, Y., Tabor, E., and Becker, Y., 1985, Cells from patients with ataxia telangiectasia are abnormally sensitive to the cytotoxic effect of a tumor promotor, phorbol-12-myristate-13 acetate, Mut.Res., 149:283.

Silber, J.R., and Achey, P.M., 1984, Excision repair participates in the Weigle reactivation of ultraviolet light-irradiated ØX174 double-stranded DNA, Mut.Res., 131:1

Su, Z.Z., Cornelis, J.J., and Rommelaere, J., 1981, Mutagenesis of intact H-1 is expressed coordinately with enhanced reactivation of ultraviolet-irradiated virus in human and rat cells treated with 2-nitronaphtofurans, Carcinogenesis, 2:1039

Su, Z.Z., Avalosse, B., Vos, J.M., Cornelis, J.J., and Rommelaere, J., 1985, Mutagenesis at putative apurinic sites in alkylated single-stranded DNA of parvovirus H-1 propagated in human cells, Mut.Res., 149:1.

Takimoto, K., 1983, Lack of enhanced mutation of UV- and γ-irradiated herpes virus in UV-irradiated CV-1 monkey cells, Mut.Res., 121:159.

Taylor, W.D., Bockstahler, L.E., Montes, J., Babich, M.A., and Lytle, C.D., 1982, Further evidence that ultraviolet radiation-enhanced reactivation of simian virus 40 in monkey kidney cells is not accompanied by mutagenesis, Mut.Res., 105:291.

Walker, G.C., 1984, Mutagenesis and inducible responses to deoxyribonucleic acid damage in E.coli, Microbiol. Rev., 48:60.

DEFENCE AGAINST SOLAR RADIATION DAMAGE TO HUMAN SKIN CELLS

Rex M. Tyrrell

Cell Mutation Group
Swiss Institute for Cancer Research
CH-1066 Epalinges/Lausanne
Switzerland

INTRODUCTION

The skin, and in particular the basal layer of the epidermis, is the main target for damage to humans by solar radiation. The physical manifestations of this injury range from acute sunburn to long term effects such as skin ageing and basal and squamous cell carcinomas. Although solar radiation in the UVB (290-320 nm) range almost certainly makes the major contribution to the carcinogenic effectiveness of sunlight, radiation at longer wavelengths in the UVA (320-380 nm) range also causes many cellular changes. Indeed, using in vitro cytotoxicity data we have calculated that under certain solar spectral irradiance conditions (i.e. low zenith angles), UVA can account for up to 80 percent of the cytotoxic effectiveness of the combined UVA and UVB regions (Tyrrell and Pidoux, 1987). There is therefore a considerable interest in determining the targets and nature of the damage induced in cells by radiation throughout the solar range and learning more about the mechanisms of cellular defence against such damage.

Human skin cells have several lines of defence against solar radiation damage, the first of which is the purely physical protection provided by the stratum corneum. A second, and major line of cellular defence mechanism is repair and in particular DNA excision repair. However, activated oxygen species are likely to be important mediators of solar radiation damage (see R. Menighini, this volume) so that cellular scavenging mechanisms are also likely to play a role in protecting cells. In this respect, evidence from this laboratory strongly implicates endogenous glutathione as an important defence mechanism (Tyrrell and Pidoux, 1986). We also have evidence for a possible fourth line of defence since we have recently demonstrated the induction of a unique stress protein by UVA radiation (Keyse and Tyrrell, in press). The purpose of the following overview is to summarise information in these various areas, focusing in particular on information obtained using cultured human skin cells.

PHYSICAL PROTECTION

Absorption of solar radiation by the stratum corneum leads to a major attenuation of the levels of UVB radiation reaching the living epidermis. Nevertheless, a small but significant percentage of high energy UVB radiation can reach the basal layer of the epidermis and even the upper layer of the dermis (Bruls et al., 1984). Attenuation of UVA radiation is much weaker and as much as 20 percent of radiation at 365 nm is transmitted to the basal layer of the epidermis.

DNA REPAIR

UVB. Many studies have demonstrated that the clone-forming efficiency of fibroblasts cultured from individuals with the classical forms of Xeroderma pigmentosum (XP, excision repair deficient) is much more UV sensitive than that of cells derived from normal individuals (see for example, Andrews et al., 1978). These experiments provide the strongest evidence that excision repair plays a major role in removal of lethal damage induced by UVC (254 nm) radiation. More recent studies (Smith and Paterson, 1982; Keyse et al., 1983, our own unpublished observations) have shown that a large sector of lethal damage induced by radiation at 313 nm is also removed by excision repair. These studies not only demonstrate the importance of excision repair for removal of UVB-induced damage but also make it evident that DNA damage is involved. Removal of pyrimidine dimers induced by radiation at 313 nm has also been measured biochemically in human fibroblasts (Niggli and Cerutti, 1983). Although repair of UVB-induced DNA strand breaks has been measured in bacteria (Miguel and Tyrrell, 1983) no clear-cut measurements are yet available in mammalian cells. Certain cell lines derived from patients with Ataxia telangectasia, known to be sensitive to ionizing radiation, are also significantly more sensitive than normal strains to radiation at 313 nm (Smith and Paterson, 1981; Keyse et al., 1985). Although the precise defect in Ataxia telangectasia is unknown, it almost certainly involves DNA metabolism.

UVA. Although excision deficient human fibroblast cell lines are more sensitive to irradiation at monochromatic UVA wavelengths than are cell lines from normal individuals (Smith and Paterson, 1982; Keyse et al., 1983; unpublished results, this laboratory), the differences are much less marked than after UVB or UVC irradiations. Studies with polychromatic UVA radiation suggest that single-strand breaks are both induced and repaired in mammalian cell DNA (Roza et al., 1985; Rosenstein, 1986) but there is little evidence to suggest that either pyrimidine dimers or strand breaks are critical cytotoxic lesions at these longer wavelengths. It is possible that repair systems themselves are damaged by the longer wavelengths and that this could account for the only slightly enhanced sensitivity of repair deficient strains. However, it is equally reasonable to suggest that DNA is only a minor target for UVA-induced cytotoxicity and that damage to other cellular components becomes critical. This is a question which remains to be resolved.

SCAVENGING AND REVERSAL OF OXIDATIVE DAMAGE

There is considerable evidence that active oxygen species may be involved in the cytotoxic action of solar UV, particularly in the UVA region. UVA radiation inactivation of mammalian cells is dependent on the presence of oxygen (Danpure and Tyrrell, 1976) as is both UVB (Miguel and Tyrrell, 1983) and UVA (Webb, 1977) inactivation of bacteria. Recent in vitro studies have shown that both hydrogen peroxide and superoxide anions can be produced by photo-oxidation of NADH and NADPH (Czochralska et al., 1984; Cunningham et al., 1985) and it has been known for some time that hydrogen peroxide is formed via the photo-oxidation of tryptophan (McCormick et al., 1976). In vitro scavenger studies with transforming DNA (Peak et al., 1981) and recent observations concerning the enhancement of UVA cytotoxicity to human cells by deuterium (S.H. Moss, personal communication; unpublished observations, this laboratory) suggest that singlet oxygen may be involved in UVA effects. There are therefore several candidates for UVA generated active oxygen species.

If hydrogen peroxide or superoxide anions are involved in UVA cytotoxicity, we would expect that the common cellular scavenging enzymes, catalase and superoxide dismutase may exert a protective role. Inhibition of catalase by aminotriazole or superoxide dismutase by diethyldithiocarbamate lead to only slight modifications in the UVA sensitivity of human fibroblast populations (this laboratory, manuscript in preparation). Addition of exogenous catalase or superoxide dismutase has no effect on UVA cytotoxicity, nor does addition of the iron chelater, desferrioxamine. Taken together, these results suggest that neither H_2O_2 nor hydroxyl radical derived from H_2O_2 via an iron-catalysed Fenton's reaction are involved in UVA cytoxicity.

In addition to scavenging enzymes, cells contain endogenous compounds with radical scavenging properties such as metalothionein and glutathione, the latter being present in very high concentrations (around 5 mM) in many cell types. In recent experiments we have shown that endogenous glutathione plays a major role in protecting human skin cells against the cytotoxic action of both UVB and UVA radiations (Tyrrell and Pidoux, 1986; Tyrrell and Pidoux, submitted). Using an overnight (18 h) incubation of cell populations in appropriate concentrations of buthionine-S, R-sulfoximine (BSO), a specific inhibitor of ɣ-glutamyl cysteine synthetase, glutathione levels can be reduced to levels below 5 percent of normal at the maximum BSO concentrations employed. By incubation of cells in various concentrations of BSO to control glutathione levels, we have been able to show that there is an extremely close correlation between levels of resistance to radiation at 365 nm and levels of glutathione. Using fibroblast populations which have been maximally depleted of glutathione, we have examined the wavelength range over which glutathione can protect against lethal damage. There is a strong effect after irradiation at two UVA (334 nm, 365 nm) wavelengths and a visible wavelength (405 nm). More surprisingly, there is also a significant protective effect of glutathione after irradiation at 302 nm and a major effect at 313 nm. These data provide a very strong indication that there is free radical involvement in the cytotoxic action of UVB. This may be particularly significant, since the UVB range is believed to be the region of peak carcinogenic effectiveness of sunlight for humans. By comparing sensitivities of excision deficient human fibroblast strains and cells maximally depleted of glutathione, it is clear that glutathione is almost as important as excision repair in protecting human fibroblasts against cytotoxic radiation damage at 313 nm.

In further studies, we have compared the protective role of glutathione in fibroblasts and epidermal keratinocytes from the same human foreskin biopsy. The keratinocytes are sensitised to radiation at 313 nm by depletion of glutathione much less than are the fibroblasts. However, the epithelial cells contain at least 3 times the levels of endogenous glutathione as the fibroblasts and even maximum BSO concentrations cannot deplete the thiol to below 20-25 percent of normal levels. Higher levels of glutathione are also observed in the epidermis as compared to the dermis of both mice and man (Connor, 1986). These studies not only support qualitatively the relationship between glutathione levels and protection against UVB radiation but also indicate that the tissue at highest risk of incurring solar radiation damage (i.e. the epidermis) is also better protected by endogenous scavengers.

There are at least two ways in which glutathione may be involved in protecting cells against radiation damage. Firstly, it may simply act by quenching free radicals directly. However, glutathione also participates in detoxification of organic hydroperoxides by acting as the unique hydrogen donor for glutathione peroxidase. If a radical quenching mechanism is involved then glutathione depletion should be compensated for by adding back a suitable exogenous thiol compound such as cysteamine, which cannot act as a hydrogen donor for glutathione peroxidase. In this type of experiment we have shown that cysteamine does completely restore resistance against radiation at 302 nm in glutathione depleted cells, at least in the low to intermediate fluence range. However, both at longer UVB (313 nm) and UVA (365 nm) wavelengths, cysteamine is unable to restore protection to glutathione depleted cells except at very high fluences. These results indicate that the mechanism of radiation inactivation changes with increase in wavelength. At the shortest solar wavelengths, free radicals appear to be involved. However, at longer wavelengths, the role of glutathione appears to be more specific. Since singlet oxygen appears to be involved in UVA cytotoxicity (see above) and lipids are a prime target for this oxygen species, it is tempting to speculate that peroxidation of membrane lipids may constitute critical damage after irradiation at longer wavelengths and that glutathione does indeed exert a more specific role by acting as the hydrogen donor for glutathione peroxidase.

IS PROTECTION AGAINST SOLAR RADIATION DAMAGE INDUCIBLE?

UVC (254 nm) radiation treatment leads to a series of inducible responses in mammalian cells which resembles the SOS response in bacteria (Radman, 1980). However, the biological responses involved are much smaller and clearly do not constitute such a major defence mechanism as in bacteria. Since UVC and UVB radiations induce major DNA lesions in common (e.g. pyrimidine dimers, 6-4 pyo adducts), it is likely that UVB irradiation of cells will lead to similar inducible responses. However except for viral reactivation measurements at short UVB wavelengths (up to 302 nm, Coohill et al., 1978) there is no experimental support for this possibility.

In bacteria, UVA and H_2O_2 appear to induce a common protective pathway (Tyrrell, 1985; Sammartano and Tuveson, 1985) and catalase is induced with similar kinetics. We have found no evidence for a similar response in mammalian cells (unpublished results). However, we have also analysed the pattern of protein synthesis in human cells after treatment with UVA (334 nm, 365 nm) and near-visible (405 nm) radiations and found that a single 32 kDa stress protein is induced in cells within 2 h of radiation treatment (Keyse and Tyrrell, in press). This protein appears to be identical, by chemical peptide mapping, with a similar protein induced by low concentrations of hydrogen peroxide. Furthermore, the fluence threshold for induction of the protein is considerably lowered by depletion of intracellular glutathione. These observations suggest that the induction response probably involves radiation-induced free radicals and could be a cellular response to oxidative stress. Although there is no functional data available on this protein, it is induced at near-maximal levels at fluences which kill only a small percentage of the population, an observation which is consistent with a protective action against damage by natural sunlight.

SUMMARY

In summary, it appears that there exist in human cells at least three (and possibly a fourth inducible) lines of defence against damage by radiation in the solar UV range. The physical barrier provided by the stratum corneum provides the first line of defence, and this is particularly important in attenuating UVB radiation. A second, and major line of defence against UVB damage is DNA excision repair. However, active oxygen intermediates are generated by both UVA and UVB radiation and our recent experiments have shown that intracellular glutathione plays a major role in protecting cells against potentially cytotoxic damage induced throughout the solar UV range. Finally, we also have evidence for the induction of a unique 32 kDa stress protein by UVA radiation and we are now seeking evidence that this protein plays a functional role in protection against solar radiation damage.

ACKNOWLEDGEMENTS

The experimental work was supported by grants from the Swiss National Science Foundation (3.108.85) and the Swiss League Against Cancer.

REFERENCES

Andrews, A.D., Barret, S.F., and Robbins, J.H., 1978, Xeroderma pigmentosum neurological abnormalities correlate with colony forming ability after ultraviolet radiation, Proc. Natl. Acad. Sci. (USA), 75:1984.

Bruls, W.A.G., Sloper, H., van der Leun, J.C., and Berrens, L., 1984, Transmission of human epidermis and stratum corneum as a function of thickness in the ultraviolet and visible wavelengths, Photochem. Photobiol., 40:485.

Connor, M.J., 1986, Depletion of cutaneous glutathione by ultraviolet radiation, Photochem. Photobiol., 43:69S.

Coohill, T.P., James, L.C., and Moore, S.P., 1978, The wavelength dependence of ultraviolet enhanced reactivation in a mammalian cell-virus system, Photochem. Photobiol., 27:725.

Cunningham, M.L., Johnson, J.S., Giovanazzi, S.M., and Peak, M.J., 1985, Photosensitized production of superoxide anion by monochromatic (290-405 nm) ultraviolet irradiation of NADH and NADPH coenzymes. Photochem. Photobiol. 42:125.

Czochralska, B., Kawczynski, G., Bartosz, G., and Shugar, D., 1984, Oxidation of excited state NADH and NAD dimer in aqueous medium involvement of O_2^- as a mediator in the presence of oxygen, Biochim. Biophys. Acta, 801:403.

Danpure, H.J., and Tyrrell, R.M., 1976, Oxygen-dependence of near-UV (365 nm) lethality and the interaction of near-UV and X-rays in two mammalian cell lines, Photochem. Photobiol. 23: 171.

Keyse, S.M., and Tyrrell, R.M., Both near ultraviolet radiation and the oxidising agent hydrogen peroxide induce a 32 kDa stress protein in normal human skin fibroblasts. J. Biol. Chem. (in press).

Keyse, S.M., McAleer, M.A., Davies, D.J.G., and Moss, S.H., 1985, The response of normal and ataxia-telangiectasia human fibroblasts to the lethal effects of far, mid and near ultraviolet radiations, Int. J. Radiat. Biol., 48:975.

Keyse, S.M., Moss, S.H., and Davies, D.J.G., 1983: Action spectra for inactivation of normal and Xeroderma pigmentosum human skin fibroblasts by ultraviolet radiations, Photochem. Photobiol., 37:307.

McCormick, J.P., Fisher, J.R., Pachlatko, J.P., and Eisenstark, A., 1976, Characterisation of a cell lethal tryptophan photooxidation product: hydrogen peroxide, Science, 191-468.

Miguel, A.G., and Tyrrell, R.M., 1983, Induction of oxygen-dependent lethal damage by monochromatic (313 nm) radiation; strand breakage, repair and cell death, Carcinogenesis, 4:375.

Niggli, H.J., and Cerutti, P.A., 1983, Cyclobutane-type pyrimidine photodimer formation and excision in human skin fibroblasts after irradiation with 313 nm ultraviolet light, Biochemistry, 22:1390.

Peak, J.G., Foote, C.S., and Peak, M.J., 1981, Protection by DABCO against inactivation of transforming DNA by near-ultraviolet light: action spectra and implications for involvement of singlet oxygen, Photochem. Photobiol., 34:45.

Radman, M., 1980, Is there SOS induction in mammalian cells, Photochem. Photobiol., 32:823.

Rosenstein, B.S., 1986, Kinetics of the induction and repair of DNA strand breaks in normal human cells exposed to solar UV radiation, Photochem. Photobiol., 41:60S.

Roza, L., van der Schans, G.P., and Lohman, P.H.M., 1985, The induction and repair of DNA damage and its influence on cell death in primary human fibroblasts exposed to UVA or UVC irradiation, Mutation Res., 146:89.

Sammartano, L.J., and Tuveson, R.W., 1985, Hydrogen peroxide induced resistance to broad spectrum near-ultraviolet (300-400 nm) inactivation in Escherichia coli, Photochem. Photobiol., 41:367.

Smith, P.J., and Paterson, M.C., 1981, Abnormal responses to mid ultra-violet light of cultured fibroblasts from patients with disorders featuring sunlight sensitivity, Cancer Res., 41:511.

Smith, P.J., and Paterson, M.C., 1982, Lethality and the induction and repair of DNA damage in far, mid or near-UV irradiated human fibroblasts: comparison of effects in normal, Xeroderma pigmentosum and Bloom's syndrome cells. Photochem. Photobiol., 36:333.

Tyrrell, R.M., 1985, A common pathway for protection of bacteria against damage by solar UVA (334 nm, 365 nm) and an oxidising agent (H_2O_2), Mutation Res., 145:129.

Tyrrell, R.M., and Pidoux, M., 1986, Endogenous glutathione protects human skin fibroblasts against the cytotoxic action of UVB, UVA and near-visible radiations. Photochem. Photobiol., 44:561.

Tyrrell, R.M., and Pidoux, M., 1987, Action spectra for human skin cells: Estimates of the relative cytotoxicity of the middle ultraviolet, near ultraviolet and violet regions of sunlight on epidermal keratinocytes, Cancer Res., 47:1825.

Webb, R.B., 1977, Lethal and mutagenic effects of near-ultraviolet radiation, in: Photochemical and Photobiological Reviews, Vol. 2 (K.C. Smith, ed.) pp 169-261, Plenum Press, New York.

THE MOUSE TAIL PHOTOTOXICITY TEST

Bo Ljunggren

Department of Dermatology
Malmö General Hospital
214 01 Malmö, Sweden

INTRODUCTION

Phototoxicity to drugs and to other compounds to which
man may be exposed can be studied in a number of systems.
Oscar Raab around the turn of the century was the pioneer in
this field using paramecia, and since then a spectrum of
organisms ranging from single cells to man have been used for
this purpose (Table 1). In vitro techniques have the advantage
of being easy to perform and inexpensive, but they don't
account for the metabolic handling of the drug, and results
obtained in vitro are often flawed by the occurrence of false
negative as well as positive reactions. These tests may give
information about the phototoxic mechanism of action and they
may be helpful in screening studies. However, their value in
predicting clinical phototoxicity is usually limited (Ljunggren and Bjellerup, 1986).

Among the animal models, the guinea pig has been used predominantly for topical phototoxicity studies, whereas for systemic
phototoxicity the mouse has been the animal of choice. Earlier
animal studies have been based on the qualitative evaluation
of erythema and sometimes edema of ears and tails. To obtain
more accurate data, which will allow dose-response studies, a
quantitative method, such as the mouse tail technique is
desirable.

Table I. Models used to study phototoxicity

In vitro:	In vivo:
Paramecia	Guinea pig
Candida albicans	Rat
Red blood cells	Rabbit
Lymphocytes	Swine
Cell cultures	Mouse
Human fibroblasts	Man

METHOD

Ordinary albino mice in groups of 5-10 animals (5 is usually a sufficient number) are administered the test drug either intraperitoneally, or, if the compound is poorly soluble or locally irritating, through a gastric tube. The technique has been described in detail elsewhere (Ljunggren, 1983). The animals are then placed in a fixation device (Fig 1) which will prevent them from escaping during the irradiation procedure. The device will also guarantee a constant UV dose to the tail during the exposure, and it will prevent systemic phototoxic reactions, which otherwise might occur, by shielding the rest of the animal.

While restrained in the fixation tubes, the tails of the mice are exposed to UV-A from a bank of fluorescent blacklight tubes (Philips TL 40W/08). Since even large doses of UV-A fail to induce tail edema in the absence of a phototoxic substance, the exposure time is kept long in order to make sure that the peak concentration of the drug in the tail skin will fall within the irradiation period. The standard exposure time we use is 5 h. After the irradiation is completed, the animals are returned to their cages and kept out of light until the evaluation takes place.

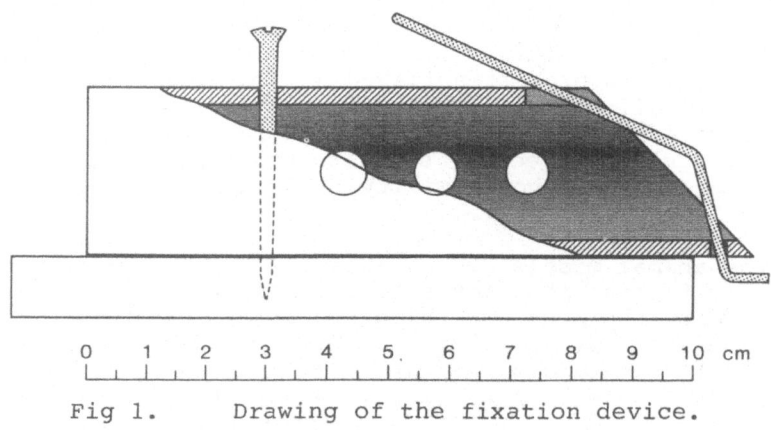

Fig 1.   Drawing of the fixation device.

After the mice have been killed, a 2 cm piece of the proximal portion of the tail is excised. The evaluation is based on the wet weight increase of the tail tissue as a consequence of the phototoxic damage. The relative wet weight is calculated following a drying procedure.

The optimal time for sacrifice varies with the phototoxic agent. With the psoralens a maximal reaction is obtained not until 72 h after irradiation, but with the majority of other phototoxic agents, like chlorpromazine, the reaction can be best evaluated at 24 h (Fig 2).

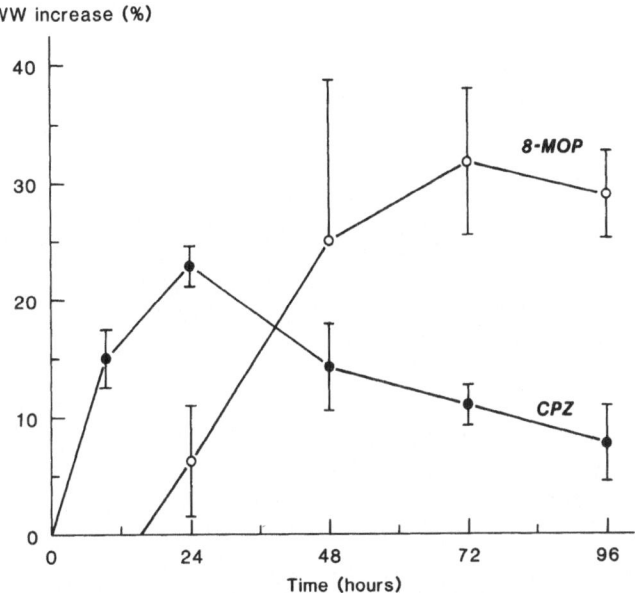

Fig.2 Time-course for phototoxic inflammation for 8-MOP and
chlorpromazine (CPZ)   ww = wet weight

The intensity of the inflammatory reaction also influences the time-course. Fig 3 shows the time-course for 3 different doses of UV-B, and it can be seen that the larger the dose, the earlier the maximal response will fall.

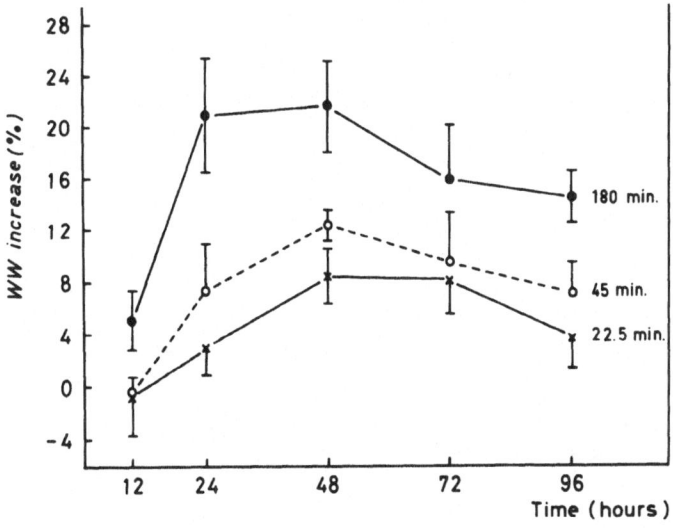

Fig.3. Time-course for 3 different doses of UV-B radiation

APPLICATIONS

Members of most groups of drugs known to cause photosensitivity reactions have been tested with the mouse tail technique (Ljunggren and Möller, 1978). Table II lists some representative compounds which are phototoxic in this model. The threshold phototoxic dose is given in the middle column. There is a wide dose range, and an occasional compound like amiodarone, is phototoxic only after repeated administration. Some compounds were negative in this test (Table III), most notably the sulfonamides and their derivatives. In this group only sulfanilamide was phototoxic. Other evidence, mainly clinical, however suggest that these photoreactions may not be of toxic origin.

Table II. Some representative compounds phototoxic in the mouse tail technique. TD = threshold dose in mg/kg. a = demethylchlortetracycline. b = given on 4 consecutive days.

Compound	TD	Adm.
Chlorpromazine	2.5	i.p.
8-methoxypsoralen	10	p.o., i.p.
Protoporphyrin	10	i.p.
Chlordiazepoxide	20	i.p.
Benoxaprofen	25	i.p.
Trimethylpsoralen	40	p.o.
Doxycycline	50	p.o., i.p.
Nalidixic acid	50	p.o.
DMCT[a]	100	i.p.
Griseofulvin	200	p.o.
Sulfanilamide	600	p.o.
Kynurenic acid	800[b]	p.o.
Amiodarone	400[b]	p.o.

Table III. Compounds negative in the mouse tail test. a = also tested with UV-B

Compound	Dose range tested, mg/kg	Adm.
Chloroquine	20-160	i.p.
Chlorothiazide[a]	600-1200	p.o.
Cyclamate	1600	p.o.
Sulfaisodimidine[a]	200-2400	p.o., i.p.
Tolbutamide	600	p.o., i.p.

The psoralens and some of the phenothiazines, particularly chlorpromazine, are among the most powerful phototoxic agents of all tested, on a weight basis. The dose-response curves for some psoralen derivatives shown in Fig 4, demonstrate the very steep gradation and the high relative wet weight increases attained of about 40%.

When a number of structurally closely related substances are available for testing, conclusions regarding structure-activity relationships may be drawn. For the phenothiazine group such studies have been performed (Ljunggren and Möller, 1977). Fig 5 is an example of a number of chlorpromazine metabolites tested for phototoxicity in the mouse. The chlorpromazine curve falls in the middle. The figure shows that one of the major metabolites in man, the chlorpromazine sulfoxide, is markedly less phototoxic than the mother compound, whereas on the other hand the two desmethylated metabolites (DDCPZ and DCPZ) are at least twice as phototoxic as chlorpromazine itself. Obviously, metabolites occurring in vivo can contribute to clinical phototoxicity.

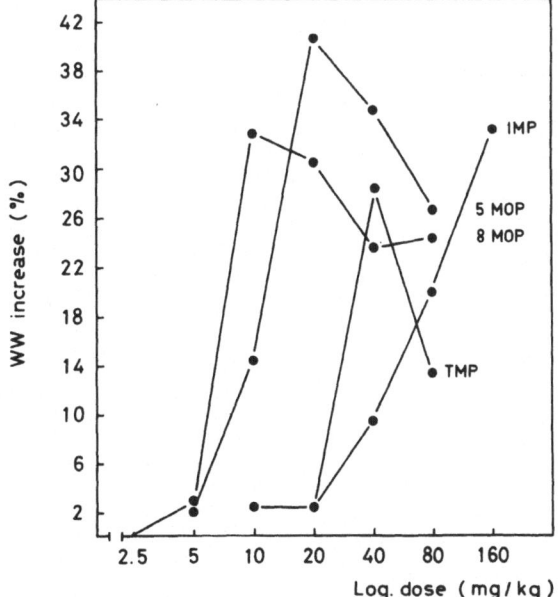

Fig 4. Dose-response curves for 4 psoralens (TMP = trimethylpsoralen, 8-MOP = 8-methoxypsoralen, 5-MOP = 5-methoxypsoralen, IMP = imperatorin).

A class of compounds recently in the focus concerning photosensitivity reactions is the NSAID group. In vivo, in the mouse, a considerable number of NSAID's are phototoxic, the propionic acid derivatives generally being the most potent. The wellknown antiphlogistic benoxaprofen which was withdrawn from the market a couple of years ago, is also quite active (Fig 6) (Ljunggren and Lundberg, 1985). Another wellknown NSAID, piroxicam, was on the other hand negative in this model.

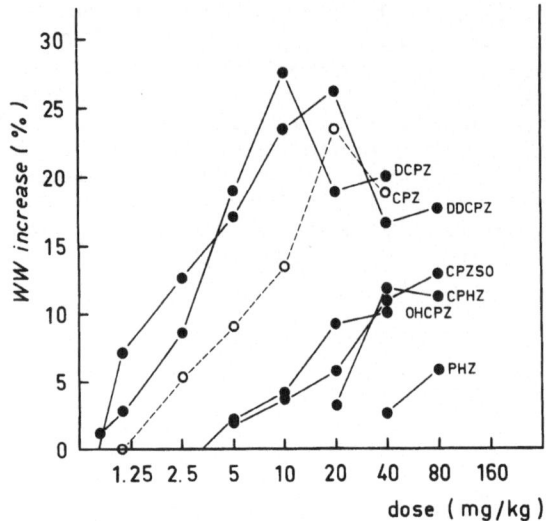

Fig 5.   Dose-response curves for 7 chlorpromazine (CPZ)
         metabolites. CPZSO = chlorpromazine sulfoxide,
         DCPZ and DDCPZ = des- and didesmethylchlorproma-
         zine respectively.

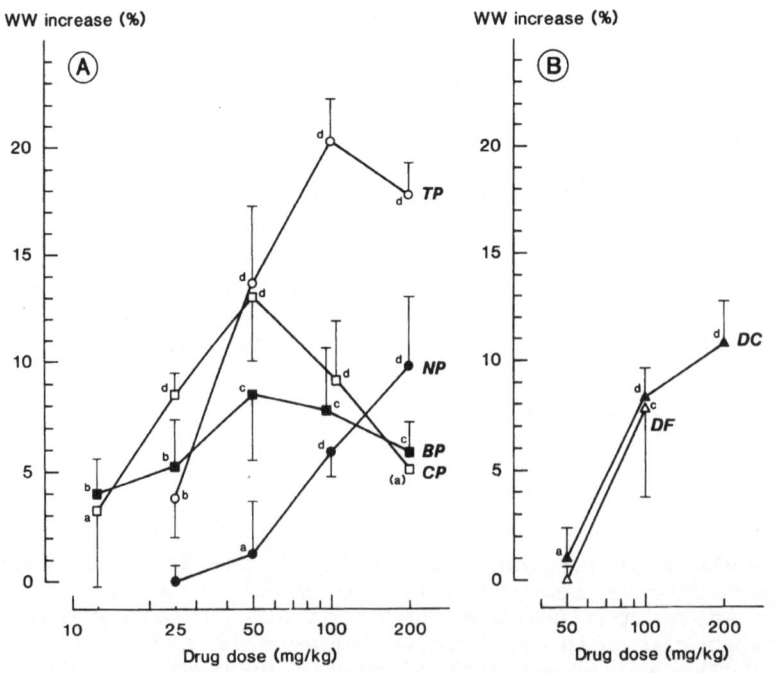

Fig 6.   Dose-response curves for 6 phototoxically active
         NSAID-derivatives. TP = tiaprofenic acid, NP =
         naproxene, BP = benoxaprofen, CP = carprofen, DC =
         diclofenac, DF = diflunisal.

CONCLUSIONS

The mouse tail technique is a sensitive, quantifiable and cost-efficient method that has successfully been applied to most known phototoxic agents. It can conveniently be adapted to the screening of large numbers of compounds, and it is thus a method well suited for predictive purposes.

REFERENCES

Ljunggren B., 1983, The mouse tail technique, Photoderma-tology, 1:96-100.

Ljunggren B., Bjellerup M., 1986, Systemic drug photosen-sitivity. Photodermatology, 3:26-35.

Ljunggren B., Lundberg K., 1985, In vivo phototoxicity of non-steroidal antiinflammatory drugs evaluated by the mouse tail technique. Photodermatology, 2: 377-382.

Ljunggren B., Möller H., 1977, Phenothiazine phototoxicity: An experimental study on chlorpromazine and its metabolites. J Invest Dermatol 68:313-317.

Ljunggren B., Möller H., 1978, Drug phototoxicity in mice. Acta Dermatovener (Stockh) 58:125-130.

PHOTOHEMOLYSIS TECHNIQUES -

AN UPDATE

Mats Bjellerup

Department of Dermatology
Malmö General Hospital
214 01 Malmö, Sweden

INTRODUCTION

Several experimental methods are available for the detection of phototoxic reactions. The action mechanisms studied with the different systems range from molecular level events to complex experimental animal reactions. For investigations on phototoxic membrane damage the photohemolysis technique is a suitable method.

METHOD

The pioneering work using the photohemolysis technique was done by Blum (1964) and Cook (1955) and was further developed by Kahn and Fleischaker (1971). A recent update by Hetherington and Johnson (1984) describes the technique in some detail. Red blood cells (RBC) irradiated with ultraviolet radiation (UVR) in the presence of a photosensitizing agent may hemolyze secondary to membrane damage. The concentration of hemoglobin in the supernatant can be measured spectrophotometrically and is proportional to the injury inflicted. Since RBC are organelle - poor and devoid of nuclei this method is thought to reflect cell membrane damage exclusively. The mechanism behind the lysis of RBC is a photodegradation of biomembranes probably synonymous with lightinduced oxidation of membrane components. These oxidative reactions may involve different components of the cell membrane such as phospholipids, cholesterol and amino acids which in the presence of radiation, molecular oxygen and appropriate sensitizer may undergo processes of lipid peroxidation and destruction of amino acid side chains (Lamola, 1976). Another, indirect, mechanism leading to lysis of RBC is the production of a toxic photoproduct as has been shown with preirradiation of chlorpromazine (Johnson, 1973).

APPLICATIONS

Comparative Studies

The photohemolysis method can be used for comparative studies of phototoxic potency within one specific group of well known photosensitizers such as the tetracyclines (Bjellerup, 1985) (fig 1), as well as for comparative studies of non related groups of sensitizers (Kahn, Fleischaker 1971).

Predictive Testing

Newly developed substances can be screened for assessing the probability of adverse photoreactions when introduced into humans. However substances that are dependent on the interaction with other than membrane components such as RNA or DNA for their damaging effect (e.g. psoralens) will not be detected.

Studies on Photochemical Mechanisms

Phototoxic reactions on the molecular level can be studied with the photohemolysis technique either indirectly by studying changes in the hemolysis rate by addition of certain quenchers or accelerators or by direct chemical analyses of membrane components. Thus, in the former case, the involvment of singlet oxygen in the process is indicated if the lysis of RBC is enhanced in a deuterium oxide environment (which prolongs the lifetime of singlet oxygen) or is slowed down by the addition of e.g. histidine (which is one of many quenchers of singlet oxygen) as reported by Swanbeck et al. (1974). Direct chemical analyses of hydroperoxide residues in ruptured RBC membranes produced by thiazide photohemolysis suggested that lipidperoxidation in RBC membranes may play a significant role in mediating the hemolysis (Matsuo et al. 1986).

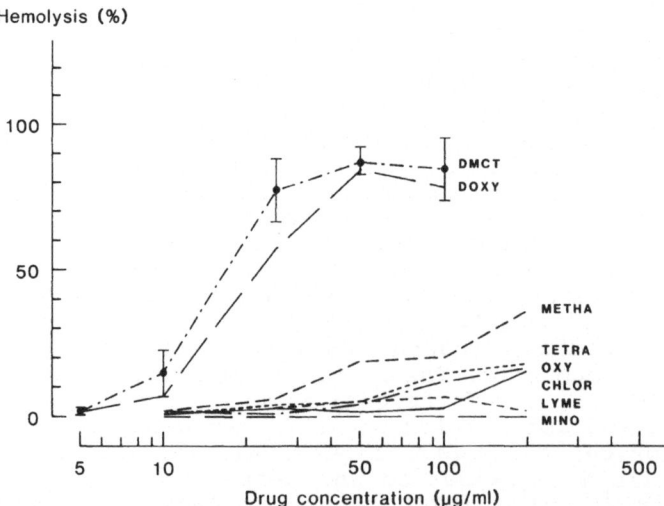

Fig.1. Dose-response curves for photohemolysis with 8 tetracycline derivatives (DMCT demethylchlortetracycline, DOXY doxycycline, METHA methacycline, TETRA tetracycline, OXY oxytetracycline, CHLOR chlortetracycline, LYME lymecycline, MINO minocycline). RBC in increasing concentrations of drug were irradiated with 72 J/cm^2 of UVA and incubated for 2 h. The values for DMCT are the mean of 10 experiments, vertical bars indicating the SD.

TECHNICAL ADVICE

There are several pitfalls that has to be considered
when applying the photohemolysis technique in order to gain
optimal results. The significant ones are discussed in the
following:

Dark Hemolysis

Some substances are hemolytically active in themselves
without the addition of UVR. Such hemolysis can be compensated
for if an appropriate dark hemolysis control is included
according to fig 2. row 4.

Irradiation Quality

Most phototoxic substances have their action spectrum in
the UVA and the photohemolysis is thus provoked by these wavelengths. Some substances however are activated by UVB only
while other, such as benoxaprofen, are activated by both UVA
and UVB (Ljunggren et al. 1983) making it necessary to test
both parts of the spectrum.

UVR Induced Colour Changes

It is important to compensate for possible colour changes
of the irradiated drug solutions. Changes in colour may cause
absorption of the wavelengths used for the spectrophotometrical

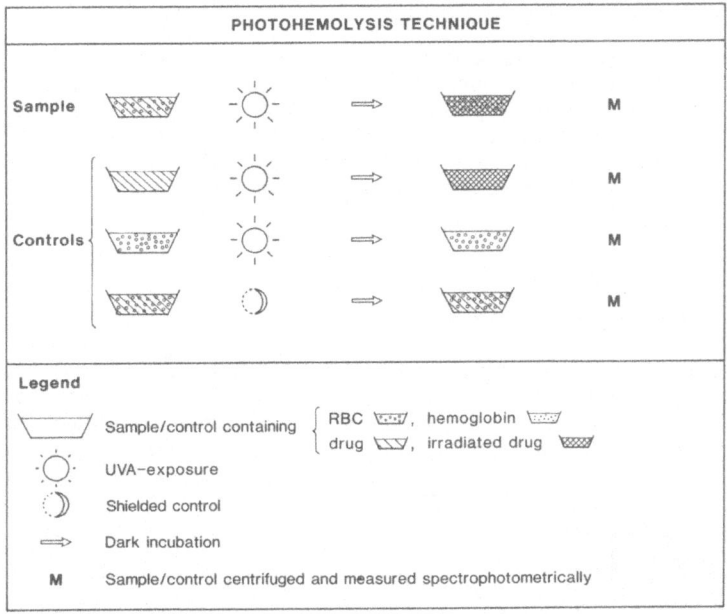

Fig. 2. Schematic illustration of the photohemolysis technique.

hemoglobin determination and thus cause falsely high estimated hemolysis. This can be overcome by an irradiated drug control without RBC as indicated in fig 2. row 2.

UVR Induced Hemolysis

UVR in itself, without the addition of sensitizer, may cause hemolysis. This is the case especially with UVB but also with very large doses of UVA and is detected with a control consisting of RBC irradiated in the absence of photosensitizer (fig 2. row 3).

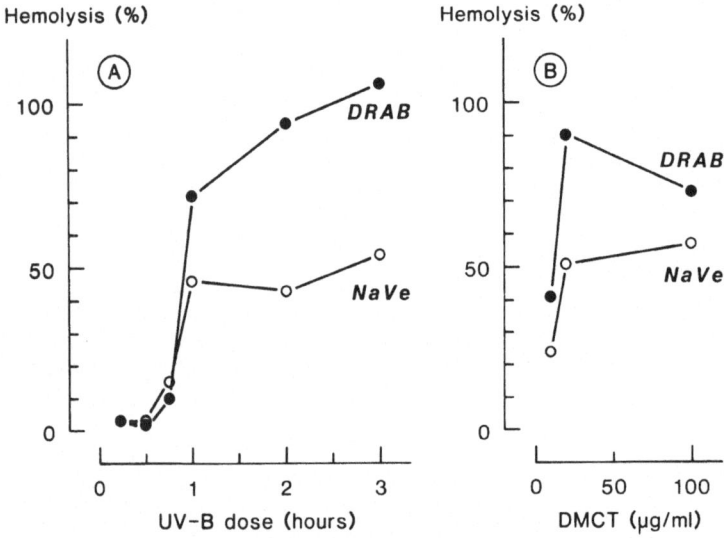

Fig. 3. Dose-response curves for photohemolysis showing the difference in hemoglobin determination with (DRAB) and without (NaVe) the use of Drabkin's method. A: RBC irradiated with UVB without drug. Incubation time 2 h. B: RBC together with 10, 25 and 100 ug/ml of the DMCT irradiated with 2 h of UVA. Incubation time 2 h.

Formation of Methemoglobin

Two methods for hemoglobin determination exist namely the oxyhemoglobin method and the cyanmethemoglobin method. With the oxyhemoglobin method blood is oxygenated by exposure to atmospheric oxygen and read at 540 nm, one disadvantage being the inability to include methemoglobin which under natural conditions only constitutes 1% of RBC hemoglobin. With the cyanmethemoglobin method using Drabkin's solution [K_3Fe (CN)$_6$ 0.2 g, KNC 0.05 g, KH_2PO_4 0.14 g diluted with distilled water to 1000 ml with pH 7.0 - 7.4] practically all forms of hemoglobin are converted to cyanmethemoglobin absorbing at 540 nm. Since UVR alone or in combination with a photosensitizer is capable of converting oxyhemoglobin to methemoglobin, having its absorption peak at 630 nm, the oxyhemoglobin method is not suitable when UVR is involved.

This problem was recognized though not solved by Kahn and Fleischaker who named it the fading of hemoglobin (1971). The difficulties however can be overcome by the use of Drabkin's solution as proposed by Johnson (personal communication 1981) and later Hetherington and Johnson (1984). Without the use of Drabkin's solution falsely low values will be obtained as has been shown for UVB hemolysis and demethylchlortetracyclin + UVA hemolysis as shown in fig 3 (Bjellerup, Ljunggren 1984).

SUMMARY

The photohemolysis technique is a suitable method for studies on phototoxic membrane damage. It is sensitive, quantifiable, cost efficient and simple. This method can be used for comparative studies on the phototoxic potency of well known sensitizers, for screening of putative sensitizers and for investigation of phototoxic mechanisms, especially those who involve lightinduced oxidation of membrane components. In order to gain optimal results several factors have to be considered such as; possible colour changes during irradiation, conversion of hemoglobin to methemoglobin during irradiation and the influence of different qualities of UVR.

REFERENCES

Bjellerup M., Ljunggren B., 1984, Studies on Photohemolysis with Special Reference to Demethylchlortetracycline, Acta Dermatovenereol (Stockh), 64:378.

Bjellerup M., Ljunggren B., 1985, Photohemolytic Potency of Tetracyclines, J Invest Dermatol, 84:262.

Blum, H.F., 1964, "Photodynamic action and diseases caused by light", Hafner Publishing Company, New York.

Cook, J.S., 1955, Some charcteristics of hemolysis by ultraviolet light, J Cellular and Comparative Physiology, 47:55.

Hetherington, A.M., Johnson, B.E., 1984, Photohemolysis, Photodermatology, 1:255.

Johnson, B.E., 1973, Chlorpromazine phototoxicity: a non-classical mode of action, Br J Dermatol, 89:16.

Kahn G, Fleischaker B., 1971, Red blood cell hemolysis by photosensitizing compounds, J Invest Dermatol, 56:85.

Lamola, A.A., 1976, Photodegradation of Biomembranes, in: "Research in Photobiology", A.Castellani, ed., Plenum Press, New York.

Ljunggren B., Bjellerup M., Möller H., 1983, Experimental Studies on the Mechanism of Benoxaprofen Photoreactions, Arch Dermatol Res, 275:318.

Matsuo I., Fujita H., Hayakawa K., Okkido M., 1986, Lipid Peroxidative Potency of Photosensitized Thiazide Diuretics, J Invest Dermatol, 87:637.

Swanbeck G., Wennersten G., Nilsson R., 1974, Participation of Singlet State Exited Oxygen in Photohemolysis Induced by Kynurenic Acid, Acta Dermatovenereol (Stockh), 54:433.

SYSTEMIC REACTIONS IN PHOTOTOXICITY

G.M.J. Beijersbergen van Henegouwen, R.W. Busker, H.de
Vries and S.A. Schoonderwoerd

Center for Bio-Pharmaceutical Sciences, State University
of Leiden
P.O.Box 9502, 2300 RA Leiden, The Netherlands

The conversion of 7-dehydrocholesterol into vitamin D_3 and that of
bilirubin into more water soluble products, upon exposure of the body
to UV-B (290-320 nm) and to visible light respectively, are two
examples of systemic photobiological effects by underdogenous compounds.
Vitamin D_3 is essential for proper bone calcification and the
photoconversion of bilirubin in the visible light therapy of neonatal
jaundice results in a decrease of brain-damaging effects. That
simultaneous exposure to (sun)light and a xenobiotic may also result in
systemic effects is indicated by results from animal experiments.

Fig. 1 Quindoxin (R_1=R_2=H); carbadox (R_1= -CH=N-NH-COOCH$_3$, R_2=H);
cyadox (R_1= -CH=N-NH-COCH$_2$CN, R_2=H) and olaquindox (OX; R_1= -CO-
NH-CH$_2$-CH$_2$OH; R_2= -CH$_3$). OX is supposed to react according to
the scheme (see also Jarrar and Fataftah, 1977).

Two examples of this research will be dealt with in this paper.
The first concerns olaquindox (OX, Fig.1) which, like carbadox and
cyadox, is a growth-promoting substance used as an additive to pig
feed. OX has two imino-N-oxide groups. The imino-N-oxide group has been
proven to be the cause of the phototoxicity of chlordiazepoxide (CDZ).

In that case systemic effects concerned a.o. a profound change of urinary metabolism, covalent binding of CDZ-fragments to liver and kidney tissue and a 25 - 30% decrease of liver/total weight (Beijersbergen van Henegouwen, 1987). The quinoxaline-1,4-dioxides are extremely photolabile. Quindoxin caused several cases of photocontact dermatitis (Zaynoun et al., 1976) and has been removed from the market. OX has been reported to be phototoxic in pigs.

The present experiments with OX have been performed as follows. On four successive days male Wistar rats (140 g) were shaved and given 60 mg/kg OX in 0.5 ml PBS suspension by oral intubation under brief ether anesthesia. The animals were housed in small metabolism cages covered with netting (mesh 2x2 cm). Four rats were kept in the dark and four others were exposed each day to UV-A (5 lamps, Philips TL 80W/ 10 R; UV-A= 340-400 nm, λ_{max}= 370 nm) for 12 h/day; light intensity at the level of the rats is 6 mW/cm^2 as measured with a UV-X radiometer (UV-products inc., San Gabriel USA). Urine, collected in light resistant

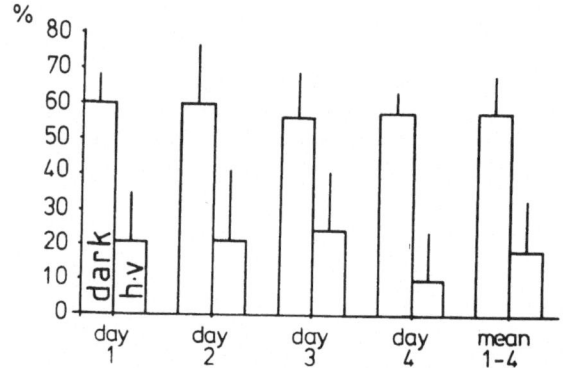

Fig. 2 Urinary excretion of OX, 60 mg/kg.day, as a % of dose applied by four rats exposed to UV-A (λ_{max}= 370 nm; I=6 mW/cm^2; 12 h/day) and four rats kept in the dark. Mean is summation of % of dose found daily divided by four (Dark= 58±11%; UV-A=19±15%; p<10^{-5} Student t test).

containers, was analyzed by HPLC (column: 200 mm x 3mm ID filled with Chromspher RP18 10μ; mobile phase 5% methanol/ phosphate buffer 0.01 M pH 3; detection: UV abs. 260 nm; retention-time OX= 10 min). The complete experiment was performed twice. After four days of treatment only the OX-treated UV-A exposed rats showed severe erythema on irradiated parts of the body, e.g. skin of the back, ears, feet and tail, oedema of feet and necrosis of ears. Systemic effects were indicated by a profound change of metabolism (Fig. 2).

Unchanged OX appeared to be eliminated via the urine for ~60% of total applied dose by non-UV-A exposed rats. This figure was only ~20% for the UV-A exposed rats. From preliminary experiments it appeared that one of the metabolites was excreted in a much higher quantity by the UV-A exposed rats (~5 times that of rats in the dark). Investigation of the identity of this metabolite, and of the possibility that covalent binding to organ tissue takes place, is in progress now. This will give more insight into the molecular mechanism underlying the phototoxic action of OX.

The widely used urinary antiseptic nitrofurantoin (NFT) is well known to cause a considerable number of hematologic reactions (Koch-Weser et al., 1971). Many of the NFT-induced hematologic reactions are related to glucose-6- phosphate-dehydrogenase (G6PD) deficiency. In some cases of hemolytic anemia caused by NFT, also high levels of MetHb were found (Stefanini, 1972). Earlier we reported on the formation of nitrite, a well known MetHb inducer, from 5-nitrofurfural (NFA) upon UV-A exposure (Busker and Beijersbergen van Henegouwen, 1987). As NFA is the major photodecomposition product of NFT, we investigated whether photolysis of NFT in vitro in human blood, or in NFT treated rats exposed to UV-A light, leads to an increase of the MetHb content (Busker et al. submitted).

The normal MetHb value is 0.5 %. The MetHb value of shaved rats, who were UV-A irradiated (intensity at 360 nm 2.5 mW.cm^{-2}, during 4 days, 12 h/day), was not significantly different from the normal value: 0.59 (s.d.=0.22; n=29). The same holds for the MetHb value of rats treated with NFT (4x12 mg p.o. on 4 consecutive days), and kept in the dark: 0.55 (s.d.=0.32; n=24). However rats exposed to UV-A after having been treated with 4x12 mg NFT p.o. showed a significant increase in MetHb%: 0.97 (s.d.=0.37; n=36; p<0.0001, Student t test). The combination of NFA (2x3 mg; i.p.) and UV-A was also capable to induce a MetHb increase with respect to treatment with either UV-A or NFA alone: 1.30 (s.d.=0.44; n=15; p<0.00001, Student t test) whereas NFA without light gave a MetHb% of 0.41 (s.d.=0.24; n=10). Further even a low dose of the secundary photoproduct of NFT, nitrite, is still active: 0.25 mg NaNO$_2$ in PBS (i.p.) led to MetHb% of 1.35 (s.d.=0.39; n=6).

The mechanism involved here was further studied in vitro. To examine whether nitrite plays a role in the MetHb formation, we assayed the [NO$_2^-$] and the [MetHb] in blood (1:10 diluted with PBS) formed upon irradiation of NFT. As can be seen in Fig. 3, only irradiation gives a considerable increase in both [MetHb] and [NO$_2^-$]. Depending on the starting concentration, 40-60 mol % NFT appeared to be decomposed in the irradiated samples. In the samples kept in the dark >95 % of NFT was unchanged. Nitrite alone, added to diluted blood in concentrations comparable to those in Fig.3 panel B, also caused an increase of [MetHb] although it could not account for all the MetHb found upon irradiation of NFT (Fig. 3 panel A). Further investigations, results not shown, revealed that NFA itself, the first photoproduct of NFT, also plays a role in the formation of MetHb in vitro.

The increase in MetHb found in rats due to NFT/UV-A is not very large. Thus the effects of light-induced MetHb formation are not expected to form a widespread problem for "normal" patients receiving NFT. However for G6PD or methemoglobin reductase deficient patients photo-induced MetHb generation may well be a cause of the dangerous blood disorders induced by NFT in these patients.

The two examples reported in this paper concern compounds of very different molecular structure. They indicate that, analogous to photobiological processes with endogeneous compounds systemic effects from the combination of (sun)light and xenobiotics may be expected as well.

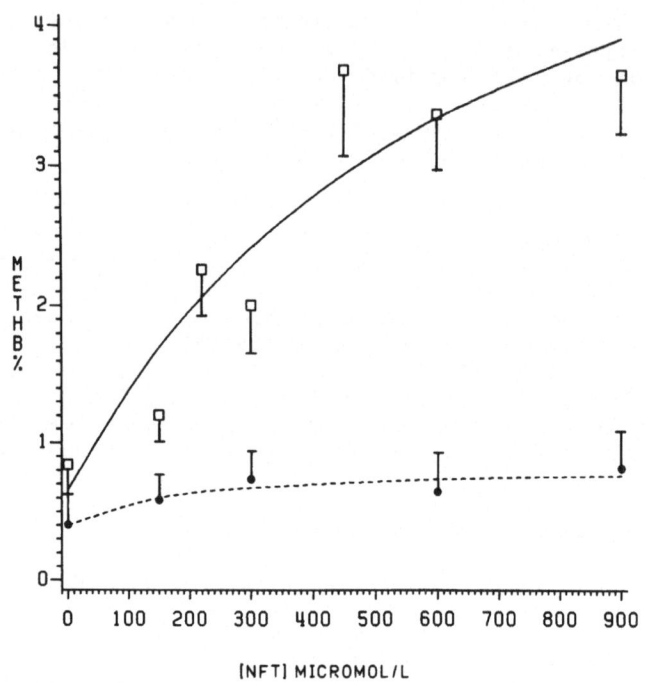

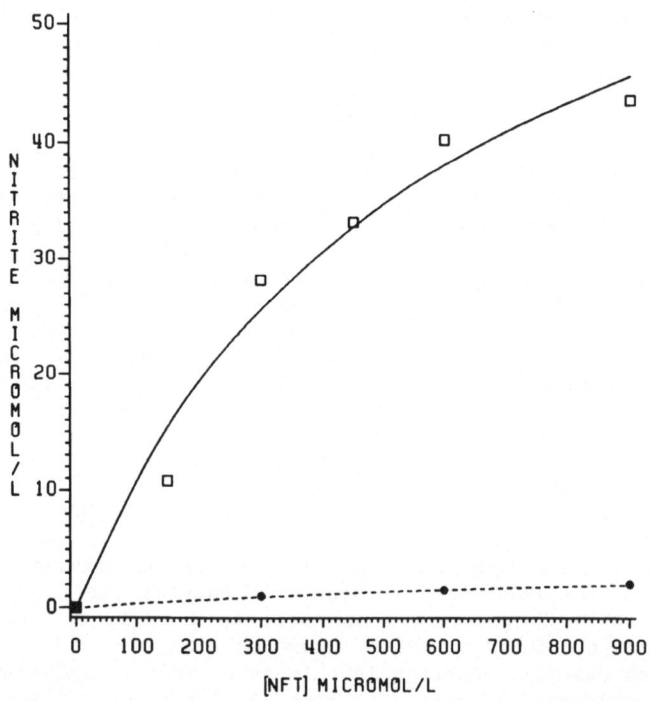

Fig. 3 Formation of MetHb (Panel A) and nitrite (Panel B) <u>in vitro</u> in diluted human blood (1:10). NFT was either irradiated (▭——▭) or kept in the dark (●----●) for 60 min. $[NFT]_0 = 100$ μM in PBS; light intensity at 310 nm: 0.04 mW.cm^{-2} and at 360 nm 0.4 mW.cm^{-2}.

REFERENCES

Beijersbergen van Henegouwen, G.M.J., 1987, In vitro and in vivo research of phototoxic xenobiotics: structure reactivity relationships, Arch. Toxicol. Suppl., in press.

Busker, R.W. and Beijersbergen van Henegouwen, G.M.J., 1987, The photolysis of 5-nitrofurfural in aqueous solutions: nucleophilic substitution of the nitro-group, Photochem. Photobiol., 45: 331-335.

Busker, R.W., Beijersbergen van Henegouwen, G.M.J., Menke, R.F. and Vasbinder, G., 1987, Formation of methemoglobin by photoactivation of nitrofurantoin in rats exposed to UV-A light; involvement of nitrofurfural and nitrite, Toxicol. Lett. submitted for publication.

Koch-Weser, J., Sidel, V.W., Dexter, M., Parish, C., Finer, D.C. and Kanarek, P., 1971, Adverse reactions to sulfisoxazole, sulfamethoxazole, and nitrofurantoin, Arch. Intern. Med., 128: 399-404.

Jarrar, A.A. and Fataftah, Z.A., 1977, Photolysis of some quinoxaline-1,4-dioxides, Tetrahedron, 33:2127-2129.

Stefanini, M., 1972, Chronic hemolytic anemia associated with erythrocyte enolase deficiency exacerbated by ingestion of nitrofurantoin, Am. J. Clin. Pathol., 58: 408-414.

Zaynoun, S., Johnson, B.E. and Frain-Bell, W., 1976, The investigation of quindoxin photosensitivity, Contact dermatitis, 2: 343-352.

MINIMIZING PHOTOTOXICITY OF XENOBIOTICS BY IN VITRO AND IN VIVO

RESEARCH OF THEIR STRUCTURAL ANALOGUES

G.M.J. Beijersbergen van Henegouwen

Center for Bio-Pharmaceutical Sciences, State
University of Leyden, P.O. Box 9502, 2300 RA
Leyden, The Netherlands

Simultaneous exposure to (sun)light and a xenobiotic may provoke a phototoxic effect. Effects often concern the skin. However, normal photobiological processes in man, such as the light-induced Vitamin D_3 production or the conversion of bilirubin with visible light, as well as animal experiments with xenobiotics (Beijersbergen van Henegouwen et al., 1987), show that systemic effects may also occur. Xenobiotics involved are present in drugs, cosmetics, food products, in chemicals used in agriculture, in the household etc. The variety in molecular structure of phototoxic compounds is immense, which implies that they can be found in virtually all classes of xenobiotics. Important objective of the research is to find the part of the molecular structure of a given xenobiotic that causes the unwanted effects. This gives the opportunity to alter the structure in such a way that the phototoxicity diminishes, whereas the desired properties of the xenobiotic, e.g. a drug, are conserved. This aim may be reached by a combination of data from 3 different research lines: a photoreactivity in vitro of the phototoxic xenobiotic and structure analogues whether or not in the presence of essential bio(macro)molecules. b phototoxicity in microbiogical test systems (bacteries, yeast, mammalian cell cultures). c phototoxicity in experimental animals. Attention should also be paid to the processes on molecular level, underwent by the xenobiotic studied e.g. to the formation of covalent bonds with biomacromolecules in vivo and to changes in metabolism as a result of exposure to light. This integration of in vitro and in vivo research has been applied to the phototoxic drug chlordiazepoxide (CDZ=A_1 in Fig.1) known under the trade name Librium[R] and some of its analogues. Amongst these are the major metabolites of CDZ, desmethyl CDZ (A_2) and demoxepam (A_3), because phototoxic effects may be caused not only by the drug itself but also by its metabolites. A_4 is the N-oxide of diazepam. (C_4). Diazepam (Valium[R]) is never reported as a phototoxic compound; its metabolites lack the N-oxide group as well, e.g. C_3.

In this review only some of the main results can be dealt with on which the conclusion was based that the N-oxide function is the cause of the phototoxicity of CDZ; more information and experimental details are given in the articles referred to.

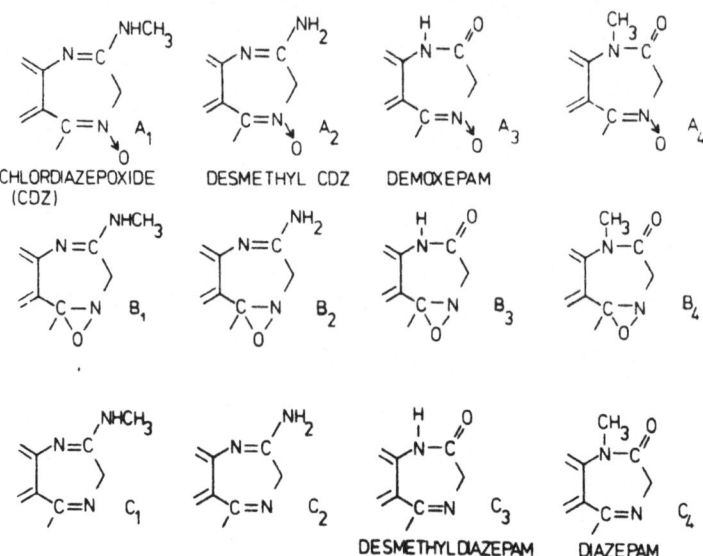

CHLORDIAZEPOXIDE (CDZ)   DESMETHYL CDZ   DEMOXEPAM

A_1 A_2 A_3 A_4

B_1 B_2 B_3 B_4

C_1 C_2 C_3 C_4

DESMETHYLDIAZEPAM   DIAZEPAM

Fig.1. Structures of 7-chloro-1,4-benzodiazepines.

a. Photoreactivity in vitro

Being present as the only compound in solution the N-oxides (A_1-A_4) isomerize for nearly 100% into their oxaziridine B_1-B_4 upon irradiation with UV-A (320-400 nm). (De Vries et al., 1983), Bakri et al., in press). Oxaziridines being unstable compounds the quantity B_1-B_4 actually found depends on the conditions applied e.g. $t_{1/2}$ at pH 7.4 and 37°C: $B_1\approx140$ min, $B_2\approx110$ min, $B_3\approx1$ min and $B_4\approx20$ h.

Also in the presence of SH group containing compounds, such as glutathione (GSH), the photoreaction of the N-oxides proceeds via the formation of an oxaziridine. This appeared from the fact that compounds B_1 and B_2 thermochemically and A_1 and A_2 photochemically react in the same way with GSH, namely by the non-enzymatic formation of a conjugate with GSH. In the case of A_1 and B_1 this conjugate has a $t_{1/2}\approx100$ min at 37°C and pH 7.4 and decomposes into the reduced form of A_1 (or B_1) namely C_1 (Cornelissen et al., 1980 a and De Vries et al., submitted).

The extent to which A_1 upon UV-A irradiation or B_1 in the dark irreversibly binds to human plasma proteins in vitro has been investigated and found to be 50% in both cases (Bakri et al., 1986); for A_4 and B_4, which react far more slowly under the same conditions, this appeared to be 30% (Bakri et al., in press). Under comparable conditions the compounds C_1-C_4 appeared to be photostable (Cornelissen et al., 1980 b, Bakri et al., 1986 and Bakri et al., in press).

b. Phototoxicity in microbiogical test systems

At this stage of the research it was preferred to get an idea of the relevance of the in vitro data obtained to the in vivo situation and not to investigate the in vitro (photo)reactions of A and B in more detail. To that aim microbiological test systems are convenient because of their speed and simplicity. In Fig. 2 part of the results obtained with Salmonella typhimurium TA 100 (Cornelissen et al., 1980 b and De Vries et al., 1983) are represented.

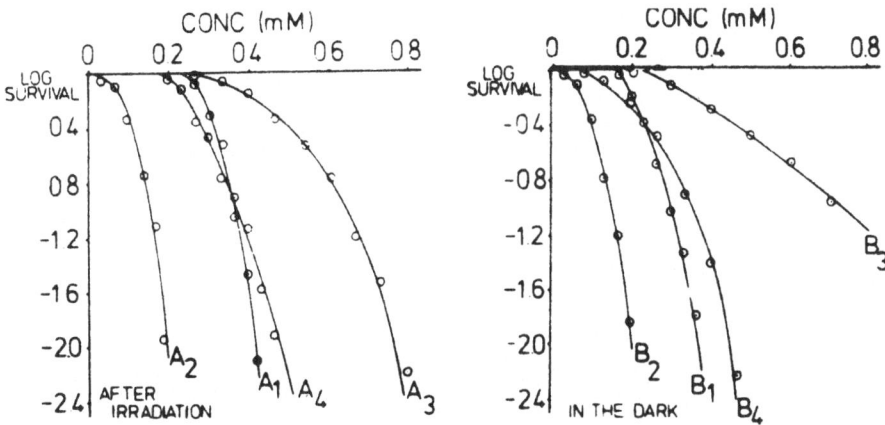

Fig.2. Concentration-dependent survival of the colony forming ability of Salmonella thyphimurium TA 100. Left panel: phototoxicity of the N-oxides A_1-A_4 (irradiation with UV-A, λ_{max}=350 nm) and right panel: toxicity of the oxaziridines B_1-B_4.

The N-oxides A_1-A_4 which form an oxaziridine as primary product, are not toxic without light (not represented) but only phototoxic in this test system. Of the two metabolites of CDZ, demoxepam (A_3) appears to be less phototoxic and desmethyl CDZ (A_2) more phototoxic than CDZ.

The toxicity curves of the oxaziridines B_1-B_4 correspond nicely with those for the phototoxicity of A_1-A_4. This indicates that formation of an oxaziridine is the cause of the phototoxicity of the N-oxides A_1-A_4. In this respect it is of importance that the reduced forms of the N-oxides (Fig.1, C_1-C_4), amongst which diazepam and its metabolite desmethyldiazepam, appeared to be non-phototoxic. (Cornelissen et al., 1980 b).

Neither with strain TA 100 nor with TA 98 Cornelissen et al., (1980 b) found photomutagenic effects. However recently De Vries et al., (submitted) reported that the oxaziridines B_1 and B_4 can induce DNA damage with E.coli K-12 # 765 and # 753.

Another interesting finding from this research of De Vries is, that (photo)conjugation of A_1, A_2, B_1 and B_2 with GSH has only a small detoxificating effect (Salmonella typhimurium TA 100). The results with bacterial test systems supported the supposition that the formation of an oxaziridine from photo-excited CDZ is the cause of the phototoxicity. Together with the data from the in vitro photoreactivity study they were considered as a firm bases for experiments with the rat. To verify whether the N-oxide group is the cause of the phototoxicity compound A_1, CDZ, was compared with C_1 (Bakri et al., 1983 and 1985) and A_4 with C_4, diazepam (Bakri et al., in press and submitted). A proper comparison can only be made if two factors are taken into account which quantitatively determine the eventual photobiological effect. The first is the concentration of the compound investigated in the irradiated organs, e.g. the skin, as a function of time and the second is the extent to which it absorbs the light in the spectral region of the lamp used. In the case of A_1 and C_1 the dose C_1 was 1.5 that of A_1 because of differences in these two factors. Rats of which the back was shaved got an intraperitoneal injection of the compound and were either kept in the dark or exposed to UV-A on five consecutive days for 8h/day (UV-A dose per day comparable to that on a sunny May day in Holland.)

Most of the reduced compound C_1 and its metabolites in the rat is excreted via the faeces. After deconjugation and extraction the mixture was submitted to TLC. Quantitative analysis of the spots proved that the metabolism of C_1 is not altered by UV-A. (see Table 1).

Table 1. Rf and percentage of each metabolite from C_1 in the extract of faeces after deconjugation. The total quantity extracted was put at 100%.

Rf	0.27	0.33	0.43	0.46	0.59	0.71	0.90
UV-A	11.7	11.9	4.1	9.2	10.5	26.8	25.6
Dark	11.8	12.3	4.1	8.7	11.0	26.4	25.5

The excretion of CDZ (A_1) and metabolites proceeds for ~ 45% via the urine.

In Table 2 the results are represented for urine collected between 10-20h. and 82-106h. after administration of CDZ.

Table 2. Percentage of N-oxymetabolites (including CDZ) and N-desoxymetabolites in the extract of urine after deconjugation. The total quantity extracted was put at 100%.

Metabolites	Urine(10-20h)		Urine(82-106h)	
	UV-A	Dark	UV-A	Dark
N-oxy	61 ± 3	87 ± 3	45 ± 3	83 ± 3
N-desoxy	17 ± 2	5 ± 2	30 ± 2	6 ± 2
Total	78 ± 3	92 ± 3	75 ± 3	89 ± 3
Unidentified	22 ± 2	8 ± 2	25 ± 2	11 ± 2

As can be seen, the percentage of N-oxy metabolites is lower for the UV-A exposed rats; this differences is even much larger after 82 h.: 45% vs 83%. The reverse was found for N-desoxymetabolites: after 82 h. an increase from 6 to about 30%. Comparable results were obtained for diazepam (C_4) and for its N-oxide (A_4). With diazepam quantitative TLC analysis of metabolites extracted from urine and faeces after deconjugation showed that there was no difference between UV-A exposed rats and those kept in the dark. With diazepam N-oxide (A_4) the extract of urine after deconjugation appeared to contain 85% N-oxymetabolites for rats kept in the dark and 67% for UV-A exposed animals, whereas these percentages were 15% and 33% respectively for the N-desoxymetabolites.

An explanation can be brought forward in the light of the reaction of oxaziridines, formed by photoisomerisation of N-oxides in the UV-A exposed skin, with SH-group containing compounds. As already mentioned CDZ upon UV-A irradiation and its oxaziridine without light react spontaneously with GSH with as the ultimate product the N-desoxyform C_1.

A further confirmation of the responsability of N-oxide function in the molecule for the phototoxic effects was obtained by investigation of covalent binding of benzodiazepine-fragments to biomacromolecules of e.g. the skin.

This was expected because of the in vitro data obtained: covalent binding to plasma proteins. With CDZ irreversible binding to biomacromolecules of the skin of back, to a lesser extent to those of the skin of the belly but also to those of liver and kidney was found with rats exposed to UV-A. (Bakri et al., 1983). With the reduced form of CDZ (C_1) irreversible photobinding in vivo was not observed (Bakri et al., 1985), what corresponds with the in vitro data. Remarkable is that the N-oxide of diazepam (A_4) did not photobind in vivo to a measurable extent. (Bakri et al., submitted). In vitro investigation already showed that its oxaziridine (B_4) has a relatively long life time and reacts far more slowly with plasma proteins than that of CDZ (B_1). It is supposed that in vivo irreversible binding of the oxaziridine B_4 to proteins is suppressed by enzymatic reaction with GSH. This would correspond with the increased percentage of N-desoxymetabolites, at the expense of N-oxymetabolites, in the UV-A exposed rats.

Table 3. Percentage of glucuroconjugated metabolites of CDZ (A_1); the total quantity of metabolites, conjugated and extracted at a certain pH, was put at 100%.

pH	Non-irradiated	UV-A	Non-irradiated/UV-A
2	81 ± 1	32 ± 1	2.5
4	49 ± 1	34 ± 1	1.4
7	70 ± 2	32 ± 3	2.2

Other remarkable differences between UV-A exposed rats and those kept in the dark concerned the conjugation of metabolites. These were observed with CDZ (A_1, Table 3, Bakri et al., 1983) but not with its reduced form (C_1, not represented, see Bakri et al., 1985) and with diazepam N-oxide (A_4) but not with diazepam (C_4) (Table 4, Bakri et al., in press and submitted).

Table 4. Percentage of urinary metabolites (40-50% of dose) of diazepam (C_4) and its N-oxide (A_4): I = non-conjugated; II = glucuroconjugates; III = other conjugates I + II + III = 100%.

		I	II	III
C_4	UV-A	18 ± 2	69 ± 2	13 ± 3
C_4	Non-irradiated	19 ± 2	70 ± 3	11 ± 2
A_4	UV-A	43 ± 2	2 ± 2	55 ± 3
A_4	Non-irradiated	65 ± 2	9 ± 3	26 ± 2

Perhaps the most remarkable fact is that the combination of CDZ (A_1) and UV-A, or its oxaziridine (B_1) alone, caused a decrease of the ratio of the weight of the liver and the total weight with 24% and 17% respectively, whereas no change was found with C_1 and UV-A (Bakri et al., 1983 and 1985).

By combination of the results from in vitro and in vivo research it could be concluded that the N-oxide function is responsible for the phototoxicity. In this respect it is interesting to mention that the reduced form of CDZ, compound C_1, shows biological activity approaching that of CDZ and the question remains whether no-phototoxic C_1 might have been an acceptable alternative to CDZ (Librium[R]) as tranquilizer. On the other hand as far as diazepam-N-oxide (A_4) and diazepam (C_4) are concerned this question does not arise because the non-phototoxic compound has been commercialized (Valium[R]).

References

Beijersbergen van Henegouwen, G.M.J., Busker, R.W., De Vries, H., Schoonderwoerd, S.A., 1987, Systemic photobiological effects from xenobiotics, Arch.Toxicol. Suppl., in press.

Bakri, A., Beijersbergen van Henegouwen, G.M.J., Chanal, J.L., 1983, Photopharmacology of the tranquilizer chlordiazepoxide in relation to its phototoxicity, Photochem.Photobiol.,38:177-183

Bakri, A., Beijersbergen van Henegouwen, G.M.J., Chanal, J.L. 1985, Involvement of the N_4-oxide group in the phototoxicity of chlordiazepoxide in the rat, Photodermatology., 2:205-212

Bakri, A., Beijersbergen van Henegouwen, G.M.J., Sedee, A.G.J. 1986, Irreversible binding of chlordiazepoxide to human plasma protein induced by UV-A radiation, Photochem.Photobiol., 44:181-185

Bakri, A., Beijersbergen van Henegouwen, G.M.J., De Vries, H., Photobinding of some 7-chloro-1,4-benzodiazepines to human plasma protein in vitro and photopharmacology of diazepam in the rat, Pharm.Weekbl.Sci.Ed., In press

Bakri, A., Beijersbergen van Henegouwen, G.M.J., Photopharmacology of diazepam-N_4-oxide in the rat, Submitted for publication

Cornelissen, P.J.G., Beijersbergen van Henegouwen, G.M.J. 1980 a Photochemical decomposition of 1,4-benzodiazepines; quantitative analysis of decomposed solutions of chlordiazepoxide and diazepam, Pharm.Weekbl.Sci.Ed., 2:39-48

Cornelissen, P.J.G., Beijersbergen van Henegouwen, G.M.J., Mohn, G.R., 1980 b, Structure and photobiogical activity of 7-chloro-1,4-benzodiazepines. Studies on the phototoxic effects of chlordiazepoxide, desmethylchlordiazepoxide and demoxepam using a bacterial indicator system, Photochem.Photobiol., 32:653-661

De Vries, H., Beijersbergen van Henegouwen, G.M.J., Wouters, R., 1983, Correlation between phototoxicity of some 7-chloro-1,4-benzodiazepines and their (photo)chemical properties, Pharm.Weekbl.Sci.Ed., 5:302-308

De Vries, H., Beijersbergen van Henegouwen, G.M.J., Bakri, A., The effect of glutathione on the oxaziridine mediated phototoxicity of chlordiazepoxide, Submitted for publication

REACTIVE OXYGEN SPECIES AND OTHER MEDIATORS OF PHOTOTOXIC DAMAGE

Eino Hietanen

Department of Clinical Physiology, Turku University Hospital
SF-20520 Turku, Finland

INTRODUCTION

Environmental factors are major causes of cancer. In addition to
xenobiotics also radiation may have importance in carcinogenesis as an
association has been demonstrated e.g. between skin cancers and exposure to
UV light. Although the exact mechanisms how environmental agents cause cancer
are partially unknown, a 3-stage model including initiation, promotion and
progression of cancer is generally accepted. Usually chemicals are metabolized into reactive intermediates reacting with cellular macromolecules to
initiate cancer or they may act as tumor promoters when the mechanisms have
not been explored. Some xenobiotics may be metabolized into free radicals
which may attack on DNA and initiate cancer or these free radicals may act
on other components such as oxygen to activate it into active forms which
may act as tumor initiators or promoters. It is well known that radiation
may cause DNA damage via the formation of hydroxyl radicals. Hydrogen peroxide
causes DNA strand breakage and liberation of DNA bases as well as chemical
alterations of bases. Superoxide anion causes single-strand scissions.

Although free radicals and various active oxygen products have very
short half lives they form secondary products which can be utilized to
monitor their effects. Lipid peroxidation (LPO) is a typical event secondary
to radical reactions which has stable end-products to be monitored via
various means. As the membrane composition is of importance regulating lipid
peroxidation the factors modulating membrane composition may also modify LPO
and even influence on the carcinogenesis.

Those phototoxic reactions will be dealt here where molecular oxygen is
involved and not the direct mechanisms. Furthermore, as it has been previously
reviewed, phototoxic chemicals together with UV/visible light are activated
into excited intermediates activating oxygen or UV light may itself cause
the formation of active oxygen intermediates. In this context the mechanisms
how active oxygen species mediate the toxicity and those factors which may
modify the toxicity will be dealt.

PHOTOTOXICITY, FREE RADICALS AND ACTIVE OXYGEN

Ultraviolet light may produce skin damage and in experimental animals
may also be carcinogenic (IARC, 1986). Yet especially combined with certain
photoactive furocoumarins, e.g. 8-methoxypsoralen and 4,5´,8-trimethylpsoralen,
UV light may have therapeutic uses in the treatment of psoriasis and other

skin diseases. Possibly both the photosensitizing and chemotherapeutic effects of psoralen are related to the photoexcitation yielding reactive intermediates to bind covalently to DNA (Pathak, 1982; Pathak and Joshi, 1984). Active psoralens may undergo reactions leading to the production of singlet oxygen and superoxide anion radicals after photoexcitation of psoralen The skin photosensitizing effect of various psoralen derivatives is related to the ability to form DNA adducts and interstrand cross links (Dall´Acqua et al., 1971), and further, there seems to be a simultaneous production of singlet oxygen and superoxide anion (Pathak and Joshi, 1984). Also other commonly used drugs may cause phototoxic reactions by similar mechanisms (Kochevar, 1987). The involvement of singlet oxygen and superoxide anion in the photosensitizing reactions has further been confirmed by utilizing various specific quenchers of active oxygen products. It was earlier believed that psoralens after the excitation by UV light reacted only according to type-I, anoxic reaction, involving photocyclo addition reactions with pyrimidine DNA bases. Although most therapeutic psoralens form bifunctional adducts, photosensitize the skin and have also carcinogenis effects, recently also some non-photosensitizing monofunctional furocoumarins have been shown to have therapeutic effects on psoriasis. Most (or all) of monofunctional and bifunctional adduct-forming psoralens seem to react according to type II, oxygen-demanding reaction, yielding to the production of singlet oxygen and superoxide anions (Pathak and Joshi, 1984). The production of singlet oxygen and superoxide anion may be related to the carcinogenic potency of psoralen derivatives.

Table 1.   Free radicals and selected cellular events

---

1. Activation of chemical carcinogens (not all) in the
   endoplasmic reticulum requires free radical reactions.

2. Pathologic ionic fluxes across cell membranes
   - Na^+/K^+ ATPase affected by lipid free radical reactions

3. Alteration of DNA; alkylation and other DNA-adduct
   formation mechanisms may be via free radical reactions.

4. Oxygen-related free radicals (generated by redox
   cycling) induce lysosomal membrane destabilization.

5. Alteration of cyclic nucleotide levels; O_2-related
   free radicals activate guanylate cyclase and inhibit
   adenylate cyclase -> influence on cAMP levels

6. Hampered cell differentiation; effects on cell membranes,
   nuclear membranes -> changes in control factors -> changes
   in membrane fluidity

Psoralens and other phototoxic drugs are metabolized or activated by the UV light into free radicals acting then on molecular oxygen and yielding to the production of various active oxygen species. The free radical derivatives may have a very short half life leading to measurable secondary events. Free radicals may alter cellular and subcellular membrane homeostasis by altering membrane permeability, fluidity and decrease membrane stability (Table 1). Many of the active oxygen species are free radicals although others have a nonradical nature (Table 2).

Table 2   Oxy-radicals and related species that have been implicated
           in cellular processes. Radicals are denoted by r and non-
           radicals by nr.

Name	Notes	Name	Notes
singlet oxygen	nr	acyloxyl	r
ozone	nr	acylperoxyl	r
nitrogen dioxide	r	aryloxyl	r
hydrogen peroxide	nr	arylperoxyl	r
hydroxyl	r	hypochlorous acid	nr
superoxide	r	semiquinone radical	r
hydroperoxyl	r	semiquinone radical	
alkoxyl	r	anion	r
peroxyl	r		

The formation of free radicals from oxygen, endogenous compounds or xenobiotics
may take place via different mechanisms depending on the original compound
in concern. Reactive oxygen species in the body can be generated by hyperbaric
oxygen tension or by radiation, e.g. UV light. Xenobiotics are metabolized
into free radicals via cytochrome P-450 dependent enzyme system while
peroxisomal proliferators typically also produce active oxygen species. In
general free radicals, including oxygen radicals, are suspected to cause
mutations, sister chromatide exchanges, chromosomal aberrations, of being
cytotoxic, cause cancer and possibly involved in ageing process. Active
oxygen species act mainly as tumor promoters enhancing carcinogenesis after
the initiation, light irradiation or chemical initiation (Cerutti, 1985)

Free radicals may themselves bind to nucleic acids covalently and form
adducts causing cancer or they may activate oxygen and further initiate LPO
yielding various active products which can also be utilized to detect
indirectly free radical reactions as many of the end products of LPO reactions
are stable (Del Maestro, 1980; Hietanen et al., 1987a). The final toxicity
whether expressed as carcinogenicity, mutagenicity, hemolysis, necrosis or
irritation, is dependent, not only on the formation of reactive intermediates
but also on the defence systems inactivating these.

Typically low molecular weight compounds such as vitamin E, vitamin C, uric
acid and glutathione may trap reactive intermediates and catalase, glutathione
peroxidase and superoxide dismutase may inactivate various active oxygen
species. There is a strong evidence that LPO, often the consequence of free
radical reactions, and the presence of active oxygen species may be involved
in carcinogenesis. Many structurally unrelated classes of prooxidant promoters
cause LPO while the suppression of LPO has an antipromotional effect. Also,
stable end products of LPO may either initiate or promote cancer. Data exist
also that UVA radiation does not cause skin pigmentation in anoxic conditions
but UVB and PUVA treatments cause pigmentation and inflammation independently
of oxygen supply which might suggest that UVB and PUVA are themselves strong
enough stimulators to cause possible free radical reactions while UVA needs
the presence of oxygen (Tegner, 1984).

Photosensitization involves both phototoxic reactions and less common
photoallergic reactions. Direct photosensitization (Type I reactions) is due
to the direct action of triplet state sensitizer on cellular macromolecules
while indirect photosensitization (type II reactions) means that sensitizer
first reacts with molecular oxygen activating it into singlet oxygen,

superoxide anion and hydroxy radicals. Some phototoxic agents require the presence of molecular oxygen which the photoxic chemical activates such as hematoporphyrin while the activated oxygen reacts with macromolecules (Pathak, 1982). In the phototoxic reactions requiring molecular oxygen the photoexcitation energy is transferred to oxygen molecules present in biological systems. The photosensitizer being activated by photoenergy transfers the energy to oxygen existing in triplet state which results in the formation of reactive singlet oxygen capable to oxidize biological substrates.

ACTIVE OXYGEN SPECIES AND CANCER

Singlet oxygen causes vasodilatation due to the endothelial damage, further, membrane and cell damages, damage to DNA, lipid peroxidation and induction of cross-linkages in proteins. All these changes may cause the initiation of cancerous transformations. Although psoralens may cause biological photosensitization independently of oxygen psoralens have also been demonstrated to be activated into free radicals by UV light in the presence of oxygen which radicals may further activate oxygen (Pathak, 1961). Hematoporphyrins have been used in the treatment of malignant tumors based on the assumption that the following photoradiation of these compounds generate oxygen species resulting the oxidation of essential cellular components and, as these compounds are taken into the tumorous cells, destroys malignant cells. When rats were pretreated with hematoporphyrins and liver microsomes were prepared a rapid enhancement of LPO was found in the presence of solar radiation and quenchers of singlet oxygen and inhibitors of the hydroxyl radical protected against increase in LPO (Das et al., 1985a, b).

There is considerable experimental evidence that UV radiation enhances skin tumors simultaneously with the increase in lipid peroxidation in epidermal homogenates (Black et al., 1985). Furthermore, the presence of polyunsaturated fats in the diet increases tumor incidence and shortens the latency of tumors to appear while the level of LPO is highly increased. The promoting activity of phorbol derivatives in skin cells is related to the chemiluminescence (Fischer and Adams, 1985). The use of various inhibitors has aided to find out specific oxygen derivatives to be produced during the carcinogenic transformation. The use of superoxide dismutase has revealed that during phorbol ester induced skin tumor promotion mainly superoxide anion is formed and responsible for the chemiluminescence. Furthermore, inhibitors of arachidonic acid patway have indicated that phorbol ester induced chemiluminescence in epidermal homogenates is mainly produced via the lipoxygenase patway (Fischer and Adams, 1985). After UV radiation of the skin ultraweak chemiluminescence is detected due to the formation of singlet oxygen in the skin, which chemiluminescence is inhibited by beta-carotene suggesting also a link between phototoxicity and active oxygen formation (Torinuki et al.,1982).

PROOXIDANT STATE AND CARCINOGENICITY

Materials and methods. Male Wistar rats were fed from weanling onwards either 30 % high fat diet or low fat diet for 40 weeks. After 2 weeks on diets each group was divided into two; another received 10 weeks N-nitro-sodiethylamine (NDEA), a carcinogen initator, and another subgroup was as a control. Rats were followed 40 weeks on diet whereafter histopathological analyses of all organs were made and during and at the end of the experiment biochemical analyses were performed. As the aim of the study was to find out the possible role of enhanced prooxidant state in carcinogenesis LPO was monitored by various methods in isolated hepatocytes prepared as described elsewhere (Moldéus et al.,1978), in liver microsomal fraction and in exhaled air (Hietanen et al.,1987b). The malondialdehyde (MDA) concentration was measured by detecting the thiobarbituric acid reactive substances (Uchiyama and Mihara, 1978); in microsomes MDA was measured after stimulation by 0.1 mM $FeCl_3$ -1.7 mM ADP and the reaction was initiated by NADPH (Slater, 1984). Chemiluminescence was measured in microsomes in the presence of NADPH. Ethane as a product of LPO was monitored by measuring the ethane in exhaled air gas chromatographically (Wendel and Dumelin, 1981).

Results. MDA production was enhanced in stimulated liver microsomes of rats fed the high fat diet whether or not given NDEA (Table 3). When rats were fed low fat diet microsomal MDA content remained at low level despite of NDEA administration. When microsomal LPO was monitored via detecting chemiluminecence high fat diet produced higher level of NADPH stimulated chemiluminescence than low fat diet whether or not NDEA was previously given (Table 3).

In isolated hepatocytes after 40 weeks experiment highly enhanced LPO in tumoroas tissue was found (Table 3). In exhaled air NDEA always produced elevated ethane exhalation even though the administration was discontinued previously whichever diet was given and also high fat diet as such enhanced ethane exhalation (Table 3).

Table 3. Lipid peroxidation in isolated hepatocytes and liver microsomes measured as malonaldehyde (MDA), chemiluminescence (CL) in microsomes calculated per cytochrome P-450 and ethane in exhaled air calculated per weight from rats fed high or low fat diet and given N-nitrosodiethylamine. In NDEA nontumorous and tumorous liver samples were separated. The tumor distribution is also given.

| | High Fat | | | Low Fat | |
| | -NDEA | +NDEA | | -NDEA | +NDEA |
		no tumor	tumor		
MDA ($umol/10^6$ cells)	0.7+0.1	0.7 <	3.3+0.6	0.5+0.2	0.5+0.1
Microsomal MDA (umol/g w.wt)	12.9+0.7*	12.5+2.3	7.2+2.4	2.2+0.2	3.2+0.6
CL (10^6 mV/nmol P-450)	7.7+0.7	9.6+1.4 <	28.7+9.0	7.0+0.6	7.1+1.7
Ethane (nmol/kg/h)	2.2+0.4* <	3.8+0.3		1.1+0.3 <	3.6+0.6
Tumors (no of rats)					
Liver	-	13		-	8
Extrahepatic	-	-		-	14
No of tumors	23	9		22	5
No of rats	23	22		22	23

< gives statistically significant change. * Significantly higher than in respective (-NDEA) low fat group. Mean + SEM given.

The tumor distribution was dependent on the dietary fat content in a way that high fat diet promoted liver tumors while in rats fed low fat diet total tumor frequency was higher and numerous extrahepatic tumors existed (Table 3).

Discussion. High fat diet enhanced prooxidant state in the liver when judged from the enhanced level of LPO in various liver fractions or exhaled air. The elevated prooxidant state apparently was related to the increased amounts of liver cancers. Analogously high cholesterol diets have enhanced skin tumor rates induced by UV light in rats simultaneously with the increase in the skin cholesterol content although in mice no such an association was found (Baumann and Rusch, 1939). In hairless mice polyunsaturated lipids have enhanced the photocarcinogenesis of the skin as compared to respective amounts of hydrogenated fats in the diet (Black et al.,1985). An association was found between dietary lipid levels and skin tumor incidence due to photocarcinogenesis while antioxidants have prevented tumor formation preventing simultaneosly LPO in the skin. Also in recent studies dietary lipid levels were related to the skin tumor rates (Black et al.,1985). The increase of the dietary unsaturated fat concentration to 4 % maximally enhanced the skin photocarcinogenicity as the number of tumors animals were carrying. The antioxidant addition to the diet prevented tumor development. Simultaneously with changes in tumor formation epidermal LPO levels were increased and LPO was inhibited by the dietary antioxidants suggesting also that in the skin photocarcinogenesis the production of active oxygen species and radical reactions measured as LPO may be involved in epidermal carcinogenesis.

Although many experimental and indirect clinical data suggest that phototoxic damage including cancerous transformations are due to the presence of active oxygen species as such possibly also secondary events, e.g. LPO products, may be involved in the cancerous transformations. A stable peroxide, benzoyl peroxide, may act as a promoter of mouse skin cancer facilitating the transformation of papillomas to cancers (O´Connell et al.,1986). Various products of LPO such as malondialdehyde may be mutagenic and cause chromosomal damage (Basu et al.,1984; Draper et al.,1986; Ueda et al.,1985).

CONCLUSION

Various animal models and cell cultures have suggested that enhanced prooxidant state, resulting increased amounts of reactive oxygen species, is involved in the phototoxic reactions, skin carcinogenesis and also in toxic and carcinogenic reactions in other organs mediating also the phototoxic effects of many drugs.

ACKNOWLEDGEMENTS

This study was supported by grants from J. Vainio Fdn (Helsinki) and from the Finnish Ministry of Agriculture and Forestry for cell culture studies.

REFERENCES

Basu, A.K.,Marnett, L.J.,and Romano, L.J.,1984, Dissociation of malondialdehyde mutagenicity in Salmonella typhimurium from its ability to induce interstrand DNA cross-links, Mut. Res., 129:39.
Black, H.S.,Lenger, W.A., Mac Callum, M, and Gergius, J., 1983, The influence of dietary lipid level on photocarcinogenesis, Photochem. Photobiol., 37:539.
Black, H.S.,Lenger, W.A.,Gerguis, J.,and Thornby, J.I.,1985, Relation of antioxidants and level of dietary lipid to epidermal lipid peroxidation and ultraviolet carcinogenesis, Cancer Res., 45:6254.
Baumann, C.A. and Rusch, H.P.,1939, Effect of diet on tumors induced by ultraviolet light, Am. J. Cancer, 35:213
Cerutti, P.A.,1985, Prooxidant states and tumor promotion, Science, 227:375.

Dall'Acqua, F.,Marciani, S.,Ciavatta, L., and Rodighiero, G.,1971,
   Formation of interstrand cross-linkings in the photoreactions
   between furocoumarins and DNA, Z. Naturforsch., 26:561

Das, M., Dixit, R.,Mukhtar, H.,and Bickers, D.R.,1985a, Role of active
   oxygen species in the photodestruction of microsomal cytochrome
   P-450 and associated monooxygenases by hematoporphyrin derivative
   in rats, Cancer Res., 45:608.

Das, M.,Mukhtar, H.,Greenspan, E.R.,and Bickers, D.R.,1985b,
   Photoenhancement of lipid peroxidation associated with the
   generation of reactive oxygen species in hepatic microsomes of
   hematoporphyrin derivative-treated rats, Cancer Res., 45:6328.

Del Maestro, R.F.,1980, An approach to free radicals in medicine and
   biology, Acta Physiol. Scand., Suppl. 492:153.

Draper, H.H.,McGirr, L.G.,and Hadley, M.,1986,The metabolism of malondialdehyde, Lipids, 21:305.

Fischer, S.M. and Adams, L.M.,Suppression of tumor promoter-induced
   chemiluminescence in mouse epidermal cells by several inhibitors of
   arachidonic acid metabolism, Cancer Res., 45:3130.

Hietanen, E.,Ahotupa, M.,Bereziat, J.C.,Bussacchini, V.,Camus, A.-M., and
   Bartsch, H.,1987a, Elevated lipid peroxidation in rats induced by
   dietary lipids and N-nitrosodimethylamine and its inhibition by
   indomethacin monitored via ethane exhalation, Tox. Path., 15:93.

Hietanen, E.,Ahotupa, M.,Bartsch, H.,Bereziat, J.-C., Bussacchini, V.,
   Camus, A.-M., and Wild, H.,1987b, Lipid peroxidation and chemically
   induced cancer in rats fed lipid rich diet, Proc. XIV Int. Cancer
   Congress, In Press.

IARC Monographs, 1986, Some naturally occurring and synthetic food
   components, furocoumarins and ultraviolet radiation, 40:379.

Kochevar, I.E.,1987, Mechanisms of drug photosensitization, Photochem.
   Photobiol., 45:891.

Moldéus, P.,Högberg, J.,and Orrenius, S.,1978, Isolation and use of liver
   cells, Methods Enzymol., 52:60.

O´Connell, J.F.,Klein-Szanto, A.J.P.,DiGiovanni, D.M.,Fries, J.A.W., and
   Slaga, T.J.,1986, Enhanced malignant progression of mouse skin
   tumors by the free-radical generator benzoyl peroxide, Cancer Res.,
   46, 2863.

Pathak, M.A.,1961, Mechanism of psoralen photosensitization and in vivo
   biological action spectrum of 8-methoxypsoralen. J.Invest. Dermatol.,
   37:397.

Pathak, M.A.,1982, Molecular aspects of drug photosensitivity with
   special emphasis on psoralen photosensitization reaction, J.Natl.
   Cancer Inst.,69, 163

Pathak, M.A. and Joshi, P.C.,1984, Production of active oxygen species
   ($^1O_2$ and O_2^{-}) by psoralens and ultraviolet radiation (329-400 nm),
   Biochim. Biophys. Acta, 798:115.

Slater, T.F.,1984, Overview of methods used for detecting lipid
   peroxidation, Methods Enzymol., 52:60.

Tegner, E.,1984, Tissue anoxia prevents inflammation and pigmentation
   caused by UVA but not by UVB or PUVA, Photodermatology, 1:311.

Torinuki, W.,Kumai, N.,Miura, T.,and Seiji, M.,1982, Effect of beta-
   carotene on ultraweak chemiluminescence of UVB-irradiated
   squalene, Tohoku J.Exp. Med., 136:459.

Uchiyama, M. and Mihara, M.,1978, Determination of malonaldehyde
   precursor in tissue by thiobarbituric acid test, Anal. Biochem., 86:274.

Ueda, K.,Kobayashi, S.,Morita, J.,and Komano, T.,1985, Site-specific DNA
   damage caused by lipid peroxidation products, Biochim. Biophys. Acta,
   824, 341.

Wendel, A. and Dumelin, E.E.,1981, Hydrocarbon exhalation, Methods
   Enzymol., 77:10.

UNDESIRED PHOTOTOXICITY: CUTANEOUS REACTIONS TO SYSTEMIC DRUGS

Olle Larkö

Department of Dermatology
Sahlgren's Hospital
Göteborg, Sweden

INTRODUCTION

Most clinical photoreactions to systemic drugs are caused by a limited number of drugs. The problem has increased in importance due to the abundant use of solaria. However, surprisingly few reactions have been reported in these situations. The majority of the photo reactions are of the phototoxic type. This article will deal with unintended harmful photobiological effects in conjunction with systemic drug therapy.

Tetracyclines

It is well known that many tetracyclines have a phototoxic potential. Photoallergic reactions seem to be very rare (Hoigné 1975). Methacycline has been shown to cause pigmentation of the skin and conjunctivae (Möller and Rausing, 1980). The action spectrum for tetracycline phototoxicity seems to be in the UVA. However, it seems that UVB induced sunburn can facilitate the development of a phototoxic tetracycline reaction (Bjellerup 1986). It seems that demethylchlortetracycline is the most photoactive compound. However, phototoxicity towards doxycycline is probably more relevant clinically due to its extensive use. In our clinical experience tetracycline itself rarely cause any problem in natural sunlight. We have quite a few patients that receive both UVB and UVA treatment for acne while receiving tetracycline.

Sulphonamides

Sulphonamides are less used nowadays compared to previously. The reaction may be of the phototoxic or photoallergic type. The action spectrum is probably in the UVA but this has not been completely established. Diuretics and antihypertensives occasionally cause problems.

Phenothiazines

Among the phenothiazines chlorpromazine most commonly cause
problems. When used topically a photoallergy can develop but following
systemic use phenothiazines cause phototoxic reactions. The action
spectrum is probably in the UVA. It seems that metabolites of
chlorpromazine may be photoactive.

NSAID

Some years ago benoxaprofen was reported to cause photoreactions.
The incidence was high as reported by Halsey and Cardoe. The action
spectrum lies mainly between 320-340 nm (Addo et al. 1982). Other
NSAIDs have also been shown to cause photosensitization. Among these
are piroxicam and azapropazone.

Amiodarone

Amiodarone is a cardiac antiarythmic drug that can cause a bluish
discoloration of exposed sites in the face. The incidence of
photosensitivity has been reported to be as high as 57%. Maximal
sensitivity seems to be in the mid-UVA (Diffey et al., 1984; Roupe et
al., 1987).

Nalidixic acid

Photobiological side effects of nalidixic acid often mimics
porphyria cutanea tarda. However, porphyrine metabolism seems to be
normal. It may be that a metabolite contributes.

Other drugs

Griseofulvin has been reported to cause photoreactions but in our
experience this is rare. Quinidine has been reported to induce a
photosensitivity with an eczematous or lichenoid histology (Marx et
al., 1983).

Phototesting

Phototesting is often difficult to do as it is often hard to
reproduce the reactions (Rosén and Swanbeck, 1982).

According to the Swedish Drug Information Committee most reports
concerning photosensitivity has been made on nalidixic acid followed by
hydrochlorotiazide. Among the tetracyclines doxycycline dominates.

In conclusion it seems important to screen out potentially
photosensitizing drugs. It is also important to inform the patients
when new such drugs are being introduced on the market. Furthermore, it
is necessary to inform patients that most sunscreens are worthless in
this situation.

References

Addo, H.A., Ferguson, J., Johnson, B.E., Frain-Vell, W., 1982.
A study of benoxaprofen-induced photosensitivity. Br J Dermatol 107
(suppl 22):17.

Bjellerup, M., 1986. Tetracycline phototoxicity. An experimental
and clinical study. Thesis, Lunds university, Malmö.

Diffey, B.L., Chalmers, R.J.G., Muston, H.L., 1984. Photobiology
of amiodarone: preliminary in vitro and in vivo studies. Clin Exp
Dermatol 9:248.

Halsey, J.P., Cardoe, W., 1982. Benoxaprofen: side-effect profile
in 300 patients. Br Med J 284:1365.

Hoigné, R., 1975. Penicillins, Cephalosporins and tetracyclines.
In: Dukes MNG, ed. Side effects of drugs. Amsterdam, Oxford:
Excerpta Medica 8:571.

Marx, J.L., Eisenstat, B.A., Gladstein, A.H., 1983. Quinidine
photosensitivity. Arch Dermatol 119:39.

Möller, H., Rausing A., 1980. Methacycline hyperpigmentation:
A five-year follow-up. Acta Derm Venereol (Stockh) 60:495.

Rosén, K., Swanbeck, G., 1982. Phototoxic reactions from some common
drugs provoked by a high-intensity UVA lamp. Acta Derm Venereol
(Stockh) 62:246.

Roupe, G., Larkö, O., Ohlsson, S.B., 1987. Amiodarone
photoreactions. Acta Derm Venereol (Stockh) 67:76.

UVB INDUCED PRODUCTION OF KERATINOCYTE DERIVED SUPPRESSOR FACTORS

Thomas Schwarz[+], Agatha Urbanski[*], Fritz Gschnait[+], and Thomas A.Luger

[+]Department of Dermatology, Hospital Vienna-Lainz, Department of Dermatology II, University of Vienna and Ludwig Boltzmann Institute for Dermatovenerologic Serodiagnosis, Laboratory for Cellbiology, Vienna, Austria

Excessive exposure to sunlight is the major etiologic factor in the development of squamous cell carcinomas, basal cell carcinomas and possibly malignant melanomas. Ultraviolet (UV) light mediated carcinogenesis is a complex process which is not completely understood. First, chronic exposure of animals to UV light shows that it acts both as a tumor initiator and promotor upon epidermal cells (EC). Secondly, it may impair the body's immune response to the altered ECs permitting the survival and growth of neoplastic epidermal cells (1). In particular mice exposed to UV light develop T-suppressor lymphocytes that prevent 1. sensitization to potent contact allergens (2) and 2. immunologic rejection of highly antigenic UV induced skin cancer (3). Immunologic alterations that occur following exposure to UV radiation can be divided into two types: local and systemic immuno-suppression.

Local alterations are those that result from a direct interaction between UV exposure and immunologic reactions taking place locally, in the irradiated area. It has been clearly demonstrated that after exposure of mice to UV light epicutaneous application of the potent allergen dinitrofluorobencene (DNFB) to exposed skin produced only a minimal contact hypersensitivity (CHS) response, while administration of DNFB at an unexposed site resulted in a normal CHS reaction (4). The induction of this immunologic tolerance correlated with a reduction of the number as well as alteration of the morphology and antigen presenting capacity of Langerhans cells at the site of UV exposure. It was subsequently shown that the immunologic tolerance was associated with the induction of hapten specific suppressor cells (5).

In addition it has been demonstrated that UV irradiation at higher doses alters certain immune responses e.g. induction of CHS at sites not exposed directly to UV light ("systemic" immunosuppression) (6). This systemic immunosuppression is also associated with the induction of antigen-specific suppressor cells (7). However, in contrast to the local suppression of CHS the distant suppression appears to be unrelated to morphologic alterations induced in Langerhans cells by UV irradiation and thus the pathomechanisms involved have not been clarified so far (8). In particular the question, how the events occurring at the irradiated site lead to an abnormal response to an antigen applied at a distant non UV exposed skin area remains to be answered. The most attractive hypothesis for explaining the systemic suppression of CHS by UVB irradiation involves a soluble mediator (9). Additional evidence consistent with this hypothesis was recently provided demonstrating that serum of UVB irradiated mice contains a factor that inhibits the induction of CHS and induces T-suppressor cells when injected intravenously into normal mice (10). However, the nature and the origin of such a mediator remains to be determined.

Keratinocyte derived immunomodulatory cytokines

Within the last decade the epidermis has been recognized as an immunoregulatory tissue. Dendritic cells within the epidermis such as the Langerhans cells may function as antigen presenting cells and thereby play an important role during the initiation of an immune response (11). In addition a recently discovered dendritic Thy 1 positive EC apparently belongs to the T-cell lineage and exhibits natural killer like activities (12). Therefore, the observation was not surprising that the keratinocytes surrounding these immunocompetent cells are endowed with the capacity to release an immunomodulating factor known as epidermal cell derived thymocyte activating factor (ETAF) which is indistinguishable antigenically, biochemically and biologically from macrophage derived IL 1 (13). ETAF, like IL 1 is a nonspecific mediator of immunity and inflammation which activates a variety of cells such as lymphocytes, granulocytes and fibroblasts, as well as osteoclasts and may also cause fever, acute phase protein production and protein degradation (14,15). In as much as the expression of mRNA both for human IL 1α and IL 1β in human keratinoctes and murine IL 1α mRNA in murine keratinocytes has recently been reported, ETAF and IL 1 apparently are identical (16,17). Chronic UV radiation in vivo as well as in vitro has been demonstrated to increase the production of ETAF/IL 1 (18). Since increased serum levels of IL 1 after UV treatment also have been shown to induce fever, this mediator at least partly may be responsible for some of the systemic inflammatory reactions after UV (18). However, under certain experimental conditions in other studies reduced ETAF/IL 1 has been described after UV treatment (19). The latter observation on the first glance appears to be more consistent with the fact that UV causes immunosuppression, however, the reason for this contradictive observations, remained to be determined so far.

In addition to the release of ETAF/IL 1 the keratinocyte has been demonstrated to produce a variety of other immunostimulating cytokines, such as a natural killer cell activity augmenting factor (ENKAF) (20), an interleukin 3 like cytokine (EC IL3) (21) and a granulocyte activating mediator (GRAM) (22). When purified to apparent homogeneity human ENKAF, EC-IL 3 and EC-GRAM coeluted, strongly suggesting that they are identical. Therefore the term epidermal cell leukocyte stimulating activity (ELSA) was proposed for this new cytokine, which is distinct from IL 1, IL 2, interferons and colony stimulating factors (23). Although murine keratinocytes have been demonstrated to express IL 3 mRNA (24) it is not clear whether human ELSA is related to human IL 3. Moreover keratinocytes have been shown to produce a cytokine which stimulates the proliferation of certain T-lymphocytes, but is distinct from IL 2 and therefore was named keratinocyte T-cell growth factor (KTGF) (25). More recently it became evident that KTGF antigenically, biologically and biochemically may not be separated from GM-CSF. Furthermore murine keratinocytes have been found to contain mRNA homologous to GM-CSF (24).

These data clearly indicate that the keratinocyte is much more than part of a mechanical barrier to the environment but is a fully immunocompetent cell which through the capacity to release immunomodulating cytokines may play an important role in the regulation of the immune response.

Keratinocyte derived suppressor factors

As the skin provides an optical barrier against UV radiation and UVB light can hardly affect cells outside the epidermis the keratinocyte has to be assumed as the main target of UV light. If soluble mediators, as mentioned above, are in fact of importance in the pathogenesis of UV induced systemic immunosuppression, the keratinocyte has to be considered as a primary source of such suppressor factors. This hypothesis indeed gains weight by the observation that the keratinocyte exhibits capacity to release a variety of immunomodulatory cytokines. In order to investigate whether keratinocytes after UV exposure release factors that influence CHS reaction to common antigens, epidermal cell (EC) suspensions were prepared from Balb/c mice and cells exposed to UVB light (Osram Ultravitalux, 20 mJ/cm^2). 24 hours after UV radiation supernatants were harvested and injected intravenously into Balb/c mice. Five days later animals were senisitized by the application of 0.5% DNFB on the abdomen and subsequently CHS reaction was induced by administration of 0.3% DNFB on the left ear. Ear swelling was measured 24 hours later. In contrast to the injection of supernatant derived from unirradiated ECs which did not affect ear swelling, application of fluid obtained from UV exposed ECs significantly suppressed CHS reaction thus indicating that ECs upon UV exposure release an immunosuppressive cytokine (26). The main source of this inhibitor appears to be the keratinocyte as this CHS blocking activity could also be detected in supernatants derived from UV exposed Pam 212 cells, a transformed murine keratinocyte cell line which is devoid of Langerhans cells, Thy 1 positive cells and melanocytes. Moreover, the release of this inhibitory factor appears to be a keratinocyte specific phenomenon since UV exposed macrophages and fibroblasts in culture failed to produce a similar factor. The inhibitor obviously only blocks the induction but not the elicitation of CHS because injection of supernantant derived from UV exposed

ECs into presensitized animals did not affect ear swelling. Upon HPLC gel filtration this CHS inhibitor exhibits a molecular weight between 20 and 50 kD clearly distinct from that of other immunomodulators such as prostaglandins and leukotrienes. Moreover, the EC derived inhibitor of CHS appears to be different from urocanic acid, a soluble compound located in the superficial layers of the epidermis, which has been suggested to be a photoreceptor for immunosuppression (27). Further studies have to clarify the mode of action of this inhibitor in particular whether it leads to the induction of T-suppressor cells or interferes with antigen presentation and processing. The capacity of keratinocytes to release an inhibitor of the induction of CHS may at least partly explain some of the events in UV mediated systemic immunosuppression. Whether the release of such a suppressor factor has in vivo implications e.g. in the suppression of CHS in men (28,29) or in the pathogenesis of skin cancer remains to be determined.

In order to gain more insight on the mode of action of this CHS inhibitor supernatants derived from UV exposed ECs were tested in vitro for a variety of other immunoinhibitory properties. When supernatants obtained from UV irradiated Balb/c ECs were subjected to HPLC gel filtration and fractions tested for IL 1 inhibition as measured in the thymocyte costimulator assay in the presence of natural murine IL 1 a significant suppression of IL 1 induced thymocyte proliferation was detected. This inhibitory activity was not observed in supernatants of unirradiated ECs and eluted at about 40 kD thus appearing in the same molecular weight range as the CHS inhibitor. The IL 1 inhibitor which was named EC-contra-IL 1 blocks natural and recombinant murine IL 1 in a dose dependent manner and exhibits an isoelectric point of 8.8 (30). EC-contra-IL 1 does not represent a nonspecific inhibitor of DNA synthesis as it does not affect the spontaneous proliferation of other murine cell lines, such as P388D1, L929, EL 4, Pam 212. The inhibitory activity of EC-contra-IL 1 appears to be specific for IL 1, since it also blocks IL 1 induced fibroblast and D10 lymphocyte proliferation and does not interfere with IL 2 and IL 3 activity. EC-contra-IL 1 is not constitutively produced by ECs, but the release can be induced either by UVB exposure or by stimulation with the tumor promoting agent PMA. Other stimuli such as lipopolysaccharide had no effect on the production of EC-contra-IL 1. This suppressor factor does not represent a preformed substance released after cellular damage but is actively synthesized, as EC-contra-IL 1 activity was not detectable after treatment of ECs with the potent protein synthesis inhibitor cycloheximide. The release of EC-contra-IL 1 does not seem to be only an in vitro observation as recently an IL 1 blocking mediator with similar biochemical characteristics as EC-contra-IL 1 could be detected in the serum of UVB irradiated mice (31). A variety of IL 1 inhibitory factors produced by distinct cell lines have been described (32-37). Although the biological in vivo function of these inhibitors is not yet clear; they may be part of a physiological feedback mechanism downregulating IL 1 activity after excessive release of IL 1. The fact that in addition to IL 1 keratinocytes do produce its own inhibitor might explain the contradictory results concerning the release of IL 1 after UV exposure by various groups in the past (18,19). Although the biological function of EC-contra-IL 1 is not yet clear, it might exhibit immuno-suppressive and anti-inflammatory activity. As EC-contra-IL 1 is induced by UVB light it thus may play an important role in the UV mediated immunosuppression. Whether EC-contra-IL 1 is involved in the generation of skin tumors cannot be answered yet, however, the capacity to induce EC-contra-IL 1 seems to be confined to tumor promoting agents such as UV light and PMA.

There seems to be a close similarity between the CHS inhibitor and EC-contra-IL 1: both suppressor factors are released by normal as well as transformed murine keratino-cytes in culture under identical conditions; they are both induced by UV light; and they share a similar molecular weight (table 1).

However, only further characterization, sequencing and gene cloning of both mediators will clarify wether these factors are identical or two distinct cytokines release by keratinocytes. Further purification will also indicate whe-ther both factors are new mediators or well-known immunomo-dulators with a so far undescribed activity. Although the in vivo function of these inhibitors has not been deter- mined the release of immunosuppressive cytokines by keratinocytes might explain systemic changes in vivo after UV exposure. Therefore the keratinocytes through the capacity to release suppressor factors seems to play an important role in the pathogenesis of UVB induced systemic immunosuppression.

Table 1.

	CHS inhibitor	EC-contra-IL 1
Source	Pam 212, Balb/c, EC	Pam 212, Balb/c, EC
Induction	UVB light	UVB light, PMA
Molecular weight	20-50 kD	40 kD
pI	n.d.	8.8

References

1. M. L. Kripke, Immunology and photocarcinogenesis. J. Am. Acad. Dermatol. 14: 149 (1986).
2. F. P. Noonan, E. Defabo, M. L. Kripke, Suppression of contact hypersensitivity by ultraviolet radiation and its relationship to ultraviolet induced suppression of tumor immunity. Photochem. Photobiol. 34: 638 (1981).
3. M. L. Kripke, Immunologic mechanisms in UV radiation carcinogenesis. Adv. Cancer Res. 34: 69 (1981)

4. G. B. Toews, P. R. Bergstresser, J. W.Streilein, Epidermal Langerhans cell density determines whether contact hypersensitivity or unresponsiveness follows skin painting with DNFB. J. Immunol. 124: 445 (1980).

5. C. A. Elmets, P. R. Bergstresser, R. E. Tigellar, P. J. Wood, J. W. Streilein, Analysis of mechanism of unresponsiveness produced by haptens painted on skin exposed to low dose ultraviolet radiation. J. Exp. Med. 158: 781 (1983).

6. J. M. Jessup, N. Hanna, E. Palszynski, M. L. Kripke, Mechanisms of depressed reactivity to dinitrochlorobenzene and ultraviolet-induced tumors during ultraviolet carcinogenesis in Balb/c mice. Cell. Immunol. 38: 105 (1978).

7. E. C. Defabo, F. B. Noonan, Mechanism of immune suppression by ultraviolet irradiation in vivo. J. Exp. Med. 157: 84 (1983).

8. W. L. Morison, C. Bucana, M. L. Kripke, Systemic suppression of contanct hypersensitivity by UVB radiation is unrelated to the VU-induced alterations in the morphology and number of Langerhans cells. Immunology 52: 229 (1984).

9. M. L. Kripke, W. L. Morison, Studies on the mechanism of systemic suppression of contact hypersensitivity by ultraviolet B radiation. Photodermatology 3: 4 (1986).

10. R. Swartz, Role of UVB-induced serum factor(s) in suppression of contact hypersensitivity in mice. J. Invest. Dermatol. 83: 304 (1984).

11. G. Stingl, K. Tamaki, S. I. Katz, Origin and function of epidermal Langerhans cells. Immunol. Rev. 53: 149 (1980).

12. N. Romani, G. Stingl, E. Tschachler, M. D. Witmer, R. M. Steinmann, E. M. Shevach, G. Schuler, The thy-1-bearing cell or murine epidermis. A distinctive leukocyte perhaps related to natural killer cells. J. Exp. Med. 161: 1368 (1985).

13. T. A. Luger, B. M. Stadler, S. I. Katz, J. J. Oppenheim, Epidermal cell (keratinocyte)-derived thymocyte activating factor (ETAF). J. Immunol. 127: 1493 (1981).

14. T. A. Luger, B. M. Stadler, B.M. Luger, B. J. Mathieson, M. Mage, J. A. Schmidt, J. J. Oppenheim, Murine epidermal cell derived thymocyte-activating factor resembles murine interleukin 1. J. Immunol. 128: 2147 (1982).

15. T. A. Luger, J. J. Oppenheim, Characteristics of interleukin 1 and epidermal cell-derived thymocyte-activating factor. Adv. Inflammation Res. 5: 1 (1983).

16. J. C. Ansel, T. A. Luger, D. R. Lowy, J. D. Mountz, Expression of IL 1 in murine keratinocytes. J. Invest. Dermatol. 87: 127 (1986).

17. T. S. Kupper, D. W. Ballard, A. O. Chua, J. S. McGuire, P. Flood, M. C. Horowitz, R. Langdon, L. Lightfood, U. Gubler, Human keratinocytes contain mRNA indistinguishable from monocyte interleukin 1α and βmRNA. J. Exp. Med. 164: 2095 (1986).

18. J. C. Ansel, T. A. Luger, I. Green, The effect of in vitro and in vivo irradiation of the production of ETAF activity by human and murine keratinocytes. J. Invest. Dermatol. 81: 513 (1983).

19. L. Gahring, M. Baltz, M. B. Pepys, R. Daynes, Effect of ultraviolet radiation on production of epidermal cell thymocyte-activating factor/interleukin 1 in vivo and in vitro. Proc. Natl. Acad. Sci. USA 81: 1198 (1984).

20. T. A. Luger, A. Uchida, A. Köck, M. Colot, M. Micksche, Human epidermal cells and squamous cell synthesize a cytokine which augments natural killer cell activity. J. Immunol. 134: 2477 (1985).

21. T. A. Luger, U. Wirth, A. Köck, Epidermal cells synthesize a cytokine with interleukin 3-like properties. J. Immunol. 134: 915 (1985).

22. A. Kapp, M. Danner, T. A. Luger, C. Hauser, E. Schöpf, Granulocyte-activating mediators (GRAM): II. Generation by human epidermal cells - Relation to GM-CSF. Dermatol. Res. in press (1987).

23. T. A. Luger, A. Kapp, M. Micksche, M. Danner, Characterization of a distinct epidermal cytokine with multiple immunoregulatory properties. J. Invest. Dermatol. 88: 504 (1987)

24. T. S. Kupper, D. L. Coleman, J. McGuire, D. Goldminz, M. C. Horowitz, Keratinocyte derived T-cell growth factor. A T cell growth factor functionally distinct from Interleukin 2. Proc. Natl. Acad. Sci. USA 83: 4451 (1986).

25. T. S. Kupper, M. Horowitz, F. Lee, D. Coleman, P. Flood, Molecular characterization of keratinocyte cytokines. J. Invest. Dermatol. 88: 501 (1987).

26. T. Schwarz, A. Urbanska, F. Gschnait, T. A. Luger, Inhibition of the induction of contact hypersensitivity by an UV-mediated epidermal cytokine. J. Invest. Dermatol. 87: 289 (1986).

27. E. C. Defabo, F. P. Noonan, Mechanism of immune suppression by ultraviolet irradiation in vivo: I. Evidence for existence of a unique photoreceptor in skin and its role in photoimmunology. J. Exp. Med. 157: 84 (1983).

28. K. Rosen, H. Mobacken, G.Swanbeck, Chronic eczematous dermatitis of the hands: A comparison of PUVA and UVB treatment. Acta Derm. Venereol. (Stockh) 67: 48 (1987).

29. P. Hersey, E. Hasic, J. Edwards, M. Bradeley, G. Haran, W. H. McCarthy, Immunological effects of solarium exposure. Lancet i: 545 (1983).

30. T. Schwarz, A. Urbanska, F. Gschnait, T. A. Luger, UV-irradiated epidermal cells produce a specific inhibitor of interleukin 1 activity. J. Immunol. 13: 1457 (1987).

31. T. Schwarz, A. Urbanska, F. Gschnait, T. A. Luger, Characterization of a suppressor of IL 1 activity in sera of UVB treated mice. J. Invest. Dermatol. 88: 517 (1987).

32. P. Miossec, D. Cavender, M. Ziff, Production of interleukin 1 by human and endothelial cells. J. Immunol. 136: 2486 (1986).

33. G. Scala, Y. D. Kuang, R. E. Hall, A. V. Muchmore, J. J. Oppenheim, Accessory cell function of human B. cells. I. Production of both interleukin 1-like activity and an interleukin 1 inhibitory factor by an EBV-transformed human B-cell line. J. Exp. Med. 159: 1637 (1984).

34. N. J. Roberts, A. H. Prill, T. N. Mann, Interleukin 1 and inhibitor production by human macrophages exposed to influenza or respiratory syncytial viruses. J. Exp. Med. 163: 511 (1986).

35. Z. Liao, R. S. Grimshaw, D. L. Rosenstreich, Identification of a specific interleukin 1 inhibitor in the urine of febrile patients. J. Exp. Med. 159: 126 (1984).

36. H. Fujiwara, J. J. Ellner, Spontaneous production of a suppressor factor by the human macrophage-like cell line U937. I. Suppression of interleukin 1, interleukin 2 and mitogen-induced blastogenesis in mouse thymocytes. J. Immunol. 136: 181 (1986).

37. K. Tiku, M. L. Tiku, S. Liu, J. L. Skosey, Normal human neutrophils are a source of a specific interleukin 1 inhibitor. J. Immunol. 10: 3686 (1986).

A SYSTEMATIC INFLUENCE OF UV RADIATION IN PHOTOCARCINOGENESIS

Jan C. van der Leun and Frank R. de Gruijl

Institute of Dermatology
University of Utrecht
The Netherlands

INTRODUCTION

Long before the days of modern photoimmunology, there was evidence
of systematic changes caused by exposure of the human body to UV radiation.
Clinicians noted that UV irradiations influenced almost any blood component which could be counted or measured.(1) A practice developed in which
a small quantity of a patient's own blood was exposed to UV radiation and
then re-injected; this was used as a treatment against infections and
allergies.(2,3) Most of this was understood vaguely at best. Many of
the recent observations are more specific and better accessible to check;
one may hope that they will also be more fruitful.

The organizers of this symposium asked us to tell about the experiments performed in our group on a systemic influence of UV radiation in
photocarcinogenesis. These experiments, on mice (4,5) were designed to
find out if the systemic effect discovered by photoimmunologists in experiments with transplantation of UV-induced tumors (6), would also change
the response of a mouse to primary tumors, induced by UV radiation in its
own skin. If so, we wanted to know whether such an influence would be
quantitatively important, and whether or not we would have to incorporate
it into the mathematical model we were developing to describe the process
of photocarcinogenesis. The study was not primarily directed, therefore,
on details of the photoimmunological process, but rather on its possible
practical implications.

EXPERIMENTAL SYSTEM

We used Skh hr 1 mice in the experiments; these are immunologically
normal, hairless albino mice. The mice were housed individually, and exposed to UV radiation from fluorescent sunlamps of the type Westinghouse
FS 40, which emit predominantly in the UV-B range. The mice were irradiated daily; the daily dose was 190 mJ/cm^2, measured over the full spectrum
of the lamp emission.

In three successive experiments, we first exposed part of the animals'
skin to UV-B radiation; the question to be answered by the experiments was,
if this pre-irradiation would facilitate the induction of tumors by UV-B
radiation in other skin areas.

EXPERIMENTS AND RESULTS

In all three experiments, the final induction of tumors was investigated using irradiations from above, with the sunlamps mounted over the cages. The pre-irradiations were different.

In experiment A, the pre-irradiations were given with the same arrangement, with the lamps above the cages. The mice had the left half of the back and the left flank shielded from the UV radiation by opaque adhesive tape. In this way, the pre-irradiations were given to the right half of the back, the right flank and the uppersides of head and tail. The pre-irradiations were given daily during 13 weeks, and led to small tumors in all of the mice. No tumors occurred on the shielded areas. After these pre-irradiations, the tapes were removed and not applied again. From then on the mice were given the same daily irradiations on the complete dorsal side. This was also done with a control group which had been treated in the same way, except for the pre-irradiations. The question to be answered was, if the tumors on the initially shielded sides would appear faster in the mice that had received the pre-irradiations.

The results of experiment A are shown in Figure 1. The tumor yield on the ordinate was defined as the average number of tumors per mouse, in this case on the left sides initially shielded. The figure shows that the tumors did appear faster on the animals pre-irradiated. The difference was statistically significant at the level $P < 0.01$.

In experiment B, the pre-irradiations were given on the abdomen. The mice were irradiated from below; this was achieved by placing the lamps under the cages. In this case, the pre-irradiations were given daily during 15 weeks, and led to small tumors on the abdomen in the majority of animals. After these 15 weeks, the dorsal sides of the mice were irradiated daily, with the lamps above the cages. The induction of tumors on the back was compared with that in a control group, which received the same dorsal exposures, but had not had the pre-irradiations.

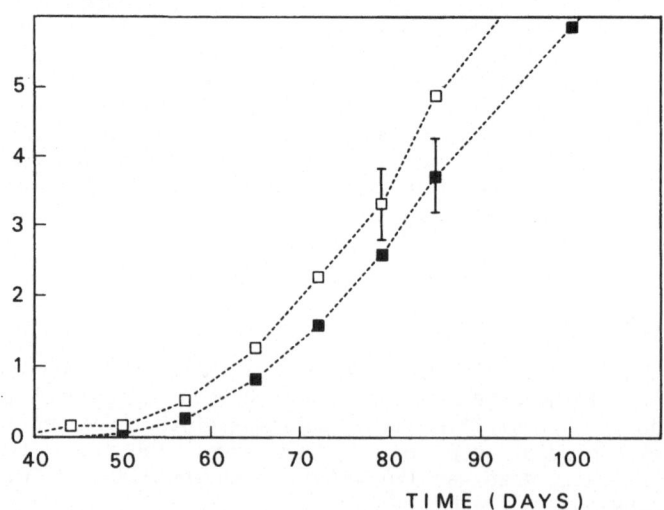

TUMOR YIELD

TIME (DAYS)

Fig.1 Tumor yield as a function of time on the skin areas initially shielded from the UV radiaiton (experiment A). Open squares represent the mice pre-irradiated on other skin areas, closed squares the controls that did not receive the pre-irradiations. Vertical bars indicate standard error of the mean.

The results of experiment B are shown in Figure 2. Again, the tumors appeared faster in the mice pre-irradiated than in the control group. The difference was even more marked than in experiment A.

Experiment C was carried out in almost the same way as experiment B. The pre-irradiations on the abdominal side were given, however, during only 4 weeks. This gave a correspondingly lower total dose of the pre-irradiations, which did not lead to visible tumors on the abdomen.

The results of experiment C are show in Figure 3, now in the form of percent of mice with tumors as a function of time. Time is on a logarithmic scale; this produces the cumulative log-normal distributions as known from the classic work by Blum.(7) Again, the mice pre-irradiated developed their tumors earlier than the control animals not pre-irradiated.

ANALYSIS

In all three experiments, the induction of tumors by UV radiation was facilitated by pre-irradiation of the mice in other skin areas. It is attractive to express this facilitation in quantitative terms, so that the results of the various experiments can be compared and the influence of the characteristics of the pre-irradiation examined. We chose to express the facilitation as an increase of the effectiveness of the radiation finally inducing the tumors. That is feasible, because the dose-effect relationship for tumorigenesis by UV radiation is known. For the same strain of mice exposed to the same type of fluorescent lamps, we found (8)

$$t_m = kD^{-0.6} \tag{1}$$

where t_m is the median tumor appearance time (defined as the time since the start of the exposures until the day at which 50 percent of the mice have developed one or more tumors); D is the daily dose of UV radiation administered and k a proportionality constant. This dose-response relationship is immediately applicable to data as presented in fig.3; it may, of course, also be applied to the data collected in experiments A and B.

Calculation on this basis shows that in experiment A (fig.1) the UV doses inducing tumors in the pre-irradiated mice had an effectiveness equivalent to a 35 percent higher dose given to animals not pre-irradiated. The corresponding increase of the dose-effectiveness in experiment B amounted to 70 percent and in experiment C to 25 percent. These results are displayed in figures 4 and 5.

Figure 4 shows the influence of the surface area of skin pre-irradiated, with approximately the same doses, in experiments A and B. The fraction of the skin surface area pre-irradiated was extimated 30 percent in experiment A and 50 percent in experiment B. The effect appears to be approximately proportional to the surface area of skin pre-irradiated.

Figure 5 shows the influence of the total dose of the pre-irradiations, given to the same area of skin in experiments B and C. The result suggests that the effect of the pre-irradiations is approximately proportional to the total dose of the pre-irradiations. An attempt was made to make the scales on the two axes of fig.5 comparable; both were related to t_m, the median tumor appearance time. The total dose of the pre-irradiations, on the horizontal axis, was expressed as a percentage of the dose accumulated at t_m. The increase of dose-effectiveness on the vertical axis was expressed as the percent increase of the daily doses on the back which, according to equation (1), would produce the same decrease of t_m as the pre-irradiations did.

TUMOR YIELD

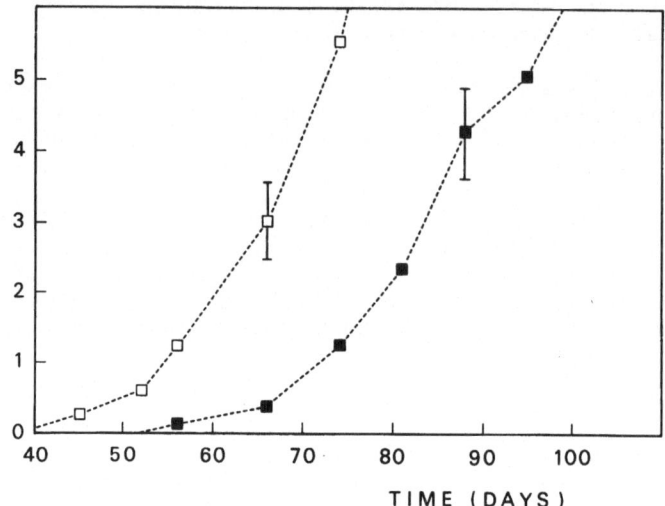

Fig.2 Tumor yield on the backs of the mice as a function of time
(experiment B). Open squares represent the mice pre-irradiated
on the abdomen, closed squares the control group which was not
pre-irradiated. Bars indicate standard errors.

PERCENT OF MICE WITH TUMORS

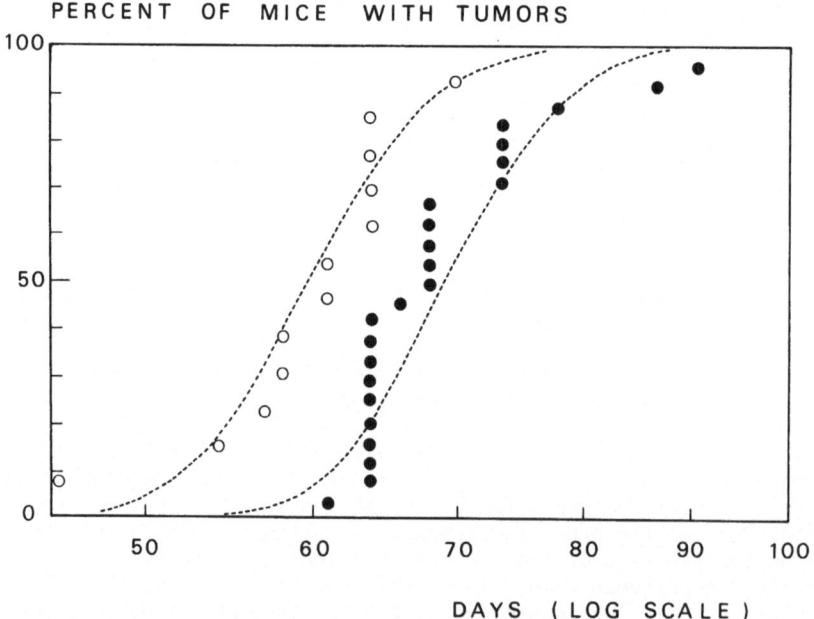

Fig.3 The percent of mice with tumors on the back as a function of time
(experiment C). Open points represent the mice pre-irradiated on
the abdomen, closed points the controls not pre-irradiated.

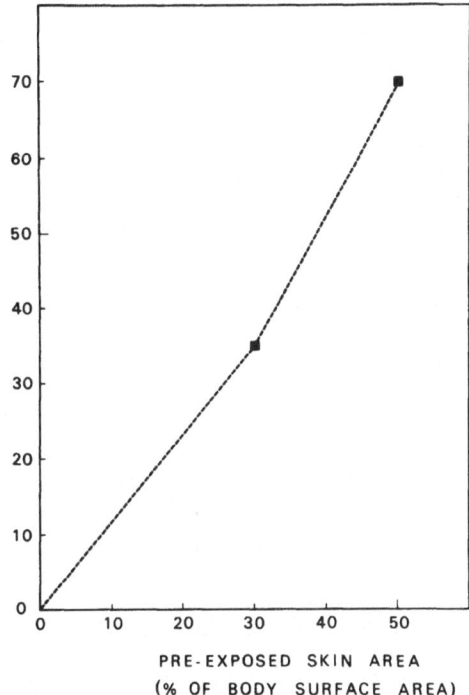

INCREASE OF DOSE-EFFECTIVENESS (%)

PRE-EXPOSED SKIN AREA
(% OF BODY SURFACE AREA)

Fig.4 Increase of dose-effectiveness for the induction of tumors due to pre-irradiation of another skin area, as it depends on the surface area of skin pre-irradiated.

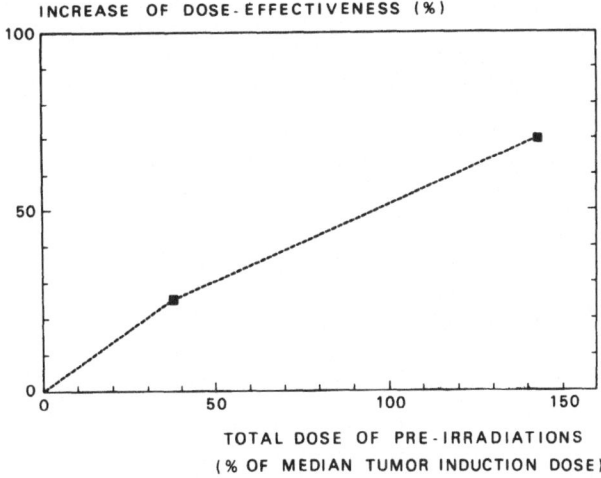

INCREASE OF DOSE-EFFECTIVENESS (%)

TOTAL DOSE OF PRE-IRRADIATIONS
(% OF MEDIAN TUMOR INDUCTION DOSE)

Fig.5 Increase of dose-effectiveness for the induction of tumors on the back as it depends on the total dose of the pre-irradiations on the abdomen.

DISCUSSION

The steepness of the relationship depicted in fig.5 suggests that the systemic effect caused by the pre-irradiations is highly efficient in facilitating tumorigenesis in another skin area. The total dose of the pre-irradiations given on the abdomen in experiment B amounted to 150 percent of the dose accumulated at t_m. This increased the tumorigenic effectiveness of the doses given on the backs of the mice by 70 percent. The effect of the pre-irradiations on the abdomen was, therefore, passed on to the back with approximately half its initial effectiveness.

This high efficiency of the systemic effect examined rules out several explanations. One could imagine, for instance, that the pre-irradiations produced a carcinogen photochemically. This carcinogen might spread through the entire body and also end up in the skin on the other side of the animal. A systemic effect operating in this way would in principle facilitate the induction of tumors in other skin area, but the carcinogen would be diluted by at least a factor of 100 by the time it arrived at the other side of the body. It is inconceivable that such a passive system would lead to a "dilution" by only a factor of 2. The high efficiency of the systemic effect examined is well compatible, however, with an active system based on the programming of suppressor T-lymphocytes, as found in photoimmunology.(9)

The experiments discussed demonstrate that the systemic, presumably immunological change caused by exposures of the skin to UV radiation indeed also facilitates the acceptance of primary tumors, induced by UV radiation in the mouse's own skin. It forms, therefore, an integral part of the process of photocarcinogenesis.

A practical consequence of the findings presented has to do with the problem of a "safe limit", which repeatedly arises in discussions on the protection of people against UV radiation. Life for regulators would be more comfortable if there were UV doses so small that they would not contribute to the risk of skin cancer. Many investigations have produced evidence that the induction of skin cancer by UV radiation is a chance process. In this conception, every exposure increases the risk: there is no threshold dose and therefore no safe limit. That leaves open the possibility of a "practical threshold dose". That could arise, for instance, from the systemic effect under discussion, assumed that it would operate also in man. Certain UV exposure regimens could be so minute that they were insufficient to induce the systemic effect. In that case, any tumors initiated in the skin by the UV radiation would be removed by the immune system.

There is no indication in the experimental results presented for even such a practical threshold. The systemic effect described is not an all-or-none phenomenon. Over a wide range of realistic doses the effect is present, to varying degrees, dependent on the surface area of skin exposed and on the doses received. It is, therefore, already included in the observational dose-effect relationship given by equation (1). This relationship has been established down to very low doses, smaller by an order of magnitude than the doses received on the average by indoor workers in the Netherlands (52° N.lat). Even at these low doses, there was no indication of any deflection of the dose-effect relationship.(8) This also implies that, if a mathematical model for photocarcinogenesis is based on such observational dose-effect relationship, there is no reason to incorporate the systemic effect discussed as a separate factor into the model.

CONCLUSIONS

The induction of tumors by UV radiation in mouse skin is facilitated by pre-irradiation of other skin areas. This indicates a systemic effect.

The systemic effect examined is highly efficient. This suggests an active process. Presumably it has the same basis as the facilitation of transplantation of UV-induced tumors, namely, the generation of suppressor cells. The effect examined also affects the acceptance of primary tumors, induced by UV radiation in the mouse's own skin.

The systemic effect examined is implicitly included in observational dose-effect relationships. It does not support the assumption of a "practical threshold dose", and a safe limit for UV exposures.

REFERENCES

1. J. Kimmig and A. Wiskemann, Lichtbiologie und Lichttherapie, in: "Handbuch der Haut- und Geschlechtskrankheiten", J. Jadassohn, ed., Ergänzungswerk V/2, p. 1986, Springer, Berlin (1959).
2. O. Glasser, Ultraviolet irradiation of blood, in: "Medical Physics", O. Glasser, ed., Vol. 1, p. 1596, The Year Book Publishers, Chicago (1944).
3. G. P. Miley, R. E. Seidel and J. A. Christensen, Ultraviolet blood irradiation therapy of apparently intractable bronchial asthma, Arch. phys. Med. 27:24 (1946)
4. F. R. de Gruijl and J. C. van der Leun, Systemic influence of pre-irradiation of a limited skin area on UV-tumorigenesis. Photochem. Photobiol. 35: 379 (1982).
5. F. R. de Gruijl and J. C. van der Leun, Follow up on systemic influence of partial pre-irradiaiton on UV-tumorigenesis. Photochem. Photobiol. 38: 381 (1983).
6. M. L. Kripke, Photoimmunology: the first decade, in: "Therapeutic Photomedicine", H. Hönigsmann and G. Stingl, eds., Curr. Probl. Derm. 15: 164, Karger, Basel (1986).
7. H. F. Blum, "Carcinogenesis by Ultraviolet Light", Princeton University Press, Princeton, N.J. (1959).
8. F. R. de Gruijl, J. B. van der Meer and J. C. van der Leun, Dose-time dependency of tumor formation by chronic UV exposure, Photochem. Photobiol. 37: 53 (1983).
9. R. A. Daynes and C. W. Spellman, Evidence for the generation of suppressor cells by ultraviolet radiation, Cell. Immunol. 31: 182 (1977).

DIFFERENTIAL EFFECTS OF VARIOUS PHYSICOCHEMICAL AGENTS ON MURINE Ia-
AND Thy-1 POSITIVE DENDRITIC EPIDERMAL CELLS

Adelheid Elbe[*], Werner Aberer[*], Nikolaus Romani[°],
Erwin Tschachler[*], and Georg Stingl

Department of Dermatology I., Immunobiology Unit
University of Vienna, Vienna, Austria
[°]Department of Dermatology, University of Innsbruck
Innsbruck, Austria

INTRODUCTION

It is now well established that epidermal cells (EC) can initiate
T-cell-dependent immune responses[1] and that this functional capacity
can be altered by various physicochemical agents. Low-dose ultraviolet
B (UV-B) irradiation of EC in vivo or in vitro impairs their sensitizing
potential in contact hypersensitivity (CHS)[2-4]. Similarly, both man
and experimental animals that had received psoralen plus UV-A (PUVA)
treatment exhibited a markedly reduced CHS response to simple chemical
haptens[4-6]. In the search for the mechanism(s) by which these various
physicochemical agents modulate EC-induced immune responses, particular
attention has been focused on the effects of these agents on Ia-bearing
Langerhans cells (LC) because the immunostimulatory capacities of EC
are critically dependent on these cells[1,7-10]. Results obtained showed
that PUVA-treatment leads to a dramatic reduction of Ia-bearing dendritic EC as assessed by immunofluorescence (IF)[11,12], and to a concom-
mitant impairment of LC-dependent immunologic functions of EC[12]. In the
case of low-dose UV-B, however, the functional defect of LC-containing
EC[8] is reportedly not paralleled by a disappearance of surface Ia
antigens[3,11].

The search for the target structure for the action of UV-B/
PUVA became more complicated when it became clear that EC other than LC
may contribute to the immunological function of the epidermis. Keratinocytes produce the Interleukin-1-like cytokine ETAF (= epidermal cell-
derived thymocyte-activating factor)[13,14] the production/secretion of
which appears to be increased[15] or at least, not adversely influenced by
UV-B[16,17].

The recently discovered population of Thy-1-positive dendritic epidermal cells (Thy-1[+]DEC)[18,19] is another candidate for participating in
EC immune functions. These dendritic cells are derived from the bone
marrow[20-22], belong to the T-cell lineage[23,24], and some evidence exists
that they deliver down-regulating signals in CHS induction[25]. In addition, both phenotypical and functional studies suggest their being related to natural killer cells[26,27]. In view of these date, we decided to
study the influence of UV-B irradiation and PUVA-treatment on both phenotypic equipment and ultrastructure of murine LC and Thy-1[+]DEC. We will
demonstrate that UV-B and PUVA-treatment not only alter the phenotypic
appearance of dendritic, bone marrow-derived EC but also provide a useful tool to remove selectively either LC or Thy-1[+]DEC from the epidermis.

Effect of low-dose UV-B radiation on Ia^+ EC and $Thy-1^+DEC$

Ears of C3H/He mice were irradiated with 100-1000 J/m² UV-B for four consecutive days by means of Sylvania F20 T12 fluorescent light bulbs, emitting a continuous spectrum from 290 - 320 nm. At defined time intervals after the last treatment, ears were cut off, mechanically split and placed on 0.5 M NH_4-SCN for 20 min at 37°C. Resulting epidermal sheets were then subjected to appropriate anti-Ia or anti-Thy-1 reagents using indirect immunolabeling techniques and the mean number of positive cells/mm² was determined in 400 x power field using a calibrated ocular grid.

Administration of UV-B doses which did not result in any gross or histological changes of the ear skin ($4 \times 100-200$ J/m²) altered neither the density nor the structural appearance of both Ia^+ EC and $Thy-1^+DEC$ when compared to untreated control animals (Figs.1,2). Exposure to UV-B doses of 4 x 400 J/m² led to erythema, edema and scaling of the irradiated ears and to a moderate decrease of both Ia^+ EC and $Thy-1^+DEC$ when assessed two days after cessation of UV-B treatment (Figs.1,2). During the following weeks, we observed a steady numerical increase of both cell systems, and after aproximatley six weeks their density was virtually indistinguishable from that encountered in control animals (Figs.1, 2). When experimental animals were irradiated with 4 x 700 - 1000 J/m², the universal tissue injury was also reflected in the Ia^+ and $Thy-1^+DEC$ which - two days after the last irradiation - were both greatly reduced in number and which appeared structurally distorted. Thereafter, we observed a continuing reappearance of Ia^+ EC similar to that seen in animals which were exposed to 4 x 400 J/m². In contrast, the reemergence of $Thy-1^+DEC$ lagged far behind, i.e. six weeks after irradiation, the density of Ia^+ EC had reached control values whereas $Thy-1^+DEC$ were still substantially reduced in number (Figs.1,2).

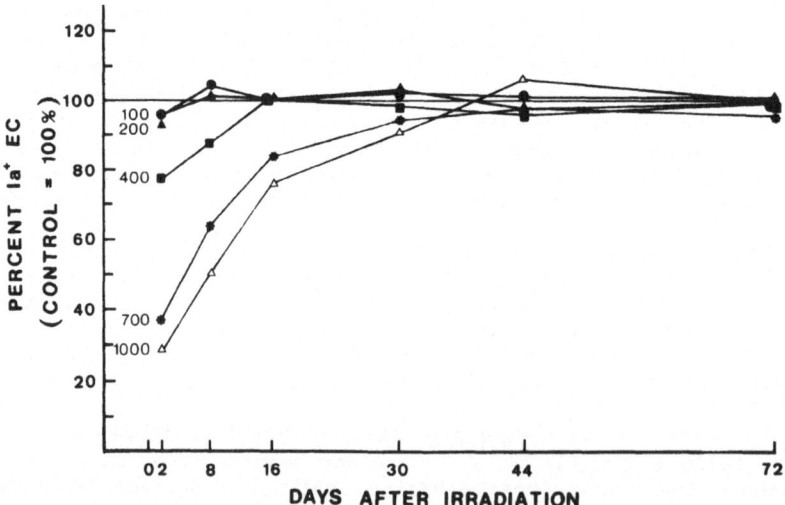

Figure 1. Effect of UV-B radiation on the number of Ia^+EC. Mice were irradiated for four consecutive days (-3 to 0) with 100 (●), 200 (▲), 400 (■), 700 (*) or 1000 (△) J/m² UV-B; they were sacrificed starting on day 2 and in time intervals of 14 days; the density of Ia^+ EC was determined in epidermal sheet preparations. UV-B led to a dose-dependent disappearance of Ia^+ EC 2 days after irradiation, which was followed by a steady and complete recovery within 6 weeks.

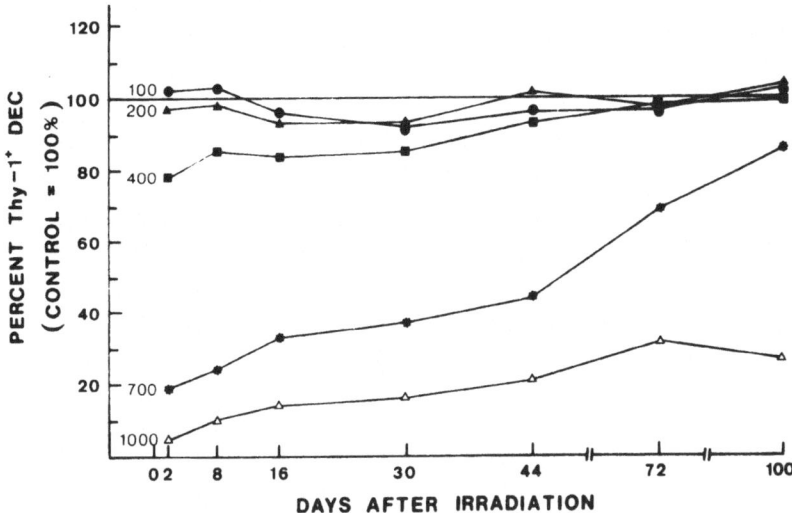

Figure 2. Effect of UV-B radiation on the number of Thy-1[+]DEC. Irradiation protocol, days of experiments, and graphic design are the same as in Figure 1. The density of Thy-1[+]DEC was determined in epidermal sheet preparations. UV-B radiation led to a dose-dependent disappearance of Thy-1[+]DEC as assessed 2 days after the last irradiation. The reemergence of Thy-1[+]DEC was very slow, and even 14 weeks after termination of high-dose UV-B treatment (4 x 1000 J/m²) the percentage of Thy-1[+]DEC was only 26 % of that found in untreated animals.

Effect of PUVA-treatment on Ia[+]EC and Thy-1[+]DEC

Two groups of mice were exposed 3 times per week for three weeks to either 8-methoxypsoralen (8-MOP) alone, UV-A alone (control groups) or irradiated with UV-A after 8-MOP application (PUVA group). 50 µl of a 0.15 % 8-MOP solution were applied to both sides of the ears. 60 min later, UV-A was administered in single doses of 4000 J/m² using Sylvania FR90 T12 PUVA/HO lamps as radiation source. These bulbs are routinely used as light source for clinical PUVA-treatment. At defined time intervals after the last treatment, Ia[+]EC and Thy-1[+]DEC were visualized and enumerated on epidermal sheets using indirect immunolabeling. In certain experiments, single EC suspensions prepared from ear skin by standard trypsinization procedures were searched for the presence of LC and Thy-1[+]DEC as identified either solely by their characteristic ultrastructural features or by additional ultrastructural immunolabeling.

A dose of 4000 J/m² applied 9 times over a period of three weeks induced a marked darkening of both sides of the ears - indicative for melanocyte activation - but did not cause any visible gross or microscopic alterations of the ear epidermis. As opposed to untreated, 8-MOP treated or UV-A treated control animals, PUVA-treatment caused an almost complete disappearance of both Ia[+]EC and Thy-1[+]DEC (Fig.3) as assessed two days after cessation of PUVA-treatment. Electronmicroscopic studies revealed, however, that the PUVA-induced disappearance of surface antigens is not necessarily paralleled by a loss of the respective cell population: whereas cells exhibiting the characteristic features of Thy-1$^+$ DEC were no longer detectable, structurally undamaged LC were readily identified, and LC numbers were not significantly reduced in PUVA-treated animals as compared to the controls. Evaluation of the reemergence kinetics of Ia$^+$- and Thy-1$^+$DEC over a time period of 22 weeks demonstrated striking differences between the two cell populations. In the case of Ia$^+$ EC (Fig.3), a measurable increase was first noted four weeks after termination of PUVA-treatment, and already after six weeks their density was virtually indistinguishable from that encountered in control animals.

In sharp contrast to Ia$^+$EC, phenotypic markers (Thy-1.2 antigen; asialo-GM1 antigen) of Thy-1$^+$DEC (Fig.3) were not detectable up to twelve weeks (day 86) after the last PUVA-treatment - not even when using

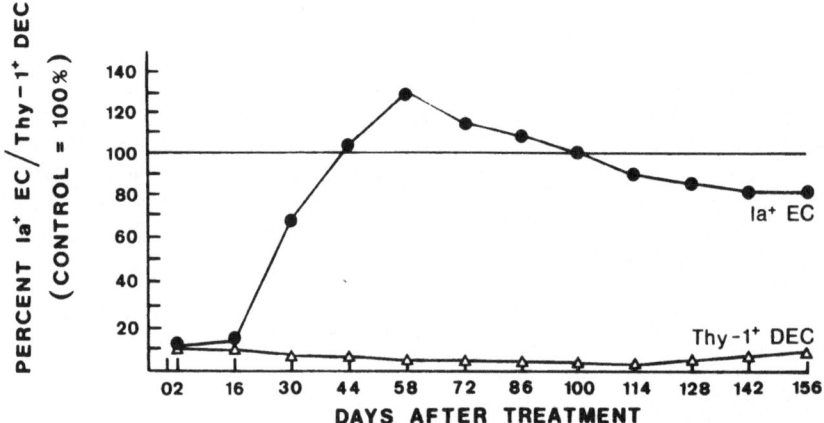

Figure 3. C3H/He mice received PUVA-treatment 3 times per week for three weeks; at various time intervals after cessation of PUVA-therapy, the density of Ia$^+$EC and Thy-1$^+$DEC was determined in sheet preparations of ear epidermis using anti-I-A^k and anti-Thy-1.2 monoclonal antibodies, respectively, in an indirect immunofluorescence technique; both Ia$^+$EC (closed circles) and Thy-1$^+$DEC (asterisks) were dramatically reduced in number on day 2; assessment of reemergence kinetics revealed that whereas the dinsity of Ia$^+$EC returned to normal values within six weeks, only negligible numbers of Thy-1$^+$DEC were encountered over the entire observation period.

highly sensitive immunoperoxidase techniques. After that time, dendritic cells exhibiting anti-Thy-1.2- and anti-asialo-GM1 reactivity reappeared in small, patchy, randomly distributed areas but their percentage never exceeded 8 % of that seen in controls, not even 22 weeks after termination of PUVA-treatment. Our further ultrastructural finding, that EC suspensions derived from animals which had received the last PUVA-treatment 45 days before were essentially devoid of cells exhibiting the morphological characteristics of Thy-1$^+$DEC[18,27] strongly suggests that the PUVA-induced non-detectability of cells bearing Thy-1.2- and asialo-GM1 antigens signifies the physical absence of the Thy-1$^+$DEC population.

DISCUSSION

We have described the effects of UV-B irradiation and PUVA-treatment on phenotype and structure of LC and Thy-1[+]DEC using UV-B/PUVA-
doses which, at a functional level, abrogate the in vivo or in vitro
sensitizing potential of EC. There was a good reason for focusing our
interest on both LC and Thy-1[+]DEC. It is well established that the
capacity of EC to induce proliferative and cytotoxic T-cell responses is
critically dependent upon Ia-positive LC[1,8-10]. Although the functional
role of bone marrow-derived Thy-1[+]DEC is not yet clear, it is now well-
known, that they belong to the T-cell lineage[23,24]. On the one hand
there exists suggestive evidence that Thy-1[+]DEC not only display morphological and phenotypic similarities to natural killer (NK) cells[27]
but - when activated - can also exhibit NK activity[26]. On the other
hand, some data indicate that Thy-1[+]DEC may be operative in the activation of suppressor mechanisms by providing down-regulating signals in
the induction of CHS[25].

Studies with either C57Bl/6 or C3H/He mice have shown that immuni-
zation via a skin area exposed to 4 x 200 J/m² and 4 x 700 J/m², respectively, results in a state of antigen-specific unresponsiveness rather
than sensitization[16,28]. Our finding that 4 x 200 J/m² exert no visible
changes in number, phenotype and structure of both Ia[+]EC and Thy-1[+] DEC
thus provides no explanation for the above in vivo observation. Based on
in vitro data that low-dose UV-B interferes with antigen uptake and/or
processing by LC[29,30] it is conceivable that 4 x 200 J/m² UV-B in vivo
affects LC only at a functional but not at a phenotypical or structural
level. Administration of 4 x 700 J/m² UV-B, however, leads to a substantial numerical decrease and structural alteration in both the Ia[+]EC and
Thy-1[+]DEC populations. It is thus conceivable that the loss of the
sensitizing capacity of EC seen after irradiation with 4 x 700 J/m² is,
at least partly, due to the numerical reduction and injury of Ia[+]EC. Our
further finding that Thy-1[+]DEC appear to be at least as, if not even
more (4 x 1000 J/m²) susceptible to UV-B argues against the possibility
that the recently described I-J[+], UV-resistant suppressor cell activating EC[31] are included in the Thy-1[+]DEC population.

Our study further demonstrates that PUVA-treatment affects LC and
Thy-1[+]DEC in a different manner. It physically eliminates the Thy-1[+]DEC
population whereas - despite their loss of IF-detectable Ia determinants
- LC remain morphologically intact. This finding differs from data by
Friedman et al.[32] who reported that PUVA-treatment of psoriasis patients
leads to an almost complete disappearance of ultrastructurally identifi-
able LC. The reason for this discrepancy is not yet clear but may be
either due to species differences or due to a different experimental
design. In partial accordance with repopulation kinetics described in
radiation bone marrow chimeras[21,22] we found that the repopulation of
the PUVA-treated epidermis with Thy-1[+]DEC occurs at a much slower rate
and much less complete than the reappearance of Ia[+]EC. Although yet
unproven, the almost complete lack of Thy-1[+]DEC even 22 weeks after
PUVA-treatment may suggest that this treatment modality leads to a
permanent damage of the putative precursor cell. Following the suggestion that Thy-1[+]DEC perform natural killer (NK) functions[26], the PUVA-
induced elimination of this cell system from the epidermis may contribute to photochemical carcinogenesis[33]. In addition, this phenomenon
may provide an excellent tool to study the functional role of Thy-1[+]DEC
in CHS. If the assumption that Thy-1[+]DEC are concerned with the delivery
of suppressor cell-activating signals[25] is correct, one would expect
from our experimental data that epicutaneous application of a contact
allergen on mice immediately after termination of PUVA-treatment would
result in a null event, i.e. neither sensitization nor tolerization
should occur. Consequently, a contact allergen introduced via a PUVA-
treated epidermis at a time when Ia[+]EC are present in normal numbers and

Thy-1$^+$DEC are virtually absent (i.e. 44 - 86 days after termination of PUVA-treatment) should result in maximal sensitization, i.e. sensitization without accompanying down-regulating signals. After reappearance of sufficient quantities of Thy-1$^+$DEC the sensitizing potential of PUVA-treated epidermis would then be equal to that of normal, untreated epidermis. Unfortunately, our observation period (= 22 weeks) was not long enough to determine at which time or whether at all such a numerical recovery of Thy-1$^+$DEC would occur. Regarding the effects of PUVA-treatment on LC, we have previously shown that the PUVA-induced numerical reduction of Ia$^+$EC was paralleled by an impairment of LC-dependent immune functions of EC[12]. Apparently, this PUVA-induced functional deficit was not due to the loss of the LC population but rather due to modulation of metabolic events in or of surface moieties on LC needed for T-cell activation. The observation that PUVA-treatment leaves the LC population physically intact raises the question whether the reappearance of Ia$^+$EC is due to "reexpression" of class II determinants by LC rendered Ia-negative by PUVA-treatment or, alternatively, due to an actual repopulation with newly immigrating cells. Although yet unproven, our observation that the shape of some reappearing Ia$^+$EC is slightly different from that of control animals is, at least partially, compatible with the latter hypothesis.

In summary, this study demonstrates that UV-B irradiation and PUVA-treatment affect the two dendritic, bone marrow-derived EC populations of the mouse in a somewhat different fashion and may explain why different immunoregulatory T-cell pathways are activated by either UV-B or PUVA-treatment.

REFERENCES

1. G. Stingl, K. Tamaki, and S. I. Katz, Origin and function of epidermal Langerhans cells. Immunol. Rev., 53: 149 (1980).
2. G. B. Toewes, P. R. Bergstresser, J. W. Streilein, and S. Sullivan, Epidermal Langerhans cell density determines whether contact hypersensitivity or unresponsiveness follows skin painting with DNFB. J. Immunol., 124: 445 (1980).
3. D. N. Sauder, K. Tamaki, A. N. Moshell, H. Fujiwara, and S. I. Katz, Induction of tolerance to topically applied TNCB using TNP-conjugated ultraviolet light-irradiated epidermal cells. J. Immunol., 127: 261 (1981).
4. K. M. Halprin, M. Comerford, S. E. Presser, and J. R. Taylor, Ultraviolet light treatment delays contact sensitization to nitrogen mustard. Br. J. Dermatol. 105: 71 (1981).
5. D. H. Lynch, M. F. Gurish, and R. A. Daynes, Relationship between epidermal Langerhans cell density. ATPase activity and the induction of contact hypersensitivity. J. Immunol., 126: 1892 (1981).
6. T. Horio, and H. Okamato, Immunologic unresponsiveness induced by topical application of hapten to PUVA-treated skin in guinea pigs. J. Invest. Dermatol., 80: 90 (1983).
7. J. W. Streilein, Skin-associated lymphoid tissues (SALT): origins and functions. J. Invest. Dermatol. 80: 12s (1983).
8. G. Stingl, L. A. Gazze-Stingl, W. Aberer, and K. Wolff, Antigen presentation by murine epidermal Langerhans cells and its alteration by ultraviolet B light. J. Immunol., 127: 1707 (1981).
9. W. Aberer, G. Stingl, L. A. Stingl-Gazze, and K. Wolff, Langerhans cells as stimulator cells in the murine primary epidermal cell-lymphocyte reaction: alteration by UV-B irradiation. J. Invest. Dermatol., 79: 129 (1982).
10. H. Pehamberger, L. A. Stingl, S. Pogantsch, G. Steiner, K. Wolff,

and G. Stingl, Epidermal cell-induced generation of cytotoxic T-lymphocyte responses against alloantigens or TNP-modified syngeneic cells: requirement for Ia-positive Langerhans cells. J. Invest. Dermatol., 81: 208 (1983).

11. J. J. Nordlund, A. E. Ackles, and A. B. Lerner, The effects of ultraviolet light and certain drugs on Ia-bearing Langerhans cells in murine epidermis. Cell. Immunol., 60: 50 (1981).

12. J. Hutterer, L. Stingl, S. Pogantsch, and G. Stingl, Effects of PUVA on epidermal Langerhans cells: immunohistologic and functional studies. Arch. Dermatol. Forsch., 275: 278 A (1983).

13. T. A. Luger, B. M. Stadler, S. I. Katz, and J. J. Oppenheim, Epidermal cell (keratinocyte) derived thymocyte activating factor (ETAF). J. Immunol., 127: 1493 (1981).

14. D. N. Sauder, C. S. Carter, S. I. Katz, and J. J. Oppenheim, Epidermal cell production of thymocyte-activating factor (ETAF). J. Invest. Dermatol., 79: 34 (1982).

15. T. S. Kupper, A. O. Chua, P. Flood, J. Mc Guire, and U. Gubler, Interleukin 1 gene expression in cultured human keratinocytes is augmented by ultraviolet irradiation. J. Clin. Invest., 80: 430 (1987).

16. J. C. Ansel, T. A. Luger, and I. Green, The effect of in vitro and in vivo UV irradiation on the production of ETAF activity by human and murine keratinocytes. J. Invest. Dermatol., 81: 519 (1983).

17. L. Gahring, M. Baltz, M. B. Pepys, and R. Daynes, Effect of ultraviolet radiation on production of epidermal cell thymocyte-activating factor/interleukin-1 in vivo and in vitro. Proc. Natl. Acad. Sci. USA, 81: 1198 (1984).

18. E. Tschachler, G. Schuler, J. Hutterer, H. Leibl, K. Wolff, and G. Stingl, Expression of Thy-1 antigen by murine epidermal cells. J. Invest. Dermatol., 81: 282 (1983).

19. P. R. Bergstresser, R. E. Tigelaar, J. H. Dees, and J. W. Streilein, Thy-1 antigen-bearing dendritic cells populate murine epidermis. J. Invest. Dermatol., 81: 286 (1983).

20. S. M. Breathnach, and S. I. Katz, Thy-1[+] dendritic cells in murine epidermis are bone marrow-derived. J. Invest. Dermatol., 83: 74 (1984).

21. P. R. Bergstresser, R. E. Tigelaar, and J. W. Streilein, Thy-1 antigen-bearing dendritic cells in murine epidermis are derived from bone marrow precursors. J. Invest. Dermatol., 83: 83 (1984).

22. N. Romani, E. Tschachler, G. Schuler, W. Aberer, R. Ceredig, A. Elbe, K. Wolff, P. O. Fritsch, and G. Stingl, Morphological and phenotypical characterization of bone marrow-derived dendritic Thy-1-positive epidermal cells of the mouse. J. Invest. Dermatol., 85: 91s (1985).

23. G. Stingl, K. C. Gunter, E. Tschachler, H. Yamada, R. J. Lechler, W. M. Yokoyama, G. Steiner, R. N. Germain, and E. M. Shevach, Thy-1[+] dendritic epidermal cells belong to the T-cell lineage. Proc. Natl. Acad. Sci. USA, 84: 2430 (1987).

24. G. Stingl, F. Koning, H. Yamada, W. M. Yokoyama, E. Tschachler, J. A. Bluestone, G. Steiner, L. E. Samelson, A. M. Lew, J. E. Coligan, and E. M. Shevach, Thy-1[+] dendritic epidermal cells express T3 and the T cell receptor gamma chain. Proc. Natl. Acad. Sci. USA., 84: 4586 (1987).

25. S. Sullivan, P. R. Bergstresser, R. E. Tigelaar, and J. W. Streilein, Induction and regulation of contact hypersensitivity by resident, bone marrow-derived, dendritic epidermal cells: Langerhans cells and Thy-1[+] epidermal cells. J. Immunol., 137: 2460 (1986).

26. J. L. Nixon-Fulton, P. R. Bergstresser, J. Hackett, Jr., V. Kumar,

and R. E. Tigelaar, Con-A stimulated Thy-1$^+$ epidermal cells exhibit natural killer (NK)-like activity. J. Invest. Dermatol., 84: 289 A (1985).

27. N. Romani, G. Stingl, E. Tschachler, M. D. Witmer, R. M. Steinman, E. M. Shevach, and G. Schuler, The Thy-1-bearing cell of murine epidermis. A distinctive leukocyte perhaps related to natural killer cells. J. Exp. Med., 161: 1368 (1985).

28. C. A. Elmets, P. R. Bergstresser, R. E. Tigelaar, P.J. Wood, and J. W. Streilein, Analysis of the mechanism of unresponsiveness produced by haptens painted on skin exposed to low dose ultraviolet radiation. J. Exp. Med., 158: 781 (1983).

29. S. Shimada, Z. Kovac, R. H. Schwartz, and S. I. Katz, Effect of ultraviolet radiation on antigen presentation. Clin. Res., 33: 531 A (1985).

30. L. A. Stingl, D. N. Sauder, M. Iijima, K. Wolff, H. Pehamberger, and G. Stingl, Mechanism of UV-B-induced impairment of the antigen-presenting capacity of murine epidermal cells. J. Immunol., 130: 1586 (1983).

31. R. D. Granstein, A. Lowy, and M. I. Greene, Epidermal antigen-presenting cells in activation of suppression: identification of a new functional type of ultraviolet radiation-resistant epidermal cell. J. Immunol., 132: 563 (1984).

32. P. S. Friedmann, G. Ford, J. Ross, and B. L. Diffey, Reappearance of epidermal Langerhans cells after PUVA therapy. Br. J. Dermatol., 109: 301 (1983).

33. P. D. Forbes, Experimental photocarcinogenesis: an overview. J. Invest. Dermatol., 77: 139 (1981).

IMMUNOGENETIC MODULATION OF

PHOTOINDUCED SKIN TUMORS

D. Cerimele, L. Contu*, M. Mulargia*, A. Ledda*, and S. Saccabusi

Institute of Dermatology
University of Sassari
*Chair of Medical Genetics
University of Cagliari, Italy

It has long been known that prolonged U.V. irradiation of human skin has an important role in the development of skin carcinomas. In man over 90% of skin cancers develop on the face which constitutes less than 10% of the total body skin surface. The probability of developing skin tumors on exposed areas is therefor 100 fold that of developing tumors on unexposed areas (Cerimele and Serri, 1972).

Both environmental and somatic factors play an important role in the development of skin tumors: prolonged exposure to the sun radiation appears to be the most important environmental factor, while the fair skin and hair and plae irises are the most common somatic factors (Fois et al. 1981).

In a series of 506 patients with cancer following immunosuppressive therapy for renal transplantation, 42% (213/506) had skin cancer: of this patients 39% had multiple skin carcinomas (16% of the patients with skin carcinomas in normal population). This suggests that the development of skin carcinomas could be under immunologic control (Penn, 1978).

In Australia the incidence of carcinomas in patients immunosuppressed after renal transplantation is even higher: 26 out of 28 (93%) patients who developed neoplasms had skin carcinomas, thus suggesting a positive inter-relationship between sun exsposure and immunosuppression in the pathogenesis of skin carcinomas (Hardie et al., 1980.).

It has been suggested that patients with skin carcinomas are at risk of developing new carcinomas (Bergstresser and Halprin 1975). In patients with more than one carcinoma the risk is even higher, perhaps because these patients have a predisposing factor, or lack some factors that would protect them against skin carcinomas. We therefore carried out an immuno-genetic investigation, to search for the presence of HLA-associated gene-tic factors in patients with multiple skin carcinomas.

Materials and Methods

43 patients with multiple skin carcinomas (MSC) and 220 healthy age matched controls, all born in Sardinia and with Sardinian ancestors, were

typed for HLA-A,B, C, DR antigens; 30 had developed only basal cell carcinomas (from 2 to 8 MBCC) and 13 had developed either only squamous cell
carcinomas, or both squamous cell and and basal cell carcinomas (from 2 to
4 MSCC).

HLA typing was performed by a standard two stage microcytotoxicity
test using enriched B cell suspensions obtained by rosetting with AET
treated sheep RBC for DR and DQ antigens. All patients and controls were
typed for 72 HLA antigens: 16 HLA-A, 34 HLA-B, 7 HLA-CW, 12 HLA-DR, 3 HLA-
DQ. For statistical analysis the Chi-square test and the relative risk
(RR) according to Woolf's method were used (Svejgard et al.,1974).

Results

Among the class I antigens the following results have been observed.
No significant modifications were observed in A antigens. (Table I).

Table I. HLA-A frequencies in 43 patients with multiple
skin carcinomas.

ANTIGEN	MSC (43) N	MSC (43) %	MBCC (30) N	MBCC (30) %	MSCC (13) N	MSCC (13) %	CONTROLS (220) N	CONTROLS (220) %	x^2 (p) MSC	MBCC	MSCC
A1	10	23.2	7	23.3	3	23.0	27	12.3	ns	ns	ns
A2	19	44.1	16	53.3	3	23.0	117	53.2	''	''	''
A3	2	4.6	1	3.3	1	7.6	12	5.4	''	''	''
A9	9	20.9	6	20.0	3	23.0	56	25.3	''	''	''
A23	0	-	0	-	0	-	14	6.4	''	''	''
A24	9	20.9	6	20.0	3	23.0	41	18.7	''	''	''
A10	5	11.6	4	13.3	1	7.6	15	6.8	''	''	''
A25	1	2.3	1	3.3	0	-	0	-	''	''	''
A26	4	9.3	3	10.0	1	7.6	15	6.8	''	''	''
A11	6	13.9	3	10.0	3	23.0	33	15.0	''	''	''
A28	1	2.3	0	-	1	7.6	7	3.1	''	''	''
A29	1	2.3	1	3.3	0	-	20	9.1	''	''	''
A30	12	27.9	7	23.3	5	38.4	86	39.1	''	''	''
A31	1	2.3	1	3.3	0	-	3	1.4	''	''	''
A32	4	9.3	2	6.6	2	15.3	37	16.8	''	''	''
A33	6	13.9	4	13.3	2	15.3	17	7.7	''	''	''

MSC: multiple skin carcinomas;
MBCC: multiple basal cell carcinomas.
MSCC: multiple squamous cell carcinomas.

The antigen B-17 was present in 23.6% of normal controls, in 9.3% of
all the patients, and in 6.6% of the patients with multiple basal cell car
cinomas and in 15.3% of the patients with multiple squamous cell carcinomas.
Among the B-17 splittings the B-58 frequencies decreased from 21% to
6.9% for all cell cancers,to 6.6% for multiple basal cell carcinomas and
to 7.6% for multiple squamous cell carcinomas while the values of B-57
were not affected (Tab.II).

Table II. HLA-B frequencies in 43 patients with multiple skin carcinomas.

ANTIGEN	MSC (43) N	%	MBCC (30) N	%	MSCC (13) N	%	CONTROLS (220) N	%	$x^2(p)$ CCM	BM	SM
B5	9	20.9	6	20.0	3	23.0	35	15.0	ns	ns	ns
B51	7	16.2	4	13.3	3	23.0	29	13.2	"	"	"
B52	2	4.6	2	6.6	0	-	4	1.8	"	"	"
B7	3	6.9	1	3.3	2	15.3	5	2.3	"	"	"
B8	2	4.6	2	6.6	0	-	9	4.1	"	"	"
B12	4	9.3	3	10.0	1	7.6	15	6.8	"	"	"
B44	3	6.9	3	10.0	0	-	12	5.4	"	"	"
B45	1	2.3	0	-	1	7.6	3	1.4	"	"	"
B13	2	4.6	1	3.3	1	7.6	7	3.2	"	"	"
B14	9	20.9	5	16.6	4	30.7	37	16.8	"	"	"
B15	1	2.3	1	3.3	0	-	6	2.7	"	"	"
B62	1	2.3	1	3.3	0	-	6	2.7	"	"	"
B63	0	-	0	-	0	-	0	-			
B16	3	6.9	3	10.0	0	-	5	2.3	"	"	"
B38	1	2.3	1	3.3	0	-	1	0.4	"	"	"
B39	2	4.6	2	6.6	0	-	4	1.8	"	"	"
B17	4	9.3	2	6.6	2	15.3	52	23.6	.014	.005	"
B57	1	2.3	0	-	1	7.6	4	1.8	ns	ns	ns
B58	3	6.9	2	6.6	1	7.6	46	21.0	.004	.003	.024
B18	20	46.5	15	50.0	5	38.4	118	53.6	ns	ns	ns
B21	6	13.9	3	10.0	3	23.0	27	12.2	"	"	"
B49	4	9.3	3	10.0	1	7.6	23	10.4	"	"	"
B50	2	4.6	0	-	2	15.3	4	1.8	"	"	"
B22	2	4.6	1	3.3	1	7.6	13	5.9	"	"	"
B55	2	4.6	1	3.3	1	7.6	13	5.9	"	"	"
B56	0	-	0	-	0	-	0	-			
B27	1	2.3	1	3.3	0	-	10	4.5	"	"	"
B35	8	18.6	4	13.3	4	30.7	42	19.1	"	"	"
B37	0	-	0	-	-	-	3	1.4	"	"	"
B40	3	6.9	3	10.0	0	-	10	4.5	"	"	"
B60	3	6.9	3	10.0	0	-	3	1.3	"	"	"
B61	0	-	0	-	0	-	7	3.2	"	"	"
B41	1	2.3	1	3.3	0	-	4	1.8	"	"	"
B53	0	-	0	-	0	-	1	0.4	"	"	"

MSC:multiple skin carcinomas; MBCC:multiple basal cell carcinomas; MSCC: multiple squamous cell carcinomas.

The frequency of Cw-3 was 7.3% in the controls, 20.3% in all cancers, 23.3% in multiple basal cell carcinomas and 15.3% in multiple squamous cell carcinomas (Tab. III).

Table III. HLA-Cw frequencies in 43 patients with multiple skin carcinomas.

ANTIGEN	MSC (43) N	%	MBCC (30) N	%	MSCC (13) N	%	CONTROLS (220) N	%	$X^2(p)$ MSC	MBCC	MSCC
CW1	3	6.9	2	6.6	1	7.6	11	5.0	ns	ns	ns
Cw2	5	11.6	4	13.6	1	7.6	32	14.5	"	"	"
Cw3	9	20.3	7	23.3	2	15.3	16	7.3	.005	.004	"
Cw4	9	20.3	5	16.6	4	30.7	39	17.7	"	"	"
Cw5	17	39.5	14	46.6	3	23.0	99	45.0	"	"	"
Cw6	1	2.3	1	3.3	0	-	16	7.3	"	"	"
Cw7	3	6.9	3	10.0	0	-	9	10.9	"	"	"

MSC: multiple skin carcinomas
MBCC: multiple basal cell carcinomas
MSCC: multiple squamous cell carcinomas

Among the class II antigens the frequency of DR-1 which was 18.1% among the controls was significantly raised to 39.5% for all the patients, to 40% for multiple basal cell carcinomas and to 38.4% for multiple squamous cell carcinomas. The frequency of DR-3, which was 53.6% among the controls was decreased to 32.5% for all the patients, but this decrease was not statistically significant. (Table IV).

Table IV. HLA-DR frequencies in 43 patients with multiple carcinomas.

ANTIGEN	MSC (43) N	%	MBCC (30) N	%	MSCC (13) N	%	CONTROLS (220) N	%	$X^2(p)$ MSC	MBCC	MSCC
DR1	17	39.5	12	40.0	5	38.4	40	18.1	.002	.005	ns
DR2	13	30.2	9	30.0	4	30.7	77	35.0	ns	ns	ns
DR3	14	32.5	10	33.3	4	30.7	118	53.6	"	"	"
DR4	10	23.2	7	23.3	3	23.0	67	30.4	"	"	"
DR5	13	30.2	10	33.3	3	23.0	51	23.1	"	"	"
DR6	4	9.3	3	10.0	1	7.6	25	11.3	"	"	"
DR7	5	11.6	2	6.6	3	23.0	16	7.2	"	"	"
DR8	1	2.3	1	3.3	0	-	6	2.7	"	"	"
DR9	_3	6.9	1	3.3	2	15.3	2	0.9	"	"	"
DR10	0	-	0	-	0	-	3	1.3	"	"	"
DRw52	26	60.4	19	63.3	7	53.8	148	67.2	"	"	"
DRw53	18	41.8	11	36.6	7	53.8	85	38.6	"	"	"
DQw1	23	53.4	15	50.0	8	61.5	123	55.9	"	"	"
DQw2	17	39.5	11	36.6	6	46.1	120	54.5	"	"	"
DQw3	29	67.4	20	46.5	9	69.2	117	53.1	"	"	"

MSC : multiple skin carcinomas
MBCC : multiple basal cell carcinomas
MSCC : multiple squamous cell carcinomas

The relative risk of developing multiple skin cancers in the total group is 3.37 for the presence of the antigen Cw-3, 2.94 for the antigen DR-1, 0.33 for the antigen B-17 ; among the patients with multiple basal cell carcinomas the relative risk is 3.88 for Cw-3, 3.00 for DR-1 and 0.23 for B-17; among the patients with multiple squamous cell carcinomas, it is 2.31 for Cw-3 2.81 for DR-1 and 0.58 for B-17 (Tab. V).

Table V. Relative risk (RR) for HLA antigens associated with multiple skin carcinomas.

ANTIGEN	MSC (RR)	MBCC (RR)	MSCC (RR)
Cw3	3.37	3.38	2.31
B17	0.33	0.23	0.58
DR1	2.94	3.00	2.81

MSC : multiple skin carcinomas
MBCC : multiple cell carcinomas;
MSCC: multiple squamous cell carcinomas

DISCUSSION

Our data demostrate that B-17 (mostly B-58) has a negative association and Cw3 and DR-1 a positive association with multiple skin cancers.It is interesting that B-58 (apparently associated with cutaneous cancer resistance in our patients) has higher frequencies in Negroid populations which are relatively unsusceptible to skin cancers, whereas its frequency is much lower in Northern European populations, which are more susceptible to sunlight-induced cutaneous carcinomas (Baur and Danilovs, 1980).

It is also worthy of note that B-17 is an antigen commonly associated with psoriasis, a disease which have a negative correlation with skin cancers (Jacobs et al, 1977).

Moreover, an increased frequency of DR-1 antigen has been observed in cases of Xeroderma pigmentosum, where a defect in DNA repair mechanisms is highly associated to sunligth-induced skin cancers (Holge et al., 1980).

HLA antigens were studied in 31 patients with multiple basal cell carcinomas, but no statistically significan association with HLA-A,B,C and DR antigens was noted (Myskowski et al., 1985).

An increase in Dr-1 from 20% among the controls to 33% among the patients was observed. The increase was not statistically significant for all the patients, but only for a subgroup (so-called other persons, excluding Irish and Ashkenazi Jewish).We believe that the discrepancy between these results and ours may be due to the differences between the samples of the patients and of the controls: all our patients and controls were of Sardinian ancestry, while the 31 patients typed by Myskowski were distributed in 12 ethnic groups, and the composition of the control group was not stated.

REFERENCES

Baur M.P., Danilovs J.A., 1980 : Population analysis of HLA-A, B,C,DR and other genetic markers, in "Histocompatibility Testing 1980" Terasaki P.I. ed. Los Angeles, VCLA, Tissue Typing Laboratory, 955.

Bergstresser P., Halprin K. 1975: Multiple sequential Skin Cancers. Arch. Dermat. 111,995.

Cerimele D., Serri F., 1972: Clinical and Histological alterations of human skin from sunlight, in: "Res. Org. Biol. Med. Chem." Vol. 2, part II 623.

Fois G.M., Campus G. V., Sanna E., Cerimele D., Scuderi N., 1981 "Fattori costituzionali ed ambientali nella patogenesi dei carcinomi cutanei". Ann. It. Dermat. Clin. Sper. 35,117.

Hardie I.R., Strong R.W., Hartley L.C.J. 1980 "Skin cancer in Caucasian renal allograft recipients living in a subtropical climate". Surgery, 87, 177.

Holge S.E., Degos L., Walford R.L., 1980 "Four chromosomal instability syndromes: Bloom's syndrome, Franconi's anemia, Werner's syndrome, and

Xeroderma Pigmentosum. In "Histocompatibility testing 1980". Terasaki P.I., ed. Los Angeles, UCLA, Tissue Typing laboratory, 730.

Jacobs P.A., Farber E.M., Nall M.L., 1977: Psoriasis and skin cancer. In "Psoriasis", FarberE.M., Cox A. Y. eds., New York, Yorke Med. Books, 350.

Myskowski P.L. Pollack M.S., Schorr E., Dupont B., Safai B., 1985 : "Human leukocyte antigen associations in basal cell carcinoma". J. Amer. Acad. dermat. 12, 997.

Penn. I., 1978: "Malignancies associated with immunosuppressive or cyto-toxic therapy". surgery (St. Louis), 83,492.

Svejgard A., Jersild C., Staub Nielsen L., Bodmer W.F., 1974: "HLA anti-gens and dissease. Statistical and genetic considerations." Tissue Antigens 4, 95.

PHOTOREACTIVITY OF DIFFERENT PSORALENS IN THE TREATMENT OF VITILIGO

Herbert Hönigsmann

Division of Photobiology, Department of Dermatology I, Univ.of Vienna, Vienna, Austria

INTRODUCTION

Vitiligo is a relatively common condition characterized by the progressive loss of normal skin coloration in certain skin areas. It is not a life-threatening disease and therefore does not require treatment unless serious disfigurement leads to social segregation and emotional distress. Treatment is directed to reverse the progressive loss of epidermal melanocytes and to reconstitute normal skin color. Although a minority of patients with vitiligo may develop transient repigmentation when exposed to sunlight or artificial UV radiation, only photochemotherapy is effective in inducing a permanent cosmetically acceptable treatment result[1-3].

TREATMENT RESULTS

Several different psoralens are in use or have been tested for photochemotherapy of vitiligo. Psoralen, 8-methoxypsoralen, 5-methoxypsoralen and 4,5',8-trimethylpsoralen have so far proved effective in reconstituting normal skin coloration[2-4]. Controlled studies in generalized vitiligo with PUVA using 8-MOP and/or TMP have yielded success rates ranging from 25 to 40 %. Repigmentation of more than 70 % usually is considered an excellent treatment response. The most elaborate study of Pathak et al in 366 patients from India documents that the most effective treatment consists of a combination of 8-MOP and TMP, and that dark-skinned people show a better response[4].

PHOTOBIOLOGICAL ACTIVITY AND MECHANISMS OF PIGMENT INDUCTION

The mode of action of all forms of photochemotherapy, be it topical or systemical, is unknown. The mechanisms involved in photochemotherapy-induced pigmentation which are currently discussed include activation of dormant melanocytes, migration of dormant and/or active melanocytes of elsewhere (e.g. hair follicle), and an increased mytotic rate of melanocytes. Finally, since vitiligo is considered an autoimmune disease, the possible alteration of immune mechanisms during PUVA treatment may also play a role in this process (for reference see[5]). With regard to the known antiproliferative action of photochemotherapy it seems to be a paradox, that an antiproliferative stimulus should induce proliferation of a certain cell type. It has been hypothesized, that the photochemical reaction would arrest the melanocyte a certain phase during the cell cycle, and by doing so making receptors to possible melanocyte-stimulating mediators accessible for a longer period of time. As yet there exists no evidence that could confirm this assumption. However, since melanocytes are functionally related to keratinocytes the proliferation could well depend upon signals from the keratinocytes[6].

Recently, khellin, a furochromone which is structurally related to 8-MOP, was introduced for photochemotherapy of vitiligo[7,8]. It was shown that khellin-UVA photochemotherapy is effective in restoring normal skin color in a substantial number of patients with vitiligo after 1 to 2 years of continuous therapy and this equals the success rate with psoralen photochemotherapy[8]. The advantage of the use of khellin is that it does not induce phototoxic skin erythema even with exposure doses of up to 100 J/cm². Therefore severe erythema reactions as they may occur in PUVA therapy, particularly in fair skinned caucasians, can be avoided. Khellin-UVA therapy can be considered save with natural sunlight as UVA source and also as home treatment with artificial UVA provided that the patients are instructed properly and are regularly monitored.

The phototoxic and genotoxic properties of khellin have been studied in various experimental systems[9-15]. Khellin was found to be 100 to 1000 fold less active as compared with 8-MOP. At the same UVA fluence 8-MOP is about 1000-fold more active than khellin in reducing the capacity of human cells to support virus plaque formation or in killing cells and increasing mutations in mammalian cells[15]. The very low mutagenic activity of khellin in vitro suggests that potential risks of long term side effects such as skin carcinoma formation may be much lower than in conventional psoralen photochemotherapy.

The different psoralens, and as already mentioned, khellin exhibit different photosensitizing properties with regard to skin phototoxicity (erythema formation) and to the induction of pigmentation in normal skin. The reactivity of these compounds after oral and topical application is shown in table 1 and table 2.

It is quite obvious that some compounds do not induce erythema reactions with therapeutic UVA irradiation doses. All psoralens dramatically enhance the tanning properties of UVA in normal skin, but khellin appears to be not more melanogenic than UVA alone[8]. However, the therapeutic

Table 1. Erythemal & melanogenic properties
of oral photosensitizers

	E	P
8-MOP	++	++
5-MOP	+	+++
TMP	-	+
khellin	~	-

Table 2. Erythemal & melanogenic properties
of topical photosensitizers

	E	P
8-MOP	++	++
5-MOP	nd	nd
TMP	+++	++
khellin	-	-

* nd = not done

success in the treatment of vitiligo appears to be similar
with all compounds as opposed to the treatment of psoriasis
which, after oral administration, responds to 8-MOP and
5-MOP only (table 3).

Table 3. Therapeutic success & drug

psoriasis	8-MOP	yes
	5-MOP	yes
	TMP	no
	khellin	no
vitiligo	8-MOP	yes
	5-MOP	yes
	TMP	yes
	khellin	yes

Pigment induction occurs UVA alone and to the same
extent with khellin plus UVA[8]. Only psoralens stimulate
melanin formation in a higher order of magnitude than UVA
alone or khellin plus UVA.

CONCLUSION

Repopulation with melanocytes of vitiligo skin occurs
independent of the drugs ability to induce pigmentation
in normal skin during photochemotherapy (table 4).

Table 4.

	pigmentation of normal skin	repopulation of vitiligo skin
UVA	+	-
psoralen + UVA	++/+++	++
khellin + UVA	+	++

This indicates that the stimulation of melanogenesis and the induction of repopulation of vitiliginous skin with melanocytes may be mediated by different mechanisms. Thus the treatment success in vitiligo may be based upon a denominator common to the psoralens in use, and to khellin, that is unrelated to erythemogenic and/or melanogenic properties of these compounds.

REFERENCES

1. J. P. Ortonne , D. B. Mosher, T. B. Fitzpatrick, "Vitiligo and other hypomelanoses of the skin", Plenum Medical Book Company, New York (1983).

2. T. B. Fitzpatrick, J. A. Parrish, M. A. Pathak, Phototherapy of vitiligo, in: "Sunlight and man", T. B. Fitzpatrick, M. A. Pathak, L. C. Harber, M. Seiji, A. Kukita eds., Tokyo University Press, Tokyo, pp 783-791 (1974).

3. J. A. Parrish, T. B. Fitzpatrick, C. Shea, M. A. Pathak, Photochemotherapy of vitiligo, Arch.Dermatol. 112:1531-1534 (1976).

4. M. A. Pathak, D. B. Mosher, T. B. Fitzpatrick, J. A. Parrish, Safety and therapeutic effectiveness of 8-methoxypsoralen, 4,5',8-trimethylpsoralen, and psoralen in vitiligo, Natl.Cancer Inst.Monogr. 66:165-173 (1984).

5. H. Hönigsmann, K. Wolff, T. B. Fitzpatrick, M. A. Pathak, J. A. Parrish, Photochemotherapy (PUVA). in: "Dermatology in General Medicine", 3rd edition, T. B. Fitzpatrick, A. Z. Eisen, K. Wolff, I. M. Freedberg, K. F. Austen eds., McGraw-Hill Inc., New York pp 1533-1558 (1986).

6. W. C. Quevedo Jr., M. C. Youle, D. T. Rover, T. C. Bienieki, The developmental fate of melanocytes in marine, in: "Structure and control of melanocyte", G. Della Porta, O. Muhlbock eds., Springer-Verlag, Berlin pp 228-241 (1966).

7. A. Abdel-Fattah, M. N. Aboul-Enein, G. M. Wassel, B. S. El-Menshawi, An approach to the treatment of vitiligo by khellin, Dermatologica 165:136-140 (1982).

8. B. Ortel, A. Tanew, and H. Hönigsmann, Treatment of vitiligo with khellin and UVA, J.Am.Acad.Dermatol. in press.

9. P. C. Beaumont, E. J. Land, S. Navaratnam, B. J. Parsons, G. O. Phillips, A pulse radiolysis study

of the complexing of furocoumarins with DNA and proteins, Biochim.Biophys.Acta 603:182-189 (1980).

10. N. Niccolai, L. Bovalini, P. Martelli, The mechanisms of interaction between furanochromones and DNA: a heternuclear Overhauser effect study on the khellin-thymidine model system. Biophys.Chem. 24: 217-220 (1986).

11. M. Rucheton, P. Jeanteur, Studies on amikhellin: I. Intercalative binding to double-stranded DNA. Biochimie 55:1415-1420 (1973).

12. E. Cassuto, N. Gross, E. Bardwell, P. Howard-Flanders, Genetic effects of photoadducts and photocross-links in the DNA of phage lambda exposed to 360 nm light and tri-methylpsoralen or khellin. Biochim. Biophys.Acta 475:589-600 (1977).

13. P. Martelli, L. Bovalini, S. Ferri, G. G. Franchi, M. Bari, Active oxygen forms in photoreaction between DNA and furochromones khellin and visnagin FEBS Lett. 189:255-257 (1985).

14. B. F. Abeysekera, Z. Abramowski, G. H. N. Towers, Genotoxicity of the natural furochromones, khellin and visnagin and the identification of a khellin-thymine photoadduct, Photochem.Photobiol. 38:311-315 (1983).

15. V. M. Hitchins, L. E. Bockstahler, P. G. Carney, C. D. Lytle, K. M. Olvey, T. J. Withrow, H. Hönigsmann, Comparison of khellin and 8-MOP as UVA photosensitizers for mammalian cells, Photochem.Photobiol. 43 (Suppl) 275 (1986).

PHOTOPROTECTIVE FUNCTIONS OF EUMELANIN: BIOPHYSICAL AND BIOCHEMICAL

PROPERTIES OF EUMELANIN AND PHEOMELANIN

Hans Rorsman

Department of Dermatology
University of Lund
Lund, Sweden

It is generally accepted that black and brown skin is well-protected against sunlight-induced inflammation and degeneration. In contrast the fair skin of certain Caucasian subjects and especially the skin of red-haired people is highly sensitive to the damaging effects of sunlight. Pathologically nonpigmented skin in vitiligo and in albinism is easily sun-burned and albino patients get sun-induced squamous cell carcinoma at an early age.

Kaidbey et al. (1979) investigated the photoprotective role of melanin and found that on an average five times as much ultraviolet light (UVB and UVA) reaches the upper dermis of Caucasians compared with that of blacks. The main site of UV filtration in Caucasians was the stratum corneum, whereas in blacks it was the Malpighian layers.

Histological studies of light-exposed skin of people with different complexions have demonstrated that marked elastosis is considerably less likely to develop in dark individuals (Kligman, 1974).

MELANIN

The protective function of dark pigmented skin against the damaging effects of sunlight is mainly due to its content of melanins, which are black, brown or yellow polymers formed by the action of the enzyme tyrosinase present in the melanocytes. The main substrates of tyrosinase are tyrosine and dopa. Tyrosinase has two different effects on these substrates, 1) tyrosine is oxygenated in 3-position with production of dopa as a result, 2) dopa is oxidized with production of dopaquinone, a highly reactive molecule. Thus, the substrate dopa is also a product of the enzyme. Furthermore, dopa is necessary as a cosubstrate for the function of tyrosinase as oxygenase. References to classical and more recent work on tyrosinase are found in a review on the biosynthesis of melanin (Rorsman and Rosengren, 1986).

The currently accepted mechanism for melanin synthesis is mainly based on the pioneering work of Raper. Studies by many other researchers have since then substantially contributed to a more detailed understanding of the chemical mechanisms involved in the formation of melanin pigments. The reviews by Mason (1967), Swan (1963, 1973), and the monograph by Nicolaus (1968) cover the chemistry of **eumelanin**. Modern views concerning **pheomelanin** are based on the studies of Prota and Nicolaus (1967) and Prota et al. (1968) and have been reviewed by Prota (1972, 1980).

Studies on melanogenesis in the past few years have led to the concept of **mixed-type melanins** which have properties in common with both eumelanin and pheomelanin (Rorsman et al. 1979; Ito et al., 1980). Fig. 2 shows the principal steps in melanin formation. According to modern concepts melanins are polymers in which all intermediary substances may be co-polymerized.

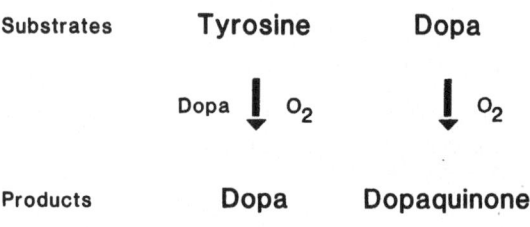

Fig. 1. Main functions of human tyrosinase.

It is generally believed that eumelanin and mixed-type melanin are more effective than pheomelanin as photoprotecting pigments.

Pathak and Fitzpatrick (1974) have reviewed the physical and chemical properties of melanin responsible for its photoprotective function which is exercised by the absorption and scattering of UV light. Absorbed energy in the UV and visible spectrum can be transformed into heat; the absorbed energy can also lead to oxidation of reduced units in the polymer.

Melanins have a stable electron spin resonance signal due to free radicals in the polymer. The signal is different for eumelanin, mixed-type melanins and pheomelanins (Hansson et al., 1979; Sealy et al., 1982a and b).

Free radicals are generated by ultraviolet light absorbed in the skin, and such radicals produced in the tissue are considered to mediate cell damage and inflammation. Production of free radicals by UV light is not dependent on melanin in the skin. However, UV light also produces enhanced ESR signals in melanin (Pathak and Stratton, 1968), and the free radicals of melanin are thought to have a protective function as traps for short-lived free radicals in other molecules.

PHOTOSENSITIZATION BY MELANIN OR PRECURSORS?

The inflammatory response to UVR in redheads is so strong that it has been suggested that pheomelanin present in the epidermis may act as a photosensitizing compound (Chedekel et al., 1977; Menon et al., 1985). The photochemistry of eu- and pheomelanins and of melanin precursors has been the subject of several studies by Chedekel and collaborators (Chedekel et al., 1979; Chedekel, 1982). However, amelanotic vitiliginous skin and albinotic skin which lack melanins also show pronounced sensitivity to sunlight (Pathak and Fitzpatrick, 1974), and it remains to be demonstrated that skin containing pheomelanin is more light-sensitive than nonpigmented vitiliginous or albinotic skin.

Studies on MED in black and white subjects clearly indicate good photoprotection of melanin against UVR (Olson et al., 1973; Kaidbey et al., 1979), but there are some reports which suggest that high melanin content in a cell may be of importance for its transformation by UVR into a "sunburn cell" (Johnson et al., 1972; Olson et al., 1974).

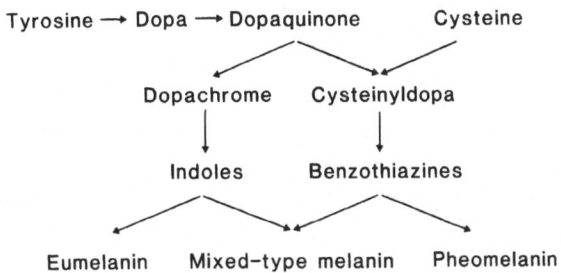

Fig. 2. Principal steps in melanin formation.

PIGMENT RESPONSE TO UV LIGHT

Long ultraviolet light (UV-A), >320 nm, produces an immediate pigment darkening (IPD) which is most easily observed in skin with good pigmentation even before the exposure to light. IPD seems to depend mainly on chemical reactions in melanosomes already present and does not represent new formation of melanin. Light and electron microscopic studies of the melanocytes after exposure of the skin to UV-A with immediate tanning have demonstrated that the most significant change is a rapid dislocation of melanosomes with the bundles of thick filaments from the perinuclear area towards the tip of dendrites. There was no change in the ultrastructure of melanogenic organelles (Jimbow et al., 1974).

Delayed tanning (DT) can be observed already on the first day after UV-A exposure, whereas short UV light (UV-B), <320 nm, gives no IPD but delayed tanning after 3 days. Rosen et al. (1987) studied the number of tyrosinase positive melanocytes after exposure to UV-A or UV-B and found a similar response with an increased number of melanocytes after 1 and 2 weeks. The increase was about 50%. Melanocytes were also more intensely dopapositive after UV exposure. In a delayed tanning reaction after a single dose of UV light (290-320 nm) a significant increase in the formation, melanization and transfer of melanosomes was observed (Jimbow et al., 1974).

In keratinocytes a redistribution of melanosomal complexes has been observed after exposure to UV-A (Lavker and Kaidbey, 1982). Thus, 18 hours after irradiation there was migration and dispersion of packaged melanosomes within keratinocytes from their normal location around the nucleus towards the periphery of the cell.

Freckles offer a model illustrating the importance of a genetically determined light-induced melanin hyperpigmentation in protection against UV light. A freckle is a sharply demarcated area of relative hyperpigmentation in paler skin. Freckles are characteristic for people with red hair, but they occur also in association with other hair colours. In freckles the melanocytes are large with numerous branching processes. The surrounding paler epidermis contains smaller less branched melanocytes. Freckles contain melanocytes with melanomes resembling those in dark hair. The paler surrounding skin has in redheads melanocytes with melanosomes similar to those in red hair (Breathnach, 1957).

Wilson and Kligman (1982) studied the light protection in freckled and surrounding skin by counting the number of epidermal sunburn cells present 24 hours after exposure of the skin to suprathreshold doses of solar simulated radiation. Equal doses of UVR produced fewer sunburn cells in freckled than in non-freckled white skin. The authors concluded that freckles provide a significant degree of protection against acute sunburn.

In recent years the tan induced by high doses of UV-A has received increased attention. Interestingly it has been observed that the induction of pigmentation by UV-A is dependent on oxygenation of the tissue at the time of light exposure. Anoxia prevents the inflammation as well as the pigment response to UV-A (Tegner et al., 1983; Auletta et al., 1986). In subjects who tan well the inflammatory response to pigmenting doses of UV-A is less pronounced than that produced by pigmenting doses of UV-B (Gange and Parrish, 1986).

PHOTOPROTECTION BY INDUCED PIGMENTATION

IPD

Immediate pigment-darkening induced by UV-A could not be demonstrated to have any photoprotective effect when the subjects were exposed to UV-B (Black et al., 1985).

DT

Kaidbey and Kligman (1978) observed a modest increase in MED of UV-B after previous repeated exposure to UV-A over an 8-week period. Roser-Maass et al.(1982) studied the effect of UV-A-induced pigmentation on the erythema produced by UV-B and on the histologic changes in the epidermis. They found less erythema, less epidermal edema and far fewer sunburn cells produced by UV-B in skin previously exposed to UV-A.

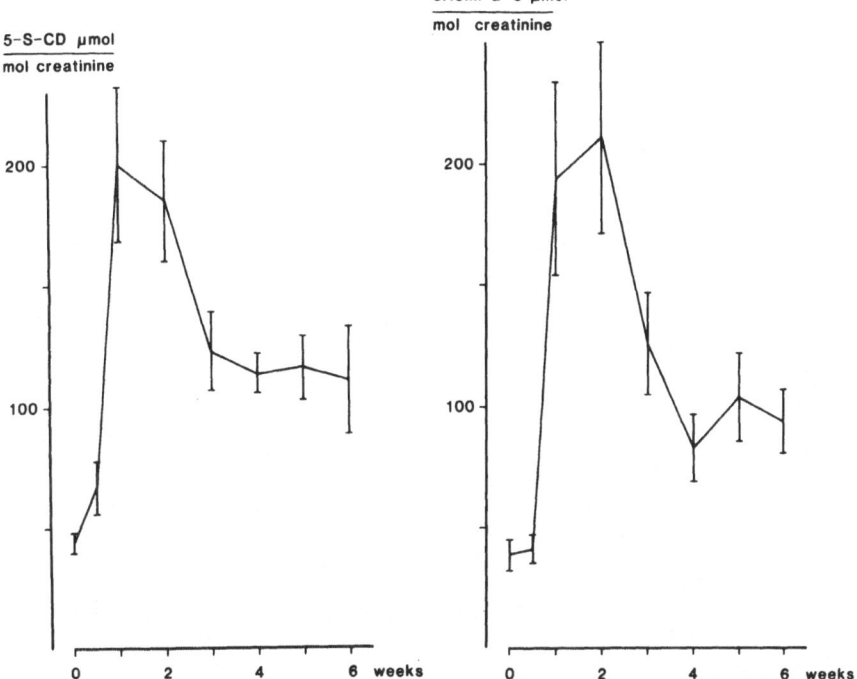

Fig. 3. The urinary excretion of the two melanocyte metabolites
5-S-cysteinyldopa (5-S-CD) and 6-hydroxy-5-methoxyindole-
2-carboxylic acid (6H5MI-2-C) during PUVA treatment.

A study by Gange et al. (1985) has demonstrated a dissociation of the
protection against erythema and against DNA damage by UVA-induced
pigmentation. This was demonstrated by comparison of the MED of UV-B in
UV-A-exposed and non UV-A-exposed skin. No significant increase of MED was
observed. In contrast the previous UV-A exposure had produced protection
for DNA measured as a reduction of endonuclease-sensitive sites in
epidermal DNA. The absence of protection against UV-B erythema in this
report may be due to the protocol used. UV-A was given 4 times over an
8-day period and UV-B was given one week after the last UV-A dose. It
should be noted that an equal tan produced by UV-B according to the same
protocol gave protection both against erythema and DNA damage induced by
UVB challenge one week after the tanning treatment.

It is evident that the experimental models and protocols used are of
the greatest importance in the evaluation of photoprotection induced by
different qualities and quantities of UV light. Methods for kinetic
analysis of the epidermal melanocyte population should be useful in the
study of light effects and photoprotection (Rosdahl, 1979). Analysis of the
urinary excretion of melanocytic metabolites provides a further possibility
to evaluate the melanocytic response to UV. The pronounced increase of such
metabolites during the first and second week of PUVA treatment seems to
illustrate the response of the initially weakly pigmented skin (Fig. 3).
The decline in the excretion of metabolites towards more normal values
thereafter probably reflects a certain protection of the pigment cells by
the induced pigmentation (Hansson et al., 1985).

Acknowledgements. Supported by the Swedish Cancer Society (Project
626-B88-16XA) and the Walter, Ellen and Lennart Hesselman Foundation for
Scientific Research.

REFERENCES

Auletta, M., Gange, R.W., Tan, O.T., and Matzinger, E., 1986, Effect of
   cutaneous hypoxia upon erythema and pigment responses to UVA, UVB,
   and PUVA (8-MOP + UVA) in human skin, J. Invest. Dermatol.,
   86:649-652.
Black, G., Matzinger, E., and Gange, R.W., 1985, Lack of photoprotection
   against UVB-induced erythema by immediate pigmentation induced by
   382 nm radiation, J. Invest. Dermatol., 85:448-449.
Breathnach, A.S., 1957, Melanocyte distribution in forearm epidermis of
   freckled human subjects, J. Invest. Dermatol., 29:253-261.
Chedekel, M.R., 1982, Photochemistry and photobiology of epidermal
   melanins, Photochem. Photobiol., 35:881-885.
Chedekel, M.R., Post, P.W., Deibel, R.M., and Kalus, M., 1977, Photochem.
   Photobiol., 26:651-653.
Chedekel, M.R., Post, P.W., and Vessell, D.L., 1979, Effects of pH and
   wavelength upon the photodestruction of phaeomelanin, in: Pigment
   Cell, Vol. 4, S.N. Klaus, ed., S. Karger, Basel, pp. 323-328.
Gange, R.W., Blackett, A.D., Matzinger, E.A., Sutherland, B.M., and
   Kochevar, I.E., 1985, Comparative protection efficiency of UVA-
   and UVB-induced tans against erythema and formation of endonuclease-
   sensitive sites in DNA by UVB in human skin, J. Invest. Dermatol.,
   85:362-364.
Gange, W., and Parrish, J.A., 1986, Photobiology of melanin pigmentation,
   in: Brown Melanoderma, T.B. Fitzpatrick, M.M. Wick, and K. Toda,
   eds., University of Tokyo Press, Tokyo, pp. 75-86.
Hansson, C., Agrup, G., Rorsman, H., Rosengren, A.-M., and Rosengren, E.,
   1979, Acta Derm. Venereol.(Stockh.), 59:453-456.
Hansson, C., Wirestrand, L.-E., Aronsson, A., Rorsman, H., and Rosengren,
   E., 1985, Urinary excretion of 6-hydroxy-5-methoxyindole-2-carbox-
   ylic acid and 5-S-cysteinyldopa during PUVA treatment, Photodermatology, 2:52-57.
Ito, S., Novellino, E., Chioccara, F., Misuraca, G., and Prota, G., 1980,
   Co-polymerization of dopa and cysteinyldopa in melanogenesis in
   vitro, Experientia, 36:822-823.
Jimbow, K., Pathak, M.A., Szabo, G., and Fitzpatrick, T.B., 1974, Ultra-
   structural changes in human melanocytes after ultraviolet radiation,
   in: Sunlight and Man, T.B. Fitzpatrick, ed., University of Tokyo
   Press, Tokyo, pp. 195-215.
Johnson, B.E., Mandell, G., and Daniels, F., 1972, Melanin and cellular
   reactions to ultraviolet radiation, Nature (New Biology), 235:147.
Kaidbey, K.H., and Kligman, A.M., 1978, Sunburn protection by longwave
   ultraviolet radiation-induced pigmentation, Arch. Dermatol., 114:
   46-48.
Kaidbey, K.H., Agin, P.P., Sayre, R.M., and Kligman, A.M., 1979, Photopro-
   tection by melanin - a comparison of black and Caucasian skin, J.
   Am. Acad. Dermatol., 1:249-260.
Kligman, A.M., 1974, Solar elastosis in relation to pigmentation, in:
   Sunlight and Man, T.B. Fitzpatrick, ed., University of Tokyo Press,
   Tokyo, pp. 157-163.
Lavker, R.M., and Kaidbey, K.H., 1982, Redistribution of melanosomal
   complexes within keratinocytes following UV-A irradiation: A
   possible mechanism for cutaneous darkening in man, Arch. Dermatol.,
   272:215-228.
Mason, H.S., 1967, The structure of melanin, in: Advances in Biology of
   Skin, Vol. 8, The Pigmentary System, W. Montagna, and F. Hu, eds.,
   Pergamon Press, Oxford, pp. 293-312.
Menon, I.A.,Persad, S., Ranadive, N.S., and Haberman, H.F., 1985, Photo-
   biological effects of eumelanin and pheomelanin, in: Pigment Cell
   1985, J. Bagnara, S.N. Klaus, E. Paul, and M. Schartl, eds.,
   University of Tokyo Press, Tokyo, pp.77-85.

Nicolaus, R.A., 1968, Melanins, Hermann, Paris.

Olson, R.L., Gaylor, J., and Everett, M.A., 1973, Skin color, melanin and erythema, Arch. Dermatol., 108:541-544.

Olson, R.L., Gaylor, J., and Everett, M.A., 1974, Ultraviolet-induced individual cell keratinization, J. Cutan. Pathol., 1:120-125.

Pathak, M.A., and Fitzpatrick, T.B., 1974, The role of natural photoprotective agents in human skin, in: Sunlight and Man, T.B. Fitzpatrick, ed., University of Tokyo Press, Tokyo, pp. 725-750.

Pathak, M.A., and Stratton, K., 1969, Effects of ultraviolet and visible radiation and the production of free radicals in skin, in: The Biologic Effects of Ultraviolet Radiation, F. Urbach, ed., Pergamon Press, Oxford, pp. 207-222.

Prota, G., 1972, Structure and biogenesis of phaeomelanins, in: Pigmentation: Its Genesis and Biologic Control, V. Riley, ed., Appleton-Century-Crofts, Meredith Corp., New York, pp. 615-630.

Prota, G., 1980, Recent advances in the chemistry of melanogenesis in mammals, J. Invest. Dermatol., 75:122-127.

Prota, G., and Nicolaus, R.A., 1967, On the biogenesis of phaeomelanins, in: Advances in Biology of Skin, Vol. 8, The Pigmentary System, W. Montagna, and F. Hu, eds., Pergamon Press, Oxford, pp. 323-328.

Prota, G., Scherillo, G., Nicolaus, R.A., 1968, Struttura e biogenesi delle feomelanine. IV. Sintesi e proprierá della 5-S-cisteinildopa, G. Chim. Ital., 98:495-510.

Rorsman, H., Agrup, G., Hansson, C., Rosengren, A.-M., and Rosengren, E., 1979, Detection of pheomelanins, in: Pigment Cell, Vol. 4, S.N. Klaus, ed., S. Karger, Basel, pp. 244-252.

Rorsman, H., and Rosengren, E., 1986, Biosynthesis of melanin, in: Brown Melanoderma, T.B. Fitzpatrick, M.M. Wick, and K. Toda, eds., University of Tokyo Press, Tokyo, pp. 63-74.

Rosdahl, I., 1979, The epidermal melanocyte population and its reaction to ultraviolet light, Acta Derm. Venereol.(Stockh), Suppl. 88.

Rosen, C.F., Seki, Y., Farinelli, W., Stern, R.S., Fitzpatrick, T.B., Pathak, M.A., and Gange, R.W.,1987, A comparison of the melanocyte response to narrow band UVA and UVB exposure in vivo, J. Invest. Dermatol., 88:774-779.

Roser-Maass, E., Hölzle, E., and Plewig, G., 1982, Protection against UV-B by UV-A-induced tan, Arch. Dermatol., 118:483-486.

Sealy, R.C., Hyde, J.S., Felix, C.C., Menon, I.A., Prota, G., Swartz, H.M., Persad, S., and Haberman, H.F., 1982a, Novel free radicals in synthetic and natural pheomelanins: Distinction between dopa melanins and cysteinyldopa melanins by ESR spectroscopy, Proc. Natl. Acad. Sci. USA, 79:2885-2889.

Sealy, R.C., Hyde, J.S., Felix, C.C., Menon, I.A., and Prota, G., 1982b, Eumelanins and pheomelanins: Characterization by electron spin resonance spectroscopy, Science, 217:545-547.

Swan, G.A., 1963, Chemical structure of melanins, Ann. N.Y. Acad. Sci., 100:1005-1016.

Swan, G.A., 1973, Current knowledge of melanin structure, in: Pigment Cell, Vol. 1, Mechanisms in Pigmentation, V.J. McGovern, and P. Russell, eds., S. Karger, Basel, pp. 151-157.

Tegner, E., Rorsman, H.,and Rosengren, E., 1983, 5-S-Cysteinyldopa and pigment response to UVA light, Acta Derm. Venereol.(Stockh.), 63:21-25.

Wilson, P.D., and Kligman, A.M., 1982, Do freckles protect the skin from actinic damage?, Brit. J. Dermatol., 106:27-32.

ENZYMIC AND LIGHT-INDUCED CONVERSION OF 5,6-DIHYDROXYINDOLE(S) TO

MELANIN(S)

Giuseppe Prota

Department of Organic and Biological Chemistry, University
of Naples, Naples, Italy

INTRODUCTION

During the last two decades, whereas knowledge of the chemistry of
phaeomelanins has been considerably added to, little progress has been
made in the investigation of the structure and biosynthesis of eumelanins
(Prota,1980). These pigments are of much greater biological interest
since they are the major determinants of skin colour differences in man
and are involved in a number of intriguing pigmentary phenomena, including
sun tanning (Fitzpatrick et al.,1979).

The pioneering studies of Raper (1927) on the tyrosinase-catalyzed
oxidation of tyrosine have delineated the biosynthesis of these pigments
as far as the formation of 5,6-dihydroxyindole (DHI,1), and it is
generally agreed that the subsequent steps involve oxidative
polymerization of indole-5,6-quinone (2) (Mason,1967). However, despite
extensive studies made over the years (Nicolaus,1968; Swan,1974), no stage
between DHI and the melanin polymer has been adequately characterized; the
polymerization process is generally rapid and the intermediates are
fugitive.

1   2   3

On oxidation of DHI in the presence of tyrosinase at pH 5.6, Mason
(1948) observed the appearance of a purple pigment, designated

melanochrome, spectroscopically identical with that formed in the second
chromogenic phase of the enzymatic oxidation of dopa. This was originally
considered to be the indolequinone 2 (first proposed by Raper as a
probable intermediate), but later evidence was reported suggesting that
melanochrome was probably a dimer or a mixture of oligomers derived from
DHI (Beer et al.,1954; Bu'Lock,1960). Subsequent investigation of the
structure of melanochrome did not proceed very far, owing to the adverse
chemical properties of the material. However, based on model studies and
partly on theoretical grounds, a number of speculative suggestions on the
mode of polymerization of 5,6-dihydroxyindole were made from time to time.
Among these, a widely accepted view was that the formation of melanin from
DHI could proceed by repeated condensation of indole-5,6-quinone (see
fig.1) between positions 3- and 7-,this being favoured on steric grounds
to the alternative linkage between positions 3- and 4- (Bu'Lock and
Harley-Mason,1951).

     As a part of our continuing studies on the chemistry of
melanogenesis, we have recently re-examined the oxidative polymerization
of DHI and some analogues, e.g. 3, to melanin(s), and have succeeded in
characterizing some of the oligomeric intermediates formed at the
melanochrome stage. In the present article a brief account of these
advances is given, together with the results of a pilot study on the
photochemistry of 5,6-dihydroxyindoles, which is of interest in relation
to the oxidation processes underlying light-induced hyperpigmentation.

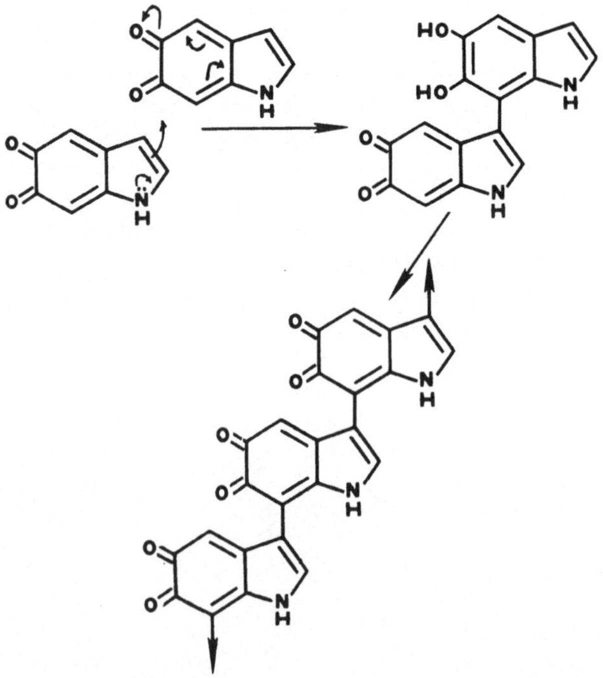

Fig. 1. Suggested mechanism of polymerization of 5,6-dihydroxyindole by
        repeated self condensation of 5,6-indolequinone at 3- and 7-
        positions (Bu'Lock, 1960).

Our approach to the structure of melanochrome was provided by the observation that formation of this intermediate from DHI is susceptible of being catalyzed by a number of metal ions, such as Cu^{2+}, Fe^{2+} and Zn^{2+}, commonly found in pigmented tissues (Prota,1987; Schultz, 1986). As a rule, the extent of the observed effects was found to be dependent upon the nature and concentration of the metal added, the pH of the medium and the partial pressure of oxygen. On this basis, a procedure was developed for the preparation of melanochrome, which involves aerial oxidation of DHI in the presence of Ni^{2+} or Zn^{2+} ions in 0.1 M hepes buffer at pH 7.5. When the characteristic purple colour of melanochrome had attained the maximum intensity (about 20 min), the reaction was stopped by addition of sodium dithionite, and the ethyl acetate-extractable fraction acetylated with acetic anhydride/pyridine overnight at room temperature. On analysis, the product thus obtained was found to consist of a mixture of DHI oligomers, the major of which could be isolated in crystalline form and identified as 5,6,5',6'-tetracetoxy-2,2'-biindolyl (4), mainly by NMR spectroscopy (Napolitano et al.,1985).

Interestingly, a different pattern of products was obtained when the oxidation of DHI was carried out under the usual conditions of melanogenesis in vitro, i.e. in phosphate buffer at pH 6.8 with mushroom tyrosinase (Napolitano and Prota, unpublished results). After work up as above, TLC and HPLC analysis of the reaction mixture revealed the presence of a different dimer, identified as 5, and no trace of the symmetrical dimer 4, which was the prevailing product in the metal-catalyzed oxidation .

Attempts to characterize other components of the melanochrome preparation were unsuccessful owing to difficulties encounterd with the chromatographic separation of the mixture. In further studies (Corradini et al.,1986) we therefore resolved to use as a substrate 5,6-dihydroxy-1-methylindole (3), since on oxidation it gave products with more favourable analytical and solubility properties. When a solution of 3 was oxidized in the presence of tyrosinase, work up of the reaction mixture at the melanochrome stage as above led to the isolation, besides a small amount of the symmetrical dimer 6 , of the 2,4-linked dimer 7 , and of a trimer which was formulated as 8 by careful analysis of its ^1H- and ^{13}C-NMR spectra.

Chemically, the structure elucidation of these oligomers is of particular interest since it reveals for the first time an unexpected tendency of the 5,6-dihydroxyindole system to undergo oxidative coupling at the 2- and 4-positions rather than at the 3- and 7-positions, as previously suggested (Bu'Lock and Harley-Mason,1951). As far as the mechanism of polymerization of DHI is concerned, available evidence suggests that the process involves the formation of semiquinone radicals, which can easily be formed by hydrogen abstraction from the phenoxyl groups at the 5- and 6-positions. Subsequent coupling of these radicals would account for the formation of all the oxidation products so far isolated, including the trimer 8 which represents a most adequate model for the structure of eumelanins from 5,6-dihydroxyindoles. Consistent with this mechanism are the results of kinetic experiments showing that, besides metal ions, UV light (see below) and high pHs catalyze the

4   R = CH₃CO-, R′=H   5 R = CH₃CO-, R=H

6   R = CH₃CO-, R′=CH₃   7 R = CH₃CO-, R′=CH₃

8 R = CH₃CO-

oxidation, whereas in the presence of superoxide dismutase the rate of the reaction is markedly lowered.

PHOTOOXIDATION OF 5,6-DIHYDROXYINDOLE(S)

The familiar darkening of human skin that follows exposure to solar radiation or UV light is known to involve two distinct photobiological processes: immediate tanning (IT) and delayed tanning (DT). While the biophysical phenomena accompanying these two complex processes have been widely explored (Pathak et al., 1978; Parrish, 1981), little is known about the molecular mechanisms accounting for light-induced stimulation of melanogenesis. In the case of IT there is evidence which suggests that the process involves photooxidation of melanin intermediates believed to be present in the skin in a colourless reduced state (Pathak et al.,1981). In this connection, it is noteworthy that DHI and other eumelanic metabolites undergo ready photodestruction on irradiation with Pyrex-filtered UV light to give brownish pigments resembling eumelanins. Pulse radiolysis and laser flash photolysis experiments showed that the process involves generation of transient radical species, such as the 5,6-dihydroxyindole semiquinones and their neutral radicals, arising from photohomolysis of the OH bond(s) (Thompson et al.,1986). However, in these and other studies the subsequent fate of these reactive intermediates as well as the nature of the photooxidation products formed remained unknown.

In an attempt to cover this gap, we have recently undertaken a systematic investigation of the photooxidative behaviour of DHI and some analogues. In preliminary experiments, it was found that in phosphate buffer at pH 7 autoxidative processes significantly compete with the primary photochemical reactions. Therefore, in subsequent studies irradiations were performed in organic solvents, in which all the indoles tested proved to be fairly stable to autoxidation. When methanol was used as a solvent, DHI as well as its 1-methyl derivative 3 decomposed more or less rapidly to give reddish to dark brown solutions, showing very complex and ill-defined chromatographic patterns. Interestingly, under the same conditions, also some hydroxyl-protected 5,6-dihydroxyindoles, such as 5,6-diacetoxyindole and 5,6-dimethoxyindole, underwent photodestruction at comparable rates to give complex reaction mixtures (d'Ischia and Prota, 1986). In the case of 5,6-diacetoxyindole, this behaviour has been shown to result from a light-induced Fries rearrangement of the 6-acetoxyl group to give 5-acetoxy-6-hydroxy-7-acetylindole, which is susceptible of being photooxidized (Chan, 1986).

Some insight into the mechanism of photopolymerization of 5,6-dihydroxyindoles was obtained by careful analysis of the products formed by irradiation of 3 in methanol with Pyrex-filtered UV light (d'Ischia and Prota, 1987). TLC analysis of the reaction mixture after reduction and subsequent acetylation revealed the presence, besides large amounts of the starting material as the acetyl derivative, of a complex pattern of products all in exceedingly minute amounts. One of these, which exhibited a characteristic blue fluorescence, was found to be identical with the dimer 7, obtained by enzymic oxidation of 3. Repeated

fractionation of the mixture on silica gel plates allowed the isolation
of five new compounds, two of which were readily identified as the isomeric
triacetoxy-1-methylindoles 9 and 10 by straightforward analysis of their
mass and NMR spectra. Another photooxygenation product, $C_{13}H_{13}NO_5$, was
negative to Ehrlich reagent and exhibited an oxindole chromophore
characterized by two absorption maxima at 252 and 281 nm. The
identification of the product as 5,6-diacetoxy-1-methyloxindole (11) was
secured by the ^{1}H-NMR spectrum showing a singlet at δ 3.51 for the
methylene group and two singlets at 6.66 and 7.08, attributable to the H-7
and H-4 protons, respectively.

The remaining two products isolated from the photooxigenation mixture
had the same molecular formula, $C_{28}H_{26}N_2O_{10}$, and were identified as the
isomeric pentaacetoxybiindolyls 12 and 13 mainly by careful analysis of
their H-NMR spectra. These exhibited in the aromatic region two sets of
pyrrolic protons (H2 and H3) and one H-7 resonance, the latter appearing
as a doublet owing to long range coupling with one of the protons at the
3-position. The unsuallly upfield resonance of one of the N-methyl groups
(δ 3.05) and the marked difference between the chemical shifts of the two
protons at the 3-positions would favour structure 12 for the less polar
isomer. This assignment was substantiated by the close similarity between
the resonances of the two N-methyl groups as well as of the H-3 and H-3'
in the NMR spectrum of the more polar isomer, consistent with the more
symmetrical structure 13.

Overall, the structures of the photooxygenation products so far
identified reveal a marked tendency of the indole to undergo, in addition
to radical coupling at the 2- and 4-positions, light- catalyzed
oxygenation at the 2-, 4- and 7-positions. This reaction conceivably
involves interaction of semiquinone radicals with triplet oxygen to give
peroxide derivatives (see fig. 2) which can partake along with the
semiquinone intermediates in the polymerization processes, thus accounting
for the extreme complexity of the photooxidation mixture.

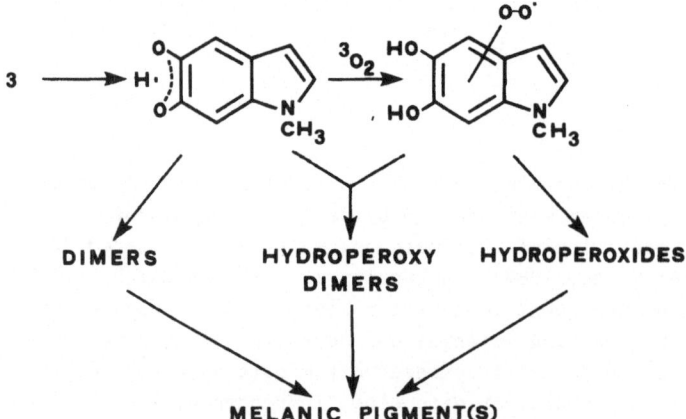

Fig. 2. Schematic outline of the photooxidation of
5,6-dihydroxy-1-metylindole.

Work is now in progress to extend this study to the photochemical behaviour of other 5,6-dihydroxyindoles in order to gain some more insight into the mechanism of light induced formation of melanin(s).

CONCLUSIONS

Since the formulation of the Raper-Mason scheme, the mechanism of polymerization of 5,6-dihydroxyindole to melanin has been extensively investigated by a number of research groups, each using different methods; however, no definite conclusion could be drawn from these studies regarding the site of reactions between the indole nuclei as well as the nature of the reacting species. The results of our studies provide now evidence that tyrosinase- catalyzed polymerization of 5,6-dihydroxyindole proceeds by a radical mechanism involving the initial attack of the semiquinone to the parent indole to give mainly a dimer with a 2-4 linkage which by subsequent coupling with 5,6-dihydroxyindole radical(s) can give rise to trimers of the type 8, and eventually to oligomers of higher molecular weight. The further conversion of these intermediates to melanin remains open to question owing to the lack of definite information regarding 1) the average molecular weight of the final polymer; 2) the state of oxidation of the 5,6-dihydroxyindole units, and 3) the possible oxidative breakdown of the 5,6-dihydroxyindole nuclei by the hydrogen peroxide formed during the polymerization reaction. However, the observed reactivity of the 5,6-dihydroxyindole system to undergo oxidative coupling at the 2- and 4-positions rather than at the 3- and 7- positions, as previously believed, provides a new and most fundamental chemical background to look into the structure of synthetic and naturally occurring eumelanins. As far as the light-induced polymerization of 5,6-dihydroxyindoles is concerned, it seems clear from the results so far obtained that the reaction follows a pathway which is basically different from that observed in the metal or enzymically promoted oxidation. This is because, just formed, the 5,6-dihydroxyindole radical can interact with molecular oxygen to form hydroperoxide intermediates which can partake in the polymerization process to give eventually oxygenated polymers. If these model studies have some bearing with the process of sun tanning, one can suggest the possibility that light-induced skin hyperpigmentation is associated with the formation of melanins having chemical and physical properties different from those enzymically formed in melanocytes.

Acknowledgements. This work was supported by grants from Consiglio Nazionale delle Ricerche, M.P.I.(Rome), and the Lawrence M.Gelb Research Foundation, Clairol (Stamford,CT).

REFERENCES

Beer, R.J.S, Broadhurst,T. and Robertson, A.,1954, The chemistry of melanins. Part V. The autoxidation of 5,6-dihydroxyindoles, J.Chem. Soc., 1947.
Bu'Lock, J.D., 1960, The formation of melanin from adrenochrome, J.Chem. Soc., 52.

Bu'Lock, J.D. and Harley-Mason, J., 1951, Melanin and its precursors. Part
    II. Model experiments on the reactions between quinones and indoles,
    and consideration of a possible structure for the melanin polymer,
    J.Chem. Soc., 703.

Chan, A., 1986, personal communication

Corradini,M.G., Napolitano, A. and Prota,G. ,1986, A biosynthetic approach
    to the structure of eumelanins. The isolation of oligomers from
    5,6-dihydroxy-1-methylindole, Tetrahedron, 42:2083.

d'Ischia, M. and Prota, G., 1986, Photooxidation of 5,6-dihydroxyindoles,
    Gazz.Chim.Ital., 116:407.

d'Ischia, M. and Prota, G., 1987, Photooxidation of 5,6-dihydroxy-1-
    -methylindole, Tetrahedron, 43:431.

Fitzpatrick, T.B., Szabo', G., Seiji, M. and Quevedo, W.C., 1979, Biology
    of the melanin pigmenting system, in:"Dermatology in general medicine",
    T.B.Fitzpatrick, A.Z.Eisen, K.Wolff, I.M.Freedberg, K.F.Austen, eds.,
    McGraw-Hill Book Co., Inc., New York, p.131.

Mason, H.S., 1948, The chemistry of melanin. III. Mechanism of the oxidation
    of dihydroxyphenylalanine by tyrosinase, J.Biol.Chem., 172:83.

Mason, H.S., 1967, The structure of melanin, in: "Advances in biology of
    skin", vol. 8, The pigmentary system, W.Montagna, F.Hu eds., Pergamon
    Press, Oxford, p. 293.

Napolitano, A., Corradini, M.G. and Prota, G., 1985, A reinvestigation of
    the structure of melanochrome, Tetrahedron Letters, 26:2805.

Nicolaus, R.A., 1968, "Melanins", E.Lederer ed., Hermann, Paris.

Parrish, J.A., 1981, Photobiology of melanin pigmentation, in: "Psoralens
    in cosmetics and dermatology", J.Cahn, P.Forlot, C.Grupper, A.Meybeck,
    F.Urbach eds., Pergamon Press, Paris, p. 17.

Pathak, M.A., Jimbow, K., Szabo', G. and Fitzpatrick, T.B. , 1978, in:
    "Photochemical and Photobiological Reviews", vol. 3, K.C. Smith ed.,
    Plenum Press, New York.

Pathak, M.A., Jimbow, K. and Fitzpatrick, T.B. , 1981, in: "Phenotypic
    Expression in Pigment Cells", M. Seiji ed., University of Tokyo Press,
    Tokyo, p. 655.

Prota, G., 1980, Recent advances in the chemistry of melanogenesis in
    mammals, J.Invest. Dermatol., 75:122.

Prota,G., 1987, Pigment cell metabolism: chemical and enzymatic control, in:
    "Cutaneous melanoma: status of knowledge and future perspective", U.
    Veronesi, N.Cascinelli, M.Santinami eds., Academic Press, London, p.233.

Raper, H.S., 1927, The tyrosinase-tyrosine reaction.VI. Production from
    tyrosine of 5,6-dihydroxyindole and 5,6-dihydroxyindole-2-carboxylic
    acid. The precursors of melanin, Biochem. J., 21:89.

Schultz, T.M., 1986, Chelation of 5,6-dihydroxyindole with divalent metals,
    Abs. XIII International Pigment Cell Conference, Tucson, Arizona,
    1986, J.Invest.Dermatol., 87:406.

Swan, G.A., 1974, Structure, chemistry and biosynthesis of the melanins,in:
    "Fortsch. Chem. Org. Naturst.", W.Herz, H.Grisebach, G.W.Kirby, eds.,
    Springer-Verlag, Wien, Vol. 31, p. 521.

Thompson, A., Koch, W.H., Chedekel, M.R., Land, E.J. and Truscott, T.G.,
    1986, Photooxidation of indolic melanogenic intermediates, Abs.XIII
    International Pigment Cell Conference, Tucson, Arizona, 1986, J.Invest.
    Dermatol., 87:407.

PHOTOPROTECTIVE ROLE OF MELANIN (EUMELANIN)

IN HUMAN SKIN

Madhu A. Pathak

Department of Dermatology, Harvard Medical School
and Massachusetts General Hospital
Boston, Massachusetts

INTRODUCTION

Natural light and UV radiation (UVR) from artificial sources has always been recognized for and endowed with health-giving powers as well as for its tonic value of generating a feeling that it is good for our psyche and contributes to the synthesis of vitamin D essential for maintaining calcium and phosphorus homeostasis and a healthy skeleton. It provides the healthy looking tan that enhances the photoprotective properties of our skin against skin cancer and photoaging.

In this presentation, we will focus our discussion on the photoprotective role of melanin against the acute effects (e.g., sunburn and phototoxicity) and the potential long-term risk of actinic changes (wrinkling and premature aging of skin) and premalignancies and malignancies (solar kertosis, basal and squamous cell carcinomas, and even melanomas).

NATURAL DEFENSES OF SKIN AGAINST SUNLIGHT

One of the most important of the many diversified functions of skin is the protection of internal organs and the integument itself against the acute and chronic damaging effects of solar radiation. To survive the insults of actinic damage resulting from direct (intentional sunbathing) or indirect sun exposure (unavoidable outdoor exposure), human skin has evolved six basic defensive mechanisms and these include:[1-3]

(1) The process of keratinization leading to the formation of compact and cohesive horny cell layer (stratum corneum) of varying thickness containing ultraviolet-absorbing proteins (amino acids of keratins); the compact horny layer not only absorbs and filters UVR but also attenuates the impinging radiation by scattering.

337

(2)   The process of genetically controlled constitutive
        melanin pigmentation in melanocytes involving the
        formation and transfer of UV-absorbing melanin in the
        epidermis in the form of melanosomes. The melanin acts
        as a UV-absorbing optical filter and a free-radical
        scavenger that shields nuclear DNA of keratinocytes and
        the dermal proteins, collagen and elastin, from
        UVR-induced harmful alterations.

(3)   The preferential accumulation of carotenoid pigment
        (β-carotene) in subcutaneous tissue allows this pigment
        to diffuse to and enrich both the epidermis and dermis
        to act as a membrane stabilizer and a quencher against
        the damaging forms of reactive O_2 species (singlet
        oxygen or $^1O_2$, $\cdot$OH, etc.) generated by UVR.

(4)   The formation and accumulation of urocanic acid, the
        deaminated product of histidine in the epidermis, which
        by virtue of its ability to undergo cis-trans
        isomerization and oxidation reaction protects the
        viable cells of the epidermis against actinic damage.

(5)   The presence of superoxide dismutase (SOD) and
        glutathione peroxidase-reductase enzyme system in the
        epidermis acting as selective scavengers for the
        reactive forms of O_2 (superoxide anion or O_2^-) generated
        by UVR and protecting the lipoprotein damage of cell
        membranes resulting from peroxidation reaction. SOD
        protects the epidermal and dermal proteins (keratin,
        elastin, and collagen) against the cross-linking
        reaction resulting from the harmful effects of reactive
        oxygen generated by UVR.[3]

(6)   The excision repair capacity of cutaneous cells to
        appropriately repair UVR-induced damage in DNA by an
        error-free DNA replicating mechanism. Xeroderma
        pigmentosum or XP, an autosomal recessive disease,
        serves as the prototype in which the defective excision
        repair due to altered endonuclease activity and
        diminished photoreactivation repair in the XP cell
        strain leads to an early onset of neoplastic changes.

        Although each of these six defensive mechanisms that
exist in mammalian skin are important and compliment each
other, normal human skin of most individuals has two major
defensive barriers that play a significant role in
protection against the harmful effects of UVR. These
include: (a) the compact multicell layers of the stratum
corneum, and (b) the presence of the melanin filter in the
viable cells of epidermis that shields the nuclear DNA from
harmful alterations. Of the two barriers to UVR, the
melanin filter is the most important, in as much as humans
with normal stratum corneum but without melanin (e.g.,
albinos) succumb to repeated UV exposures with early onset
of chronic solar damage (actinic elastosis or
dermtoheliosis) and skin cancer despite the capacity of the
albino skin to respond normally by the process of
hyperplasia and subsequent thickening of the stratum
corneum. The central factor in assessing the relative
importance of natural defenses of skin against UVR exposure
is the presence or absence of a melanized epidermis (i.e.,

brown or black skin) and the genetic capacity of the individual to develop a melanized epidermis (light, moderate, or dark tan).

For limitations of space, only the photoprotective property of melanin will be discussed.

PHOTOPROTECTIVE ROLE OF MELANIN

Not all people in the world share an equal risk for the development of dermatoheliosis (photoaging or changes in connective tissue components of the dermis, including the vascular system, collagen and elastin tissue leading to wrinkling and thining of skin) and development of skin cancer. We will first document nature's well controlled experiment to illustrate the unique photoprotective role of melanin.

The world population contains over 2.5 billion nonwhite people (individuals of Skin Types IV, V, and V) whose unexposed skin of the body (e.g., buttock) is brown or black due to the presence of a large amount of melanin. These individuals live in hot, sunny regions but are virtually resistant to the deleterious effects of UVR, while the remaining white population in excess of one billion living in Europe, the United Kingdom, North America, South Africa, Australia, etc., have white skin in the unexposed areas (e.g., buttock skin due to a low amount of melanin) and exhibit all the acute and chronic problems associated with exposure to sunlight [e.g., sunburn, keratoses, dermatoheliosis or changes associated with photoaging, such as wrinkling, freckling, lentigenes, and telangiectasia, and skin cancer). The nearly 700 million people of the Middle East (Saudi Arabia, Lebanon, Iran, etc.), India, and Pakistan, although classified as Caucasians, have moderately brown skin of unexposed areas and are not easily susceptible to sun-induced damage (dermatoheliosis) even though the intensity of harmful UVR in this region is significantly high throughout the year.[2]

Phenotypic characteristics associated with skin cancer in well documented studies[2,4] include humans with a fair complexion, light eyes, and light hair color (usually blond or red hair, blue eyes, and freckles) and who sunburn easily and repeatedly and exhibit a poor ability to tan. Individuals of Celtic descent (e.g., Scottish, Irish, and Welsh) are particularly vulnerable to skin cancer. By way of contrast, skin cancer is relatively rare in brown and darkly pigmented people; basal cell epitheliomas are uncommon while the squamous cell carcinomas, if observed, generally occur on lower extremities and do not appear related to UVR. The incidence of melanoma in the non-white world population is also very low (less than 0.9 per 100,000) compared to an average incidence of 4.5 per 100,000 in whites. In Australia, the incidence of melanoma in the fair-skinned population is 25.0 per 100,000.[4,5]

In pigmented persons, there exists a unique light-absorbing and UV-filtering system usually distributed as a supranuclear cap of melanin-laden melanosomes in the basal and suprabasal cells that minimizes the impact of

photons on viable cells of the epidermis.[2] The white
population possesses a lesser amount of this protective
filter, known as eumelanin, while brown- and black-skinned
people are endowed with a significant amount of melanin that
acts as a major barrier for the penetration of harmful UV
rays. The protective role of melanin in the epidermis is
attributable to its presence in two distinct forms:
(a) particulate form referred to as melanosomes, and
(b) nonparticulate, colloidal or amorphous form.[6] In the
stratum corneum, most of the melanin is usually in a
nonparticulate amorphous form, although in certain brown and
black-skinned individuals one will find a few melanosomes in
a particulate form scattered randomly in nonviable horny
cells. The amorphous form of melanin is derived from the
particulate form of melansomes by the process of enzymatic
degradation due to the presence of hydrolytic and
proteolytic enzymes associated with melanosome complexes or
in the outer membrane layer of these discrete organelles.
This amorphous form of the melanin undergoes rapid oxidation
reaction and is usually manifested as immediate pigment
darkening (IPD) reaction when the skin is exposed to solar
radiation or to UVA radiation (320 - 400 nm) from artificial
light sources.[9] This IPD reaction, to a limited extent,
constitutes a photoprotective mechanism and is usually more
prominent in light- or dark-skinned individuals of Skin
Types III through VI. The remainder of the viable epidermal
cells (basal and malpighian cells) contain melanin-laden
melanosomes in the particulate form. The particulate form
of melanin not only absorbs but also scatters the impinging
radiation.

How Melanin Protects the Skin

     The photoprotective role of melanin is accomplished by
the physical and biochemical properties[1,6] of the polymer
listed below on the following page:

(1) A chemically prepared melanin derived from the
     autooxidation of 3,4-dihydroxyphenylalanine (Dopa), and
     melanin isolated from human hair, melanoma cells, and
     pigmented skin of black guinea pigs reveals significant
     absorption characteristics in UV (200 - 400 nm) and
     visible spectrum (400 - 760 nm), and it can act as an
     effective filter to screen out harmful UVR. Melanized
     epidermis acts as a cloak to shield the viable cells of
     the epidermis and dermis by reducing the transmission
     of damaging UVB and UVA radiant energy. This is best
     documented in nature's ongoing experiment in humans
     living in the equatorial regions of the world.
     Sun-induced non-melanoma skin cancer on the habitually
     exposed areas of the face and upper extremities is very
     rare in the Blacks of equatorial Africa and Aborigenes
     of Australia, New Guinea, and Southern India living in
     tropical areas with high solar UVB and UVA flux; on the
     other hand, Black albinos living in South Africa or
     fair-skinned Australians living in proximity to
     Australian Aborigenes develop solar keratoses and skin
     cancer at an early age in life.[2,5]

(2) There is an apparent inverse relationship between skin
     sensitivity (reactivity) to UVR and melanin content.

Table 1 provides the laboratory determined values for minimal erythema response (MED) in over 100 individuals representd by a minimum of 10 individuals in one Skin Type ranging from Types I through VI. Richly melanized skin exhibits higher MED values than poorly melanized skin. When compared to amelanotic skin of albinos, the sun protection factor (SPF) value for melanin in the unexposed skin of the lower back and buttock ranges from 2 in individuals of Skin Type II to about 6 in individuals of Skin Type VI.

TABLE 1. MED VALUES IN NORMAL INDIVIDUALS OF SKIN TYPES I THROUGH VI BASED ON SUNBURN AND SUNTANNING HISTORY

SKIN TYPE	UNEXPOSED BUTTOCK SKIN	SUN SENSITIVITY & PIGMENT RESPONSE*	UVB MED (MJ/CM^2)
I	white	always burn easily, tan little or none	20 - 30
II	white	always burn easily, tan minimally with difficulty	25 - 35
III	white	always burn moderately, tan average (light brown)	30 - 50
IV	brown	burn minimally, exhibit IPD, tan easily (moderate brown)	50 - 75
V	moderate brown	burn with difficulty and minimally, exhibit intense IPD and tan	60 - 90
VI	dark brown, black	insensitive, never burn, tan profusely	100 - 200

* Based on first 30 to 45 minutes of sun exposure after winter season or without previous sun exposure

The second most important property by which melanin exerts photoprotection involves scattering. Melanin-laden melanosomes, present as supranuclear caps in keratinocytes, attenuate the impinging UVR by scattering. This scattering involves any process within the epidermis that deflects electromagnetic radiation from a straight line path and results in the attenuation of radiation. This increases the total absorbing path through which the photons of UV must pass. For melanin particles with dimensions of the order of wavelength of the UVR (0.3 !m or 300 nm), the impinging photons may be scattered according to the Rayleigh relation (scattering is inversely proportional to the fourth power of the incident light). Maximum scattering occurs when the wavelength of light approaches the particle size. For particles of melanosomes larger in size than the wavelength of the incident light (e.g., melanosomes which are 0.6 - 1.0 μM in size), the scattering relationship becomes quite complex, but nonetheless dominates in deviating the course of penetrating photons. The UV photons in brown and darkly pigmented skin are significantly attenuated by scattering in a forward direction.

(3)   Melanin also protects skin cells against the harmful
      effects of UVR and visible radiation through the
      process of absorption and dissipation of the absorbed
      energy as heat. In this regard, it is of interest to
      know why during the summer months one sees more of
      fair-skinned individuals (Types I - III) sunbathing on
      beaches than dark-skinned individuals (Types V - VI).
      In contrast to fair-skinned individuals endowed with a
      low concentration of melanin, who enjoy sunbathing
      without any discomfort, the high concentration of
      melanin in the darkly pigmented skin makes sunbathing a
      less enjoyable event. The heat generated in the skin
      by absorption of radiant energy causes discomfort to
      the pigmented individual. In this regard, it is of
      interest to point out the hypothesis of McGinnes and
      Procter[8,9] that melanin in the epidermal cells may
      serve as a device by which it may convert the energy of
      the excited states into heat by a phenomenon known as
      photon-phonon conversion. This hypothesis implies that
      a melanin polymer can act as an amorphous semiconductor
      in which the coupling of phonons (i.e., the vibrational
      modes of melanin polymer to its excited electronic
      states) plays a major role in the dissipation of energy
      absorbed from the impinging radiation.

(4)   An additional way by which melanin exerts a
      photoprotective effect in vivo is by way of utilizing
      the absorbed energy into a harmless photochemical
      reaction. Dermatologists and photobiologists are
      familiar with the phenomenon of immediate pigment
      darkenig or IPD reaction[7] in which the exposed skin
      becomes darker during irradiation. In this reaction,
      the absorbed UVR energy (300 - 400 nm) induces
      immediate oxidation in the melanin polymer through the
      generation of semi-quinone free radicals.[10,11]
      Although recent observations on IPD reaction reported
      by Honigsmann[12] suggest that this photochemical event
      does not provide a protective effect on the degree of
      sunburn reaction or on the induction of epidermal DNA
      damage in the form of thymine dimers[11], one must be
      very cautious in leading to conclusions based on
      negative findings. DNA damage by UVB radiation is
      generally not oxidative; it results from the direct
      absorption of photons and does not involve reactive
      oxygen. Oxidative damage is distinct (e.g., membrane
      lipid peroxidation, cytochrome P-450 oxidation, etc.).
      IPD reaction will undoubtedly minimize such an
      oxidative damage in epidermal cells.

(5)   One of the important properties of melanin by which it
      exerts photoprotection in human skin is its ability to
      act as a free radical scavenger for minimizing the
      harmful effects of other free radicals generated by
      UVR. Melanin exists in human skin as a stable free
      radical. As a stable free radical, melanin by its
      ability to undergo immediate oxidation and reduction
      reaction, can act as a biologic electron exchange
      polymer and minimize the impact of the impinging
      photons on the other vulnerable cell constituents
      (e.g., cell membranes, oxidative enzymes, etc.). The
      free radicals in melanin are quite stable and the

unpaired electrons seem to be limited to localized regions of the polymer and are stabilized by a large number of resonance structures within the polymer. Because of the unpaired electrons in the melanin polymer, it may in effect serve as a one-dimensional semi-conductor, where any bound protons serve as electron traps. A free flow of charge in the form of electrons is then possible through the melanin.[13] It is known that UV irradiation increases the unpaired spin concentration in biological tissue such as skin.[10,11] The trapping of free radicals which could disrupt the metabolism of living cells is feasible in the presence of stable free radicals in melanin polymer. It appears that melanin can also act as a scavenger of superoxide anion (O_2^-) or $\cdot OH$ radicals. In this regard, it acts as a pseudo-dismutase. Recent studies definitely indicate that UVR generates free radicals and reactive oxygen species [singlet oxygen ($^1O_2$ and O_2^-)], both in vitro and in vivo.[14] The consequences of free radical generation by UVR to a living system such as human skin are numerous and complex. Free radicals and reactive oxygen species can induce: (a) strand session, (b) DNA-protein cross-links, (c) cross-linking of proteins (e.g., in collagen and elastin), (d) inactivation of enzymes, (e) peroxidation of membrane lipids, and (f) oxidation of sulfhydryl groups causing alterations in the structural and functional state of proteins, etc. These reactive oxygen species are known to be toxic to the viability of cells. Melanin-containing cells are certainly less vulnerable to such oxidative effects of reactive oxygen.

SUMMARY

In this presentation, the photoprotective role of melanin in human skin against the acute effects (e.g., sunburn) and the potential long-term risks of actinic changes (wrinkling and premature aging of the skin) and premalignancies and malignancies (solar kertosis, basal and squamous cell carcinomas, and even melanomas) are discussed. Melanin, present in the epidermis in a particulate form as melanosomes and in an amorphous form as colloidal melanin provides protection to the viable cells in one or more of the following forms: (a) it acts as a filter to screen out harmful UV; (b) it absorbs UVR and dissipates the absorbed energy into heat; (c) melanosomes, which are bigger than the wavelengths impinging on the skin, scatter and attenuate UVR; (d) melanin as a stable free radical and as a biological electron-exchange polymer acts as a quencher of unpaired electrons in reactions involving oxidations and reductions; and (e) melanin also acts as a pseudo dismutase and a scavenger of superoxide anion generated by UVA radiation.

ACKNOWLEDGEMENT

This work was supported by NIH grant 5-R01-CA-05003-29 awarded by the US National Cancer Institute, Department of Health, Education, and Welfare, Bethesda, MD, USA.

REFERENCES

1. Pathak, M.A., The role of natural photoprotective agents in human skin, in: "Sunlight and Man," M.A. Pathak, L.C. Harber, M. Seiji, and A. Kukita, eds., Univ. Tokyo Press, Tokyo, pp. 725-750 (1974).

2. Pathak, M.A., Fitzpatrick, T.B., Greiter, F.J., and Kraus, E.W., Preventive treatment of sunburn, dermatoheliosis, and skin cancer with sun protective agents, in: "Dermatology in General Medicine," T.B. Fitzpatrick, A.Z. Eisen, K. Wolff, et al., eds., 3rd ed., McGraw-Hill Book Company, New York, pp. 1507-1522 (1987).

3. Carraro, C., and Pathak, M.A., Characterization of superoxide dismutase from mammalian skin epidermis, J. Invest. Dermatol. 90:31-36 (1988).

4. Kopf, A.W., Kripke, M.L., and Stern, R.S., Sun and malignant melanoma, J. Am. Acad. Dermatol. 11:674-684 (1984).

5. Urbach, F., Photocarcinogenesis, in: "The Science of Photomedicine," J.D. Regan and J.A. Parrish, eds., Plenum Press, New York, pp. 261-292 (1982).

6. Pathak, M.A., Jimbow, K., Szabo, G., and Fitzpatrick, T.B., Sunlight and melanin pigmentation, in: "Photochemical and Photobiological Reviews," K.C. Smith, ed., Plenum Press, New York, pp. 211-239 (1976).

7. Pathak, M.A., Immediate and delayed pigmentary and other cutaneous responses to solar UVA radiation (320 - 400 nm), in: "The Biological Effects of UVA Radiation," F. Urbach and R.W. Gange, eds., Praeger Publishers, New York, pp. 156-167 (1986).

8. McGinness, J.E., and Proctor, P.H., The importance of the fact that melanin is black, J. Theor. Biol. 39:677 (1973).

9. McGinness, M.C., Corry, P., and Procter, P., Amorphous semiconductor switching in melanin, Science 183:853-855 (1974).

10. Pathak, M.A., Photobiology of melanogenesis: Biophysical aspects, in: "Advances of Biology of Skin; the Pigmentary System," W. Montagna and F. Hu, eds., Vol. 8, Pergamon Press, Oxford, pp. 387-420 (1967).

11. Pathak, M.A., and Stratton, K., A study of the free radicals in human skin before and after exposure to light, Arch. Biochem. Biophys. 123:468-476 (1968).

12. Hongismann, H., Schuler, G., Aberer, W., Romam, N., and Wolff, K. Immediate pigment darkening phenomenon: A reevaluation of its mechanisms, J. Invest. Dermatol. 87:648-652 (1986).

13. Mason, H.S., Ingramm, D.J.E., and Allen B., The free radical property of melanins, Arch. Biochem. Biophys. 86:225-230 (1960).

14. Pathak, M.A., Reactive oxygen species and free radicals in sunlight-induced skin reactions, J. Am. Oil Chemists Soc. 64:630 (1987).

MELANOGENIC POTENTIAL OF VARIOUS FUROCOUMARINS IN NORMAL AND VITILIGINOUS SKIN

Madhu A. Pathak and M. Dalle Carbonare

Department of Dermatology, Harvard Medical School
and Massachusetts General Hospital
Boston, Massachusetts

INTRODUCTION

The herbal use of psoralens (furocoumarins) for the treatment of leukoderma (vitiligo) in India, Egypt, and many other far Eastern countries dates back to the time before the birth of Christ.[1] The roots of psoralen therapy for vitiligo can be traced as far back as 1,500 B.C. in Atharva Veda in which the early Ayurvedic therapy practiced by ancient Hindu priests is described. Extracts of Bavachee seeds or Psoralea Corylifolia and other plants containing psoralens were used in folk remedies by people of various cultures of India. Even in the late twentieth century, folk remedies of this form are being used in India, Pakistan, China, and the Far East. The powdered seeds of such plants, primarily belonging to Umbelliferae, Rutaceae, and a few to Leguminosae family, are given orally in tablet form or applied topically in paste form to the white skin. These plants are known to contain photoreactive and melanin-stimulating furocoumarins including psoralen, 8-methoxypsoralen (8-MOP), 5-methoxypsoralen (5-MOP), isopsoralen, etc.[2] Similarly, the ancient herbal use of Ammi Majus Linneaus, an umbelliferous plant, that grows wildly along the coast of the Nile in Egypt and containing well characterized photoreactive psoralens (e.g., ammoidin or 8-MOP, ammidin, or 8-isoamelaneoxypsoralen, majudin, or 5-MOP) has been carefully documented by the Egyptians.[1-4]

The development of topical or oral psoralens for stimulating melanin pigmentation in modern times has gradually evolved from the experimental observations made by several clinicians and investigators in Egypt, USA, Europe, and other countries.[4] Rarely has the issue of relative effectiveness of various psoralens, both naturally occurring and synthetically prepared, been addressed in a controlled study.[5] This brief report provides data based on laboratory investigations and clinical observations carried by the author in collaboration with Professor Thomas B. Fitzpatrick of our Department for over the past 30 years. The conclusions are also based on laboratory investigations in pigmented guinea pigs, in normal human volunteers (light brown, dark brown, and black skin), and in patients with vitiligo undergoing PUVA (oral psoralens + UVA) therapy. In this report, the pigment stimulating properties of linear and nonlinear furocoumarins (psoralens) and furanochromones are presented. A mechanism of the mode of increased pigmentation by psoralens is outlined.

MATERIALS AND METHODS

Compounds listed in Table 1 were applied topically to the skin of the backs of normal individuals and the epilated back skin of guinea pigs in concentrations ranging from 2.5, 5, 10, 25, 50, 100, and 500 μg/6 cm^2 skin area. Generally, compounds that were known to be phototoxic and evoked skin photosensitization reaction in the form of erythema, edema, and vesiculation response after exposure to UVA radiation were applied in low concentrations (2.5 - 25 μg/6 cm^2). Compounds that were nonphotosensitizing or nonphototoxic were applied in high concentrations ranging from 25 to 500 μg/cm^2. Each compound has been tested in at least three human volunteers and in two or more pigmented guinea pigs. Approximately 45 minutes after the psoralen application, the treated skin was exposed to UVA radiation. To minimize a severe phototoxic reaction, we kept the exposure dose of UVA at 2.0 to 4.0 J/cm^2 for photosensitizing psoralens in order to achieve a moderate degree of erythema reaction (+ to ++ pink-red color without edema). For nonphototoxic furocoumarins and furanochromones, UVA exposure dose was increased to 6 to 12 J/cm^2. All human subjects and test animals were treated and irradiated only once, and the pigmentation response of the exposed sites was graded on Days 7 and 10 after exposure and compared with constitutive color of the control site. The control sites included: (a) untreated skin, (b) chemically treated skin but not exposed to UVA, and (c) chemically nontreated skin exposed to UVA. Based on the constitutive pigmentation of the control sites, the melanogenic activity of the psoralen-treated and UV-exposed skin was assessed by the following criteria:

$$0 = \text{no change in skin color}$$
$$\underline{+} = \text{minimal visible pigmentation}$$
$$\overline{+} = \text{light brown tan}$$
$$++ = \text{moderate brown tan}$$
$$+++ = \text{dark brown tan}$$
$$++++ = \text{black or deep brown tan}$$

The rated potency of the compound to stimulate melanogenesis is based on the minimal melanogenic dose calculated in microgram/cm^2, and the relative degree of melanin pigmentation observed at various concentrations.

RESULTS

The relative melanin-stimulating activity for various psoralens (linear psoralens and nonlinear isopsoralens) and furanochromones is shown in Table 1. It is apparent that linearly annulated psoralens stimulated melanin pigmentation more effectively than the nonlinear isopsoralens. Until recently, it was believed that psoralens which were capable of evoking skin photosensitization responses (erythema, edema, and vesiculation) and induced interstrand cross-links with DNA upon irradiation were more melanogenic than the nonphotosensitizing monofunctional furocoumarins that formed psoralen-DNA photoadducts with single-stranded DNA but did not cause interstrand cross-links. This general concept, although valid for most chemicals that were evaluated prior to 1980 has recently undergone gradual but definite change in our interpretation and conclusion. The monofunctional isopsoralens and psoralens that are nonphotosensitizing were believed to be ineffective in stimulating melanogenesis. Newer findings indicate they do stimulate melanogenesis quite significantly. This particular property of psoralens involving the stimulation of skin pigmentation without manifestation of phototoxic reaction has not been reported earlier. It is now feasible to stimulate melanogenesis without evoking skin damaging reactions associated with phototoxic psoralens, involving erythema, edema, vesiculation, and desquamation. Melanogenesis can be effectively stimulated by nonphotosensitizing furocoumarins and furanochromones. This may generate a major therapeutic use for phototoxic psoralens. Future therapy for vitiligo patients could include the topical or systemic use of nonphotosensitizing furocoumarins or furanochromones.

TABLE 1. MELANIN-STIMULATING ACTIVITY OF VARIOUS FUROCOUMARINS (PSORALENS AND ISOPSORALENS AND FURANOCHROMONES)

COMPOUND TESTED	CONCENTRATION ($\mu g/cm^2$)	UVA DOSE (J/cm^2)	RELATIVE MELANIN STIMULATING ACTIVITY	RELATIVE PHOTOSENSITIZING ACTIVITY	TYPE OF PSORALEN
psoralen	2.5 - 20	2 - 4	++++	very strong	linear
8-MOP	2.5 - 25	2 - 4	+++	strong	linear
5-MOP	2.5 - 25	2 - 4	++ / +++	moderate	linear
4,5',8-TMP	2.5 - 5	1 - 2	++++	very strong	linear
5,8-dimethylpsoralen	25 - 100	6 - 8	- / ±	absent	linear
8-isoamelenoxypsoralen	25 - 500	6 - 8	+	low	linear
8-hydroxypsoralen	25 - 500	12.0	± / +	absent	linear
4,5'-dihydro-8-MOP	50 - 250	6 - 8	-	absent	linear
5-MOP	50 - 500	6 - 8	± / +	absent	nonlinear
3-carbethoxypsoralen	50 - 250	6 - 12	+	low	nonlinear
isopsoralen (angelicin)	50 - 1,000	6 - 12	- / ±	absent	nonlinear
5-methylangelicin	50 - 500	6 - 8	+ / ++	low	nonlinear
4,5'-dimethyl-angelicin	25 - 250	6 - 8	++	low / absent	nonlinear
4',4',6-TMP	25- 250	6 - 8	+++	low	nonlinear
4.4',8-trimethyl-5'-aminomethylpsoralen	2.5 - 5	2 - 4	+++	very strong	linear
4,4',8-trimethyl,4'-aminomethylpsoralen	2.5 - 5	2 - 4	+++	very strong	linear
khellin	25 - 500	6 - 8	+++	absent	furanochromone
visnagin	25 - 500	6 - 12	+	absent	furanochromone

linear psoralen = linearly annulated tricyclic psoralen;
nonlinear psoralen = nonlinear, annular, and tricyclic isopsoralen

When tested topically in human subjects, the potent
hotosensitizing agents such as of 4,5',8-trimethylpsoralen
4,5',8-TMP), water soluble TMP derivatives (e.g.,
,4',8-trimethyl,4'-aminomethylpsoralen and 4,4',8-trimethyl,
'-aminomethylpsoralen) and the parent psoralen were very
ffective in stimulating melanin pigmentation at low
oncentrations of the furocoumarin (< 2.5 µg/6 cm^2) and low
xposure dose of UVA radiation (2 J/cm^2). This is
llustrated below in:

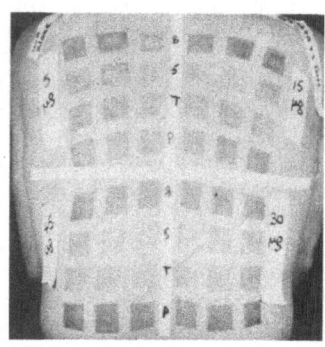

Figure 1. A human volunteer (Skin Type III)
topically treated with 5, 15, 30, and 45 µg/cm^2 of
8-MOP, 5-MOP, TMP, and PSO (psoralen). In each
quandrant, the first row received 8-MOP, second
row 5-MOP, third row TMP, and fourth PSO. Maximum
stimulation of pigmentation was observed on Day 10
with PSO, TMP, and 8-MOP.

Based on these observations, our present concepts for
melanin-stimulating activity of psoralens is as follows:[5-8]
    1. Those psoralens that photoconjugate with epidermal
DNA and produce extensive interstrand cross-links,
particularly in DNA of melanocytes, are distinctly more
capable of stimulating melanin pigmentation than other
psoralens that are primarily monofunctional and produce only
single-strand DNA adducts. These interstrand cross-linking
psoralens also cause phototoxicity (induce skin
photosensitization in the form of erythema and edema) and
produce reactive oxygen species [singlet oxygen ($^1O_2$),
superoxide anion ($O_2^{\cdot}$), or $^{\cdot}OH$ radicals]. The production of
interstrand cross-links in DNA and the generation of
reactive oxygen causes damage to DNA as well as to cell
membranes (Table 2). This leads to subsequent mitoses and
cell proliferation (see section on mechanism of
pigmentation).
    2. Those psoralens that produced minimal amounts of
interstrand cross-links (e.g., 8-isoamyleneoxypsoralen and
5,8-dimethoxypsoralen) in DNA and minimally produced
reactive oxygen species upon irradiation are very weak in
stimulating melanogenesis.
    3. Those isopsoralens (e.g., methylated angelicins)
and psoralens (3-carbethoxypsoralen) that were nonphototoxic
but showed enhanced photoconjugation with DNA in the form of
single-strand monofunctional adducts and also generated
reactive oxygen species were found to stimulate melanin
pigmentation appreciably. Those psoralens and isopsoralens
that were nonphotosensitizing but photoconjugated poorly

with DNA and were weak in producing reactive oxygen species were essentially poor in stimulating melanin pigmentation.

TABLE 2. PRELIMINARY DATA ON A COMPARATIVE STUDY INVOLVING PHOTOPROTECTIVE PROPERTIES OF 8-MOP, ANGELICINS, AND KHELLIN

Property	8-MOP	Di- or Trimethylangelicin	Khellin
structure			
skin phototoxicity	strong	absent	absent
melanin pigmentation stimulation:			
guinea pig	moderately strong	strong	strong
human	strong	moderate	moderate
topical	strong	strong	strong
oral	strong	weak	moderate
reactive O_2 production:			
$^1O_2$	moderate ++	strong +++	weak +/+
O_2^{-}	moderate ++	strong +++	weak +/+
photoconjugation with DNA:	strong	strong	low
cross-links	yes	no	uncertain
monoadducts	yes	yes	yes
distinguishing features	stimulates pigmentation with phototoxic reaction	stimulates pigmentation w/o phototoxic reaction	stimulates pigmentation w/o phototoxic reaction

MELANIN-STIMULATING ACTIVITY OF FURANOCHROMONE

The furocoumarins (e.g., psoralen, 8-MOP, and angelicin) belong to a group of heterocyclic compounds that are considered to be derivatives of coumarin (benz-α-pyrone 1,2-benzopyrine). The fusion of a pyrone ring with a benzene nucleus can give rise to a class of heterocyclic compounds known as benzopyrones (see Figs. 2a and 2b). The two distinct types of recognized benzopyrones are benzo-α-pyrone, commonly referred to as coumarins (Fig. 2a), or benzo-γ-pyrone, commonly referred to as chromones (Fig. 2b). The examples of benzo-α-pyrone or coumarins are psoralen, 8-MOP, etc. The examples of benzo-γ-pyrone or chromones are khellin and visnagin.

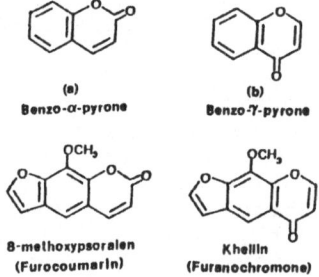

(a) Benzo-α-pyrone (b) Benzo-γ-pyrone

8-methoxypsoralen (Furocoumarin) Khellin (Furanochromone)

Figure 2. Chemical structure of benzo-α-pyrone (a) and benzo-γ-pyrone (b). Also the structure of furocoumarin (8-MOP) (c), and furanochromone (khellin, d) are shown.

Khellin, a compound isolated from the seeds of Ammi Visnaga Lam (Umbelliferae), has been recently reported to be a promising photochemotherapeutic agent for stimulating hyperpigmentation and repigmentation of skin in patients with vitiligo.[9,10] Unlike, 8-MOP (isolated from Ammi Majus Linn, Umbelliferae), khellin is a furanochromone.[9] First, the Egyptian investigators, Abdel-Fatah et al.,[9] and subsequently the Austrian investigators, Ortel et al.[10,11] reported the preliminary results on the effects of oral administration of khellin on the stimulation of melanin pigmentation in patients with vitiligo. Their observations indicated the oral administration of khellin (100 mg/70 kg) followed by exposures to sunlight or UVA radiation can stimulate satisfactory repigmentation of vitiligo macules in at least 40% of patients. After about 2 to 10 months of continuous therapy, up to 70% of the original vitiligo areas exhibited good pigment response.[10,11] We have investigated several photochemical properties, including the pigment-stimulating activity of khellin and visnagin. Khellin was found to be a good stimulator of melanin pigmentation in normal volunteers and in the skin of brown-haired guinea pigs (Figs. 3 and 4). Visnagin was not a remarkably impressive compound for stimulating new melanogenesis. Our recent data based on a comparative study involving photoreactive properties of linear psoralens (8-MOP), nonlinear isopsoralens (angelicins), and khellin

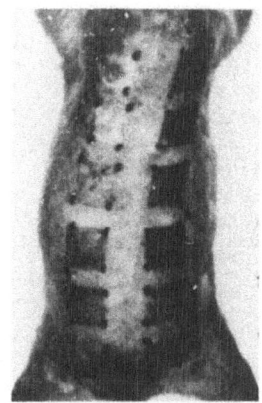

Psoralen 25 μg	5-MeA 50 μg
8-MOP 25 μg	6,4,4'TMeA 50 μg
5-MOP 25 μg	Khellin 100 μg
Khellin 50 μg	6,4,4' TMeA 100 μg
3-CP 50 μg	5-MOP 50 μg
Control	UV-A 6 J

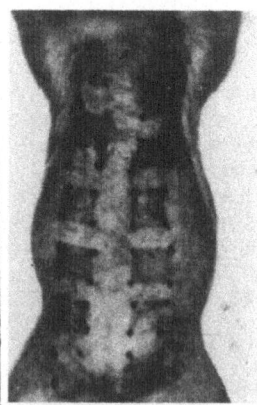

Psoralen 12.5 μg	5-MeA 100 μg
8-MOP 12.5 μg	6,4,4'TMeA 100 μg
5-MOP 12.5 μg	5-MOP 25 μg
Khellin 100 μg	Khellin 200 μg
3-CP 00 μg	8-MOP 25 μg
Control	UV-A 6 J

Figures 3 and 4. Intense pigmentation stimulated by topically applied methylangelicins (isopsoralens) and khellin. 5-MeA = 5-methylangelicin; 6,4,4'-MeA, 6,4,4',-trimethylangelicin; 3-CP = 3-carbethoxypsoralen. Note pigmentation stimulated by 8-MOP, psoralen (PSO), and 5-MOP is distinctly less than khellin, 3-CP, and methylangelicins at Day 5, when intense erythema (phototoxic reaction) observed at skin sites treated with PSO, 8-MOP, and 5-MOP retarded the process of melanogenesis. Khellin and methylangelicins were nonphototoxic but stimulated intense pigmentation within 72 hours.

(furanochromones) are summarized in Table 2. Figures 3 and 4 illustrate the strong pigment-stimulating properties of certain methylangelicins and khellin. These chemicals were applied topically (50 - 100 $\mu g/cm^2$), and the treated skin was subsequently exposed to UVA (6 J/cm^2). Khellin and methylangelicins were nonphototoxic and stimulated neomelanogenesis significantly within 72 hours. The photosensitizing 8-MOP and 5-MOP, on the contrary, caused erythema and because of this phototoxicity, the hyperpigmentation response was delayed in onset and lagged behind the pigment response stimulated by nonphotosensitizing khellin and methylangelicins.

DISCUSSION: MODE OF HYPERPIGMENTATION BY PSORALENS

The mechanism by which psoralens in the presence of UVA radiation (PUVA) stimulate hyperpigmentation of normal skin or repigmentation of vitiligo skin remains mostly unknown or speculative.[5,8,12,13] An increase in the number of melanin-producing melanocytes in normal skin and vitiliginous skin has been reported by several investigators.[14] This may help in understanding the mode of increased pigmentation stimulated by PUVA. It appears increased melanin pigmentation is related to the photoconjugation of psoralens with pyrimidine bases in DNA to form both monofunctional photoadducts (single-strand adducts of psoralen to pyrimidine bases) and subsequently interstrand cross-links (bifunctional adducts) between psoralen and pyrimidine bases of both strands of DNA.[7,14] This important event reflects damage to melanocytic DNA that retards the cell division during the repair phase of photodamaged DNA in the preexisting functional melanocytes at the dermo-epidermal juction and hair bulbs. Within 24 to 48 hours, the melanocytes appear to undergo repair of their photodamaged DNA by excision repair mechanism involving the pyrimidine-psoralen adducts.[5,12,13] Melanocytic DNA replication and DNA replication in other epidermal and dermal cells (kertinocytes and fibroblasts) follow. At this stage, we have observed an important event of mitoses in melanocytes. This suggests melanocytes undergo replication and proliferation. This event appears to stimulate certain other cellular activities in photodamaged melanocytes. The melanocyte-stimulating hormone (MSH) receptor system appears to be activated. The activity of cell surface MSH receptors increases in G_2 phase of the cell cycle and leads to increased functional activity of cyclic AMP (adenosine 3',5'-monophosphate). The resulting mitoses in preexisting melanocytes leads to new, functional melanocytes with the increased production of new melanosomes and increased levels of tyrosinase activity. Increased cell population and enhanced tyrosinase activity of melanocytes leads to increased formation of melanin in melanosomes. Thus, psoralen plus UVA irradiation influences the normal pigmentation of the skin in one or more of the following ways:

    1.   There is a photoaddition reaction of psoralen to the DNA of melanocytes and keratinocytes.

    2.   The melanocyte-stimulating hormone (MSH) receptor system appears to be activated.

    3.   This is followed by the repair of the photodamaged DNA by excision repair process. Events of cell mitoses in

melanocytes and keratinocytes are observed.

4. There is an increase in the number of functional melanocytes as a result of a direct mitotic replication of melanocytes and also possibly the activation of the dormant or resting melanocytes existing prior to irradiation in the epidermis or skin appendages (e.g., sweat ducts and hair follicles) by the activation of the MSH receptor system and the action of cyclic AMP. The melanocyte population is doubled or tripled in number within three to seven days after PUVA treatment.

5. The number of melanized melanosomes in proliferating melanocytes is increased as the result of increased synthesis of melanosomal proteins and melanin-forming tyrosinase. Melanocytic hypertrophy manifested by the enlargement of perikarya and increased arborization of dendrites is also consistently observed.

6. The rate of transfer of melanin-laden melanosomes from melanocytes to keratinocytes increases as the result of an increased turnover and upward movement of keratinocytes. Even the number of early and intermediate stages of melanosomes (stage I premelanosomes and partially melanized stage II and III melanosomes) is increased.

There is enhanced migration of the activated melanocytes from skin appendages (e.g., sweat ducts and hair follicles) to the dermo-epidermal junction. This is usually evident as a phenomenon of perifollicular repigmentation of vitiliginous macules in patients with vitiligo undergoing repigmentation.

RELATIVE EFFECTIVENESS OF FUROCOUMARINS AND FURANOCHROMONES IN REPIGMENTATION OF VITILIGO

The restoration of normal melanin pigmentation in vitiliginous areas is the goal and burning desire of most patients. The search for an effective treatment for achieving uniform repigmentation or "cure" among these vitiligo patients is intense, constant, often frantic, and endless.

At the present time, photochemotherapy involving the topical application or oral administration of a psoralen derivative in combination with exposure to ultraviolet radiation, either from sunlight or from artificial sources remains the only effective method which offers a hope of achieving repigmentation of vitiliginous macules. The most frequently recommended treatment of vitiligo involves the ingestion of a psoralen (either 8-MOP or 4,5',8-TMP) followed by a series of controlled exposures to solar radiation or to artificial sources emitting 320- to 400-nm UVA radiation.[5,12] The therapy involves two to three treatments weekly for 3 to 12 months or longer. The most limiting factor of therapy is the slow rate of response and the potential concern of phototoxicity.

Relative effectiveness of furocoumarins and furanochromones in the repigmentation of vitiligo are summarized in Table 3. Based on our clinical experience, 8-MOP (methoxsalen) is an effective agent for repigmenting brown- and dark-skinned individuals of Skin Types IV, V, and

VI (e.g., Orientals, Hispanics, Indians, Pakistanis, and Blacks). The recommended drug dose is 0.3 to 0.6 mg/kg. When administered at high doses (0.8 to 1.2 mg/kg), 8-MOP can cause nausea and severe phototoxic reactions. When given orally, trixosalen or 4,5',8-TMP is a less phototoxic agent than 8-MOP and is recommended for fair-skinned individuals of Skin Types II, III, and IV for indoor as well as outdoor PUVA therapy. Unsubstituted psoralen is not an effective agent in repigmenting vitiliginous skin. It is quite phototoxic and often gives poor results. It should be recognized that phototoxicity manifested by erythema and edema contributes to Koebner reaction and visually recognizable damage to epidermal cells. The damaged melanocytes tend to become less viable and gradually degenerate. Melanocytes undergo mitoses and cell proliferation when phototoxic reaction is minimal. It should be noted trioxsalen or 4,5',8-TMP is a potent skin photosensitizing agent which promotes melanogenesis quite effectively when it is applied topically; however, when given orally, TMP is much less erythemogenic and stimulates melanogenesis quite moderately.[15] This indicates TMP is either poorly absorbed when given orally or metabolized rapidly to a nonphotosensitizing moiety. Studies reported by Pathak et al.[14] indicated TMP is rapidly biotransformed to a nonphotosensitizing entity. Orally administered TMP, therefore, is prescribed usually to fair-skinned individuals of Skin Types I, II, III, and IV; it does not cause nausea and acute phototoxic reactions. Bergapten or 5-MOP is less phototoxic than 8-MOP and when used in higher doses (1.0 - 1.2 mg/kg) than 8-MOP (0.6 mg/kg), appears to be an effective agent for stimulating a moderate to good pigment response.[5] 5-MOP has one advantage over 8-MOP, like TMP, it also does not cause nausea. Additional control studies are necessary to determine whether orally administered khellin is as good or better than 8-MOP or 5-MOP. It is not known whether topically applied khellin will stimulate repigmentation in vitiligo macules. Khellin is a furanochromone that is moderately effective in stimulating pigment response.

Many physicians, particularly in India, Pakistan, the Middle East, including Egypt, attempt to treat large areas

TABLE 3. RELATIVE EFFECTIVENESS OF ORAL FUROCOMARINS AND FURANOCHROMONES IN REPIGMENTATION OF VITILIGINOUS SKIN

COMPOUND	DOSE (MG/KG)	THERAPEUTIC RESPONSE[c]	REMARKS
psoralen[a]	0.3	poor	
	0.6	fair	phototoxic
	1.0	fair	phototoxic
8-MOP[a]	0.3	moderate	
	0.6	good	phototoxic
	1.0	fair	phototoxic
5-MOP[b]	0.6	fair	less
	1.2	good	phototoxic[d]
4,5'8-TMP[a]	0.3	poor	less
	0.6	moderate	phototoxic[d]
	1.2	good	
khellin[b]	1.2	moderate	non-
	1.4	good	phototoxic

of vitiligo with topical psoralens. Topical treatment with
8-MOP or other psoralens should be confined to small
vitiligo macules (2.0 - 7.5 cm in diameter) and should be
carried out by the physician using a low concentration of
the psoralen (0.1%) and low dose of UVA (1 - 2 J/cm^2).

[a] Based on several clinical trials carried out by us
between 1970-86 and involving over 600 patients.
[b] Data based on studies carried out by Honigsmann and
associates (10).
[c] poor = $\leq$ 20% patients exhibited repigmentation
moderate = 30 to 50% patients exhibited repigmentation
good = $\geq$ 60% patients exhibited good repigmentation
response
[d] less phototoxic than 8-MOP

SUMMARY

In this lecture, I have attempted to review the
relative effectiveness of various linearly annulated
psoralens, nonlinear isopsoralens, and furanochromones in
stimulating melanin pigmentation in mammalian skin (human
and guinea pig skin). Of various psoralens tested, 8-MOP,
5-MOP, and orally administered 4,5',8-TMP appear to be
potent agents for stimulating cutaneous pigmentation in
normal individuals and vitiligo patients. Khellin, a
furanochromone, was also found to be an interesting,
nonphototoxic agent that stimulated melanin pigmentation
quite significantly. Earlier observations had indicated
that bifunctional psoralens and isopsoralens evoking
cutaneous phototoxicity were most active in stimulating
pigmentation, and the monofunctional psoralens which evoked
no cutaneous phototoxicity were essentially inactive in
stimulating pigmentation. Contrary to these earlier
concepts, recent studies indicate that evidence of
phototoxicity should not be the only criteria to judge the
compounds ability to stimulate pigmentation. Compounds such
as khellin and several methylangelicins that produced no
evidence of phototoxicity were quite remarkable in
stimulating skin pigmentation. It appears compounds that
photoconjugate with DNA, generate reactive oxygen species
(e.g., $^1O_2$, O_2^-, and $\cdot OH$) after irradiation with UVA, and
activate MSH receptor sites on melanocytes tend to stimulate
melanin pigmentation quite appreciably.

The therapeutic effectiveness of various psoralens and
khellin has been briefly reviewed. Studies based on the
stimulation of melanin pigmentation by phototoxic and
nonphototoxic furocoumarins and furanochromones suggest
that patients with vitiligo should be treated with
suberythemogenic (nonphototoxic) doses of UVA radiation.

An attempt has been made to elucidate the mode of
hyperpigmentation stimulated by furocoumarins and
furanochromones.

ACKNOWLEDGEMENT

This work was supported by NIH grant 5-R01-CA-05003-28
awarded by the US National Cancer Institute, Department of
Health, Education, and Welfare, Bethesda, MD, USA.

REFERENCES

1. Fitzpatrick, T.B., and Pathak, M.A., Historical aspects of methoxsalen and other furocoumarins, J. Invest. Dermatol. 32:229-231 (1959).

2. El Mofty, A.M., Vitiligo and psoralens, Pergamon Press, Oxford (1968).

3. Pathak, M.A., Daniels, F., and Fitzpatrick, T.B., The presently known distribution of furocoumarins (psoralens) in plants, J. Invest. Dermatol. 39:225-239 (1962).

4. Fitzpatrick, T.B., and Pathak, M.A., Research and development of oral psoralen and longwave radiation, Photochemotherapy: 200 BC - 1982 AD, in: "Photobiologic, Toxicologic, and Pharmacologic Aspects of Psoralens," M.A. Pathak and J.K. Dunnick, eds., Monograph 66, U.S. Dept. Health and Human Service, National Institutes of Health, Bethesda, MD, pp. 3-11 (1984).

5. Pathak, M.A., Mosher, D.B., and Fitzpatrick, T.B., Safety and therapeutic effectiveness of 8-methoxy-psoralen, 4,5',8-TMP, and psoralen in vitiligo, Ibid., pp 165-173.

6. M.A. Pathak, Mechanisms of psoralen photosensitization reactions, Ibid., pp. 41-46.

7. Dall'Acqua, F., Furocoumarin photochemistry and its main biological implications, in: "Current Problems in Dermatology: Therapeutic Photomedicine," H. Honigsmann and G. Stingl, eds., Karger, Basel, 15:137-163 (1986).

8. Honigsmann, H., Wolff, K., Fitzpatrick, T.B., and Pathak, M.A., Oral photochemotherapy with psoralens and UVA (PUVA): Principles and practice, in: "Dermatology in General Medicine," T.B. Fitzpatrick, A.Z. Eisen, and K. Wolff, et al., eds., 3rd. ed., McGraw-Hill Book Co., New York, pp. 1533-1588 (1987).

9. Abdel-Fattah, A., Aboul-Enein, M.N., Wassel, G., et al., An approach to the treatment of vitiligo by khellin. Dermatologica 165:136-140 (1982).

10. Ortel, B., Tanew A., and Honigsmann, H., "Current Problems in Dermatology: Therapeutic Photomedicine," H. Honigsmann and G. Stingl, eds., Karger, Basel, 15:265-271 (1986).

11. Ortel B., Tanew, A., Honigsmann, H., and Wolff, K., UVA photochemotherapy of vitiligo, Photochem. Photobiol. 39:52 (1984).

12. Mosher, D.B., Pathak, M.A., Fitzpatrick, T.B., Vitiligo: Etiology, pathology, diagnosis, and treatment, in: "Dermatology in General Medicine: Update," T.B. Fitzpatrick, A.Z. Eisen, K. Wolff, K.M. Freedberg, and A.F. Austin, eds., McGraw-Hill Book Co., New York, pp. 205-225 (1983).

13. Pathak, M.A., Zarebska, Z., Mihm, M.C., et al., Detection of DNA-psoralen photoadducts in mammalian skin, J. Invest. Dermatol. 86:308-315 (1986).

14. Pathak, M.A., Kramer, D.M., and Fitzpatrick, T.B., Photobiology and photochemistry of furocoumarins (psoralens), in: "Sunlight and Man," M.A. Pathak, L.C. Harber, M. Seiji, and A. Kukita, eds., Univ. Tokyo Press, Tokyo, pp. 335-368 (1974).

15. Pathak, M.A., Marciani, M.S., Guotto, A., and Rodighiero, G., A study of the relationship between photo-photosensitizing and therapeutic activity of 4,5',8-TMP and its major metabolite 4,8-dimethyl, 5'-carboxypsoralen, J. Invest. Dermatol. 81:533-539 (1983).

GUANYLATE CYCLASE ACTIVITY AND PHOTOTRANSDUCTION IN THE ROD OUTER SEGMENT

Isidoro Pepe and Isabella Panfoli

Istituto di Cibernetica e Biofisica CNR, Genova, Italy

INTRODUCTION

The mechanism of visual transduction in vertebrate rods has recently been delineated in its general outlines (Lamb, 1986; Stryer, 1986; Pugh and Cobbs, 1986). In the dark, a positive current carried by Na ions flows into the rod outer segment (ROS) through cation-specific channels that are opened by their binding to cyclic GMP (cGMP) molecules. Light activates an enzymatic cascade that initiates with the photoexcitation of rhodopsin and - through an intermediate step involving a Guanyl nucleotide binding protein, trasducin - terminates with the activation of a phosphodiesterase (PDE) that hydrolyzes cGMP. This cascade has all the necessary features such as rapidity (ms) and amplification (10^5) to be considered central to phototransduction.

The drop of cGMP concentration inside the photoreceptor leads to the closure of the cation-specific channels and to the block of the Na influx. The consequent hyperpolarization of the plasma membrane constitutes the rising phase of the neuronal signal transmitted to the synapse. In order to restore the normal membrane potential, cGMP concentration should increase rapidly. This obligatory step involves the activation of guanylate cyclase that synthesizes cGMP from GTP.

We have found that light stimulates cyclase in disrupted rod outer segments of the toad retina, provided that free Ca concentration is kept low. In the dark cyclase activity remains lower and largely independent of the Ca concentration. These results are discussed in view of a regulatory role of Ca in phototransduction.

METHODS

Preparations. Rod outer segments were obtained from dark adapted eyes of toads Bufo-Bufo, enucleated under dim red light. The eyes were hemisected under infrared light and retinae were removed and gently shaken in 35% of sucrose (w/w) in Ringer solution which did not contain $CaCl_2$ (115 mM NaCl, 2.5 mM KCl, 1 mM $MgCl_2$, buffered at pH 7.5 with 10 mM Hepes and tetramethylammonium hydroxyde). Rod outer segment suspension was centrifuged at 6000 rpm (3000 g). The pellet was discarded and the supernatant diluted with Ringer solution and centrifuged at 4000 rpm for 10 min. The resulting supernatant was discarded and the pellet, which contained rod outer segments from 4 retinae, was hypotonically shocked in 100 μl of 5 mM Mops pH 7.1 containing 5 mM dithiolthreitol. The obtained disks membranes suspension was homogenized with a 100 μl-pipette with a plastic tip (0.2 mm diameter orifice) to prevent disks from aggregating.

Guanylate cyclase assay. Guanylate cyclase was assayed following the methods of Pannbacker (1973) and Fleischman (1982) with minor modifications. The reaction mixture contained 100 mM Mops pH 7.1, 140 mM KCl, 20 mM NaCl, 5 mM $MgCl_2$, 2 mM GTP, 10 mM phosphocreatine, 1 mg/ml creatine phosphokinase, 0.4 mM Zaprinast (2-o-Propoxyphenyl-8-azapurin-6-one: a specific inhibitor of PDE), 4 mM cGMP, 3 μCi of [^{14}C]GTP (450 mCi/mmol) and 10 μCi of [^{3}H]cGMP (15 Ci/mmol). The reaction was initiated by adding a volume of rod outer segment homogenate to an equal volume of reaction mixture with a final protein concentration of about 1-2 mg/ml. The mixture was incubated at room temperature (about 24 °C) and aliquots of 20 μl were collected at different times. Reactions were stopped by adding 5 μl of 100 mM EDTA and boiling for 2 min. After centrifuging, 10 μl of the supernatants were applied to a poly-ethyleneimine-cellulose thin layer plate (20x20 cm) which was then developed in one dimension, first with distilled water and, after drying, with 0.2 M LiCl for 7 cm and then with 1 M LiCl. 5'-GMP, GDP and Guanosine were added as internal standard to the samples. Nucleotides and nucleosides were separated and visualized under short wave UV light. The spots were cut out and transferred to scintillation vials; 0.5 ml of 0.7 M $MgCl_2$/1 M Tris-HCl pH 7.4 (10/2, v/v) were added to each vial to elute the compounds; 3 ml of Instagel were added and samples were counted in a LKB liquid scintillation spectrometer. Counting efficiency was about 34% and 20% for ^{14}C and ^{3}H respectively.

Protein concentration was determined with a method based on Coomassie blue staining (Asim Esen, 1978).

RESULTS

The formation of [^{14}C] cGMP from [^{14}C] GTP in ROS homogenates in the dark and after a flash of light is shown in the lower part of Fig. 1. Each point was corrected for the cGMP hydrolysis by phosphodiesterase, which is shown in the upper part of the Fig. 1.

Guanylate cyclase activity was calculated from the rate of formation of [^{14}C] cGMP/mg protein. In the experiment shown in Fig. 1, the percentage of cGMP formed/min (slope of the straight line in the dark) was 0.39 and the initial amount of GTP was about 240 nmol in a sample containing 0.36 mg protein. The cyclase activity in the dark therefore, amounts to $0.39\times10^{-2}\times240/0.36=2.6$ nmol of cGMP formed/min/mg protein.

After a flash of light, this value increased by about 3 fold provided the concentration of Ca^{2+} was kept low (about 10^{-7} M in the experiment shown in Fig. 1). A flash of the same intensity produced an increase in cyclase activity of about 5 fold (see Fig. 2) when Ca^{2+} concentration was further decreased to 10^{-8} M, but was ineffective when calcium concentration was raised to 10^{-4} M.

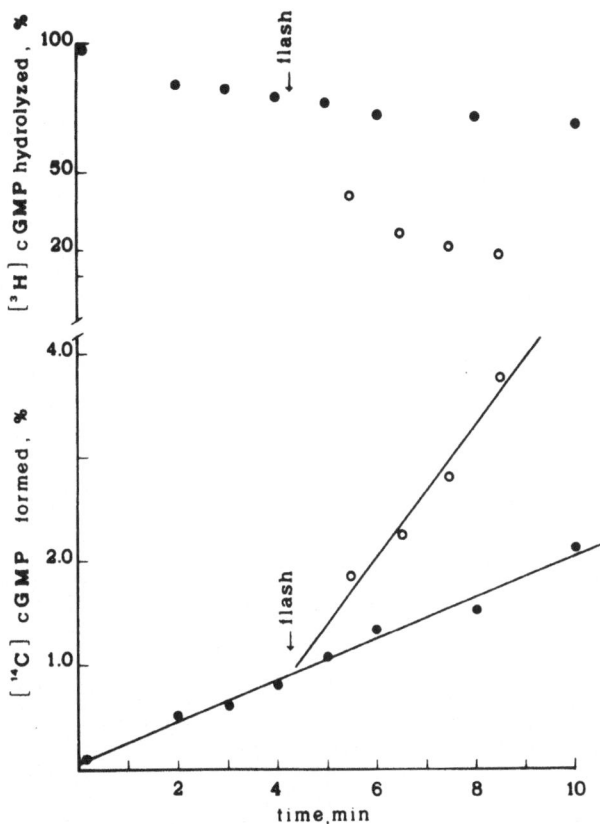

Fig. 1. Effect of light on guanylate cyclase at low calcium
concentration. ROS were incubated with the enzyme
assay (see methods) in the dark for about 10 min.
After 4 min. incubation, the sample was divided in
two: one remained in the dark (●) and the other was
illuminated (O) with a flash of light bleaching
8×10^{-4} % of rhodopsin. Ca ion concentration was kept
at about 10^{-7} M with EGTA (2 mM) and $CaCl_2$ (1 mM).

By contrast, the cyclase activity in the dark was not sensitive to
the change in calcium concentration as shown by the cGMP formation before
the flash in Fig. 2.

The stimulation of the cyclase by light might be due to the rapid
clearing away of its product by phosphodiesterase. However, the cyclase
activity, when measured in the presence of a range of concentrations of
cGMP wider than that occurring in the usual experiments, gave almost identical values, as shown in the Table.

DISCUSSION

We have reported (Pepe et al., 1986 a;b) that guanylate cyclase in
the rod outer segments of toads has a basal activity in the dark of about
3 ± 1 nmoles of cGMP formed/min/mg protein, a value that is largely inde-

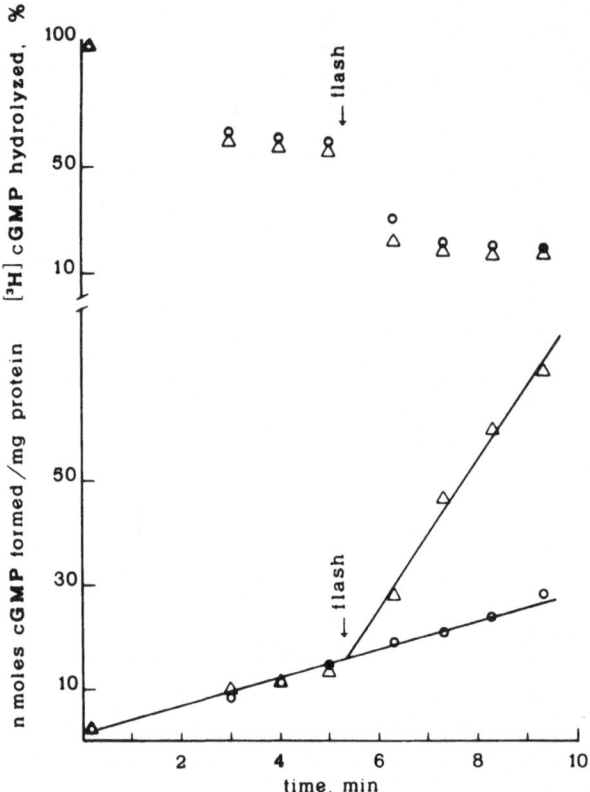

Fig. 2. Effect of light on guanylate cyclase at two different
Ca ion concentrations. Two samples incubated in the
dark at two different Ca ion concentrations: 10^{-4}M (O)
and 10^{-8}M ($\triangle$) were illuminated with a flash of light
bleaching 8×10^{-4}% of rhodopsin.

pendent of calcium ions concentration. Light enhances this activity upon
lowering Ca^{2+} from 10^{-4}M to 10^{-8}M. As shown in Fig. 2, after a flash of
light bleaching 8×10^{-4}% of rhodopsin, cyclase activity increased by 5-fold
at 10^{-8}M Ca^{2+}. Therefore the stimulation of cyclase activity by light seems
to be modulated by a variation in free Ca ions concentration. This would
be consistent with the physiological decrease of intracellular free Ca
concentration in the rod outer segments after a flash of light (McNaughton
and Cervetto, 1986). The drop of Ca inside the photoreceptor would result
in the activation of guanylate cyclase, thus restoring the cGMP concentration and the normal membrane potential with the reopening of cation-specific
channels. Cyclase activity would be eventually driven back to its minimal
value by the restored Ca influx through the opened channels.

The pathway through which light stimulates cyclase remains to be
investigated. The cyclase activation by light seems to decrease as the inhibition of phosphodiesterase is increased. For example, cyclase was activated
by 10-fold at 10^{-8}M Ca^{2+} by a flash of light bleaching 7×10^{-4}% of rhodopsin
in the presence of IBMX (isobutylmethylxantine) (Pepe et al., 1986 a;b),
whereas the cyclase activation by light was only 5-fold in the experiment

Table 1. Guanylate Cyclase Activity at Different
cGMP Concentrations.

cGMP concentration (mM)	cyclase activity	
20	3.9	4.1
10	4.5	2.7
4	3.7	4.2
0.2	4.1	3.7
0.1	4.3	4.6

Cyclase activity (nmol cGMP formed/min/mg protein)
was measured in duplicate samples incubated for 8
min in the dark at about 10^{-8} M Ca ion concentration.

of Fig. 2 where a stronger inhibition of phosphodiesterase was obtained in
the presence of Zaprinast. Moreover the activation of cyclase by light at
low calcium concentration decreases to zero upon increasing the concentration of Zaprinast up to a complete inhibition of phosphodiesterase (data not
shown), suggesting a direct coupling between phosphodiesterase and cyclase.
This possible interaction between the two antagonist enzymes would be more
complex than the mere activation of cyclase by the removal of its product
by light-activated phosphodiesterase.

REFERENCES

Asim Esen, 1978, A simple method for quantitative, semiquantitative and
        qualitative assay of protein, Analyt.Biochem., 89: 264.
Fleishman, D., 1982, Localization and assay of guanylate cyclase, in:
        "Methods in Enzymology", 81: 522, L. Packer ed., Academic Press,
        New York.
Lamb, T.D., 1986, Transduction in vertebrate photoreceptors: the roles of
        cyclic GMP and calcium, Trends Neurosci.,9: 224.
McNaughton, P.A., and Cervetto, L., 1986, The role of calcium in the light
        response, Photobiochem. Photobiophys., 13: 399.
Pannbacker, R.G., 1973, Control of guanylate cyclase activity in the rod
        outer segment, Science, 182: 1138.
Pepe, I.M., Panfoli, I., and Cugnoli, C., 1986a, Guanylate cyclase in rod
        outer segments of the toad retina, FEBS Lett.,203: 73.
Pepe, I.M., Boero, A., Vergani, L., Panfoli, I., and Cugnoli, C., 1986b,
        Effect of light and calcium on cyclic GMP synthesis in rod outer
        segments of toad retina, Biochim.Biophys.Acta, 889: 271.
Pugh Jr., E.N., and Cobbs, W.H., 1986, Visual transduction in vertebrate
        rods and cones, Vision Res., 26: 1613.
Stryer, L., 1986, Cyclic GMP cascade of vision, Ann.Rev.Neurosci.,9:224.

THE RESPONSE OF THE LIMULUS PHOTORECEPTOR TO THE ABSORPTION OF SINGLE PHOTONS[*]

Hennig Stieve

Institut für Biologie II
RWTH Aachen, Kopernikusstr. 16
D-5100 Aachen

The ventral photoreceptor of the horseshoe crab Limulus (Fig. 1)responds to very dim flashes of light with so-called "bumps", elementary excitatory events.

One bump is evoked by the absorption of one photon. Bumps evoked by identical stimuli vary greatly in size and delay (Fig. 2) .

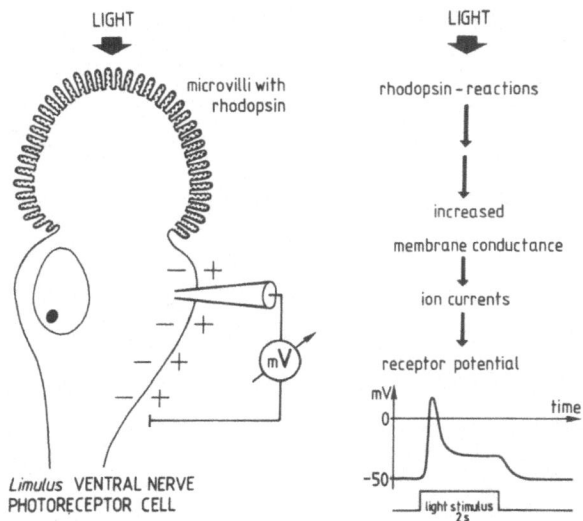

Fig. 1 Functional diagram of the ventral photoreceptor cell of Limulus. It has a diameter of 50 to 100 µm. The cell membrane of its distal lobe has glove-finger-like protrusions, the microvilli, which contain the visual pigment rhodopsin. Such a cell contains about 10^6 microvilli ca. 1 µm long and 100 nm in diameter; they contain altogether a total of about 10^9 rhodopsin molecules. (Stieve 1985)

---

[*]Dedicated to Ernst Florey, Konstanz, on the occasion of his 60. birthday anniversary.

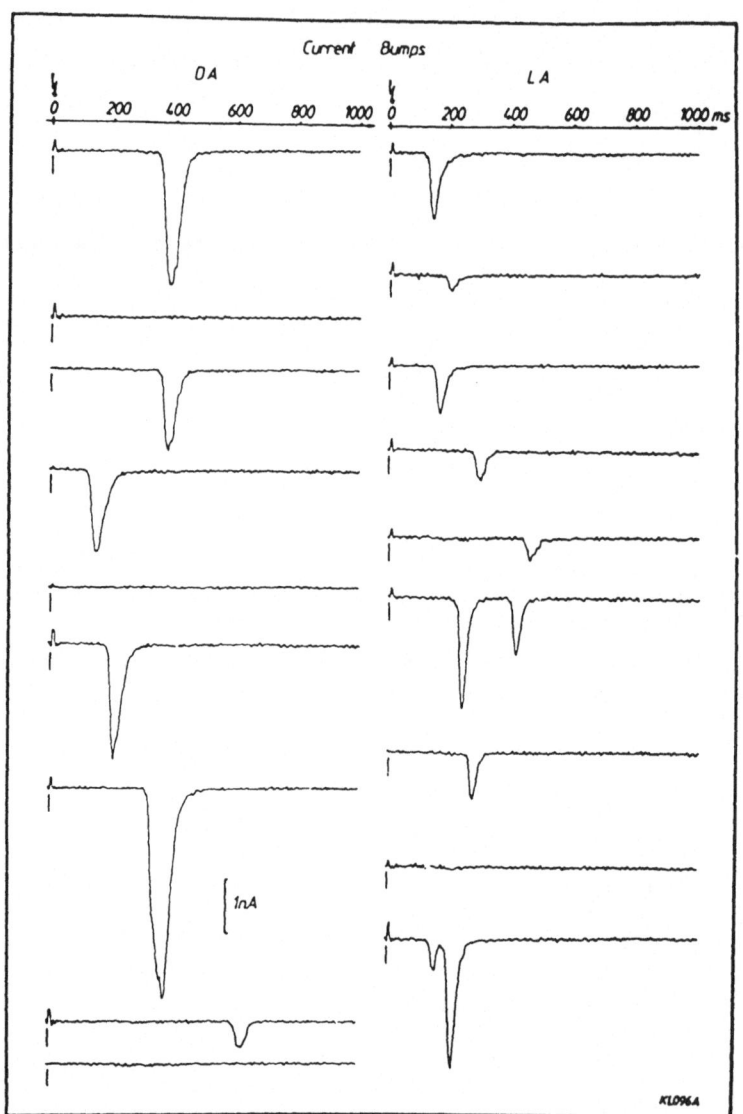

Fig.2 Current bumps measured from a Limulus ventral nerve photoreceptor cell under voltage clamp conditions. Left column: dark-adapted photoreceptor (DA); right column: photoreceptor weakly light-adapted (LA) by a conditioning light flash, 2 s prior to the bump-evoking flash. Bump-evoking flash: E_e ca. 6×10^7 photons cm^{-2}, duration 50 µs, repetition time 10 s; light-adapting flash: E_a ca. 2×10^9 photons cm^{-2}, duration 10 ms both flashes 450 nm, membrane potential constant -40 mV; 15° C. On the average bumps recorded in the dark-adapted state are larger and have a longer latency as compared to bumps which are recorded under weakly light-adapted conditions (Stieve 1985).

Bumps are assumed to be based on the concerted opening of up to 10^3-10^4 single light-activated sodium channels of 10-20 pS conductance. The general bump shape is shown in Fig. 3.

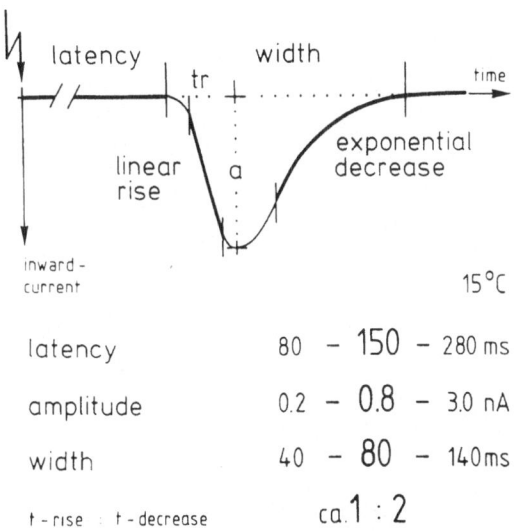

	15°C		
latency	80 –	150	– 280 ms
amplitude	0.2 –	0.8	– 3.0 nA
width	40 –	80	– 140 ms
t - rise : t - decrease		ca. 1 : 2	

Fig. 3 Time course of a current bump (schematically) with specification of the variation of some shape parameters. (Stieve 1986)

One can measure several bump parameters and determine that 4 so-called "primary" parameters are not correlated with each other: the latency, the slope of the bump rise, the rise time and the time constant of the exponential bump decay. In contrast to this, the size of the bump, determined either by the amplitude or the current-time-integral, is a "secondary" parameter, depending strongly on duration and slope of the bump rising phase, but is not correlated with the bump latency.

Latency: the rate-determining step for the duration of the latency is neither a straightforward diffusion nor a simple Michaelis-Menten process. The frequency distribution of bump latencies is such that a number of processes can be excluded. It demands for a special process such as cooperative enzyme activation. A bump-generating mechanism compatible with our present experimental knowledge about bumps is summarized in Fig. 4 and 5.

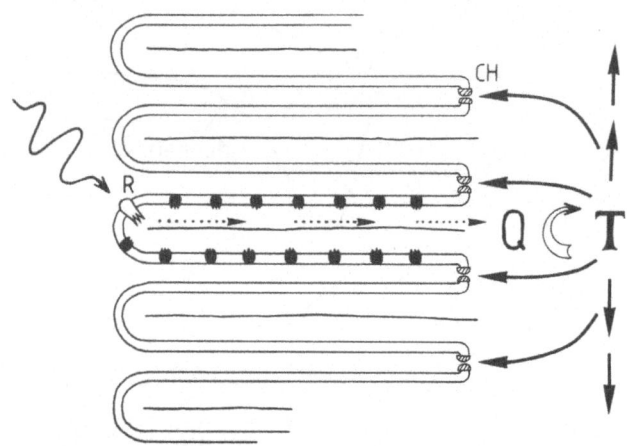

Fig. 4 Schematical diagram of proposed mechanism of bump
generation in Limulus ventral nerve photoreceptor. A light-
activated rhodopsin molecule R in a microvillus starts the
activation of an enzyme (cascade) which finally leads to the
activation of a transmitter source Q. This may be situated
at the basis of this microvillus. The transmitter source
there-upon produces, activates or releases many transmitter
molecules which are built from precursors. The transmitter T
diffuses along the bases of the microvilli and is bound by the
ion channels CH which plausibly may be situated close to the
bases of the microvilli. Following transmitter binding the
ion channels are transiently opened. Thus develops a more or
less circular "bump-speck" which in the dark adapted photoreceptor cell should have a diameter of about 4 μm. On the
average a bump-speck of a light-adapted photoreceptor cell
is smaller. The bump amplitude is proportional to the number
of simultaneously opened ion channels and should be thereby
more or less proportional to the area of the bump-speck.
(Stieve 1986)

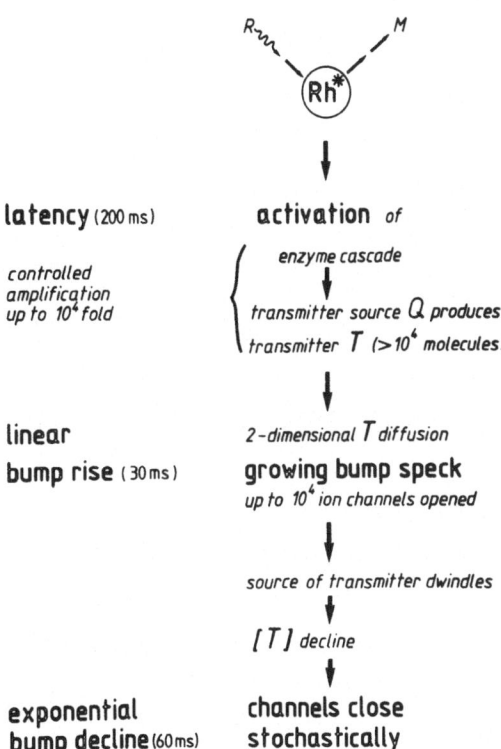

Fig. 5 Flow chart of the proposed mechanism of bump generation in the Limulus photoreceptor cell (schematically).R, M inactive forms of rhodopsin; Rh* activated state of rhodospin; [T] concentration of transmitter molecules. (Stieve 1986)

From bumps to macroscopic receptor current

With larger and larger stimulus intensity one can observe that bumps sum up to form the "macroscopic" receptor current. If one tries to synthesize a macroscopic receptor current by summation of bumps following identical, very weak light flashes, one obtains a membrane current signal which shows resemblance with the macroscopic receptor current signal after a brighter flash but has some characteristic differences. These differences are due to an automatic gain process, which in effect mainly reduces the size of the signal and shortens it; due to a feed-back-mechanism the signal "eats its own tail" (Stieve, Schnakenberg, Kuhn, Reuss 1986)

The time course $\underline{b}(t)$ of the bump sum (obtained by linear summation of many bump responses to dim flashes) is much slower than that of the directly observed macroscopic receptor current $J(t)$ evoked by a bright flash. The bump sum $\underline{b}(t)$ underlying the macroscopic response evoked by a bright flash

has to be modified (convoluted with an attenuation function
a(t)) to result in the macroscopic receptor current J(t). The
underlying bumps have to be attenuated the more the later
they follow the flash. By convoluting the bump sum b(t) with
the current-time-integral of the receptor current we can
simulate surprisingly naturalistic time courses for macroscopic receptor currents and their variation with stimulus
intensity and state of adaptation. Unexpectedly we also can
mimic the components C1 and C2 observed in the receptor
current. We propose that the intracellular transient increase
in calcium-ion-concentration (which roughly represents the
time-integral of the receptor membrane current) may be the
natural attenuation function a(t).

If one plots the intensity dependence of the membrane current
signal in a double logarithmic plot one obtains curves like
Fig. 6.

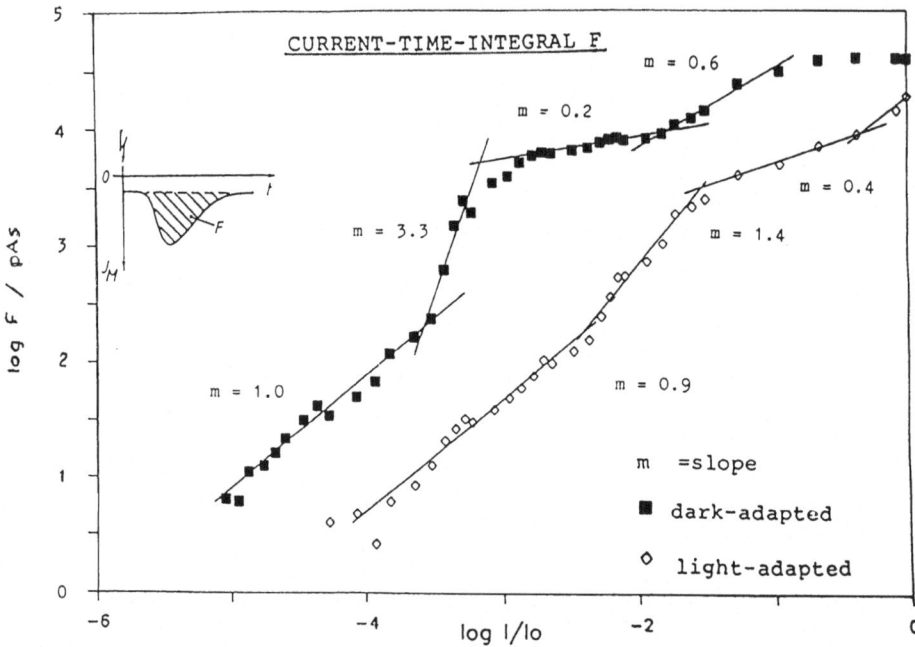

Fig. 6 The intensity dependence of the light-evoked receptor
current of the Limulus ventral nerve photoreceptor under
voltage clamp conditions at two defined states or adaptation:
moderate light adaptation (LA) and considerable dark adaptation (DA).
The light energy vs response-characteristics (current-time-
integral F) shows four regions with different slopes in the
double logarithmic plot:
a) a linear section (slope about 1)
b) a supralinear section (slope 2-4)
c) a sublinear section (slope <0.5)
d) a second sublinear section (slope >0.5)
The "heel" of the dark adapted curve, that is the point of
intersection of sections a) and b) corresponds to about 30
photons absorbed by the photoreceptor cell, the "knee" (point
of intersection of sections b) and c)) about 100 photons.
(Stieve and Schloesser, 1987).

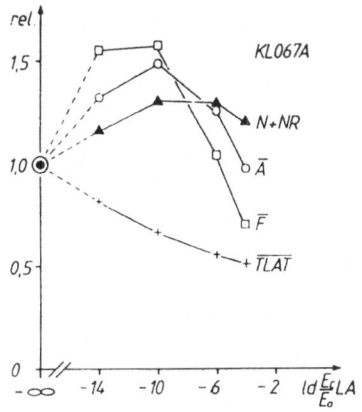

Bump Parameters in Dependence of
Strength of Light Adaptation

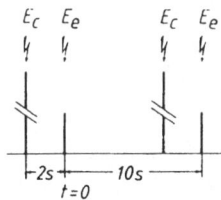

$E_o(10 ms), ca. 10^{11} Phot./cm^2$

$E_e(50 \mu s) const., ca. 1 \cdot 10^8 Phot./cm^2$

Fig. 7 Average bump parameters dependent on strength of light
adaptation demonstrating facilitation and adaptation. Ordinate:
value of bump parameter normalized with respect to the value
recorded in the dark-adapted state of Limulus ventral photoreceptor. Reference values: bump latency TLAT 178 ± 8 ms, bump
current-time-integral F 35.1 ± 3.3 pAs, bump amplitude
A 0.83 ± 0.05 nA, bump frequency N + NR 201. Abscissa ld (log
of base 2) of normalized energy of the conditioning flash;
540 nm light, 15°C; below, stimulus regime. N: number of
single and first bumps; NR: number of bumps riding on top of
preceding bumps. (Stieve 1986)

This is in accordance with our model. The average latency of
the bumps evoked by the constant test flashes, however, is
decreased (monotonically) with increasing conditioning preillumination, again demonstrating that the latency determining
processes are quite different from those determining bump
size and number of bumps. However, the energy of the conditioning test flash is in the linear section a) of the response vs stimulus dependence (Fig. 6). Additionally surprising is that not only the bump size is increased by this weak
preconditioning illumination but also the quantum efficiency
(i.e. the number of bumps evoked by the constant test flashes)
is increased. These two latter observations are not explained
by our model.

The dependence starts with a linear slope ($m_1=1$) which is followed by a supralinear slope ($m_2=3-4$) and again by two sublinear slopes ($m_3 <0,5$; $m_4 >0,5$). Slope 1 is expected if one assumes that the number of the bumps rises linearly with a stimulus energy. The supralinear slope indicates that bumps are not longer independent upon each other, but are interacting in some kind of positive cooperativity. The two sublinear slopes are due to response interactions like negative cooperativity and negative feed back (automatic gain control), in effect reducing the amplification.

A simple model to explain supralinearity is that in some process of the transduction chain more than one, probably four, ligands have to be bound to a molecule. This molecule could be, but need not be, the light-modulated ion channel. If for simplicity we assume a scheme with only two ligands, we come to the following scheme:

$$Ch + T \rightleftharpoons ChT_1$$
$$ChT_1 + T \rightleftharpoons ChT_2$$

After a weak illumination we find three types of channels. Many locked channels Ch, some unlocked channels ChT_1, and few open channels ChT_2. With increase in stimulus intensity the number of ChT_2 raises supralinearly in a certain intensity range. The slope of up to four could be explained by four ligands. Such a process seems reasonable: Cook, Hanke, Kaupp 1987 find in the case of the light-modulated sodium channel of the vertebrate rod photoreceptor that 4 cGMP molecules have to be bound until the channel is opened.

If such a mechanism as explained above is present in the Limulus photoreceptor the same phenomenon should also be demonstrable with sequential illumination.

We tested this hypothesis by administering a conditioning flash of variable light intensity I_c two seconds prior to a test flash of constant intensity I_t as in Fig. 2, left column.

In Fig. 7 it can be seen that the response to the constant test stimulus is larger after certain very weak conditioning stimulus intensities than with no or with stronger conditioning preillumination.

Acknowledgements

I thank H. Gaube, H.T.Hennig and I.Wicke for considerable help with the manuscript. The work was supported by the Deutsche Forschungsgemeinschaft, SFB 160.

References

1. Cook,N.J., Hanke,W., Kaupp,U.B. (1987)
   Identification, purification and functional reconstruction of the cyclic GMP-dependent channel from rod photoreceptors
   Proc. Natl. Acad.Sci. 84, 585-589

2. Stieve,H. (1985)
   Phototransduction in Invertebrate Visual Cells.
   The Present State of Research-Exemplified and Discussed
   Through the Limulus Photoreceptor Cell
   Neurobiology: 346-362

3. Stieve,H. (1986)
   Bumps, the Elementary Excitatory Responses of Invertebrates
   The Molecular Mechanism of Photoreception, Dahlem Konferenz
   1985, 199-230

4. Stieve,H.,Schnakenberg,J.,Kuhn,A., Reuss,H. (1986)
   An automatic gain control in the Limulus photoreceptor
   Fortschritte der Zoologie 33, 367-376

5. Stieve,H.,Schloesser,B. (1987) Abstract
   The intensity dependence of the Limulus photoreceptor
   current shows consecutive regions of linear, supralinear
   and sublinear slope
   in: "New Frontiers in Brain Research",
   Georg Thieme Verlag, Stuttgart,New York (1987)

ADVANCED TECHNIQUES IN PHOTOBEHAVIORAL STUDIES:

LIGHT SCATTERING TECHNIQUES

Donatella Petracchi

CNR – Istituto di Biofisica

Via S.Lorenzo, 26 – 56100 PISA (Italy)

INTRODUCTION

Light is a powerful tool in gathering information on cells and microorganisms. Spectroscopic studies have been widely used, and techniques such as time-resolved fluorimetry or flash photolysis make it possible to study the interaction between light and biological materials. It is also possible to study cell structures directly by using quantitative microscopy (microspectroscopy and microfluorimetry with elaboration and enhancement of the microscopic images). Moreover, light can be used to detect cell movement and cell behavioral responses, and the topic of the present paper will be this use of light.

The simplest use of light in this field has been nephelometry: by measuring the attenuation of a light beam crossing a sample of suspended cells, it is possible to evaluate the cell density. The use of more than one beam permits the measurement of the cell distribution inside the sample. This method has been widely used in studying photoresponses.

More recent experimental approaches based on the measurement of the light scattered by the sample allow us to measure motion parameters of single microorganisms and populations. There is a basic difference between methods which measure the Doppler shift of the scattered light and methods which measure modulations in light intensity. The former must use light with a long temporal coherence – laser light-, while the second can utilise a traditional light source.

When a monochromatic plane wave is scattered by a moving object its frequency is shifted by the Doppler effect; the Doppler shift is expressed by:

$$\delta f = (2V_{eff}/\lambda) \sin \Theta/2$$

where Θ is the scattering angle, λ is the wavelength and V_{eff} is the component of the scatterer velocity along the bisector of the angle α (Fig. 1). Now if the light which falls on the sample is a monochromatic plane wave (laser light), and if we collect part of the unperturbed beam, together with the scattered light, on the photodetector (Fig. 2), we can write the electric field on the photodetector as:

$$E(t) = E_0 \cos(2\pi f_0 t) + E_1 \cos(2\pi f_1 t)$$

where f_1 is equal to $f_0 + \delta f$ and δf is the Doppler shift. As the photodetector is a quadratic detector, the photocurrent will be:

$$i(t) = (E_0^2 + E_1^2)/2 + 2E_0 E_1 \cos(2\pi \, \delta f t)$$

Thus, the Doppler effect and the heterodyne detection supply a sinusoidal photocurrent whose frequency is related to the velocity of the scatterer, or, more precisely, to its effective component. When a sample of swimming microorganisms is placed on the path of the laser beam, heterodyne detection produces a photocurrent which is the sum of many independent sinusoidal contributions. The spectral analysis of the photocurrent makes it possible to measure the weight of different sinusoidal components, and, therefore, the distribution of a component of the velocity in the population.

There are some limitations when such a method is used on big non-spherical microorganisms. The physical theory holds, in fact, for scatterers smaller than the wavelength of the light, and isotropic scattering would be necessary to assure a strict correspondence between spectral distribution and velocity distribution. Nevertheless, the results obtained (Ascoli et al. 1978; Chen and Hallett,1982) show that in the spectra of the photocurrent the distribution of cell velocity in the sample can be separated from the other effects. It would be interesting to apply an improved heterodyne detection (Ascoli et al., 1980; Ascoli and Frediani, 1983), able to measure the sign of the velocity component, to a systematic study of phototaxis, i.e. of the oriented motion of microorganisms towards the actinic light source or away from it.

A second effect can be revealed through the spectral analysis of the light scattered from a sample of motile cells. This effect depends on the fluctuations of the number of cells which are inside the illuminated area (Schaefer and Berne, 1971). Each cell which enters (or leaves) the illuminated region produces a step increase (or decrease) in the scattered light. Each resulting transient in the photocurrent makes a contribution to the spectrum; this contribution is at a maximum at zero frequency. So the fluctuation in cell number produces a zero-centered band; the amplitude of this band depends on the mean number of the cells which cross the illuminated area and, therefore, on cell motility. If cells stop altogether, the zero-centered band decreases drastically; a decrease in the zero-centered band could be used to study stop reactions.

The number fluctuations can be revealed analyzing the signal in a more direct way (Pfau et al., 1983). The count of the cells which enter or leave the illuminated area is in fact possible at a low density of the sample; so a measure of the cell motility can be obtained.

Another effect which can be revealed measuring the light scattered from swimming microorganisms depends on the non

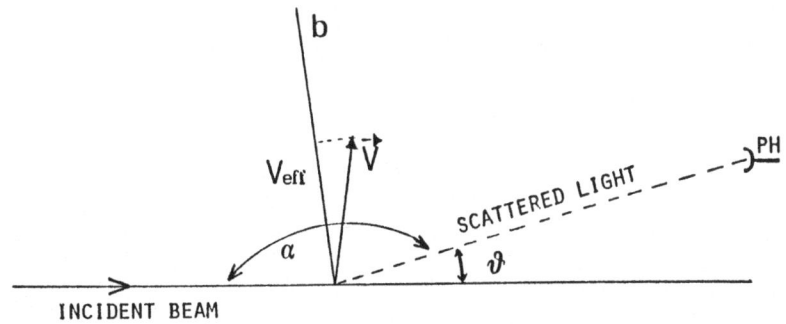

Fig. 1. The effective component of the velocity of the scatterer is along the bisector of the angle α.

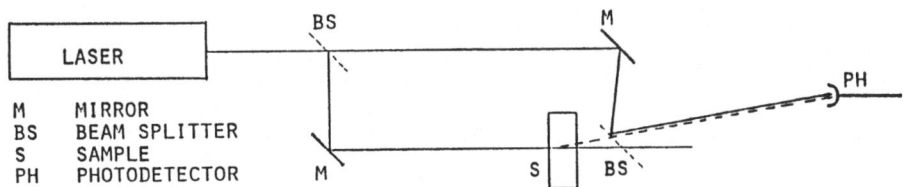

M   MIRROR
BS   BEAM SPLITTER
S   SAMPLE
PH   PHOTODETECTOR

Fig. 2. Hetherodyne Spectrometer.

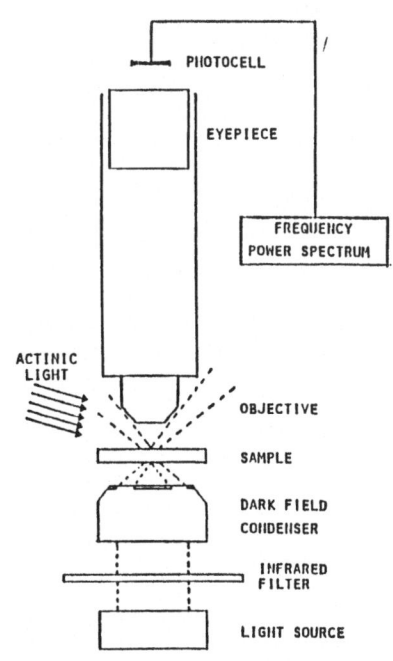

Fig. 3. Scheme of the optical apparatus to detect Dynamic Anisotropic Light Scattering.

isotropic scattering diagram of the microorganisms. The
light scattered from a cell on a photodetector depends on
the scattering angle and on the orientation of the cell body
with respect to the incident beam; then, as the cell rotates
and wiggles during its motion, the light scattered on the
photodetector turns out to be modulated in intensity.
Euglena gracilis is a typical example; it is an elongated
microorganism and its scattering diagram is far from
isotropic. In this case the modulations of the intensity of
the scattered light are very strong. To minimize them when
using Doppler velocimetry it has been useful to orient the
cells in the sample in a particularly suitable direction
(Ascoli et al. 1978). Almost all microorganisms scatter
light in an anisotropic way. Haematococcus pluvialis,
Chlamydomonas reinhardii, which is smaller, and even many
bacteria have anisotropic scattering diagrams, so that in
their motion the light scattered at a given angle changes
over time as the relative orientation between cell body,
photodetector and light beam changes. Now the orientation of
cells varies continually. Using a dark field microscope we
can observe slow modulations in light intensity, but variations due to flagellar beating are detectable too. These
light intensity modulations are simply due to the anisotropy
of the scattering; so they can be detected by using a common
incoherent light source. All that is necessary is to prevent
the incident beam falling on the photodetector. Fig. 3 shows
the very simple experimental set up used to measure modulations in the intensity of the scattered light (see also Angelini et al., 1986). The sample is placed on a dark field
microscope and the not actinic (infrared) component of the
microscope light falls on the sample and is scattered. The
light collected by the objective is transmitted to a
photocell placed at the output pupil of the eyepiece. So
there isn't any image of the microscopic field and all the
light scattered by the sample towards the objective falls on
the photodetector. With swimming microorganisms the
photocurrent so produced varies almost sinusoidally in time.
The signal detected in this way from a single Euglena
gracilis is shown in Fig. 4. Two different frequencies are
present in this signal: the higher (30-40 Hz) corresponds to
the flagellar beating and the lower (1-2 Hz) to the cell
body rotation. In other microorganisms the lower frequency
does not appear so clearly, but the flagellar beating
frequency and its distribution in the population are always
easy to measure. As for the heterodyne detection of Doppler
shift, the spectral analysis of the photocurrent must be
used to measure flagellar beating frequency distribution in
a sample.

STABILITY OF THE SPECTRA AND SIGNAL-TO-NOISE RATIO

In all these experimental approaches the signal produced
by the light on the photodetector is due to a summation of
many independent contributions and the spectral analysis of
the signal eventually yields the distribution of a motion
parameter. To handle such signals, it is necessary to allow
for the fact that a sum of many independent oscillators is
a random variable. The reason for this is that by summing
sinusoids with the same amplitude A with random phases the
resulting signal can have any value between zero and NA,
where N is the number of sinusoids summed up. Fig. 5 shows

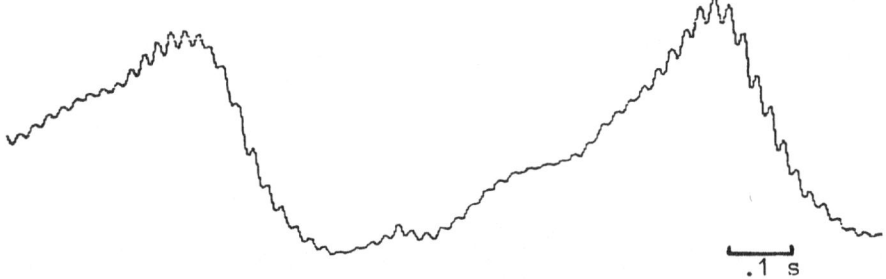

Fig. 4. Photocurrent due to light scattered from a single
swimming Euglena (redrawn from Angelini et al., 1986).

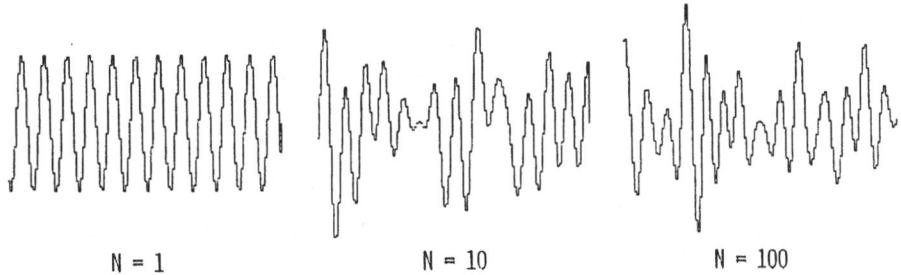

N = 1   N = 10   N = 100

Fig. 5. Sum of N sinusoids divided by $\sqrt{N}$. Frequencies (between
20 and 40 Hz) and phases are random.

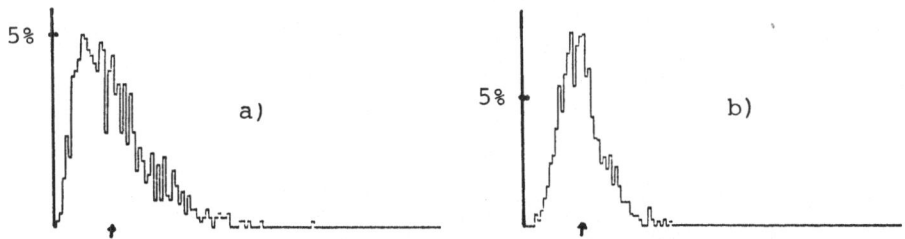

Fig. 6. Distribution of the power spectrum area. Thousand
signals, each obtained summing 10 sinusoids with random
phases and frequencies, were analysed. The frequency
distribution was gaussian around 30 Hz with a variation
coefficient .1 in a) and .3 in b). In abscissa the area
and in ordinate the percentage of signals with that
area are reported. The ratio σ/M of the area distribution results .59 in a) and .36 in b). Arrows = mean area.

that the root mean square of such a signal increases with $\sqrt{N}$ and Fig. 6 shows the distribution of the spectrum area. The variability in the spectrum area is lower for a large band signal than for a signal confined within a narrow band (compare Fig. 6a and 6b)

The aim of this discussion is to point out that the variability in a single determination of a spectrum does not depend only on biological variability; it is based on statistical factors. Thus to have stable spectral values we must average some (at least 10, but 30 would be better) independent and equivalent spectra.

Now I will determine the signal-to-noise ratio. Let us call $i(t)$ the photocurrent produced by the light scattered by a single microorganism towards the photodetector; we will have:

$$i(t) = \beta \; I_o(1+M \; cos 2\pi f t)$$

where β depends on the efficency of the photodetector, I_o is the mean light intensity scatterd by a single microorganism towards the photodetector while M (modulation depth) and f depend on the cell motion. When there are N microorganisms in the field we will have, for statistical resons,

$$i(t) = \beta \; I_o \; N + \beta \; M \; I_o \sqrt{N} \; cos 2\pi f t$$

and we are interested in the mean square of the modulated part in the photocurrent.

Any photocurrent is affected by shot noise, due to the quantum nature of light, and its mean square in our case it is expressed by:

$$i_n^2 = 2 \; \beta \; N \; e \; I_o \; \Delta f$$

where e is the electric charge of the electron and Δf is the bandwidth of the amplifier. Then the signal-to-noise ratio in the spectrum would be:

$$r = \beta^2 \; M^2 \; I_o^2 \; N/(4 \; \beta \; e \; I_o \Delta f \; N)$$

which increases with the intensity of the analysing light and appears to be independent of N. Neverthless when the noise of the amplifiers or the noise due to light scattered from non motile microorganisms is predominant the signal-to-noise ratio increases with N and it is better - to reveal a signal - working with many microorganisms in the microscopic field.

EXPERIMENTAL APPROACH

The question now arises: how can these scattering techniques be used in studying the photoresponses of microorganisms? We must give light stimuli to the sample and we must prevent the light stimuli being detected by the photocell used to collect the analyzing light; in its turn the light used to measure motility must be in a spectral range where microorganisms are not sensitive. One point to be remembered is that we must average a number of equivalent spectra, that is to say spectra with the same delay with respect to the delivery of the light stimulus. To do this several data acquisitions are stored in the computer, and

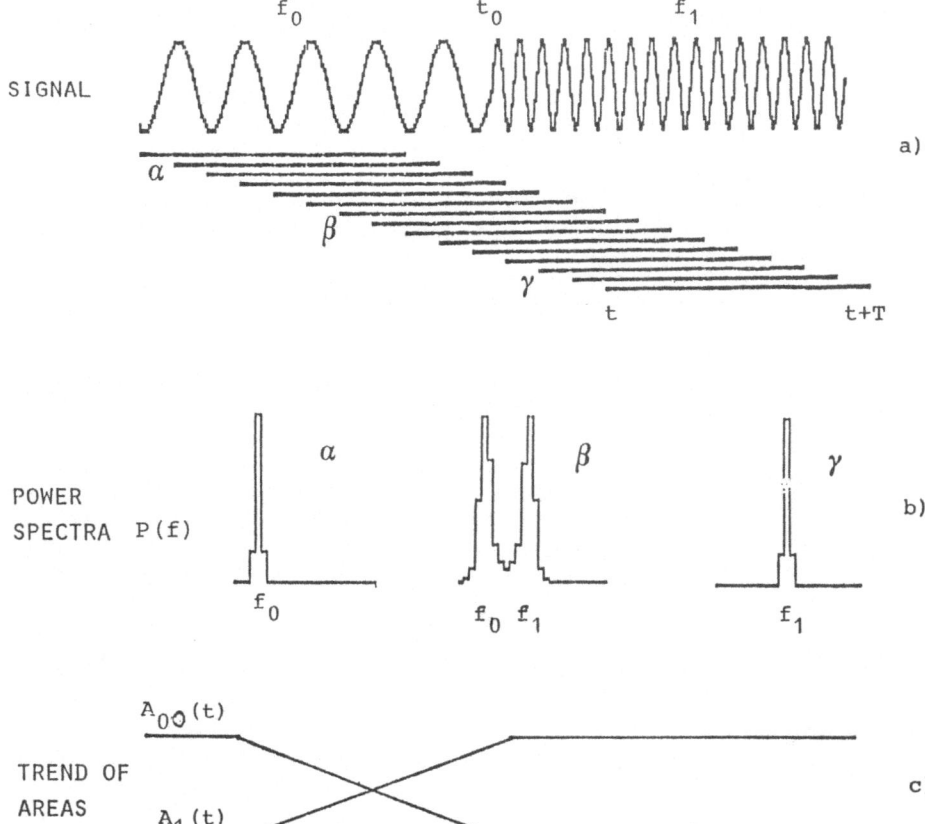

SIGNAL

f_0 t_0 f_1

a)

α

β

γ

t $t+T$

POWER
SPECTRA $P(f)$

α β γ

b)

f_0 f_0 f_1 f_1

TREND OF
AREAS

$A_{00}(t)$

$A_1(t)$

T

c)

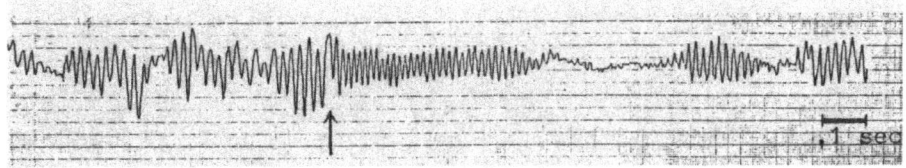

.1 sec

the spectra at same distance from the light stimulus are averaged. What is interesting in studying photoresponses is to detect, after a light stimulus, a change in a cell parameter and to follow its evolution in time with a sufficient temporal resolution (for instance, how long do cells take to respond to light, and how long do cells take to adapt to the light level ?). At first sight spectrum analysis may not appear to be a good tool for studying the time evolution of a motion parameter. In fact, to do a spectrum with a given frequency resolution δf, we must analyze a time span of the signal with duration $T = 1/\delta f$. In other words, the time equal to $1/\delta f$ is the time necessary to determine a spectrum with frequency resolution δf. So it seems that spectrum analysis is useful only in studying slowly varying processes, and without computers it would actually be very difficult to use spectral analysis to measure something in a time shorter than the time span T. The diagrams in Fig. 7 show what to do to measure the time evolution of a spectrum with a resolution independent of its frequency resolution.

Let us consider a signal which at a time t_0 changes its frequency from f_0 to f_1. Altough this is just an example, it is what happens in some photoresponses. By computing many successive spectra it is possible to detect when the signal has changed its frequency. What is necessary is a correct reading of the results: an instantaneous change in the signal frequeny yields a change in the area of the spectrum which rises to its maximum in the time span T (Fig.7). Changes in the signal amplitude can be followed with the same technique. This way of handling the signal can be achieved by the storage in a computer memory of many succesive non independent spectra which are averaged with the corresponding ones acquired in later runs.

It would be interesting to use the heterodyne detection of Doppler shift to measure the velocity distribution in a phototactic sample and to monitor how this distribution changes when the actinic light is turned on in order to measure how much time elapses before cells start tracking the light direction. As phototaxis destroys the random distribution of cells inside the sample the experiment would have to be planned in such a way to conserve the random distribution of cells.

AN EXPERIMENTAL CASE

To give an example, I will now illustrate the changes in flagellar beating frequrency that occur after the stimulation of Haematococcus pluvialis with actinic light. The typical sequence is that after a flash - if it is strong enough - Haematococcus stops moving, then moves backwards, and, after a time, moves forwards again in a new direction. In a sample with a very low cell density photocurrent due to a single microorganism has been revealed by dark field detection (Fig.8). Typically, the flagellar beating frequency increases stepwise after a light stimulus; this occurs every time a cell stops, moves backwards and starts moving forward again.

Measurements made using samples with higher densities make it possible to employ the spectrum analysis of the photocurrent to measure the distribution of flagellar beating frequency in the sample. In Chlamydomonas it is possible to observe the same behavioral response and measure a change in flagellar beating frequency, too. We have sometimes observed in <u>Chlamydomonas</u> stopping without subsequently moving backwards, and in this case the flagellar beating frequency does not increase.

Fig. 9b shows the shift in flagellar beating frequencies that occurs in <u>Haematococcus pluvialis</u>; here all the cells, as we observed on the video, stopped and reversed their motion before assuming a new forward direction. In Fig. 9a, on the other hand not all of the population stopped and we see in the spectrum that not all of the cells beat at higher frequency.

Altough the effect is all-or-nothing for the single microorganism, the response of a population increases gradually: if the intensity of light stimulus is too low there is'nt any response; increasing the intensity and/or the duration of the light flash some cells stop and for further increses all cells in the sample respond to the stimulus (Angelini et al. 1986).

It is reasonable to assume that the percentage of cells, R, which respond to light can be expressed by:

$$R = A_1/(A_0+A_1)$$

where A_0 is the area of the unperturbed peak in the spectrum and A_1 is the area of the new peak. By measuring R as function of flash intensity or duration we can draw the dose-effect curves.

It has been shown (Angelini et al. 1986) that R depends on the light dose (light dose = flash intensity X flash duration). Neverthless the reciprocity between light intensity and flash duration holds only for flash durations not exceeding 60-80 msec, which suggests an accumulation process with a time constant of about 20-30 msec. This is the time constant of the slower process in the sensory pathway.

The important point in these results is that the amount of the increase in the flagellar beating frequency is independent of the stimulus parameters. An increase can only occur fully, or not at all, as happens with an all-or-nothing effect. As its occurrence depends on the light dose, we assume that the change occurs when the light absorbed by the photoreceptor exceeds a given threshold. The typical graded rise of the dose-effect curve indicates that the threshold for the response is different in different cells.

These results have been the basis for experiments concerning the pigments involved in photoreception in these microorganisms. <u>Haematococcus pluvialis</u> is a phototactic microorganism, very similar to <u>Chlamidomonas reinhardii</u>; it has been assumed (Feinleib, 1980) that these microorganisms have a directional antenna which allows them to track the light direction. The idea is that the photoreceptor is placed strictly close to the stigma and during the motion the stigma shades the photoreceptor giving information on the light direction. More recently Foster and Smyth (1980) suggested that interference effects inside the stigma are responsible for light modulation on the photoreceptor. Now, as we have just said, in our experiments the threshold for response does not assume the same value in different cells. Apparently the threshold for frequency change is different in different cells and/or the cells may be oriented in a more or less favourable way with respect to the light source. This last hypothesis means that the variability of

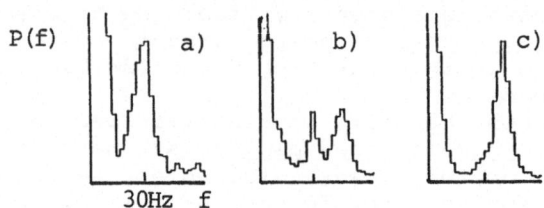

Fig. 9. Distribution of flagellar beating frequency in a population of <u>Haematococcus pluvialis</u>; a) before the flash,
b) and c) after flashes of different intensity.

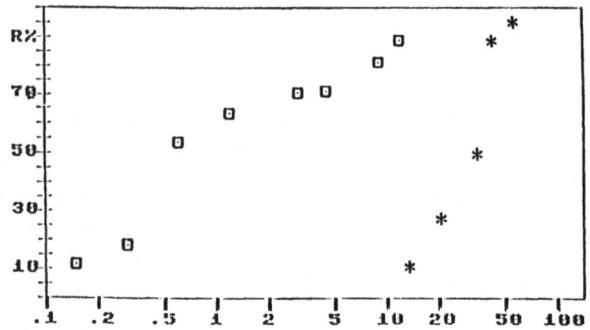

Fig. 10. Dose-effect curves at 450±25 nm ▢ and 600±25 nm * measured in the same sample. In ordinate R, defined in the
text, in abscissa fluence rate in arbitrary units.

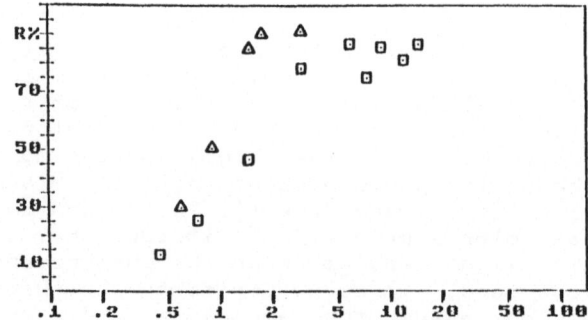

Fig. 11. Dose-effect curves at 450±25 nm obtained with only one
flash lamp ▢ and with two identical and opposite flash
lamps Δ.

the threshold between different cells depends on geometrical factors. More precisely,the light which falls on the photoreceptor of the given cell depends on the relative orientation between photoreceptor, stigma and light source. In this case we could obtain some information about the pigment involved in photoresponses from the shape of the dose-effect curves.

By measuring R for different light intensities and for different wavelengths of the stimulating light, we have drawn the dose-effect curve for different wavelengths. The first observation is that the light intensity at which the effect begins to appear and the shape of these curves both depend on the wavelength. So the individual variability of the cells is not the predominant effect.

In Fig. 10 dose-effect curves at 450 and 600 nm are shown; they were measured in a ten days sample and the slopes are very different. The slope at 450 nm is always lower than at the other wavelengths, but the difference depends on the age of the culture. In a two days sample slopes at different wavelengths do not differ very much.

Fig. 11 shows a different comparison: the two curves are both at the same wavelength (450 nm+25 nm), but they have been obtained by using a different form of illumination: in the first curve the light source was a lamp, in the other curve two identical lamps facing each other were used to stimulate the sample. We can suppose that the slope of the curve obtained when two such lamps are used depends mainly on cell variability. In fact, by simulating the experiments on the computer and using the more strongly directional antenna which has been proposed (Foster and Smyth, 1980), we see that a two-lamps curve is not very different from the curve of a receptor which is not shaded at all

Thus the different slope in Fig. 11 should allow us to extrapolate the value of the optical density of the stigma; further analyses are necessary to do this. A difference in slope can be detected also with the filter centered at 500 nm. At 600 nm there is no difference between curves obtained with one or two lamps and at 550 nm the difference is slight. A point of interest is that at this wavelength Haematococcus pluvialis is able to do phototaxis (Litvin et al., 1978), what means that a small modulation in the driving signal can produce phototaxis.

REFERENCES

Angelini, F., Ascoli, A., Frediani, C. and Petracchi, D. 1986. Transient photoresponses of a phototactic microorganism, Haematococcus pluvialis, revealed by light scattering. Biophys. J. 50:929

Ascoli, C., Barbi, M., Frediani, C. and Mure', A. 1978 Measurements of Euglena motion parameters by laser light scattering. Biophys. J. 211:585

Ascoli, C., Barbi, M., Frediani, C., Petracchi, D. and Ristori, T. 1980. Quasi-elastic light scattering for studying the motion of flagellated microorganisms. Optica Acta 271:1203

Ascoli, C. and Frediani, C. 1980. Quasi-elastic light scattering in the measurement of the motion of flagellated algae. in: "Light Scattering in Liquids and Macromolecular Solutions" V. Degiorgio, M. Corti and M. Giglio eds., Plenum Publishing Corp., London.

Chen, S.H. and Hallett, F.R. 1982. Determination of motile behavior of prokaryotic and eukaryotic cells by quasi-elastic light scattering. Q. Rev. Biophys. 15:131

Feinleib, M.E. 1980. Photomotile responses in flagellates. in: "Photoreception and Sensory Transduction" F. Lenci and G. Colombetti eds., Plenum Publishing Corp., London.

Foster, K.W. and Smyth, R.D. 1980. Light antennas in phototactic algae. Microbiol. Rev. 441:572

Litvin, F.F., Sineshchekov, U.A. and Sineshchekov V.A. 1978. Photoreceptor electric potential in the phototaxis of the alga Haematococcus pluvialis. Nature (Lond.) 271:478

Pfau, J., Nultsch, W. and Ruffer, U. 1983. A fully automated and computerized system for simultaneous measurements of motility and phototaxis in Chlamydomonas. Arch. of Microbiol. 135:259

Schaefer, D.W. and Berne, B.J. 1975. Number fluctuations spectroscopy of motile microorganisms. Biophys. J. 15:785

ADVANCED TECHNIQUES IN PHOTOBEHAVIORAL STUDIES:

COMPUTER-AIDED STUDIES

Donat-P. Häder

Fachbereich Biologie - Botanik
Karl-von-Frisch Str.
D-3550 Marburg
Fed. Rep. Germany

INTRODUCTION

Photomovement of motile microorganisms can be studied and quantified using two conceptually different approaches. Firstly, the behavior of single cells in response to the stimulus of interest can be studied. While yielding a precise and descriptive information of the behavior this technique has the disadvantage of being cumbersome and time consuming especially when high numbers of responses need to be observed in order to allow for statistical treatment of the data (Nultsch and Häder, 1987).

The alternative strategy is to integrate over the responses of many individuals by observing the net movement of a whole population (Nultsch, 1975). Though being advantageous with respect to the number of integrated responses the disadvantage of this technique is that changes in motility, competing photoresponses or reactions to other stimuli present might alter the result in an unpredictable manner.

Population techniques were the first to be automated: The change in absorbance due to the cumulative movement of the population has been recorded during the exposure of the cells in a cuvette to e.g. lateral light (Feinleib and Curry, 1967; Nultsch and Throm, 1975).

In recent years a number of technical innovations has made it possible to automate measurements and quantify photomovement of individual organisms which formerly required manual or semi-automatic assessment (Shipton, 1979; Skarnulis, 1982). Digital processing of video images is a tool for the automated tracking of microorganisms (Gualtieri et al., 1985). Due to the enormous amount of calculations, image analysis used to be restricted to the processing power of large mainframe computers and even then processing was limited to static images (Pavlidis, 1982).

The increasing power and processing speed of modern micro-computers in combination with the recent development of high resolution video digitizers capable of operating in the ms range made it possible to finally attempt image analysis of moving objects in real time and apply this technique to the tracking of motile microorganisms (Häder and Lebert, 1985; Dusenbery, 1985).

STRATEGY OF IMAGE ANALYSIS

Essentially, image processing requires digitization of an image from any video source in a matrix of points (pixels) organized in a grid of rows and columns. Each pixel is defined by a digital value which represents one of the possible gray levels the system is capable to discriminate. In the simplest case each pixel may be either zero (dark) or one (white); in more sophisticated digitizers 64 or 256 different gray levels may be defined. The host processor has access to the electronic memory in which the image is stored in digital form and the program is designed to locate and analyze objects of interest within the image. A number of mathematical treatments allows manipulation of the image before analysis (Dawson, 1987; Asmus,1987). E.g., Laplace filtering can be used to enhance edges of objects while uniform areas are reduced. For this purpose each pixel is calculated by multiplying its original value with 8 and substracting the digitized gray levels of its eight direct neighbors. By using different kernels, different manipulations of the image can be introduced. Smoothing is achieved by substituting each pixel by averaging over its original value and that of its eight neigbors.

HARDWARE

The system described in this article uses the video image recorded by a CCD camera (Philips, 0600/00) which has several advantages over conventional video cameras including a broader range of allowed fluence rates, higher sensitivity and a high spatial resolution of 512 x 512 pixels (Fig. 1). The camera is mounted on top of a conventional microscope (Olympus, BH2). The organisms are viewed in dark field technique in order to increase the image contrast which is especially necessary for non-colored, non-photosynthetic organisms, such as ciliates or bacteria. Other microscopical techniques may be used as well. The built-in light source produces the monitoring beam in combination with an infrared passing filter (RG 780 nm, Schott and Gen., Mainz, FRG) in order to prevent the organisms from being disturbed by the monitoring radiation.

The video image is digitized in a Matrox (PIP 512) single-board digitizer which plugs into the slot of a standard IBM PC/XT/AT or compatible. The microocomputer is equipped with 640 KB RAM (read and write memory), a color graphic card (CGA), two 5 ¼" floppy disks plus two 5 ¼" hard disk drives with 10 and 20 MB storage capacity, respectively. The digitizer with a spatial resolution of 512 x 512 pixels and 256 possible gray levels allows to manipulate gain and offset of the video image; thus brightness and contrast can be influenced by the operator. After digitization each pixel value is manipulated according to one of eight selectable input LUTs (look up table) before storing in the electronic memory. In this process each value is mapped to a new one; e.g. each value below a predefined threshold can be mapped to zero (dark) and each value above the threshold to 255 (white). Likewise a certain range of gray values could be spread out to allow a better distinction of similar gray levels.

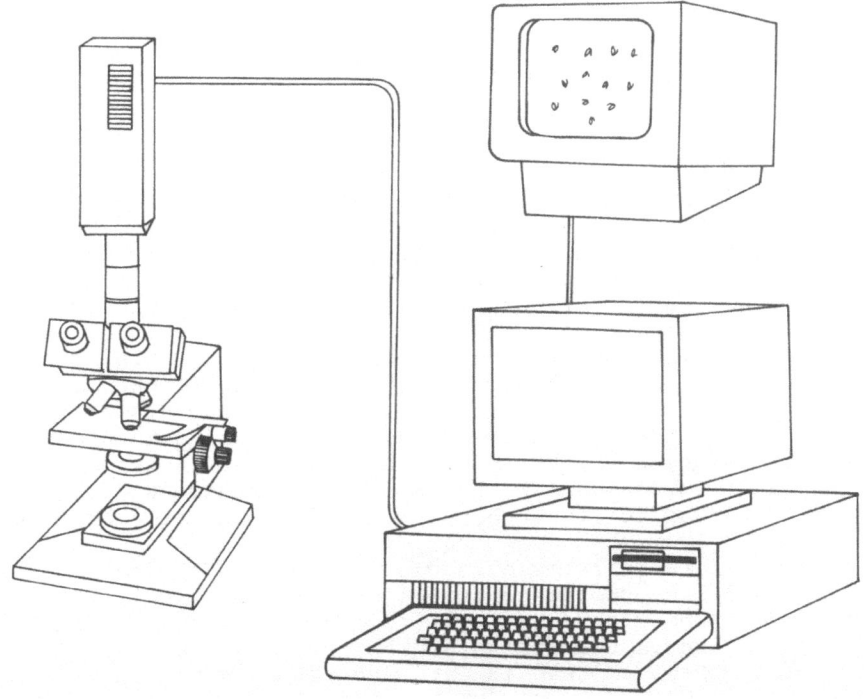

Fig. 1. Hardware system to track motile microorganisms in real time.

The electronic image stored in memory is displayed on a high resolution color monitor (NEC Multisync) after mapping each pixel value in one of eight sets of three output LUTs each, which allows e.g. to map certain gray levels to specific colors. This false color technique facilitates visualization of the image since the organisms can be mapped in one color on a different color background. In addition, overlays such as geometric masks or lettering can be produced electronically as an overlay before video output.

The host computer has access to the memory and can both read and write indvidual pixels on a random basis. Thus, it can manipulate the image as well as analyze it and extract the relevant information concerning size and position of the organisms of interest.

SOFTWARE

The analysis program is written in 8088/8086 assembly language since any of the higher programming languages are far too slow to provide the necessary high processing speed. The program is menu-driven and displays a number of options on screen. The operator may define a disk file in which the data will be stored; in addition, a headline allows to record the relevant data of the specific experiment. Then video gain and offset as well as the threshold are selected. Since the program runs cyclically each option can be chosen at any time. The search algorithm

387

starts at a random position on screen and scans either horizontally or vertically until it detects an object distinguished from the background by the predefined threshold. Using standard procedures the outline of the object is calculated (Pavlidis, 1982) and indicated by a prominent line on screen in a different color. Next the center of gravity (centroid) is calculated and stored in memory.

Now the next video image is called and digitized and the program tries to locate the new position of the same organism starting from the original centroid coordinates. This process is repeated a predefined number of times. The initial and final centroid coordinates define a trajectory vector of the moving organism which can be calculated in terms of an angle by which the organism deviates from the stimulus direction, e.g. the light direction. Several safety tests determine if the object has the correct size in order to distinguish organisms from debris and to exclude organisms which collide during movement. In addition, the speed of the object is calculated in order to determine if it corresponds with the swimming speed to exclude a passive drift and to distinguish motile from immotile organisms.

The raw data in terms of deviation angles are stored in the disk file and the program continues calculation until it reaches a predefined number of organisms (typically 1000) or until the operator terminates the analysis. The raw data are used by a number of programs written in Turbo Pascal (Borland Inc., Scotts Valley, CA) or CBASIC (Compiler Systems, Inc.) which allow mathematical and statistical treatment of the data. First of all, circular histograms are calculated and plotted as hard copy by binning the measured angles in 64 sectors and normalizing them (Häder and Lebert, 1985). Next the histograms can be subjected to a Fast Fourier Analysis in order to extract the key features such as number of orientation directions (modality) and preferred direction of movement. In addition, a Fast Fourier Synthesis can be applied to smooth the data and reject noise in the histograms (Häder and Lipson, 1986). The degree or orientation, the directedness, can be quantified using the Raleigh test (Batschelet, 1981; Mardia, 1972) or related tests for circular statistics (Häder et al., 1981). This hardware and software system has successfully been used to study photoorientation in various flagellates including *Euglena* (Häder, 1986; Häder et al., 1986), *Astasia* (Mikolajczyk et al., 1985), *Cyanophora* (Häder, 1985), *Cryptomonas* (Häder et al., in press) and *Gyrodinium*.

PHOTOTACTIC ORIENTATION OF *EUGLENA GRACILIS*

The phototactic orientation of the green unicellular flagellate, *Euglena gracilis*, has been quantified using the techniques described above in a horizontal cuvette in a lateral beam (Häder, 1987). At low fluence rates below 1.5 W m^{-2} a weak positive phototaxis was found (Fig. 2a), while at higher fluence rates a more pronounced negative phototaxis was detected (Fig. 2b). The directedness and the sense of orientation strongly depends on the culture age (Häder et al., 1987).

GRAVITAXIS IN *EUGLENA GRACILIS*

In addition to a weak positive and a strong negative phototaxis, the flagellate *Euglena* shows a pronounced negative gravi-

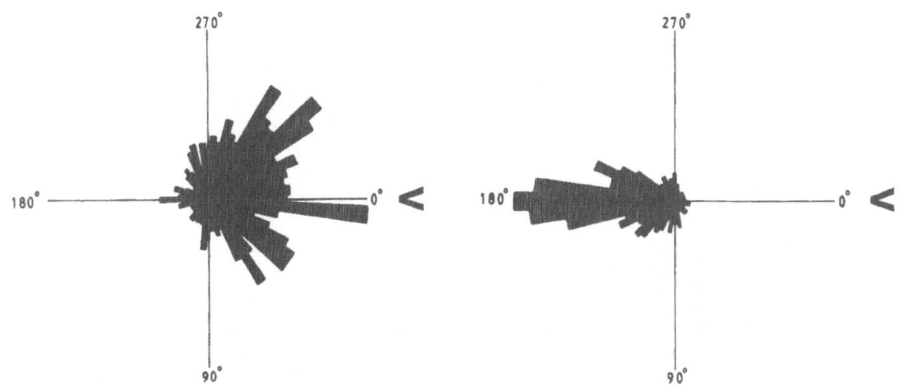

Fig. 2. Circular histograms of Euglena gracilis showing positive
phototaxis at low fluence rates (left) and negative at
higher (right).

taxis. (It should be mentioned, however, that the mechanism of
gravitactic orientation may be strictly passive and due to an
asymmetric center of gravity in the cell rather than an active
orientation with respect to the gravitational field of the
earth). In the absence of a photic stimulus, the cells move upward in a vertical cuvette with a high degree of orientation.
The ecological consequence is, that the organisms in their natural habitat use negative gravitaxis (plus positive phototaxis)
to move upward until this movement is counterbalanced by the
antagonistic effect of the negative phototaxis which commences
when the organisms experience too bright a light intensity near
the surface. The result is an accumulation of organisms in a ho-
rizon with a moderate fluence rate. This layer, however, is
found at an about 20 times higher fluence rate than indicated by
the compensation point between positive and negative phototaxis
as measured in a horizontal cuvette due to the strong component
of negative gravitaxis (Häder, 1987).

MECHANISM OF PHOTOTACTIC ORIENTATION IN *EUGLENA GRACILIS*

The original hypothesis was based on the periodic shading
of the photoreceptor (the paraflagellar body, a swelling at the
basis of the extending flagellum) by the stigma (Mast, 1911;
Jennings, 1904). According to this hypothesis, phototactic ori-
enta-tion is the result of repetitive photophobic responses
(Diehn, 1969). However, this hypothetic mechanism has been re-
jected recently for phototaxis in *Euglena* since e.g. contrary to
the expectation, stigmaless cells show an obvious phototactic
orientation. In addition, when exposed to two equal light sources oriented perpendicular to each other, the organisms should
be expected to move on the resultant; however, the population
was found to split into two components moving to either light
source (Häder et al., 1986) which is not compatible with the
shading hypothesis.

Furthermore, Doughty and Diehn (1982; 1983; 1984) have demonstrated using a number of inhibitors, ionophores and antago-
nists that the sensory transduction chain of photophobic responses involves a ouabain-sensitive Na^+/K^+ ion exchange pump and

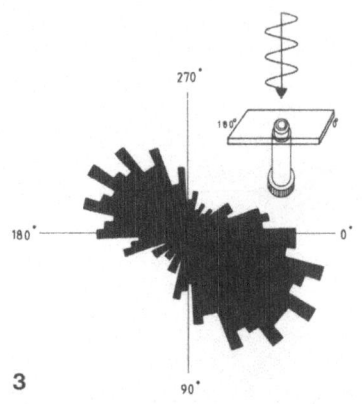

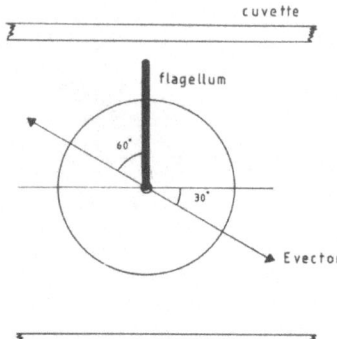

Fig. 3. (left) In a vertical, linearly polarized beam the cells
        orient 30° clockwise of the plane of the electrical
        vector.

Fig. 4. (right) Seen from the front end the absorbing vector of
        the dichroically oriented photoreceptor pigments deviate
        60° counterclockwise from the flagellar plane.

gated Ca^{2+} channels. If phobic responses were the basis for phototactic orientation these elements should be found also in the
sensory transduction chain of phototaxis. However, none of the
drugs had any appreciable effect on the phototactic orientation
even at unspecifically high concentrations (Häder et al., 1986).

In contrast, the cells detect the direction of the light
source by a dichroic orientation of the photoreceptor pigments
(Häder, 1987). In a horizontal cuvette, the cells orient at an
angle of about 30° clockwise with respect to the electric dipole
transition moment when irradiated with linearly polarized light
from above (Fig. 3). Using polarized beams of light in horizontally and vertically oriented cuvettes the three-dimensional
orientation of the absorbing vectors of the photoreceptor pigments could be analyzed: Seen from the front end of the cell,
the absorbing vector is oriented 60° counterclockwise from the
flagellar plane (Fig. 4) and 25° clockwise of the long axis of
the cell (Häder, 1987).

CONCLUSION

The hardware and software system described here allows to
measure and quantify oriented movement of individual organisms
in a population with respect to external stimuli fully automatically, objectively and in real time. Since the system processes
about 1000 angular values in about 15 min, a large number of
tracks can be analyzed in a short time which facilitates statistical analysis. The data obtained in the green flagellate
Euglena contradict the classical shading hypothesis which needs
to be replaced by a model based on the dichroic orientation of
the photoreceptor pigments.

REFERENCES

Asmus, J. F., 1987, Digital image processing in art conserva-
    tion, *Byte* 12, No. 3:151.

Batschelet, E., 1981, "Circular statistics in biology," Acad. Press, London.

Dawson, B. M., 1987, Introduction to image processing algorithms, *Byte* 12, No.3:169.

Diehn, B., 1969, Phototactic response of *Euglena* to single and repetitive pulses of actinic light, *Exp. Cell. Res.* 56:375.

Doughty, M.J., and Diehn, B., 1982, Photosensory transduction in the flagellated alga, *Euglena gracilis*. III. Induction of Ca^{2+}-dependent responses by monovalent cation ionophores, *Biochim. Biophys. Acta* 682:32.

Doughty, M.J., and Diehn, B., 1983, Photosensory transduction in the flagellated alga, *Euglena gracilis*. IV. Long term effects of ions and pH on the expression of step-dow photobehaviour, *Arch. Microbiol.* 134:204.

Doughty, M.J., and Diehn, B., 1984, Anion sensitivity of motility and step-down photophobic responses of *Euglena gracilis. Arch. Microbiol.* 138:329.

Dusenbery, D. B., 1985, Using a microcomputer and videocamera to simultaneously track 25 animals, *Comput. Biol. Med.* 15:169.

Finney, J. L., 1982, Monte Carlo techniques, *Biochem. Soc. Trans.* 10:305.

Feinleib, M. E., and Curry, G. M., 1967, Methods for measuring phototaxis of cell populations and individual cells, *Physiol. Plant.* 20:1083.

Gualtieri, P., Colombetti, G., and Lenci, F., 1985, Automatic analysis of the motion of microorganisms, *J. Microscopy* 139:57.

Häder, D.-P., 1985, Photomovement in *Cyanophora paradoxa*, *Arch. Microbiol.* 143:100.

Häder, D. P., 1986, Effects of solar and artificial UV irradiation on motility and phototaxis in the flagellate, *Euglena gracilis*, *Photochem. Photobiol.* 44:651.

Häder, D. P., 1987, Polarotaxis, gravitaxis and vertical phototaxis in the green flagellate, *Euglena gracilis*, *Arch. Microbiol.* 147:179.

Häder, D.-P., Colombetti, G., Lenci, F., and Quaglia, M., 1981, Phototaxis in the flagellates, *Euglena gracilis* and *Ochromonas danica*, *Arch. Microbiol.*, 130:78.

Häder, D.-P., and Lebert, M., 1985, Real time computer-controlled tracking of motile microorganisms. *Photochem. Photobiol.* 42:509.

Häder, D.-P., Lebert, M. and DiLena, M. R., 1986, New evidence for the mechanism of phototactic orientation of *Euglena gracilis, Curr. Microbiol.* 14:157.

Häder, D.-P., Lebert, M. and DiLena, M. R., Effects of culture are and drugs on phototaxis in the green flagellate, *Euglena gracilis, Life Sci. Adv. - Ser. B*, in press.

Häder, D.-P. and Lipson, E., 1986, Fourier analysis of angular distributions for motile microorganisms, *Photochem. Photobiol.* 44:657.

Häder, D.-P., Rhiel, E., and Wehrmeyer, W., Phototaxis in the marine flagellate, *Cryptomonas maculata, J. Photochem. Photobiol.*, in press.

Jennings, H.S., 1904, Reactions to light in ciliates and flagellates. In: "Contributions to the study of the behavior of microorganisms," Carnegie Inst. Washington, pp. 29-71.

Mardia, K.V., 1972, "Statistics of Directional Data," Acad. Press, London.

Mast, S.O., 1911, "Light and Behavior of Organisms," John Wiley & Sons, New York, Chapman & Hall, London.

Mikolajczyk, M., Häder, D.-P., and Nultsch, W., 1985, Photodynamically induced chemoresponses of the colorless flagellate, *Astasia longa*, in the presence of riboflavin, *Arch. Microbiol.* 142:397.

Nultsch, W., 1975, Phototaxis and photokinesis, in: "Primitive Sensory and Communication Systems," M. J. Carlile, ed., Academic Press, London, New York, San Francisco, 29 - 90.

Nultsch, W., and Häder, D.-P., 1987, Photomovement of motile microorganisms II, *Photochem. Photobiol.*, in press.

Nultsch, W., and Throm, G., 1975, Effect of external factors on phototaxis of *Chlamydomonas reinhardtii.*, I. Light, *Arch. Microbiol.* 103:175.

Pavlidis, T., 1982, "Graphics and Image Processing," Springer, Berlin, Heidelberg, New York.

Shipton, H. W., 1979, The microprocessor, a new tool for the biosciences, *Ann. Rev. Biophys. Bioeng.* 8:269.

Skarnulis, A. J., 1982, A computer system for on-line image capture and analysis, *J. Microsc.* 127:39.

THE PHOTOMOTILE RESPONSES IN BLUE-GREEN ALGAE

Wilhelm Nultsch

Fachbereich Biologie - Botanik
der Philipps-Universität
D-3550 Marburg, FRG

INTRODUCTION

As in many other microorganisms, in blue green algae, nowadays mostly called cyanobacteria, three types of photomotile responses occur:
1. Photokinesis, i.e. an effect of light on the speed of movement, in cyanobacteria mostly an acceleration or even a starting of movement.
2. Phototaxis, i.e. a movement towards the light source (positive) or away from it (negative), as a result of the perception of light direction.
3. Photophobic responses, in cyanobacteria mostly stops followed by a reverse of movement. They are caused by temporal changes of the fluence rate, dI/dt, either by an increase (step up) or by a decrease (step down).

Mainly two species of cyanobacteria have been used for photomovement studies during the last decennia, Phormidium uncinatum and Anabaena variabilis, the first belonging to the Oscillatoriaceae, the second to the Nostocaceae. Both families display a different movement behaviour: the Oscillatoriaceae rotate around their long axis during movement, the motile forms and stages of the Nostocaceae do not. This has consequences also for the photobehaviour, especially phototaxis.

As it has been shown in the past, photokinesis in cyanobacteria is the result of an energy supply from either cyclic or pseudo-cyclic photophosphorylation (cf. Nultsch, 1973). Step-down photo-phobic response are mediated by both photosystems and are coupled to the linear electron transport (c.f. Nultsch, 1975; Nultsch and Häder, 1979, 1980, 1987; Häder, 1979). This paper deals especially with phototaxis.

PHOTOTAXIS

In cyanobacteria, phototaxis can be brought about in two different ways: either by a change of the autonomous rhythm of reversal under unilateral illumination or by an active steering mechanism.

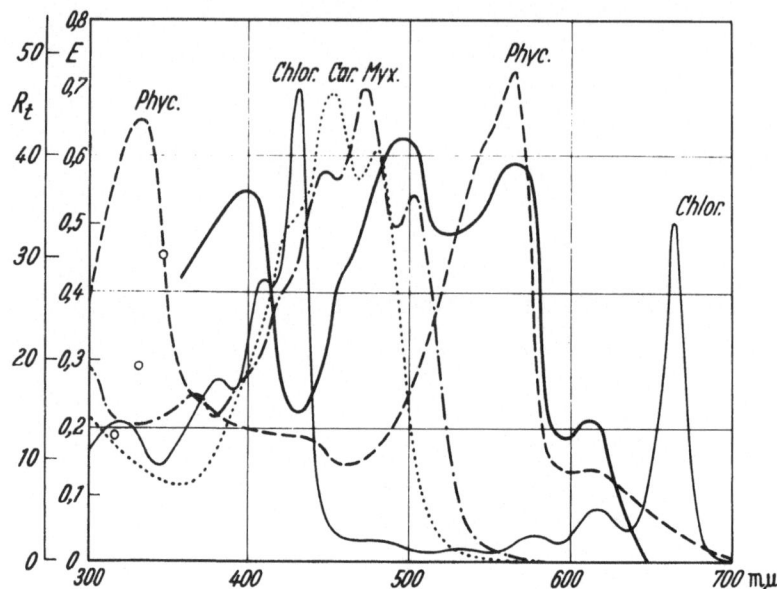

Fig. 1 Phototactic action spectrum of Phormidium autumnale (heavy solid
line) and absorption spectra of phycobiliproteins (Phyc),
chlorophyll a (Chlor), ß-carotene (Car) and myxoxanthophyll
(Myx). After Nultsch, 1961

Change of Autonomous Rhythm

This type is realized in all Oscillatoriaceae investigated so far,
and probably in all cyanobacteria which rotate around their long axis
during movement. Under diffuse light, the trichomes display an
alternating forward and backward movement without preferring any
direction. Under unilateral illumination, this autonomous rhythm is
changed: in those trichomes which are in a more or less parallel position
to the light beam, movement toward the light source is prolonged while
movement away from it is shortened in case of positive phototaxis. It
must be emphasized, however, that the individuals are not able to change
their movement direction actively. Only if they come by chance into a
position more or less parallel to the light direction, phototaxis can
occur (Drews, 1959; Nultsch, 1961). As a result, with time the population
approaches the light source.

The action spectra of positive phototaxis in the Phormidium species
measured so far (Fig. 1) show that UV and visible radiation up to 640 nm
is effective (Nultsch 1963, 1975). The UV peak in the action spectrum is
indicative of flavins, while the maximum between 450 and 510 nm points to
carotenoids. The third maximum around 560 and the shoulder at 620 nm
demonstrate that the pycobiliproteins C-phyccerythrin and C-phycocyanin
are also involved in photoperception. Though the phycobiliproteins are
photosynthetic pigments, phototaxis of Phormidium is certaninly not

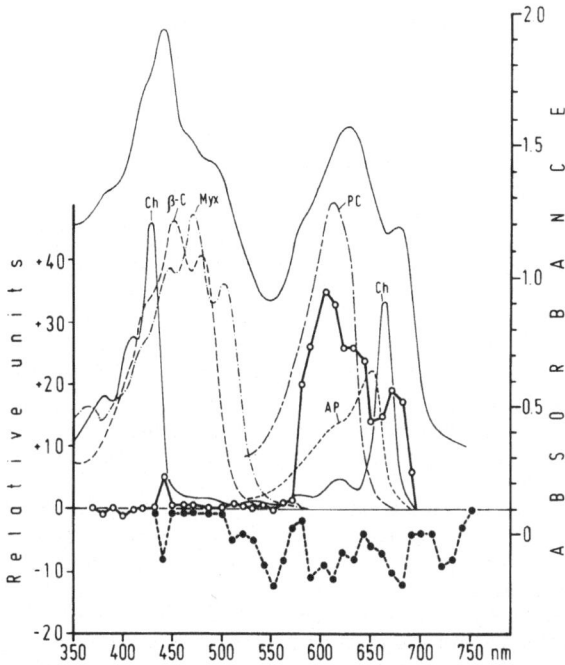

Fig. 2. Action spectra of positive (open circles, heavy solid line) and
negative phototaxis (filled circles, heavy dashed line) of A.
variabilis. For comparison the in vivo absorption spectrum
(fine solid line) and the absorption spectra of solutions of
chlorophyll a (Ch), ß-carotene (ß-C), myxoxanthophyll (Myx), C-
phyco-cyanin (PC) and allophycocyanin (AP) are shown. Abscissa:
wavelenght in nm; ordinates: phototaxis in relative units
(left) and absorbance (right). Modified after Nultsch et al.
(1979).

coupled with photosynthesis, since
1. red light above 640 nm strongly absorbed by chlorophyll a is entirely
ineffective,
2. a minimum is found in the action spectrum in the blue range absorbed
by the Soret band of chlorophyll a and
3. DCMU does not inhibit phototaxis. Thus, the reaction chain of this
type of response is still unclear.

Active steering Mechanism

The active steering mechanism is realized in the Nostocaceae
investigated so far, but probably also in other motile cyanobacteria
whose trichomes do not rotate. Apparently, these organisms make use of
the so-called one-instant mechanism detecting the spatial light
absorption gradient between the irradiated and the shaded side of the
trichome, as shown by Nultsch and Wenderoth (1983) for both positive and
negative responses of Anabaena variabilis with the aid of the partial

illumination technique. This gradient is mainly due to the absorption of the phototactically effective light by the photosynthetic pigments located in the thylakoids which are preferably arranged in the periphery of the cells. Once a trichome is unilaterally illuminated, it bends toward the light source at low but away from it at high fluence rates (Nultsch et al., 1979), resulting in positive and negative phototaxis, respectively. Only at the transition point, i.e. the fluence rate at which the phototactic response becomes negative, the trichomes show random orientation. As a result, even a population moves either toward the light source or away from it, but spreads circularly at the transition point.

The phototactic action spectra of the Phormidium species and Anabaena variabilis (Fig. 2) have in common that light absorbed by the phycobiliproteins is effective (Nultsch et al., 1979). Since A. variabilis lacks C-phycoerythrin, the main maximum coincides with the absorption maximum of C-phycocyanin, but phycoerythrocyanin and allo-phycocyanin are also involved in light perception as indicated by the strong effectiveness of 580 nm and the shoulder around 650 nm, respectively. Unlike the Phormidium species, however, even light absorbed by chlorophyll a is effective. This is indicated by a second maximum in the red around 675 nm and a small but significant peak at 440 nm. Apparently, the photosynthetic pigments with the exception of the carotenoids are the photoreceptor pigments of phototaxis. However, as DCMU does not inhibit phototaxis, the perception of the light direction is not a phototsynthetic effect.

Surprisingly, the action spectra of positive and negative phototaxis are not identical. Though the same ranges of wavelengths causing positive phototaxis are also effective in negative phototaxis, in addition light between 500 and 560 nm, entirely ineffective in positive phototaxis, causes negative phototactic reactions, whereas above 700 nm only negative responses have been observed, even at very low fluence rates. Obviously, at least one other photoreceptor of unknown nature is involved in negative phototaxis.

Reaction Sign Reversal

As mentioned above, the gradient between the illuminated and the shaded side gives the directional information. Since this gradient exists at any fluence rate, a second perception mechanism is necessary to distinguish low and high fluence rates and, hence, to give the information whether the sign of the phototactic response should be positive or negative. Based on the observation that 1 mM sodium azide, a potent quencher of singlet molecular oxygen ($^1O_2$), reverses the phototactic reaction sign from negative to positive even at high fluence rates (Fig. 3), Nultsch et al. (1983a) have suggested that the phototactic reaction sign is regulated by a hypothetical sign reversal generator which is controlled by an active oxygen species, probably singlet oxygen or one of its reaction products. It is wellknown that energy-rich oxygen species such as singlet molecular oxygen ($^1O_2$), superoxide (O_2^{-}) and the hydroxyl radical ($\cdot OH$) can be formed inside reactions of photosynthesis (Halliwell, 1981). Since the hydroxyl radical will never pass through membranes, as it reacts immediatley after its generation with the first membrane molecule it meets, and superoxide does not cross biomembranes easily unless it can pass through specific channels (Halliwell and Gutteridge,1984), singlet oxygen which is able to migrate through biomembranes very fast (Gorman et al., 1976) seemed to be the most promising candidate, though its lifetime is rather short.

Therefore, Schuchart and Nultsch (1984) and Nultsch and Schuchart (1985) carried out a series of experiments in order to find further support for this hypothesis. Since azide is not an absolutely specific $^1O_2$ quencher, as it also quenches triplet states (Hasty et al., 1972; Winter et al., 1981) and, in addition, affects several other biochemical processes, other $^1O_2$ quenchers have been tested. Unfortunately, many of them such as furan derivatives and DABCO (1,4-diazabicyclo-(2.2.2) octane) are more or less cytotoxic in the concentrations necessary for quenching. The most convincing effects were obtained with preparations of solubilized carotenoids such as ß-carotene, canthaxanthin and a C_{30} ester (=ethyl-ß-apo-8'-carotenoat), and especially the water soluble crocetin (Table 1) which shift the transition point to higher fluence rates by about one order of magnitude (Nultsch and Schuchart, 1985).

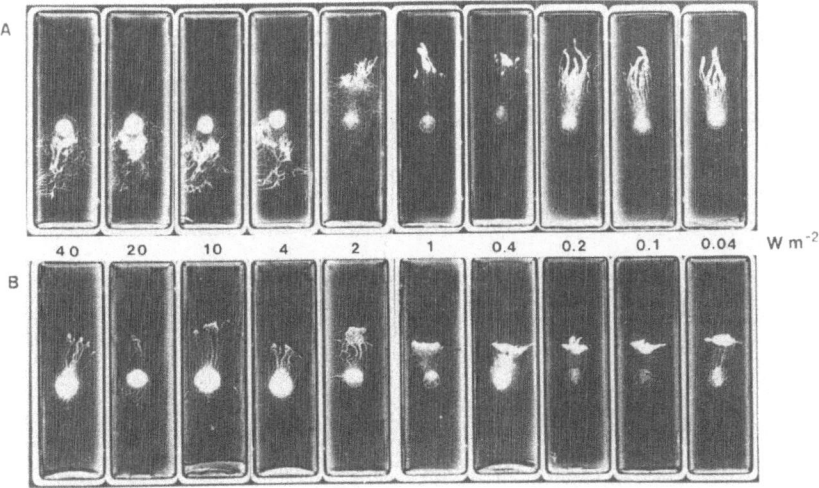

Fig. 3   Photograms of phototactic reactions of <u>A. variabilis</u> in white
light in the absence (A) and in the presence (B) of 1 mM NaN_3.

Another way is to prevent the generation of singlet oxygen. As it is known that the quantum yield of $^1O_2$ production is the less the lower the pH value is, the effects of different pH values on phototaxis have been studied. The results show (Table 2) that low pH values shift the transition point of phototaxis to higher fluence rates compared to alkaline pH values (Nultsch and Schuchart, 1985). The same effect, i.e. a shift of the transition point to higher fluence rates, can be obtained by additon of DCMU which blocks oxygen production and, hence, also the generation of active oxygen species (Nultsch et al., 1979).

Finally, a volatile substance as $^1O_2$, or a volatile reaction product, should be removed by gassing. In fact gassing with nitrogen, argon, air and even $^3O_2$ shifts the transition point to higher fluence rates. This is apparently the effect of the removal of a volatile substance by the gas stream, most probably the removal of singlet molecular oxygen.

Table 1. Effect of Carotenoids on the Phototactic
Transition Point in A. variabilis.

Carotenoid	Transition Point Wm^{-2}	Double bonds
Control	7	-
Crocetin	40	7
C_{30}-ester	66	9
ß-carotene	54	11
Canthaxanthin	70	11

Table 2. Effect of pH Value on the Phototactic Transition
Point of A. variabilis.

pH	Transition Point Wm^{-2}
5.5	> 120
6.0	15
7.0	5.7
8.0	4.5

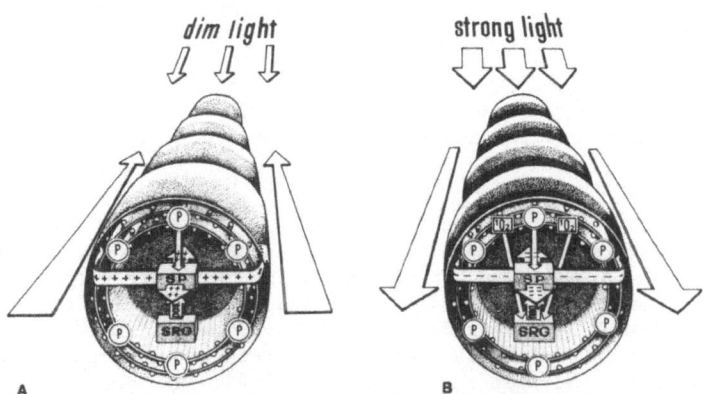

Fig. 4. Three-dimensional model of the photo-tactic reaction chain of
Anabaena variabilis, if unilaterally irradiated with dim (**A**)
and strong (**B**) light in a position parallel to the light beam.
SRG Sign reversal generator, SP signal processor, + and -
arrows = signal transmission. O_2 singlet molecular oxygen. The
arrows outside the trichomes symbolize the motive force on both
flanks, leading to forward or backward movement. The
photoreceptor molecules P are located in the thylakoids and in
the phycobilisomes, respectively. After Nultsch and Schuchart,
1985

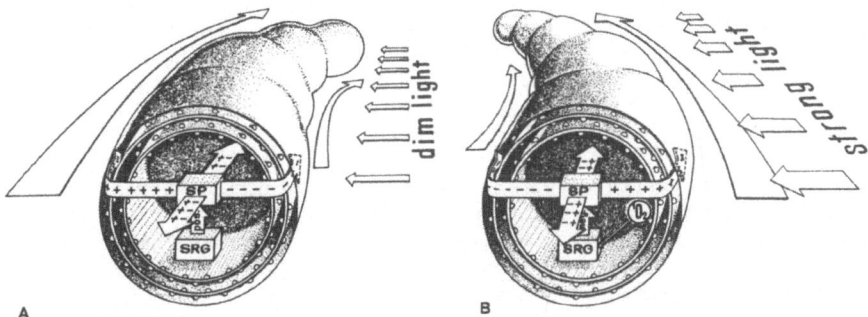

Fig. 5   Model of the phototactic reaction chain as in Fig. 4, but
trichomes irradiated from the side with dim (A) strong light
(B), respectively. For further explanations see Fig. 4.
After Nultsch and Schuchart, 1985

Based on these results a model of the sign reversal mechanism has been
proposed. As mentioned above, the light gradient perceived gives the
directional information (front-rear, left-right and vice versa). At low
fluence rates the $^1O_2$ production is low or even zero, because the
carotenoids present in the thylakoids quench $^1O_2$. Therefore, the sign
reversal generator gives a positive signal, and the input from the
processor to the motor apparatus is "forward" in trichomes being in a
parallel position (Fig. 4) or "bend towards the light" in trichomes being
in a perpendicular position to the light direction (Fig. 5). In the first
case the trichomes go ahead, in the latter case they turn toward the
light source until they are parallel to the light beam.

At high fluence rates, however, the quenching capacity is exceeded.
Consequently, $^1O_2$ is produced and activates the sign reversal generator
which reverses the signal given to the processor to negative, whereas the
directional information remains unchanged. The signal processor gives the
opposite input to the motor apparatus, and the trichomes oriented
parallel to the light beam move backwards and those in a position
perpendicular to the light beam bend away from the light source (Nultsch
and Schuchart, 1985).

DISCUSSION

From the ecological point of view this phototactic behaviour seems
to be quite reasonable. Since the cyanobacteria which arose on earth
about 2.3 billion years ago "invented" the oxygenic phototsynthesis, in
which water serves as electron donor for CO_2 reduction and molecular
oxygen is released to the atmosphere, they are largely responsible for
the substitution of the original reducing atmosphere on earth by an
oxidizing one. The development of an oxygen rich atmosphere created a
serious problem for the cyanobacteria themselves, as for other organisms,
because the already mentioned energy-rich, aggressive oxygen species can
also be generated by the photosynthetic apparatus, especially $^1O_2$. At
lower fluence rates $^1O_2$ may not be generated or, if produced in small
amounts, is quenched by the carotenoids. At higher fluence rates,
however, as in direct sun light, this photoprotective mechanism might be
insufficient in A. variabilis which is a shade plant and, hence, adapted

to low fluence rates. In order to avoide photodamage, the above described
control mechanism enables the organisms to escape dangerous light
conditions and to search out those light conditions in the biotope which
are favorable for photosynthesis but do not cause photodamage.

REFERENCES

Drews, G., 1959, Beiträge zur Kenntnis der phototaktischen Reaktionen der
    Cyanophyceen. Arch. Protitenk. 104:389
Gorman, A. A., Lovering, G., Rodgers, M. A. J., 1976, The photosensitized
    formation and reaction of singlet oxygen, $O_2^*(^1\Delta)$, in aqueous
    micellar systems. Photochem. Photobiol. 23:399
Häder, D.-P., 1979, Photomovement, in: Encyclopedia of Plant Physiolgy
    NS, Vol 7, W. Haupt, M.E. Feinleib ed., Springer, Berlin,
    Heidelberg, New York, pp 268-309
Halliwell, B., 1981, Chloroplast metabolism: The structure and function
    of chloroplasts in green leaf cells. Oxford University Press, Oxford
Haliwell, B. and Gutteridge, J. M. C. 1984, Oxygen toxiticity, oxygen
    radicals, transition metals and disease. Biochem. J. 219:1
Hasty, N., Merkel, P. B., Radlick, P., Kearns, D. R., 1972, Role of azide
    in singlet oxygen reactions: reaction of azide with singlet oxygen.
    Tetrahedron Lett. 1:49
Nultsch, W., 1961, Der Einfluß des Lichtes auf die Bewegung der
    Cyanophyceen. I. Phototaxis von Phormidium autumnale. Planta 56:632
Nultsch, W., 1962, Phototaktische Aktionsspektren von Cyanophyceen. Ber.
    Dtsch. Bot. Ges. 75:443
Nultsch, W., 1973, Relation between photomotion and photosynthesis. In:
    Primary Molecular Events in Photobiology. Checcucci, A., Weale, R.
    M. (eds.). Amsterdam, London, New York: Elsevier, pp. 245-273
Nultsch, W., 1975, Phototaxis and photokinesis. In: Primitive Sensory and
    Communication Systems (Edited by M.J. Carlile), Acad. Press, London,
    New York, pp 29-90
Nultsch, W. and Häder, D.-P., 1979, Photomovement of motile
    microorganisms. Photochem. Photobiol. 29:423
Nultsch, W. and Häder, D.-P., 1980, Light perception and sensory
    transduction in photosynthetic prokaryotes. In: Structure and
    Bonding Vol. 41, Molecular Structure and Sensory Physiology. (Edited
    by P. Hemmerich), Springer, Berlin, Heidelberg, pp. 111-139
Nultsch, W. and Häder, D.-P., 1987 Photomovement in motile microorganisms
    II. Photochem. Photobiol. (in the press)
Nultsch, W. and Schuchart, H., 1985, A model of the phototactic reaction
    chain of the cyanobacterium Anabaena variabilis. Arch. Microbiol.
    142:180
Nultsch, W. and Wenderoth, K., 1983, Partial irradiation experiments with
    Anabaena variabilis (Kütz). Z. Pflanzenphysiol. 111:1
Nultsch W., Schuchart, H. and M. Höhl, 1979, Investigations on the
    phototactic orientation of Anabaena variabilis. Arch. Microbiol.
    122:85
Nultsch W., Schuchart, H. and König, F., 1983, Effects of sodium azide
    on phototaxis of the blue-green alga Anabaena variabilis and
    consequences to the two-photoreceptor systems- hypothesis. Arch.
    Microbiol. 134:33
Schuchart, H. and Nultsch, W., 1984, Possible role of singlet molecular
    oxygen in the control of the phototactic reaction sign of Anabaena
    variabilis, J. Photochem. 25:317
Winter, G., Shioyama, H. and Steiner, U., 1981, Electron transfer
    quenching of dye triplets by NO_2^- and N_3^-. A spin-orbit coupling
    effect on the radical yield. Chem. Phys. Lett. 81:547

THE PHOTOMOTILE RESPONSES OF UNICELLULAR EUKARYOTES

Ewa Mikolajczyk

Department of Cell Biology
Nencki Institute of Experimental Biology
Warsaw, Poland

INTRODUCTION

The main goal of this paper is to give a short overview of recent research progress on photobehavior in three different groups of unicellular organisms: colored, photosynthetic flagellates (e.g., Euglena), colored, non-photosynthetic ciliates (e.g., Stentor), and colorless flagellates and ciliates (e.g., Astasia and Paramecium), respectively.

There are three main light-induced reactions in protists (for terminology see Diehn et al., 1977): photophobic responses, phototaxis, and photokinesis. The last, however, can hardly be classified as a photomotile response because, in photokinesis, the intensity of light influences the velocity of cell movement (Haupt, 1983). Thus, the light is used rather as an additional energy source for movement instead of serving as a signal, as it does for the photophobic and phototactic response. Positive and negative phototactic reactions involve oriented cellular movement toward or away from a light source, whereas step-up and step-down photophobic responses are reactions to increases or decreases in light intensity.

The green flagellate, Euglena gracilis, exhibits both step-up and step-down photophobic responses (Colombetti et al., 1982b; Colombetti and Lenci, 1983), as well as positive and negative phototaxis (Häder et al., 1981, 1986; Colombetti et al., 1982a; Lenci et al., 1983; Häder, 1985a, 1986). The blue-green colored ciliate, Stentor coeruleus, shows a step-up photophobic response and a negative phototaxis (Song et al., 1980a,b; Kim et al., 1984). Previously considered to be insensitive to light, the colorless organisms Astasia longa and several species of Paramecium (P. tetraurelia, P. multimicronucleatum, and P. arcticum) have recently been shown to respond to light stimuli, although only the step-up photophobic response has been reported in these cells thus far (Iwatsuki and Naitoh, 1982, 1983; Suzaki and Williamson, 1983; Mikołajczyk, 1984b; Reisser and Häder, 1984).

In another interesting species, P. bursaria, the ability to respond to light depends on the presence of an endosymbiotic unicellular green alga, Chlorella vulgaris (Saji and Oosawa, 1974; Cronkite and VandenBrink, 1981; Iwatsuki and Naitoh, 1981; Niess et al., 1981). The aposymbiotic ciliates (i.e., free of algae) show a step-up photophobic response, by which they can disperse from an irradiated area, whereas the ciliates with algal symbionts are able to respond also to a step-down photostimulation and to accumulate in an irradiated field (Saji and Oosawa, 1974; Iwatsuki and Naitoh, 1981; Reisser and Häder, 1984). The mechanism for such photo-accumulation and photodispersion is shown in Figure 1, using <u>Euglena gracilis</u> as an example. Photoaccumulation is brought about by a step-down

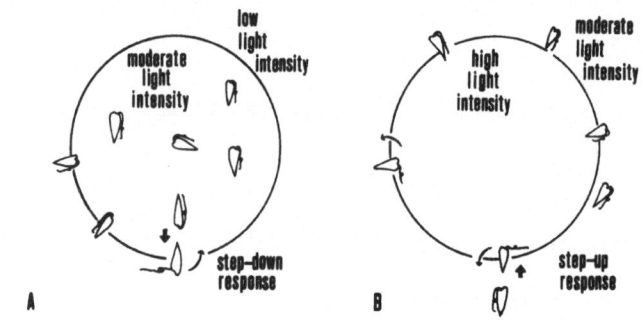

Fig. 1. Diagrammatic representation of the mechanism of photoaccumulation (A) and photodispersion (B) in <u>E. gracilis</u>. Arrows indicate: thick initial direction of swimming, thin - direction of cellular response.

response, whereby the cells move from a region of relatively high light intensity to one of lower light intensity, then turn back immediately to the more brightly illuminated field (Fig. 1 A). In contrast, photodispersion is brought about by a step-up photophobic response of cells in an area of lower light intensity; when they touch the border between areas of low and high light intensities, they turn back to the area of lower illumination. Other cells, which have left a more brightly illuminated field will not enter it again, since they, too, will show a step-up photophobic response as soon as they touch the border. After several minutes, the area of higher light intensity will thus be devoid of cells (Fig.1 B). It is suggested that the photoreceptor pigment for the step-down response is provided by the endosymbionts, while the phototransduction chain for this response is a basic feature of the host cell (Reisser and Häder, 1984). Moreover, because an activation of Ca^{2+}-channels seems to be involved in the step-down response, it is possible that the algae may increase the excitability of the <u>Paramecium</u> membrane (Cronkite, 1986).

Thus, the step-up photophobic response is apparently the only one common to all species of protists mentioned above. It seems, therefore, that this response is the primary one and perhaps the more important for cellular survival.

EXPRESSION OF THE PHOTORESPONSE

Although the expression of photophobic responses obviously differs in ciliates and flagellates, it is characterized in both types of protists by a cessation of forward movement caused by flagellar reorientation or by a change in the direction of ciliary beating (Fig. 2 A-E). For example, in the flagellates <u>E. gracilis</u> or <u>A. longa</u>, the photophobic response is expressed as pivoting (Fig. 2 B), whereas in the ciliate <u>S. coeruleus</u>, it is expressed either as swimming backward when the ciliary beat is reversed (Fig. 2 D), or as a turning aside, when only individual rows of cilia respond to the light stimulus (Fig. 2 E). In both groups of organisms, photophobic responses lead to rapid changes in the direction of movement.

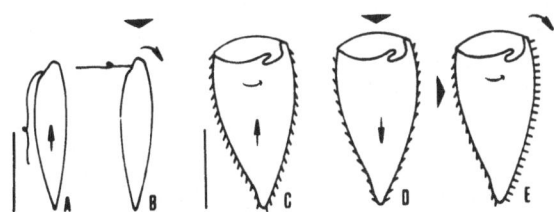

Fig. 2. Diagrammatic expression of the photophobic responses of E. gracilis (A, B) and S. coeruleus (C-E). In Euglena: flagellum trails alongside body when cell is swimming forward (A); flagellum reorients to a position perpendicular to the long axis of the cell in response to light stimuli, resulting in cellular pivoting (B). In Stentor: normal orientation of cilia when cell is swimming forward (C); ciliary reversal in response to light results in backward swimming (D); individual rows of cilia respond to a lateral light stimulus, resulting in cell turning aside (E). Arrows indicate: large - direction of light stimulation, medium - direction of cell movement, small - direction of rotation of the cell around its lateral axis. Bars indicate speciment length: Euglena 25 μm, Stentor 160 μm.

PIGMENTS RESPONSIBLE FOR PHOTOBEHAVIOR

The problem of the nature of the photoreceptor pigments and their localization seems to be solved for some organisms (e.g., Euglena gracilis and Stentor coeruleus), while for others it remains enigmatic (e.g., Paramecium spp. and Astasia longa).

Euglena. For some time it has been generally accepted that flavins in the PFB are the photoreceptor pigments involved in all light-induced motile responses in E. gracilis (Checcucci et al., 1976a,b). Recently, Lenci and co-workers (1983) reported the effects of substances (like KJ and MnCl₂) that react with excited states of flavins on oriented movement of Euglena cells away from a lateral light source, thus providing experimental evidence in support of the view that flavins are responsible for truly oriented movement, at least for negative phototaxis, in this flagellate. It has also been suggested that, in addition to flavins in the PFB, there may be another pigment, i.e., an "accessory" one, responsible for the step-up photophobic response in E. gracilis; that pigment may be a primitive one that existed before the development of the more highly specialized photoreceptor system with the PFB and stigma (Mikołajczyk, 1984a).

In other species of Euglena, too, flavins are apparently not the only photoreceptor pigment. For example, E. mutabilis reportedly differs in major aspects of its photobehavior (Häder and Melkonian, 1983; Melkonian et al., 1986); while E. gracilis shows both positive and negative phototaxis, E. mutabilis does not show a negative phototaxis, although it does respond to step-up and step-down photophobic stimuli. It is thus assumed that more than one type of photoreceptor molecule exists in <u>E. mu-</u><u>tabilis</u>, since the spectral sensitivities for both positive phototaxis and the step-down photophobic response show a number of peaks throughout the entire visible spectrum (Fig. 3). Although a flavin-type pigment may well be involved in photoresponse in <u>E. mutabilis</u>, it is also unlikely that it is the only photoreceptor molecule. The nature of the presumed additional photoreceptor molecules is unknown, however.

<u>Stentor</u>.   In <u>S. coeruleus</u> stentorin, with hypericin as a chromophore, was found to act as a photoreceptor pigment responsible for photophobic and phototactic responses (Wood, 1976; Walker et al., 1979; Song et al., 1980a,b; Song, 1981). The spectral sensitivity of the step-up photophobic response in comparison with an absorption spectrum of stentorin is shown in Figure 4. The pigment has been shown by electron microscopy to be located in rows just below the plasmalemma, near the ciliary basal bodies (Huang and Pitelka, 1973).

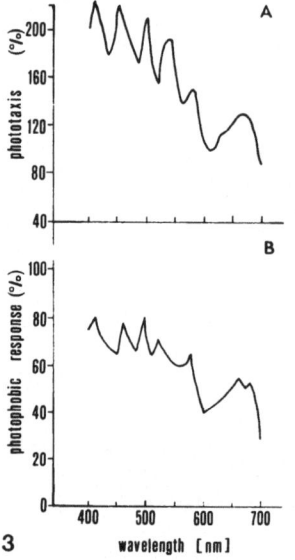 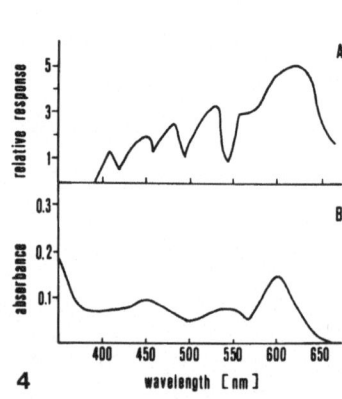

Fig. 3. Spectral sensitivity of the phototactic orientation (A) (after Häder and Melkonian, 1983) and step-down photo-phobic response (B) of <u>E. mutabilis</u> (after Melkonian et al., 1986).

Fig. 4. Spectral sensitivity of the step-up photophobic response in <u>S. coeruleus</u> (A) (after Song et al., 1980a) and the absorption spectrum of stentorin isolated from <u>S. coeru-</u><u>leus</u> (B) (after Song, 1981).

<u>Paramecium spp. and Astasia longa</u>.   To date, our knowledge about the photoreceptor molecules of <u>Paramecium</u> comes from its action spectrum, which extends beyond 680 nm and in which two prominent peaks at 520 and 680 nm are evident (Fig. 5) (Iwatsuki and Naitoh, 1982, 1983).

In the colorless flagellate, <u>A. longa</u>, the nature of the photorecep-tor pigments is also unknown, but they are not thought to be flavins (Suzaki and Williamson, 1983). The spectral sensitivity of the step-up photophobic response in comparison with the absorption spectrum of a flavo-enzyme, d-amino acid oxidase, is shown in Figure 6. Since <u>A. longa</u>

is considered to be an apochlorotic mutant of E. gracilis, the nature of the "accessory" pigment from the latter and the photoreceptor molecules responsible for the step-up photophobic response of the colorless A. longa cells could be similar or perhaps identical (Mikołajczyk, 1984b). The photoreceptor pigment molecules in Astasia are located mostly in the anterior part of the cell, since it is reportedly the most sensitive to light stimuli (Suzuki and Williamson, 1983).

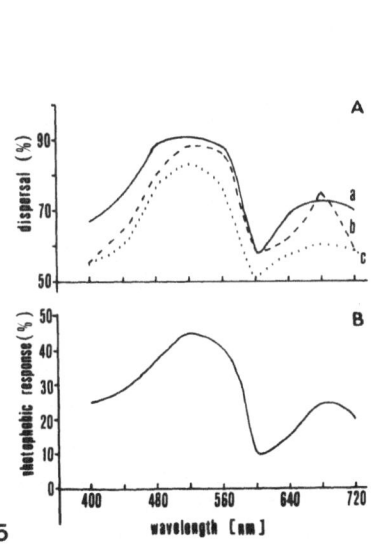

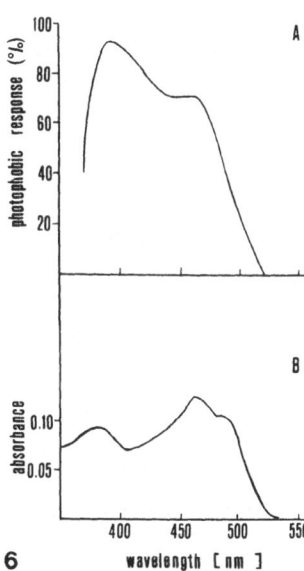

Fig. 5. Spectral sensitivity of photodispersion in three different species of Paramecium (A): a - P. caudatum, b - P. multimicronucleatum, c - P. tetraurelia (after Iwatsuki and Naitoh, 1983) and step-up photophobic response of P. multimicronucleatum (after Iwatsuki and Naitoh, 1983).

Fig. 6. Spectral sensitivity of the step-up photophobic response of A. longa (A) (after Suzaki and Williamson, 1983) and absorbance of a flavo-enzyme, d-amino acid oxidase (B) (after Colombetti et al., 1975).

MECHANISMS OF THE LIGHT-ORIENTED MOVEMENT

Until the early 1980s, such phenomena as photoaccumulation and photodispersion were identified with phototaxis in E. gracilis. Studies of the cell's truly oriented movement with respect to light direction were made possible by the use of a computer-controlled video analysis system (Häder, 1985b; Häder and Lebert, 1985; Häder and Lipson, 1986).

Is there a correlation between photophobic responses and phototaxis?
Although it was generally accepted that phototaxis in E. gracilis is brought about by repetitive photophobic responses (step-up in negative phototaxis and step-down in positive phototaxis) (Diehn et al., 1975; Checcucci et al., 1976b; Creutz and Diehn, 1976; Colombetti et al., 1982b) this is no longer believed to be so. Also, there is still disagreement about the mechanism of the phototactic movements and their relationship to photophobic responses. The experimental results of

Colombetti and co-workers (1982a) indicated, for example, that phototaxis in E. gracilis can be independent of the photophobic response. The existence of two different, possibly independent responses would require the presence of two different pigments and/or two different transduction chains. If flavins in the PFB are truly responsible for all photoresponses in E. gracilis, then differences in the mechanisms of phototactic and photophobic responses could be accounted for by differences in the photosensory transduction chains.

For several years, it was believed that the photophobic response is not involved in phototaxis in Stentor coeruleus (Song et al., 1980a; Colombetti and Lenci, 1983). According to later studies by Prusti and co-workers (1984), however, negative phototaxis in this organism is brought about by repetitive step-up responses. More recently, Iwatsuki and Song (1985) hypothesized that the sensory transduction mechanism for these two photoresponses is probably different, although the photoreceptor pigment, stentorin, is the same.

The suggested independence of phototaxis from the photophobic response in Euglena is not an unique phenomenon. There are organisms, such as the chrysophycean flagellate, Ochromonas, in which phototactic orientation is reportedly not mediated by the photophobic responses (Häder et al., 1981).

Shading hypothesis and dichroic orientation of photoreceptor pigments.
Some of the latest studies on the molecular mechanism of light-oriented movement in Euglena gracilis come from the work of Häder and co-workers (1986). They reported that phototaxis in this flagellate can be explained by the periodic-shading hypothesis in conjunction with a dichroic orientation of the photoreceptor pigments in the PFB.

Shading hypothesis. According to the shading hypothesis (Diehn, 1969, 1973), during the rotation of the cell in lateral light the stigma periodically shades the presumptive photoreceptor (PFB), thus modulating the light absorbed by the PFB. In the case of positive phototaxis, each time a shadow falls on the PFB, the cell perceives a step-down signal, the flagellum reorients to a position perpendicular to the long axis of the cell, and the organism turns toward the light source. The same mechanism functions in negative phototaxis, when the reorientation of the flagellum and a turning response occur each time the photoreceptor is exposed to the light, with the cell thus perceiving a step-up signal. Such small course corrections continue until the cells swim directly toward or away from the light source.

The so called two-instant mechanism proposed by Feinleib (1980) for phototactic orientation is based on the shading hypothesis. The essential feature of this mechanism is that the direction of light must be perceived by comparing the light absorbed in one perceptive region of the cell at two instants in time.

Dichroic orientation of photoreceptor pigments. The absorption of light by the photoreceptor molecules depends on the orientation of the organism with respect to the direction of the light source. On the basis of experiments with polarized light, Creutz and Diehn (1976) proposed an alternative to the periodic shading hypothesis, involving dichroic orientation of the photoreceptor pigment. They postulated that the photoreceptor molecules are oriented in the PFB in E. gracilis in such a way that their transition moments are aligned perpendicularly to the long axis of the cell. Then, absorption is maximal, when the anterior end of the cell is oriented toward the light beam, and minimal when cells are swimming away from the light source, with the PFB shaded by the posterior end of the cell.

Using polarized light to study oriented movement in E. gracilis, Häder (1987) has recently determined the three-dimensional dichroic orientation of the photoreceptor molecules, and has suggested that cellular oriented movement may be mediated by two different sets of photoreceptor pigments oriented at different angles or by different molecular transition moments of the same molecules.

The perception of light direction in negative phototaxis in Stentor coeruleus is most probably based on a one-instant mechanism (Song et al., 1980a; Song, 1981; Colombetti and Lenci, 1983; Prusti et al., 1984), involving the comparison of the light absorbed in two photoreceptive regions of the cell at one instant in time (Feinleib, 1980). According to this mechanism, a step-up signal induces in Stentor a step-up photophobic responce, which is expressed as a cessation of ciliary beating in the row of cilia adjacent to the row of illuminated photoreceptor pigment vesicles. Cessation of the ciliary movement occurs, but with a latency sufficient to allow a 180° rotation of the cell away from the light source (Fig. 2 E).

Is the electric potential involved in photoresponse? In Euglena gracilis recent experiments with externally applied electric fields showed no effect on oriented movement of that flagellate (Häder et al., 1986). This result is very important since it provides strong experimental evidence that the electric potential is not involved in photo-orientation in Euglena. No light-induced potential changes were detected by intracellular measurements either, as found by Diehn, Chen and Mikołajczyk (unpublished data).

In the case of Stentor coeruleus, changes in the membrane potential play an important role in photoresponses. The primary electric potential change in Stentor is apparently due to a proton flux from the pigment granules to the cytoplasm. A transient increase in the intracellular H^+ has been proposed to open or activate Ca^{2+} channels in the cell membrane causing the amplification of the primary electric potential. The increase in cytoplasmic and/or intraciliary Ca^{2+} concentration reportedly evokes ciliary reversal (Wood, 1976; Song et al., 1980a, 1981, 1983; Song, 1981, 1983; Walker et al., 1981; Kim et al., 1984; Prusti et al., 1984).

In colorless species of Paramedium the light-induced action potential produced by a photoreceptor system, presumably located in the cell membrane, causes ciliary reversal (Iwatsuki and Naitoh, 1983).

CONCLUSIONS

What is the role of photophobic and phototactic responses in the life of colored and colorless protists? It seems that for photosynthetic organisms, photophobic responses are used mostly for rapid changes in the direction of movement in response to transient drastic changes in illumination. They are responsible also for the photodispersion from and photoaccumulation in an irradiated field. In natural environments Euglena sustains optimal photosynthetic conditions by maintaining a delicate balance between positive and negative phototaxis (Häder, 1985a). In this regard, Häder (1985a, 1986) has shown that an UV-B band (270-320 nm), either as artificial or solar irradiation, has a strong inhibitory effect on negative response to UV and strong white light in E. gracilis. Since negative phototaxis results in movement of cells into deeper water, the serious consequences of the cessation of this response are that cells must remain near the surface, where they can be damaged by light.

For non-photosynthetic cells like Stentor coeruleus, step-up photophobic response and negative phototaxis lead to photoaccumulation of the cells in shaded areas of the environment. The step-up photophobic response in colorless protists, as the only possible response to light stimulation, protect cells against harmful, dangerous light and allows them to stay in an environment of lower, non-lethal light conditions. Photophobic responses thus seems to be involved in the oriented movement (i.e., phototaxis) of some protists (i.e., Stentor), but some recent data suggest that they are probably not involved in others (e.g., Euglena).

In an effort to identify precisely the site and nature of the photoreceptor, and to understand the role of the PFB in the photoreceptive process in Euglena, scientists have tried for many years to isolate the emergent flagellum with attached PFB. The task was unusually difficult, and only in 1986 did Gualtieri and co-workers finally succeed. The isolation of the PFB is a crucial development for further studies of photoreception in this flagellate, including the precise chemical and structural characterization of this organelle.

Despite the recent progress in understanding the photobehavior of protists, most problems concerning photoreception, the phototransduction chain, and the mechanism of oriented movement, even in such extensively investigated organisms as Euglena and Stentor, still remain to be solved.

ACKNOWLEDGEMENT

The author is indepted very much to Prof. Patricia Walne for valuable help.

REFERENCES

Checcucci, A., Colombetti, G., Ferrara, R., and Lenci, F., 1976a, Action spectra for photoaccumulation of green and colorless Euglena: evidence for identification of receptor pigments, Photochem. Photobiol., 23:51.
Checcucci, A., Colombetti, G., Ferrara, R., and Lenci, F., 1976b, Further analysis of the mass photoresponses of Euglena gracilis Klebs (Flagellata Euglenoidina), Monitore Zool. Ital., (N.S), 10:271.
Colombetti, G., Häder, D.-P., Lenci, F., and Quaglia, M., 1982a, Phototaxis in Euglena gracilis: effect of sodium azide and triphenylmethyl phosphonium ion on the photosensory transduction chain, Curr. Microbiol., 7:281.
Colombetti, G., and Lenci, F., 1983, Photoreception and photomovements in microorganisms, in: "The Biology of Photoreception", D. J. Cosens, and D. Vince-Price, eds., Society for Exp. Biol. Symp., Cambridge.
Colombetti, G., Lenci, F., Diehn, B., 1982b, Responses to photic, chemical, and mechanical stimuli, in: "The Biology of Euglena", D. E. Buetow, ed., Academic Press, New York, 3.
Colombetti, G., Lenci, F., McKeller, J. F., and Phillips, G. O., 1975, Light-induced effects in a flavoprotein: d-amino acid oxidase, Photochem. Photobiol., 21:303.
Creutz, C., and Diehn, B., 1976, Motor responses to polarized light and gravity sensing in Euglena gracilis, J. Protozool., 23:552.
Cronkite, D. L., 1986, Relations between photobehavior and ionically stimulated swimming behavior in Paramecium bursaria, J. Protozool., 33:52.
Cronkite, D., and VandenBrink, S., 1981, The role of oxygen and light in guiding "photoaccumulation" in the Paramecium bursaria-Chlorella symbiosis, J. Exp. Zool., 217:171.

Diehn, B., 1969, Phototactic response of Euglena to single and repetitive pulses of actinic light: orientation time and mechanism, Exp. Cell Res., 56:375.

Diehn, B., 1973, Phototaxis and sensory transduction in Euglena, Science, 181:1009.

Diehn, B., Feinleib, M. E., Haupt, W., Hildebrand, E., Lenci, F., and Nultsch, W., 1977, Terminology of behavioral responses in motile microorganisms, Photochem. Photobiol., 26:559.

Diehn, B., Fonseca, J. R., and Jahn, T. L., 1975, High speed cinematography of the direct photophobic response of Euglena and the mechanism of negative phototaxis, J. Protozool., 24:492.

Feinleib, M. E., 1980, Photomotile responses in flagellates, in: "Photoreception and Sensory Transduction in Aneural Organisms", F. Lenci, and G. Colombetti, eds, Plenum Press, New York and London.

Gualtieri, P., Barsanti, I., and Rosati, G., 1986, Isolation of the photoreceptor (paraflagellar body) of the phototactic flagellate Euglena gracilis, Arch. Microbiol., 145:303.

Häder, D.-P., 1985a, Effects of UV-B on motility and photobehavior in the green flagellate, Euglena gracilis, Arch. Microbiol., 141:159.

Häder, D.-P., 1985b, Computer - aided studies of photoinduced behaviors, in: "Sensory Perception and Transduction in Aneural Organisms", G. Colombetti, and F. Lenci, eds., Plenum Press, New York and London.

Häder, D.-P., 1986, Effects of solar and artificial UV irradiation on motility and phototaxis in the flagellate, Euglena gracilis, Photochem. Photobiol., 44:651.

Häder, D.-P., 1987, Polarotaxis, gravitaxis and vertical phototaxis in the green flagellate, Euglena gracilis, Arch. Microbiol., 147:179.

Häder, D.-P., Colombetti, G., Lenci, F., and Quaglia, M., 1981, Phototaxis in the flagellates, Euglena gracilis and Ochromonas danica, Arch. Microbiol., 130:78.

Häder, D.-P., and Lebert, M., 1985, Real time computer-controlled tracking of motile microorganisms, Photochem. Photobiol., 42:509.

Häder, D.-P., Lebert, M., and Di Lena, M. R., 1986, New evidence for the mechanism of phototactic orientation of Euglena gracilis, Curr. Microbiol., 14:157.

Häder, D.-P., and Lipson, E. D., 1986, Fourier analysis of angular distributions for motile microorganisms, Photochem. Photobiol., 44:657.

Häder, D.-P., and Melkonian, M., 1983, Phototaxis in the gliding flagellate, Euglena mutabilis, Arch. Microbiol., 135:25.

Haupt, W., 1983, Photoreception and Photomovement, Phil. Trans. R. Soc. Lond. B., 303:467.

Huang, B., and Pitelka, D., 1973, The contractile process in the ciliate, Stentor coeruleus. I. The role of microtubules and filaments, J. Cell Biol., 57:704.

Iwatsuki, K., and Naitoh, Y., 1981, The role of symbiotic Chlorella in photoresponses of Paramecium bursaria, Proc. Jpn. Acad., 57:318.

Iwatsuki, K., and Naitoh, Y., 1982, Photoresponses in colorless Paramecium, Experientia, 38:1453.

Iwatsuki, K., and Naitoh, Y., 1983, Behavioral responses in Paramecium multimicronucleatum to visible light, Photochem. Photobiol., 37:415.

Iwatsuki, K., and Song, P.-S., 1985, Deuterium oxide (D_2O) enhances the photosensitivity of Stentor coeruleus, Biophys. J., 48:1045.

Kim, I. H., Prusti, R. K., Song, P.-S., Häder, D.-P., and Häder, M., 1984, Phototaxis and photophobic response in Stentor coeruleus action spectrum and role of Ca^{2+} fluxes, Biochim. Biophys. Acta, 799:298.

Lenci, F., Colombetti, G., and Häder, D.-P., 1983, Role of flavin quenchers and inhibitors in the sensory transduction of the negative phototaxis in the flagellate, Euglena gracilis, Curr. Microbiol., 9:285.

Melkonian, M., Meinicke-Liebelt, M., and Häder, D.-P., 1986, Photokinesis and photophobic responses in the gliding flagellate, Euglena mutabilis, Plant Cell Physiol., 27:505.

Mikołajczyk, E., 1984a, Photophobic responses in Euglenina. 1. Effects of excitation wavelength and external medium on the step-up response of light- and dark-grown Euglena gracilis, Acta Protozool., 23:1.

Mikołajczyk, E., 1984b, Photophobic responses in Euglenina. 2. Sensitivity to light of the colorless flagellate Astasia longa in low and high viscosity medium, Acta Protozool., 23:85.

Niess, D., Reisser, W., and Wiessner, W., 1981, The role of endosymbiotic algae in photoaccumulation of green Paramecium bursaria, Planta, 152:268.

Prusti, R. K., Song, P.-S., Häder, D.-P., and Häder, M., 1984, Caffeine-enhanced photomovement in the ciliate Stentor coeruleus, Photochem. Photobiol., 40:369.

Reisser, W., and Häder, D.-P., 1984, Role of endosymbiotic algae in photokinesis and photophobic responses of ciliates, Photochem. Photobiol., 38:673.

Saji, M., and Oosawa, F., 1974, Mechanism of photoaccumulation in Paramecium bursaria, J. Protozool., 21:556.

Song, P.-S., 1981, Photosensory transduction in Stentor coeruleus and related organisms, Biochim. Biophys. Acta, 639:1.

Song, P.-S., 1983, Protozoan and related photoreceptors: molecular aspects, Ann. Rev. Biophys. Bioeng., 12:35.

Song, P.-S., Häder, D.-P., and Poff, K. L., 1980a, Phototactic orientation by the ciliate, Stentor coeruleus, Photochem. Photobiol., 32:781.

Song, P.-S., Häder, D.-P., and Poff, K. L., 1980b, Step-up photophobic response in the ciliate Stentor coeruleus, Arch. Microbiol., 126:181.

Song, P.-S., Tapley, K. J., and Berlin, J. D., 1983, The photoreceptor in Stentor coeruleus, in: "The Biology of Photoreception", D. J. Cosens, and D. Vince-Price, eds., Exp. Biol. Symp., Cambridge Univ. Press.

Song, P.-S., Walker, E. B., Auerbach, R. A., and Robinson, G. W., 1981, Proton release from Stentor photoreceptor in the exited states, Biophys. J., 35:551.

Suzaki, T., and Williamson, R. E., 1983, Photoresponse of a colorless euglenoid flagellate, Astasia longa, Plant Sci. Lett., 32:101.

Walker, E. B., Lee, T. Y., and Song, P.-S., 1979, Spectroscopic characterization of the Stentor photoreceptor, Biochim. Biophys. Acta, 587:129.

Walker, F. B., Lee, T. Y., and Song, P.-S., 1981, The pH dependence of photosensory responses in Stentor coeruleus and model system, Biochim. Biophys. Acta, 634:308.

Wood, D., 1976, Action spectrum and electrophysiological responses correlated with the photophobic response of Stentor, Photochem. Photobiol., 24:261.

HYDRA PHOTORESPONSES TO DIFFERENT WAVELENGTHS

Cloe Taddei-Ferretti°, V. Di Maio, S. Ferraro°° and
A. Cotugno

Istituto di Cibernetica del C.N.R., 80072 Arco Felice (NA)
Italy
°° Istituto di Zoologia, Università, 16126 Genova, Italy

Keywords. Hydra attenuata, photoreception, photoresponse, wavelength
effect, phase response curve.

Summary

Hydra is photosensitive, although missing known organized photoreceptive cellular structures. Its contraction-relaxation activity is periodic.
Short or long wavelength visible light has opposite effects (respectively
long/short main contraction-relaxation period during steady stimulation, low/
high efficiency in interrupting a contraction in progress, long/short reaction time until the next contraction after a pulse stimulation given just
after a contraction, decrease/increase of the above reaction time as long as
the time of the pulse application after a contraction increases) and they
interact on each other (as tested with two stimuli of different wavelengths
in close succession). A photocycle mechanism could be hypothesized.

1. Introduction

Hydra is photosensitive [5,12,13,14,15,16,18,19,20,21,22] although
missing known organized photoreceptive structures; either neurons or epithe-
liomuscular cells are the candidate elements for its photoreceptive function,
although not yet identified as bearing molecular structures (free in the cytoplasm or membrane bound) sensitive to light radiation. In other organisms
photosensitivity both of neurons internal to the body of transparent animals
and of epithelial tissues has been observed [1,24].
    The photosensitivity can be studied through the effect of light stimuli
on the contraction-relaxation behaviour, which is periodic (mean period of
the order of few minutes), and to which a periodic sequence of bioelectric
events is associated. It is known that: 1) after a positive/negative white

---

° Author to whom correspondence should be addressed.

light step the mean contraction-relaxation period increases/decreases [14, 20] ; 2) a contraction already in progress can be interrupted by a positive or negative step or pulse white light stimulus [14,20] ; 3) the reaction time until the next contraction after a pulse stimulus is not a constant but depends both on the polarity of the pulse (white light or darkness pulse) and on the phase of the stimulus application during the contraction-relaxation period (regardless of the contraction interruption efficiency) [20] ; 4) Hydra is red blind [14] ; 5) carotenoids (astaxanthin, cantaxanthin and xantophyll) have been found in Hydra and their level decreases with the intensity increase of permanent light [6] ; 6) various and partially contradictory are the results obtained by testing the reactivity to different wavelength light stimulation [5,16].

2. Materials and methods

Aposymbiotic Hydra magnipapillata was cultured a) in continuous darkness, b) in continuous 400 or 550nm light, c) with 12h white light and 12h darkness. The potentials were picked up by using a polyethylene suction Ag/AgCl electrode, a Tektronix 5031 Oscilloscope and a YEW 3047 pen recorder. Steady or 7.5s pulse light stimulation at different wavelengths was obtained by a Philips 6824, 12V, 100W lamp and Balzers filters, 5 cm diameter, 50nm band, maximum transmittance at 360, 400, 450, 500, 550, 600, 650, 700nm; the light transmitted was made equal in all cases (except 360nm) by means of neutral filters.

We define: P, contraction-relaxation mean period; PHI, phase along P beginning at the onset of a contraction and having normalized P in darkness as 1; RT, reaction time after a light pulse until the onset of the subsequent contraction; THETA, reaction time after a light pulse expressed in the same unit as PHI; SCND-THETA, reaction time after a second light pulse given 15s after the start of a first pulse, having normalized the mean reaction time after the first pulse as 1; WL, peak wavelength of the stimulus.

3. Results

1) P, tested during two hours in continuous darkness on animals cultured with the b) Method, was higher (5min 4s) for animals cultured under 400nm light during 3 months, lower (3min 45s) for animals cultured under 550nm during the same period.

2) The lowest P, tested during 45min under steady conditions at a definite wavelength illuminance on animals cultured with the c) Method, was obtained with 550nm (2min 9s. In the neighbouring bands: 5min 28s at 500nm and 2min 38s at 600nm).

3) The efficiency in interrupting a contraction already in progress, in animals cultured with the a) Method, was highest for 550nm pulses, which are effective also after the fourth single bioelectric event of a contraction (Fig. 1 and 2); 700nm pulses were effective only occasionally after the first event; 360 and 400nm pulses only after the first event; 450nm pulses also after the second one; 500, 600 and 650nm pulses also after the third one.

4) In animals cultured with the a) Method, the mean THETA after a pulse administered just after the onset of a contraction was higher after a 450nm pulse, lower in the 500-650nm range; it was low also at 360nm (in this case only the illuminance of the transmitted light was lower); the variance

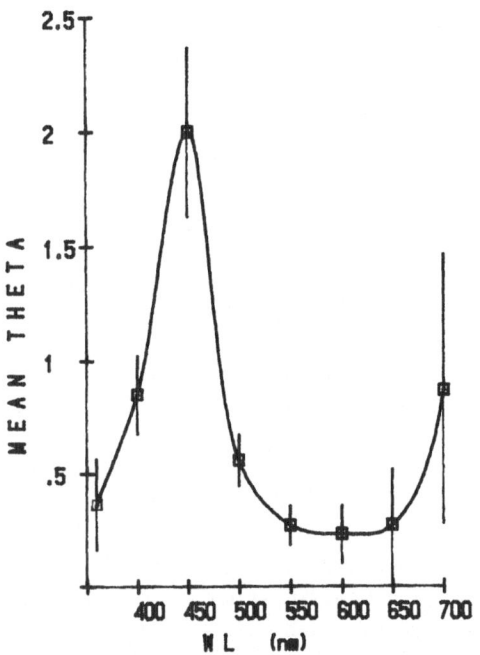

Fig. 1 - Mean THETA vs WL of a pulse efficient in interrupting a contraction immediately after the start of the 1st bioelectric event of the contraction.

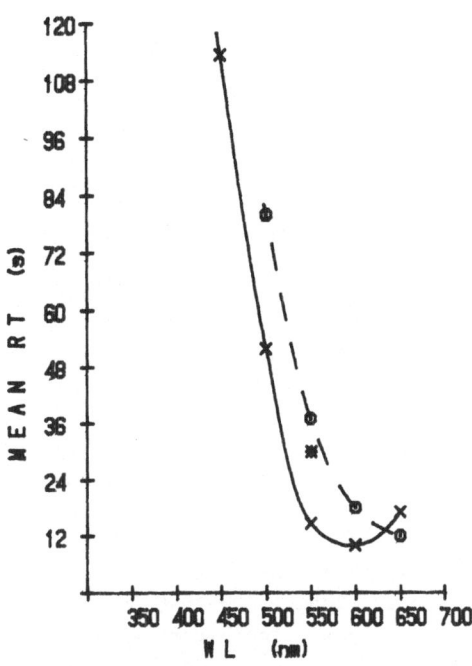

Fig. 2 - Mean RT vs WL of a pulse efficient in interrupting a contraction after the 2nd (X), 3rd (0), 4th (*) single bioelectric event of the contraction.

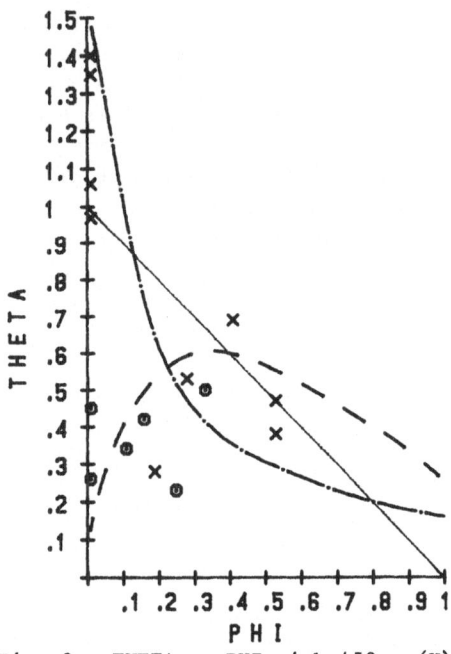

Fig. 3 - THETA vs PHI with 450nm (X) and 550nm (0) pulses; - - best fit for white light pulses,[20] ,— · — for darkness pulses from white light[20]; --- subthreshold stimuli.

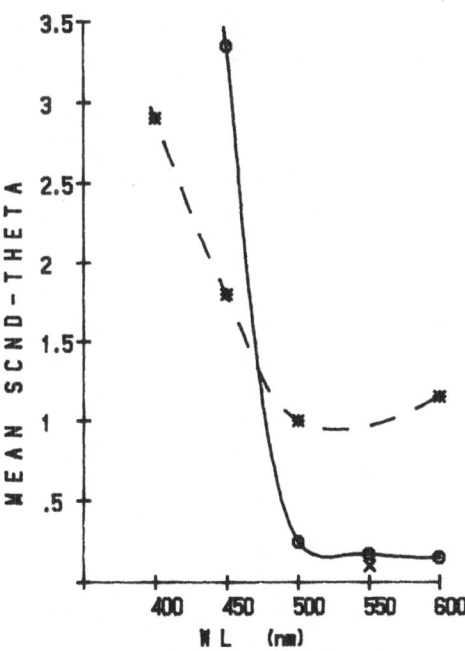

Fig. 4 - Mean SCND-THETA vs WL of a 2nd pulse after the 1st 400nm (0), 450nm (X), 550nm (*) pulse.

decreased at 500 and 550nm while it was highest at 700nm (Fig. 1).

5) With a) culture Method, THETA after a pulse administered at different PHIs varies with PHI (regardless of the efficiency of the pulse in interrupting a contraction if administered during a contraction in progress), in ways which depend on the stimulation pulse WL: the trend of the curve expressing such variations is opposite for 450nm and 550nm (Fig. 3).

6) In animals cultured with the a) Method, tested with two stimuli of different WL in close succession, the analysis of the mean SCND-THETA revealed that higher increase/decrease of the reaction time after the second pulse of low/high WL, compared to the reaction time after the first pulse, was obtained if the WL of the first pulse had been lower.

4. Discussion

From the analysis of the Results it is clear that:

1) Chromatic light has a long-lasting effect (Result 1: different behaviour in darkness of animals previously cultured for a long period under different chromatic light).

2) Different effects are obtained by short (400-450nm, and maximally 450nm) and long (500-550-600nm, and maximally 550nm) WL stimuli, which exhibit respectively inhibitory and excitatory effects. Such effects were observed both in steady conditions [Result 1: the above long-lasting effect producing a higher/lower P when animals were previously cultured under low/high WL steady light. Result 2: higher/lower P during continuous illumination at low/high WL; also Ellis (1970), by testing the growth rate, observed an inhibitory/excitatory effect of continuous blue/green light] and after a transient [Result 3: lower/higher efficiency of a pulse of various WLs in interrupting a contraction in progress , i.e. possibility of interrupting it only at the beginning/also later with low/high WL pulse. Result 4: occurrence of the next contraction after a reaction time higher/lower than the time until the next contraction in undisturbed conditions when low/high WL pulses are administered just at the beginning of a contraction. Result 5: opposite trend of the phase response curve when the above reaction time is tested at different PHIs after a pulse of low/high WL light; it has to be noted that: i) the different values obtained for low/high wavelength stimuli and corresponding to the early PHIs of Fig. 2 and to the earliest PHI of Fig. 1 are consistent with the data of Fig. 3; ii) the trend of the phase response curve obtained with 450/550nm pulses has a close resemblance respectively with that previously [20] obtained with darkness/white light pulses (cfr Fig. 3); also Ellis (1970) has observed that the effect of green light was similar to that of white light, and at this point it has to be noted that 550nm lies in the WL range which is less attenuated by clear water where Hydra normally lives [3]; iii) the Singer et al. (1963) data cover the zone of later PHIs not tested by us at present and are consistent with the above i) comment, as they show lower/higher reaction time for low/high WL pulses; on the contrary Ellis's (1970) different data on the reaction time are not confrontable as the PHI of the stimulus application is not reported].

3) The above inhibitory/excitatory effects of short/long WL counteract each other [Result 6: increase/decrease of the reaction time after a second pulse of low/high WL administered in close succession after the first one, compared to the reaction time after the first pulse alone. The fact that the above increase/decrease effect has different values depending on the first pulse WL, i.e. it is higher/lower when the first pulse had been at lower/

higher WL, is due to the fact that the second pulse arises at a new PHI -
obtained as effect of the first pulse [20] - which is different, and thus
to which corresponds a different sensitivity to a definite stimulus, depending on the first stimulus WL].

It is to be noted that, although data related to 360nm pulse stimulation were obtained at a lower illuminance, as they show a much lower response time with respect to that obtained with 450nm pulse stimulation,
they are giving in rough approximation a suggestion that the pigment responsible of the photoresponse involved in the modulation of the periodic
contraction-relaxation behaviour of Hydra should be a carotenoid [2]; this
assumption is consistent with the presence in Hydra of carotenoids with a
light-dependent concentration level [6].

Hydra color discrimination, as tested by the reported experiments,
could be explained either in terms of a dual system of two different photostable pigments or of one photoconvertable pigment with two reciprocally
reversible states of different chemical structure which are wavelength
dependent [23]; a photocycle compatible with the photochromic mechanism
seems to be preferably suggested in the present case, if one takes into
account the fact that 450nm and 550nm WL stimulation (or, more generally,
450nm stimulation and lower or higher than 450nm stimulation) have effects
which differ not only quantitatively but also from the qualitative point of
view [24], as the opposite (e.g., inhibition/excitation, phase delay/advance
of the system controlling Hydra periodic behaviour) types of reaction have
revealed. Antagonizing effects of light of two different WLs have been
already observed in bacteria [7,17], invertebrates [8,9,10,11] and verte-
brates [4]. In the present case stimulation with two very short pulses at
different WL very close each other and especially bichromatic irradiation
with background and pulse stimuli of different WL could give insight into
the problem.

References

1 P. A. V. Anderson and G. O. Mackie, Electrically coupled photosensitive
  neurons control swimming in a jellyfish, Science, 197 (1977) 186-188
2 G. Colombetti and F. Lenci, Identification and spectroscopic characterization of photoreceptor pigments. In F. Lenci and G. Colombetti (eds),
  Photoreception and sensory transduction in aneural organisms, Plenum
  Press, New York London 1980, pp. 173-188
3 H. R. Condit and F. Grum, Spectral energy distribution of daylight, J.
  Opt. Soc. Amer., 54 (1964) 973
4 W. D. Eldred and J. Nolte, Pineal photoreceptors; evidence for a vertebrate visual pigment with two physiologically active states, Vision Res.,
  18 (1978) 29-32
5 V. L. Ellis, The spectral sensitivity of Hydra carnea L. Agassiz (1850),
  J. Zool., 48 (1970) 63-68
6 M. Frigg, Vorkommen und Bedeutung der Carotinoide bei Hydra, Z. Vgl.
  Physiol., 69 (1970) 186-224
7 E. Hildebrand and N. Dencher, Two photosystems controlling behavioural
  responses of Halobacterium halobium, Nature (London), 257 (1975) 46-48
8 R. Menzel, Spectral sensitivity and color vision in Invertebrates. In
  H. Autrum (ed.), Handbook of Sensory Physiology, VII/6 Vision in Invertebrates, A. Invertebrate photoreceptors, Springer Verlag, Berlin 1979,
  pp. 503-580

9 B. Minke, S. Hochstein and P. Hillmann, Antagonist process as source of visible-light suppression of afterpotential in Limulus UV photoreceptors, J. gen. Physiol., 62 (1972) 787-791

10 J. Nolte and J. E. Brown, Ultraviolet-induced sensitivity to visible light in ultraviolet receptors in Limulus, J. gen. Physiol., 59 (1972) 186-200

11 K. Ohtsu, UV-visible antagonism in extraocular photosensitive neurons of the Anthomedusa, Spirocodon saltatrix (Tilesius), J. Neurobiol., 14 (1983) 145-155

12 L. M. Passano and C. B. McCullough, The light response and the rhythmic potentials in Hydra, Proc. Natl. Acad. Sci. USA, 48 (1962) 1376-1382

13 L. M. Passano and C. B. McCullough, Pacemaker hierachies controlling the behaviour of hydras, Nature, 199 (1963) 1174-1175

14 L. M. Passano and C. B. McCullough, Co-ordinating system and behaviour in Hydra. I. Pacemaker systems of the periodic contractions, J. Exp. Biol., 41 (1964) 643-664

15 L. M. Passano and C. B. McCullough, Co-ordinating system and behaviour in Hydra . II. The rhythmic potential system, J. Exp. Biol., 42 (1965) 205-231

16 R. H. Singer, N. B. Rushforth and A. L. Burnett, The photodinamic action of light in Hydra , J. Exp. Zool., 154 (1963) 169-173

17 J. L. Spudich and W. Stoeckenius, Photosensory and chemosensory behavior of Halobacterium halobium, Photobiochem. Photobiophys., 1 (1979) 43-53

18 C. Taddei-Ferretti and S. Chillemi, Modulation of Hydra attenuata rhythmic activity. V. A revised interpretation, Biol. Cybern., 56 (1987) 225-235

19 C. Taddei-Ferretti and L. Cordella, Modulation of Hydra attenuata rhythmic activity: Photic stimulation, Arch. Ital. Biol., 113 (1975) 107-121

20 C. Taddei-Ferretti and L. Cordella, Modulation of Hydra attenuata rhythmic activity: Phase response curve, J. Exp. Biol., 65 (1976) 737-751

21 C. Taddei-Ferretti, S. Chillemi and A. Cotugno, Modulation of Hydra attenuata rhythmic activity. IV. The mechanism responsible for the rhythmic activity, Exp. Biol., 46 (1987) 133-140

22 C. Taddei-Ferretti, L. Cordella and S. Chillemi, Analysis of Hydra contraction behaviour. In G. O. Mackie (ed.), Coelenterate ecology and behavior, Plenum Press, New York London 1976, pp. 685-694

23 B. Traulich, V. Hesse and G. Wagner, States of light adaptation in photo-attracted Halobacterium halobium cells, Intern. Conf. on Sensing and response in microorganisms, Rehovot (1985)

24 M. Yoshida, Extraocular photoreception. In H. Autrum (ed.), Handbook of Sensory Physiology, VII/6 Vision in Invertebrates, A. Invertebrate photoreceptors, Springer Verlag, Berlin 1979, pp. 581-640

416

ABSENCE OF COOPERATIVITY IN THE PRIMARY PHOTOINDUCED REACTION OF BATHOINTER-
MEDIATE FORMATION IN THE PHOTOCYCLE OF BACTERIORHODOPSIN AT ROOM TEMPERATURE

A.K. Dioumaev, V.V. Savranskii, N.V. Tkachenko, and V.I. Chukharev

Institute of General Physics
Moscow 117942, USSR

INTRODUCTION

On absorption of light by the bacteriorhodopsin chromophore, its spectrum is bathochromically shifted, corresponding to the formation of an intermediate K. Spectroscopically this photoinduced state does not differ greatly from the initial state (its spectral shift is small as compared to the FWHM of its absorption), making it difficult to estimate the percentage of excited molecules. The reaction is photoreversible and photoexcitation leads to a photostationary mixture of parent (bR) and photoproduct (K) states, in which the K/bR ratio is determined by the excitation parameters. Analysis of this photostationary state and the kinetics of its formation under different excitation conditions allows determination of the quantum yield $Ø1$ of the direct (bR $\rightarrow$ K) and $Ø2$ of the backward (K $\rightarrow$ bR) reactions, intermediate K spectrum, symmetry of the transition moment in bR and K states and its alternation in phototransition, as well as several conclusions on the mechanism of intermediate K formation, for example, cooperativity.

In the simplest case (a trivial reaction scheme bR $\leftrightarrow$ K, absence of cooperativity and photoselection effects, a simple mechanism when quantum yields are wavelength-independent, and excitation flashes are shorter than K decay time), the number (percent) of photoinduced K state formation [K] as a function of excitation energy [I] is given by the formula:

$$[K] = K \text{ sat} \times \{ 1 - \exp(-BxI) \} \qquad (1)$$
$$\text{where} \quad B = Ø1 \times \sigma\, bR + Ø2 \times \sigma\, K \qquad (2)$$
$$\sigma i - \text{ is the intermediate cross section}$$

The photostationary mixtures of bR and K (at cryogenic temperature) were studied as reported (reference 1), where two peculiarities of the primary reaction in bacteriorhodopsin were noted:

1. Saturation in experiments was achieved more slowly than predicted by formula 1, and this deviation increased with increasing doses of absorbed excitation. This led to an assumption of "negative" cooperativity in the backward reaction (K $\rightarrow$ bR).

According to reference (1), this assumption was confirmed during evaluation of experimental data. The authors assumed that, with the use of some additional data, it was possible to estimate independently the number of molecules in K state in the photostationary mixture and hence the spectrum of K. The quantum yields for the direct (Ø1) and backward (Ø2) reactions were calculated using the spectrum.

2. Unrealistic values strongly dependent on wavelengths were obtained for Ø2.

According to reference (1) these two results (deviation from formula 1 and Ø2 = f(λ)) contradict the simple photoreaction scheme, and a cooperative model for the primary act of the bacteriorhodopsin photocycle was introduced to explain them. This model postulated an intratrimer energy transfer between a photoexcited molecule bR* and a neighbouring molecule in K state with probability 0.43. Such a transfer results in transition of both molecules into bR state. This path is competitive to the usual path in which the K state is formed from the photoexcited state bR*, with probability equal to the value of quantum yield Ø1.

The effects of such cooperativity become evident only with approaching saturation, while the initial quasilinear region of the dependence [K] = f(I) is pratically not influenced.

The problem of the presence or absence of intratrimer energy transfer is important for an understanding of the mechanism of primary reactions in the bacteriorhodopsin photocycle, as well as for correct calculation of quantum yields and spectrum of K state.

MATERIALS AND METHODS

All our experiments were carried out at 23°C and pH 8 on an aqueous suspension of light-adapted purple membranes of **Halobacterium halobium.** Excitation was obtained by light pulses of different wavelengths of aprox. 20 nsec duration with energy of up to 30 mJ/cm. The kinetics of transient absorption changes at 632.8 nm, associated with K state formation, were recorded and the difference in absorption between bR and K was calculated using a multiexponential fitting programme.

RESULTS AND DISCUSSION

Formula 1 is valid only for optically thin samples, while in reference (1) a sample with OD = 0.52 was used. Unhomogeneous distribution of excitation energy inside the sample leads to a slightly different form of dependence of [K] on I (4):

$$[K] = K \text{ sat } x \{1 - \exp(\smallint -B \text{ x } I(x) \text{ x } dx)\} \qquad (3)$$

An illustration of this is given in Fig. 1. Deviation from exponent thus appears from the "non-ideal" experimental conditions (OD ≠ 0) even in a simple reaction scheme, and cannot be considered as an argument in favor of cooperativity.

Another source of deviation from the exponent in formula 1 may be photoselection effects (see Ref. 2).

In our experiments (Refs. 3 & 4) dependence [K] = f(I) was measured for nine wavelengths of excitation in the range 530-630 nm, using circular polarized excitation to eliminate photoselection problems. Analysis of these data (both initial quasilinear and saturation regions) enabled us to calculate quantum yields and K state spectrum (without using any addi-

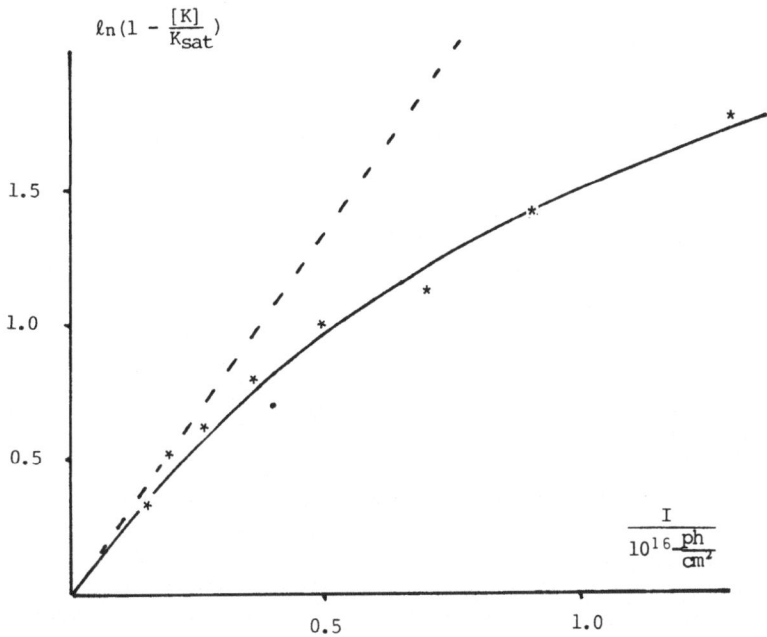

Fig. 1. Influence of increasing optical density on "percentage-dose"
dependence" (dependence of relative concentration of K,
[K]/Ksat, vs. incidental density of actinic flash) for different optical densities of the sample. [K]/Ksat = A/Asat,
A - transient increase of absorption at 632.8 nm, Asat -
its saturation value. * - experimental values for excitation
at 560 nm of the sample with optical density of OD = 0.95.
Solid line obtained by fitting parameters Asat and B in
formula 3 to experimental data in case of optically thick
sample (OD = 0.95). Broken line is the same dependence recalculated for the case of optically thin sample (OD → 0)
using same set of parameters B & Asat.

tional data, e.g. percentage of phototransformed molecules or spectra of
subsequent intermediates, etc.) for any assumed photoreaction scheme of
bathointermediate formation (for details, see reference 4).

Using the simple model without cooperativity, we obtained wavelength-
independent quantum yields (Ø1 = 0.31 ±0.10, Ø2 = 0.93 ± 0.14) and K state
spectrum with an extinction of 50 000 ± 6 000 $M^{-1}cm^{-1}\cdot l$ at its maximum at
580 nm at room temperature and pH 8.

Using the spectrum of bR and the calculated spectrum of K a differential spectrum between these two states may be calculated. On the other
hand, a differential spectrum (in arbitrary units) may be directly
measured by using, for istance, an optical multichannel analyser (OMA).
Comparison of these two differential spectra produces an experimental test
for the model suggested in our calculations (see Fig. 2).

In a set of experiments linear polarized excitation was used

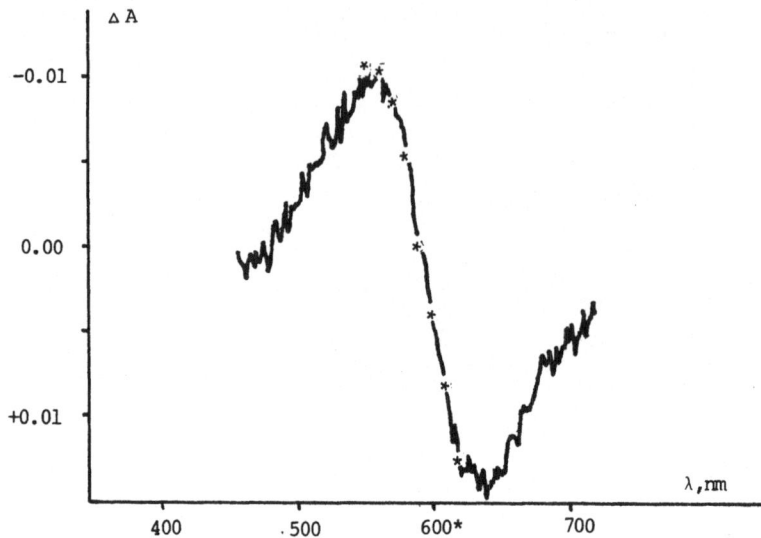

Fig. 2. Differential spectrum between bR and K states: * - values cal- culated using method of Ref. 4 for a reaction scheme without cooperativity; solid line - directly measured by optical multi- channel analyser (5) 80 nanoseconds after actinic flash, OD=0.45, averaging - 2000, room temperature, 6 Hz repetition rate (repro- duced by permission of L. Keszthelyi). This differential spec- trum was originally measured in arbitrary units and normalized to our calculated one. Most published differential spectra between bR and K were recorded 300-500 nanoseconds after excit- ation flash when 20-30% of photoexcited molecules are already in L state and hence cannot be used for comparison.

(polarized at different angles φ to probing beam polarization) to probe the symmetry of transition dipole moments in both bR and K states by measuring $[K](\varphi) = f(I)$ dependence. The three chromophores in bacterio- rhodopsin trimer are not parallel and any intratrimer energy transfer should lead to deviation from the dependence of $[K](\varphi) = f(I)$, predicted for pure dipoles in bR and K states. The angular dependence curves for parameters characterising the $[K](\varphi) = f(I)$ dependence were consistent with dipole approximation for both bR and K states; their dipoles proved to be parallel within 5 degrees, non-dipole contribution was less than 7% (both limits set by the accuracy of our experimental data) (see Fig. 3).

CONCLUSIONS

We have presented three arguments against cooperativity:

1. The non-exponentiality of percentage-dose dependence in K state formation can be readily explained by experimental conditions (OD >> 0), and cannot be used as an argument against the simple photoreaction mechanism;
2. The differential spectrum calculated for the model without cooperativity with wavelength-independent quantum yields coincides with that measured within experimental error limits;

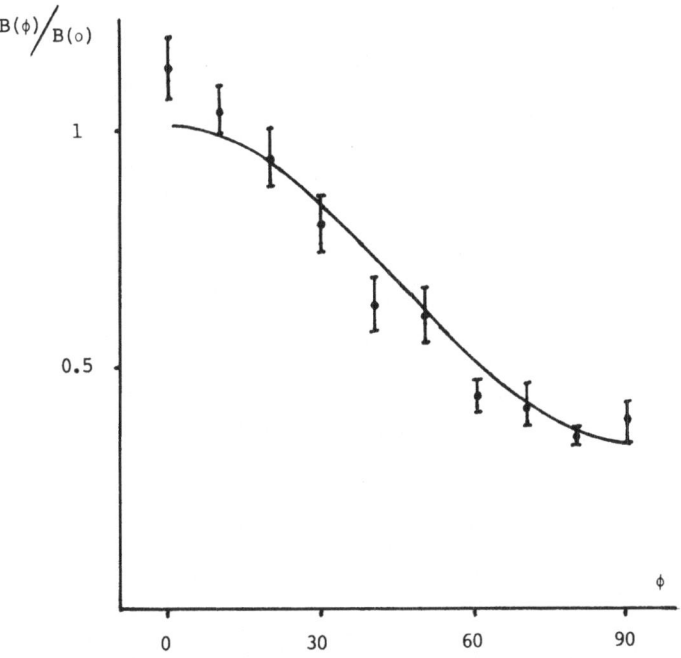

Fig. 3. Parameter B(φ) (see formula 2) - of "percentage-dose"
dependence vs. angle φ between polarizations of excitation and probe beams. Filled circles with error
bars - experimental values obtained for excitation at
532 nm. Solid line - angular dependence predicted for
pure colinear dipoles in bR and K states (B(φ)/B(0)=
(2+cos2(φ)/2). Any turn (for an angle ψ) of transient dipole moment in K as compared with bR would lead
to shift in maximum of this curve for a double angle
(2 ψ), in accordance with:

$$B(\varphi)= \emptyset 2x \, \sigma K + \emptyset 1x \sigma \, bR + 0.5x \, \{\emptyset 2x \, \sigma Kxcos(2\varphi -2 \psi) + \emptyset 1x \, \sigma \, bRxcos2\varphi\}$$

3. No changes in anisotropic response, associated with energy transfer between chromophores with non-colinear transition dipole
moments, were detected in our photoselection experiments.

We seem to have good reason to state that, at room temperature the
"primary" photochemistry of the bacteriorhodopsin photocycle may be described by the simplemechanism with wavelength-independent quantum yield
$\emptyset 1 = 0.31 \pm 0.10$ and $\emptyset 2 = 0.93 \pm 0.14$.

The initial report (1) in favor of cooperativity seems to be due
either to the specific influence of low temperature when excitation of
the frozen protein leads to creation of a bathointermediate different
from room temperature K state, or to errors in estimation of the fraction
of bacteriorhodopsin molecules phototransformed to K state.

REFERENCES

1. J. B. Hurley and T. G. Ebrey, <u>Biophys. J.</u> 22:49-66 (1978).
2. J. F. Nagle, S. M. Bhattacharjea, L.A.Parodi and R. H. Lozier, <u>Photochem. Photobiol</u>. 38:331-339 (1983).
3. A. K. Dioumaev, V. V. Savransky, N. V. Tkachenko and V. I. Chukharev, Abstr. P65 of the 1st European Congress of Photobiology, Grenoble, France (1986).
4. A. K. Dioumaev, V. V. Savransky, N. V. Tkachenko, V. I. Chukharev, Preprint 29 of General Physics Institute (1986).
5. L. Keszthelyi and J. Posfai, personal communication (1986).

THE STRUCTURE OF ANTENNA POLYPEPTIDES FROM PHOTOTROPIC BACTERIA

AND CYANOBACTERIA

Herbert Zuber

Institut für Molekularbiologie und Biophysik
Eidg. Technische Hochschule, ETH-Hönggerberg
CH - 8093 Zürich, Switzerland

INTRODUCTION

In photosynthesis light energy is converted into a photochemical
potential by means of the cooperative action of the light-harvesting
antennae and the reaction centers. The antenna system, which is composed
of various antenna complexes, funnels excited-state energy via a heterogeneous and direction-oriented energy transfer system to the reaction center
with little energy dissipation. It is generally assumed that energy migrates from pigment molecule to pigment molecule within $\sim 10^{-13}$ S and
reaches the reaction centers by random walk between the pigments within
$\sim 10^{-11}$S. It may also be postulated that practically all photons absorbed
form mobile electronic excited singlet (S_1) states called excitons in the
antenna complexes. The formation of larger (localized) excitons by strong
exciton coupling between pigments and/or Förster type energy migration by
weak coupling is discussed. To understand the physical mechanism and the
theoretical basis of the directed energy transfer to the reaction center,
knowledge of the structure and, particularly, general structural principles
of the antenna systems are important. Such general principles are best
derived from comparison of diverse antenna structures. Of special interest
is the question as to which structural elements within the wide variety of
antenna systems are invariant and are therefore of functional importance.

Fundamental to the formation of the functionally active antenna complexes (pigment-protein complexes) and finally of the whole antenna are
the structure and organization of the antenna polypeptides (1, 2). Structural and functional analysis revealed the following general features of the
whole antenna complexes with respect to the role of the antenna polypeptides:

1.) All pigment molecules are specifically bound to polypeptides and form
defined antenna complexes with a precise number of pigments (2-7
pigments per polypeptide in bacteria, cyanobacteria).

2.) The polypeptides determine the position, orientation, spacing and
environment (spectral shift) of the pigment molecules.

3.) The specific arrangement (aggregation) of a large number of relatively
small polypeptides (bacteria: 5-7 kD, cyanobacteria: 18-25 kD) forms
the basis of the three-dimensional structure of the antenna complexes,

i.e., the three-dimensional arrangement of the pigments to form the energy transfer system.

4.) A considerable number of antenna complexes associate via polypeptide-polypeptide interactions, thus forming the energy transfer system of the whole antenna.

5.) A system for heterogeneous energy transfer within the whole antenna is formed by the combination of various types of antenna complexes with pigments exhibiting different absorption maxima.

6.) There are intramentrane and extramembrane antenna systems (complexes), the two types differing considerably in the structure of the antenna polypeptides (membrane proteins, globular, fibrillar proteins).

7.) The number of pigments per reaction center varies between 25-1000.

In recent years we determined the primary structures of a number of antenna polypeptides (1) from purple (3-10) and green (11-13) photo-synthetic bacteria, cyanobacteria, red algae and cryptophyceae (2, 14-19). As a result of adaptation to diverse environmental conditions these organisms show a multitude of antenna structures with different antenna polypeptides and pigments. Light energy, as the most important environmental factor, determines the type of the pigment and of the polypeptide, i.e., the spectral range of the organisms. There is a clear distinction between the spectral range of the antenna system from oxygenic photosynthetic organisms such as cyanobacteria, algae and plants (500-700 nm) and from anoxygenic organisms such as bacteria (700-1000 nm). This corresponds to the large differences between the antenna polypeptides of cyanobacteria and red algae with respect to the extramembrane antenna of the phycobili-some (phycobiliproteins) or the intramembrane antenna of these organisms or of plants, and the antenna polypeptides of the extramembrane chlorosomes (green bacteria) and the intramembrane antenna of purple and green bacteria. On the other hand, light intensity determines the size of the antenna system and of the heterogeneous energy transfer system, i.e., the number of antenna complexes, pigments and polypeptides. Comparing the primary struc-tures of the antenna polypeptides of these various photosynthetic organisms, general structural principles of their antenna systems became apparent. It revealed also structural details such as specific domain structures (active sites) within these polypeptides which could be correlated to the functional (spectral) properties of the antenna complexes. These data are most impor-tant for the functional interpretation of the three-dimensional structure of these complexes obtained by present day or future X-ray structure analysis. In the following I will discuss some examples of these general structural principles found in the antenna polypeptides and complexes of photosynthetic bacteria and of cyanobacteria.

PURPLE PHOTOSYNTHETIC BACTERIA

The simplest antenna system is found within the photosynthetic mem-brane of bacteria. In purple photosynthetic bacteria it is located mainly in the intracytoplasmic membrane which is formed by the differentiation of the cytoplasmic membrane. By this process vesicular, lamellar, rod-shaped and tubular structures are formed (Rhodospirillaceae, Chromatiaceae). However, the antenna system of Rc. gelatinosus and Rs. tenue is located within the cytoplasmic membrane without special membrane extrusion. It is to be expected that this morphological variety of photosynthetic membranes will also be reflected in structural variations of antenna complexes and antenna polypeptides. On the basis of the near infrared absorption maxima

between 800 - 1000 nm, the purple bacteria can be classified into three
groups. Group I: Rs. rubrum, Rp. viridis (870 nm, 1015 nm), group II:
Rb. sphaeroides, Rb. capsulatus, Rs. gelatinosus (800, 850, 875 nm) and
group III: Rp. acidophila, Chromatium vinosum (800,820, 850, 880 nm).
These absorption maxima can also be considered representative of the
antenna complexes in these organisms: Group I has one complex: B 870 or
B 1015, group II possesses two complexes: B 875, B 800-850 and group III
has three complexes: B 880, B 800-850, B 800-820. Therefore all three
bacteria groups differ with respect to the extent (size) of their energy
transfer system in the neighbourhood of the reaction center. The B 870
complex (exception: B 1015) found in all bacteria is directly connected
to the reaction center (core complex) and the B 800-850 and B 800-820
complex surround the B 870 complex. Energy migrates from B 800-820 to
B 800-850, to B 870/890 and to the reaction center. Which typical
structures, structural differences and general structural principles can
be detected in these complexes on the basis of the primary structures of
the antenna polypeptides? Firstly, biochemical and sequence analysis
showed two types of polypeptides, named α- and β-polypeptides, in all of
these complexes, and 2-3 BChl or 1-2 carotenoids non-covalently bound.
The polypeptide pair (heterodimer) is the basic structure element of the
antenna complexes. We determined and compared the primary structures of
a number of α- and β-polypeptides from Rs. rubrum, Rp. viridis, Rb.
sphaeroides, Rb. capsulatus, Rp. acidophila, Rc. gelatinosus (3 - 10). The
antenna polypeptides consist in general of 50-60 amino acid residues, they
are sequence homologous and, aligned for maximum homology, show a typical
three-dimensional structure with polar, charged N- and C-terminal domains
and a central hydrophobic domain. The most homologous residue is His found
in the hydrophobic domain. It is most probably the main binding site for
BChla (b) molecules (via Mg-atom), as also Raman spectroscopy revealed. A
second His residue is present in all β-chains near the N-terminal domain.
It may form the binding site for the second (monomeric) BChl. The domain
structure of the antenna polypeptides indicates a transmembrane orientation
of these polypeptides: the hydrophobic domain forming a helix structure is
located within the membrane and the N- and C-terminal domains at the polar
head region or at the membrane surface. The concept of the α-helix structure
was supported by IR, UV and CD spectroscopy, and that of the transmembrane
orientation by IR-dichroism studies and by digestion experiments with
proteolytic enzymes. In intact vesicles with the cytoplasmic site outside,
only the N-terminal domains accessible from the cytoplasmic site were
split off. The functional (spectral) characteristics (properties) of the
antenna complexes correspond to this antenna polypeptide structure. It
reveals primarily an energy transfer between BChl 800 and B 850 (separation
of 20 Å, fluorescence measurements) and an exciton interaction between the
BChl 850 pair (distance 10-20 Å, strong CD-signal). Linear dichroism and
fluorescence polarisation measurements show that the BChl 850 are perpendicular to the membrane, whereas BChl 800 lies approximately in the
membrane plane. The α- and β-polypeptide pair (heterodimer) is formed by
specific polypeptide-polypeptide interactions via the amino acid side
chains at the α-helix surface or the N- and C-terminal domains. This forms
the basis for the tight packing and twisting of the helix pair. By polypeptide-polypeptide interactions the twisted helix-pair and the N- and C-
terminal domains can further aggregate in the lipid or aqueous environments
respectively, forming large arrays of α-, β-polypeptides. Most effective
from the structural and functional point of view would be cyclic hexamers
with 12 polypeptides in the case of the peripheric B 800-850 complex or
dodecamers in the case of the larger B 870 or B 1015 core complexes in the
environment of the reaction center. The dodecameric structure of B 1015 was

confirmed in Rp. viridis by electron microscope and image reconstruction studies. In these clusters of cyclic complex structures, four polypeptides (tetramers, 2α- 2β-) interact strongly. In the cyclic structures the BChl molecules are optimally spaced yielding effective energy migration within and between the antenna complexes. Heterogeneous energy transfer is created by combining the cyclic complexes with different absorption maxima.

The information content for the folding and aggregation of the antenna polypeptides and the interaction of the pigment molecules with each other or with the polypeptides (pigment environment) is already stored in their primary structure. Here the question arises if the corresponding structure elements representing this information content are detectable by comparing the primary structures of a series of antenna polypeptides. It is reasonable to assume that structurally or functionally important amino acid residues are conserved even in antenna polypeptides from phylogenetically distant bacteria. The existence of these amino acid residues is already indicated by the overall sequence homologies between the various α- and β-polypeptides. The α- and β-polypeptides which have different functions within the complexes show the lowest homology (7-13%), whereas the sequence homologies between the α- or β-polypeptides of the various complexes are higher (13-28%, polypeptide specific structure - function). The highest sequence homology is found between α- or β-polypeptides of the same complex type (28-78%, complex specific structure - function). With respect to the specific structure and function of the hydrophobic and the C- and N-terminal domain, a typical distribution of amino acid residues can be observed: 1.) Clusters of polar Ser and Thr residues are found at the C-terminal border between the hydrophobic and C-terminal domain of the α-chain (structural role). 2.) Clusters of acid amino acid residues (β-chains) and of Arg residues (α-chain) are present in the N-terminal domain. Interesting data on the structural - functional importance of certain amino acid residues became apparent by compairing the α- or β-polypeptides of antenna complexes (B 870, B 1015 or B 800-850, 800-820) of a larger number of purple bacteria (incl. Chloroflexus aurantiacus). This data reveals antenna complex specific subdomain structures, particularly with conserved aromatic and charged residues and typical amino acid residues such as Pro, Gly (folding), Thr, Asn (polarity) and Ala, Val, Leu (hydrophobic inter-actions). Typical for the B 870 (B 1015) core complex are clusters of aromatic residues (mainly Trp) in the N-terminal and C-terminal domain in the neighbourhood of charged residues, and in the hydrophobic domain. It is appealing to postulate that these clusters interact functionally with the BChl (carotenoids) causing the typical spectral properties (red shift) of this energy transfer system. In this sense, both chains interact. In contrast, in the B 800-850 (B 800-820) complex, these clusters are much less pronounced (lower content of aromatic residues) or missing. In the C-terminal domain the Arg residues close to the aromatic residues are missing. This may be the cause for the different spectral properties of the B 800-850 compared to B 870. Interestingly the only significant structural difference between B 800-850 and B 800-820 seems to be the substitution of the functionally important C-terminal Trp residue for Leu or Thr (α-chain). Typical differences with respect to conserved Arg, Asp and Glu residues and Pro, Asn, Thr, Ala, Val, Leu seem also to be correlated to the specific structural - functional properties of these complexes.

GREEN PHOTOSYNTHETIC BACTERIA

Green photosynthetic bacteria (Chlorobiaceae, Chloroflexaceae) are able to live at low light intensity which could be expected to influence the antenna structure. Thus, in addition to intramembrane antenna in the cytoplasmic membrane, there are also large extramembrane antennae, the chlorosomes, located on the inner surface of the cytoplasmic membrane. Their location and ultrastructure has been investigated by electron microscopy. In the case of the green gliding bacterium Chloroflexus aurantiacus, these studies and biochemical and biophysical analyses revealed three antenna complexes forming the whole antenna system: 1.) The intramembrane antenna complex B 808-866 in the environment of the reaction center (core complex) containing 20-25 BChla per reaction center. 2.) the extramembrane antenna complexes located in the chlorosomes, absorbing at 740 nm (B 740), correspondingly to the BChlc content. They contain approximately 10000 BChlc molecules. These complexes form particles visible in the electron microscope as rod-shaped elements. 3.) The base-plate BChla-complex which is located between the chlorosome and the cytoplasmic membrane (B 790). In the whole antenna system energy migrates in a heterogeneous energy transfer system from the chlorosome via the base plate to the B 808-866 complex and to the reaction center.

We isolated the α- and β-antenna polypeptides from the intramembrane antenna B 808-866 and determined their primary structures (12, 13). They consist of 44 and 51 amino acid residues respectively, and show a three-domain structure like the antenna polypeptides of purple bacteria. As antenna polypeptides they contain the central His residue for BChla binding. The relatively high sequence homology, particularly in the neighbourhood of His, to the α- and β-polypeptides of B 870 or B 1015 of purple bacteria (27-39%), means that these antenna polypeptides belong to the core complex B 808-866 of Chloroflexus located close to the reaction center. On the basis of these homologies, it can be postulated that the intramembrane antenna complexes of purple and green bacteria are phylogenetically related. We could also isolate and sequence the antenna polypeptide (BChlc binding polypeptide) from the B 740 antenna complexes of the chlorosome (11). It does not show a three-domain structure, but, following secondary structure predictions, it may possibly form an α-helix. In this α-helix structure 7 Gln and Asn residues are arranged asymmetrically and in a helical distribution and may be the possible binding sites for the 7 BChlc molecules. In this conformation the BChlc molecules can specifically interact via the carbonyl function of ring V and the hydroxyethyl group of ring I of the tetrapyrrole. This type of interaction, specific for BChlc (d, e) with its typical spectral properties, was shown with BChl model compounds, without the polypeptide matrix. A cluster of BChlc molecules, most probably exciton-coupled over the Qy-transition dipoles, is formed in this basic α-helix unit. It is further hypothesized that in the rod-shaped elements (antenna complexes) the α-helices are associated, over homodimers, in aggregates of 12 polypeptides (dimensions 52 Å x 60 Å). On the basis of these aggregates and the fibrillary arrangement of the rod-shaped elements, it can be assumed that the BChl molecules have their transition dipoles oriented more or less parallel to the long axis of the chlorosome, as was postulated on the basis of spectroscopic studies.

CYANOBACTERIA AND RED ALGAE

Cyanobacteria and red algae have three antenna systems each with characteristic features: first, as oxygenic organisms like the higher plants, they contain the two specific antenna systems of photosystem I and II located within the thylakoid membrane (not discussed here). Secondly, a large part of the light-harvesting capacity (about 50%) is provided by peripheral, extramembrane antennae, the phycobilisomes, which absorb light energy between 450-650 nm. They are arranged regularly on the surface of the thylakoid membrane, linked mainly to photosystem II. The phycobilisome (Mr 7-15 x 10^6) contain 300 - 800 phycobilin-pigment molecules covalently bound via thioether bonds to cysteine residues of the phycobiliproteins which form the antenna complexes of the phycobilisomes. Four main types of these watersoluble antenna complexes can be differentiated on the basis of their different absorption maxima: 1.) Phycoerythrin (PE, 545-567 nm) 2.) Phycoerythrocyanin (PEC, 568 nm) 3.) Phycocyanin (PC, 617-620 nm) and Allophycocyanin (AP, 650-670 nm). In the phycobilisome they form a heterogeneous energy transfer system from PE (PEC) → PC → AP → photosystem II. All phycobiliprotein antenna complexes are composed of α- and β-polypeptides (16-20 kDa). This α- and β-polypeptide pair (heterodimer) is, as in the antenna of photosynthetic bacteria, the basic structural - functional unit of the antenna system. Each polypeptide carries a typical number of pigment molecules. A characteristic feature, also with respect to the function, is their aggregation tendency to form trimeric $(\alpha/\beta)_3$ or hexameric $(\alpha/\beta)_6$ antenna complexes. These cyclic complexes are the basic elements of the whole phycobilisome antenna which have, in many cases, hemidiscoidal structures, as was found by electron microscopy. They appear as disc-shaped particles in these phycobilisomes.

In the following I will concentrate on the phycobilisomes and phycobiliproteins of the cyanobacterium Mastigocladus laminosus on which we have been working. In these phycobilisomes the trimeric or hexameric phycobiliproteins are specifically arranged to form a heterogeneous direction-oriented energy transfer system: 1.) Aggregates of AP trimers are located in the central core region and PC and PEC hexamers in the rod region (stacks of hexamers). On the basis or their varying absorption maxima, energy migrates from outside from the PEC via PC to the AP in the core region and from there to the photosystem II within the membrane. 2.) This specific arrangement of the phycobiliproteins is mediated and organized by the linker polypeptides (8-30 kDa, 10-20%) of the proteins of the phycobilisome). They bind specifically in the central cavity of the CP and PEC trimers and hexamers to form phycobiliprotein-linker polypeptide complexes. In this way they also modulate the spectral properties of these complexes to form a continuous energy transfer system. 3.) The central AP core region traps (collects) the light-energy. For this reason, specific trapping antenna complexes with extreme long wave length absorption are present, for instance the AP_B-containing antenna complex and the large 84 kDa polypeptide/β-16.2 kDa-complex. Also the core complexes are modulated by the 8.9 kD linker polypeptides.

We determined the primary structures of AP, CP, PEC and of part of the linker polypeptides (14 - 19). We could show that not only these phycobiliproteins are sequence homologous (21-67%) and phylogenetically related, building up the phycobilisome continuously during evolution, but interestingly also the phycobiliproteins and the linker polypeptides. As in the case of the antennae of photosynthetic bacteria, the primary structure data of the phycobiliproteins reveal details of the molecular structure and function of these antenna complexes. Sequence homology indicates similar folding, aggregation and functional properties of these three phycobiliproteins. The typical covalent binding sites of the blue or red bilin pigments could be located by primary structure analysis: 1.) The binding site at cysteine 84, most important for energy transfer and found in all α- and β-polypeptides. 2.) The variable binding site in position 155 of the β-polypeptide (PC, PEC and also PE) and additional binding sites found only in PE (Fremyella diplosiphon) in position 50/61 and 143a. Most important for the spectral (functional) properties of the various phycobiliproteins is the pigment polypeptide environment (about 6 amino acid residues at the α 84 and β 84 pigment). The conserved Asp and Arg residues (close to position 84) should be of functional significance, as the aromatic residues (together also with Arg residues), particularly in the environment of the α 84 and β 84 pigments seem to be. They are preferred in the long wave length absorbing AP and especially in AP^B (with Trp), like in the bacterial antenna.

As has been demonstrated, the basic units in the phycobiliprotein antenna complexes are the α-β heterodimers (monomers) which associate to form cyclic trimers and hexamers. This was revealed by X-ray structure analysis of phycocyanin trimers of Mastigocladus laminosus in a collaborative work with T. Schirmer, R. Huber, W. Bode, Max-Planck Institute, Martinsried (19). In this analysis the structure of the CP-trimers, the folding of the α- and β-polypeptide chains in the α-β heterodimer and the position of the pigments α 84 and β 150 could be determined. The α- and β-polypeptides of the heterodimer (monomer) show a practically identical secondary structure containing 8 α-helices. This α-helical structure is very similar to the α-helical structure of myoglobin, demonstrating an interesting phylogenetic relationship of an old cyanobacterium, the first oxygen producer, with a younger organism consuming oxygen. The α 84 and β 84 pigments are bound at equivalent binding sites, also with respect to the three-dimensional structure, which is in the instance of α 84 or β 84 practically identical to the non-covalent binding site of the haem in myoglobin. On the other side of the complex, the β-155 pigment is bound in a different environment. Through the formation of CP trimers arranging the a-β heterodimers in form of a water-wheel (diameter 110 Å, thickness 30 Å, central hole 35 Å), the bilins α 84 and β 84 are brought into proximity (distance 20 Å center - center). α 84 lies on the periphery, β 84 protrudes into the central hole of the trimer. Of interest is the polypeptide environment and the interacting amino acid residues (spectral properties). Strikingly all three pigments α 84, β 84 and β 155 exhibit very similar geometries showing a common principle of pigment-polypeptide interaction. Many of the amino acid residues, predicted as interaction sites in the primary structure, are found in the three-dimensional structure as interaction sites. Particular important interaction sites are the Asp residues (α 84 - Asp 87, β 84 - Asp 87, β 155 - Asp 39), the nitrogens of the pyrroles being within hydrogen-bonding distance of one of the carboxylate oxygens. The propionic acid side-chains of the tetrapyrroles form salt-bridges with Arg or Lys residues, or protrude into the water. The pigments α 84 and β 84 stretch in a cleft between helices E and F. They are partly exposed to solvent in the heterodimer, but are completely covered by the adjacent heterodimer in the trimeric state. In the trimers the α- and β-pigments are energetically coupled (possible exciton coupling). M. Mimuro in our laboratory could show by steady state absorption, circular dichroism, fluorescence and fluorescence polarisation spectroscopy that energy transfer occurs from the sensitizing α 84 pigment in one heterodimer to the fluorescent β 84 pigment in the adjacent heterodimer and from the sensitizing β 155 pigment to β 84 in the same heterodimer. The position and orientation of the bilin pigments result in directed energy transfer within the trimer towards the β 84 pigment in the center of the trimer. By association of two trimers hexamers are formed, increasing the energy transfer system by coupling a larger number of pigments. The strongest coupling occurs between two β 155 and two α 84 pigments (distance 37 Å) for inter-trimer energy transfer. The hexamers aggregate further to form stacks of hexamers in the rod region of the phycobilisome. It is to be expected that the strongest coupling occurs between the β 84 pigments of adjacent hexamers located in the central hole of the hexamer. In this region energy can migrate via random walk along a one-dimensional array of β 84 pigment molecules to the central core region of the phycobilisomes. This energy migration will be direction oriented if the β 84 pigments have different absorption maxima (heterogeneous energy transfer).

In conclusion the following statements on general structural principles in antenna systems can be made (1):

1.) The pigments are bound specifically to (in most cases) small polypeptides. The pigment organization is based on the specific organization of the polypeptides.

2.) The pigment molecules in the antenna complexes and antenna systems are highly ordered and form pigment clusters for functional reasons.

3.) Pigment clusters are important for heterogeneous, directed energy transfer between antenna complexes. They are spatially separated and have different absorption maxima to minimize random walk to the reaction center.

4.) Pigment clusters are also found within antenna complexes in the form of localized excitons or energy traps of pigment pairs or pigment oligomers (direction of energy transfer within the complexes).

The experimental work performed in our laboratory and reported here was supported by the Eidg. Technische Hochschule and by the Swiss National Foundation projects No. 3.286-0.82 and 3.207-0.85.

REFERENCES

(1) H. Zuber, Structure and function of light-harvesting complexes and their polypeptides, Photochem. Photobiol. 42: 821 (1985).

(2) H. Zuber, Structural organization of tetrapyrrole pigments in light-harvesting pigment-protein complexes, in: Optical properties and structure of tetrapyrroles, Walter de Gruyter, Berlin, p. 425 (1985).

(3) R.A. Brunisholz, P.A. Cuendet, R. Theiler, and H. Zuber, The complete amino acid sequence of the single light-harvesting protein from chromatophores of Rhodospirillum rubrum G-9[+], FEBS Lett. 129:150 (1981).

(4) R.A. Brunisholz, F. Suter, and H. Zuber, The light-harvesting polypeptides of Rhodospirillum rubrum. I. The amino acid sequence of the second light-harvesting polypeptide B 880-β (B 870-β) of Rhodospirillum rubrum S 1 and the carotenoidless mutant G-9[+]. Aspects of the molecular structure of the two light-harvesting polypeptides B 880-α (B 870-α) and B 880-β (B 870-β) and of the antenna complex B 880 (B 870) from Rhodospirillum rubrum, Hoppe-Seyler's Z. Physiol. Chem. 365:675 (1984).

(5) R.A. Brunisholz, V. Wiemken, F. Suter, R. Bachofen, and H. Zuber, The light-harvesting polypeptides of Rhodospirillum rubrum. II. Localization of the amino terminal regions of the light-harvesting polypeptides B 870-α and B 870-β and the reaction centre subunits L at the cytoplasmic side of the photosynthetic membrane of Rhodospirillum rubrum G-9[+], Hoppe-Seyler's Z. Physiol. Chem. 365: 689 (1984).

(6)   R.A. Brunisholz, F. Jay, F. Suter, and H. Zuber, The light-harvesting
        polypeptides of Rhodopseudomonas viridis. the complete amino acid
        sequence of B 1015-α, B 1015-β and B 1015-γ, Biol. Chem. Hoppe-Seyler
        366:87 (1985).

(7)   R. Theiler, F. Suter, V. Wiemken, and H. Zuber, The light-harvesting
        polypeptides of Rhodopseudomonas sphaeroides R-26.1. I. Isolation,
        purification and sequence analyses, Hoppe-Seyler's Z. Physiol. Chem.
        365:703 (1984).

(8)   R. Theiler, and H. Zuber, The light-harvesting polypeptides of Rhodo-
        pseudomonas sphaeroides R-26.1. II. Conformational analyses by
        attenuated total reflection infrared spectroscopy and possible mole-
        cular structure of the hydrophobic domain of the B 850 complex, Hoppe-
        Seyler's Z. Physiol. Chem. 365:721 (1984).

(9)   R. Theiler, F. Suter, H. Zuber, and R. J. Cogdell, A comparison of the
        primary structures of the two B 800-850 apoproteins from wild-type
        Rhodopseudomonas sphaeroides strain 2.4.1 and the carotenoidless
        mutant strain R 26.1, FEBS Lett. 175:231 (1984).

(10)   R. Theiler, F. Suter, J.D. Pennoyer, H. Zuber, and R.A. Niederman,
        Complete amino acid sequence of the B 875 light-harvesting protein
        of Rhodopseudomonas sphaeroides strain 2.4.1, FEBS Lett. 184:231
        (1985).

(11)   T. Wechsler, F. Suter, R.C. Fuller, and H. Zuber, The complete amino
        acid sequence of the bacteriochlorophyll c-binding polypeptide from
        chlorosomes of the green photosynthetic bacterium Chloroflexus
        aurantiacus, FEBS Lett. 181:173 (1985).

(12)   T. Wechsler, R. Brunisholz, F. Suter, R.C. Fuller, and H. Zuber,
        The complete amino acid sequence of a bacteriochlorophyll a binding
        polypeptide isolated from a cytoplasmic membrane of the green photo-
        synthetic bacterium Chloroflexus aurantiacus, FEBS Lett. 191:34 (1985).

(13)   T.D. Wechsler, R.A. Brunisholz, G. Frank, F. Suter, and H. Zuber,
        The complete amino acid sequence of the antenna polypeptide B 806 -
        866 β from the cytoplasmic membrane of the green bacterium Chloro-
        flexus aurantiacus, FEBS Lett. 210:189 (1987).

(14)   G. Frank, W. Sidler, H. Widmer, and H. Zuber, The complete amino acid
        sequence of both subunits from the cyanobacterium Mastigocladus
        laminosus, Hoppe-Seyler's Z. Physiol. Chem. 359:1491 (1978).

(15)   W. Sidler, W. Gysi, E. Isker, and H. Zuber, The complete amino acid
        sequence of both subunits of allophycocyanin, a light-harvesting
        protein-pigment complex from the cyanobacterium Mastigocladus
        laminosus, Hoppe-Seyler's Z. Physiol. Chem. 362:611 (1981).

(16)   P. Füglistaller, F. Suter, and H. Zuber, The complete amino-acid
        sequence of both subunits of phycoerythrocyanin from the thermophilic
        cyanobacterium Mastigocladus laminosus. Hoppe-Seyler's Z. Physiol.
        Chem. 364:691 (1983).

(17)   P. Füglistaller, R. Rümbeli, F. Suter, and H. Zuber, Minor polypep-
        tides from the phycobilisome of the cyanobacterium Mastigocladus
        laminosus, Hoppe-Seyler's Z. Physiol. Chem. 365:1085 (1984).

(18)   W. Sidler, B. Kumpf, F. Suter, W. Morisset, W. Wehrmeyer, and H. Zuber,
        Structural studies on cryptomonad biliprotein subunits. Two different
        α-subunits in Chroomonas phycocyanin-645 and Cryptomonas phycoerythrin-
        545. Biol. Chem. Hoppe-Seyler 366:233 (1985).

(19) T. Schirmer, W. Bode, R. Huber, W. Sidler, and H. Zuber, X-ray crystallographic structure of the light-harvesting biliprotein C-phycocyanin from the thermophilic cyanobacterium Mastigocladus laminosus and its resemblance to globin structures, J. Mol. Biol. 184:257 (1985).

STRESS AND ADAPTATION IN PHOTOSYNTHESIS

Gunnar Öquist

Department of Plant Physiology
University of Umeå
S-901 87 Umeå, Sweden

INTRODUCTION

Plants are exposed to a large number of abiotic as well as biotic
stress factors that affect the organization and function of photosynthesis,
and we know of a large number of responses at different organizational
levels of photosynthesis that may be considered as adaptive responses to
tolerate or avoid stress (Levitt, 1980). Because this is a congress for
photobiologists I have chosen to discuss how the function of photosynthesis
is inhibited by excessive excitation of plants with visible light, and how
plants have acquired different mechanisms to avoid or tolerate photoinhibition of photosynthesis. Photoinhibition of photosynthesis is a well
known phenomenon and it is also established that it may occur under natural
field conditions (Öquist et al., 1987; Neale, 1987). Photoinhibition has
been studied very extensively over the last 10 years. The present status
of knowledge has recently been reviewed (Powles, 1984). Mechanisms by which
plants avoid or tolerate photoinhibition is much less studied and suggested
protective mechanisms should at present be considered as working hypotheses.

PHOTOINHIBITION OF PHOTOSYNTESIS

It is generally assumed that photoinhibition of photosynthesis may occur
under conditions when the pigment antenna absorbs light in excess to what can
be orderly dissipated in photosynthesis (Osmond, 1981). At the leaf level,
photoinhibition results in a decrease of the quantum yield of CO_2-uptake or
O_2-evolution. Moderate photoinhibition is fully reversible after the plants
have been shaded or transfered to lower light levels. The time needed for
full recovery varies between species from about an hour in unicellular algae
(Lidholm et al., 1987) to several hours in higher plants (Ögren et al.,
1984).

Mechanism of photoinhibition

It is well established that photosystem II is a major site of photoinhibition (Powles, 1984). Since photoinhibition causes a decrease of variable chlorophyll fluorescence even at 77K, where only the pure photochemical
reactions of photosystem II are operative, the site of inhibition has been
localized to the reaction center (Fork et al., 1981). Although still
hypothetical, it appears today that there may be at least two principle
mechanisms responsible for photoinhibition of photosynthesis in intact

433

plants. There is an increasing number of evidences that the herbicide binding D_1- or Q_b-protein is one target in photoinhibition (Ohad et al., 1984). The feasibility of this hypothesis is supported by the recent suggestion that the D_1- and D_2-proteins are components of the reaction centers of photosystem II in analyogy with the M- and L-proteins of photosynthetic bacteria (Deisenhofer et al., 1985). To explain the concomitant loss of variable chlorophyll fluorescence, damaged reaction centers may be converted to quenchers of fluorescence. However, there are also evidences that photoinhibition of photosynthesis in vivo may be caused by increased thermal dissipation of excited chlorophylls at the antenna (Demmig and Björkman, 1987) or the reaction center (Öquist, unpublished data) levels, without destruction of the photochemical competence of the reaction centers. This would quench variable chlorophyll fluorescence and lower the photochemical efficiency under light limiting, but not light saturating, photosynthetic assay conditions (see below).

It should also be mentioned that sites in photosynthesis other than related to the function of photosystem II may be affected by excessive excitation (Krause and Cornic, 1987).

In neither intact plants nor in isolated chloroplasts or thylakoid membranes is the presence of oxygen a prerequisit for the occurrence of photoinhibition of photosystem II. However, the mechanisms for photoinhibition may be somewhat different, as well as the magnitude of inhibition, in the presence and absence of oxygen (Krause and Cornic, 1987). However, the function of oxygen is complex as it has the potential both to contribute to damages to photosynthesis if reactive oxygen species are formed, and to protect from photoinhibition by serving as electron acceptor in a Mehler type reaction or through the function of photorespiration in C_3-plants (see below).

Susceptibility of photoinhibition

Plants acclimated to low light, or genotypically adapted to shade, are especially prone to photoinhibition when exposed to light levels above that of the growth regime (Powles, 1984). However, other stress factors such as low temperatures (Ögren et al., 1984), high temperatures (Bongi and Long, 1987) and drought (Ögren et al., 1984) may predispose photosynthesis to photoinhibition under relatively low light levels that would not induce photoinhibition under favourable environmental conditions. The hypothesis that photoinhibition occurs when the reaction centers receive excitation energy in excess to what can be orderly dissipated through photosynthesis (Osmond, 1981) also apply in this case; light comes in as a secondary stress factor when the rate of photosynthesis is limited by other environmental stress factors.

PROTECTION TO PHOTOINHIBITION

Since we can envisage numerous field situations when plants are exposed to light, often for prolonged periods of time, under conditions when the rate of photosynthesis is restricted or fully inhibited by unfavourable climatic conditions, we may ask if plants have specific mechanisms to avoid or tolerate excessive excitation of photosynthesis and photosystem II. I will in this communication concentrate on possible, qualitative protective mechanisms at the subcellular levels although plant morphology and leaf anatomy may also adapt to minimize light absorption under high light conditions.

Avoidance of photoinhibtion

By accepting the hypothesis that photoinhibition of photosynthesis occurs under conditions when the rates of transfer of excitation energy from the light-harvesting antennae to the photochemical reaction centers are in excess to the rates of energy transfer from the reaction centers to the electron transport chain and the carbon reduction cycle (Osmond, 1981), we may consider that protective mechanisms related to a decrease of the excitiation pressure of the reaction centers would be means by which plants avoid photoinhibitory conditions to be created.

High light responses in the function of photosynthesis. It is well established that plants acclimated or genotypically adapted to high light have a reduced capacity to absorb photons but a high capacity to orderly dissipate photons in photosynthesis (Björkman, 1981). This is achieved by high light adapted plants having a relatively small light absorbing chlorophyll antenna and high capacities of both the photosynthetic electron transport and the reductive carbon cycle. We may generalize that within certain limits set by the genotype most species have the ability to adjust in consert both the antenna size (and photon absorption) and the capacity of the biochemical steps of photosynthesis, thereby avoiding excess excitation of photosynthesis and photoinhibition under increased light exposure.

Photorespiration and photoinhibition. Photorespiration of C_3-plants has a very high energy demand and it was early hypothesised that photorespiration may contribute to protect from photoinhibition under conditions of high light and limiting CO_2. Support for the hypothesis is given in experiments by Cornic and Powles and Osmond (for reviews see Powles, 1984; Krause and Cornic, 1987) showing that leaves of C_3-plants suffer photoinhibition when illuminated in CO_2-free air with a low oxygen pressure to suppress photorespiration. Full protection to photoinhibition was, however, obtained if the oxygen content was increased allowing the photorespiratory pathway to operate. The carbon turnover obtained at the CO_2-compensation point is often sufficient for full protection. The protective role of photorespiration appears to be twofold: 1) the photorespiratory pathway fueled by endogenous carbon reserves consumes NAD(P)H and ATP during photorespiratory evolution of CO_2; 2) refixation of photorespiratory evolved CO_2 via the carbon reduction cycle serves to increase the turnover of NADPH and ATP generated in photosynthesis.

The Mehler reaction and photoinhibition. The possibility of oxygen functioning as a Hill oxidant in a Mehler type reaction in vivo by accepting electrons on the reducing side of photosystem I has been discussed (cf. Krause and Cornic, 1987). The finding that photoinhibition of photosystem II may be promoted under anaerobosis in both isolated chloroplasts and intact leaves (Krause et al., 1987), but drastically decreased by low oxygen levels, suggests that oxygen may excert protection by serving as an electron acceptor draining electrons from the electron transfer chain. A benificial function of this would, however, be coupled to the need of scavenging systems such as superoxide dismutase and catalase to prevent accumulation of harmful superoxide radicals. It is demonstrated that these scavangers may protect from photoinhibition in isolated chloroplasts (Barényi and Krause, 1985). However, a Mehler reaction would also confer protection to photoinhibition by maintaining a high pH-gradient across the thylakoids (see below).

Thermal dissipation and photoinhibition. By the use of isolated, intact spinach chloroplasts Krause and Behrend (1986) have demonstrated that photoinhibition was considerably diminished when the chloroplasts were in a low fluorescent state related to a high proton gradient across the thylakoid membranes. The protection was abolished when uncouplers were added to eliminate the proton gradient. The hypothesis is that photoinhibition is partly avoided by an increased thermal dissipation of excitation energy, when the energy-dependent, or pH-dependent, quenching is high. It has furthermore been shown (Krause and Laasch, 1987; Weis and Berry, 1987) that the quantum

435

yield of photosynthesis is negatively correlated with the energy-dependent quenching of chlorophyll fluorescence. It is suggested that the negative correlation observed between the quantum yield of photosynthesis and the energy-dependent quenching indicates a regulatory mechanism that allows a tuning of the thermal dissipation of absorbed light to the energy requirement of photosynthesis. Such a regulatory mechanism would induce a decrease in the photochemical yield of photosystem II under conditions when photosynthesis approaches light saturation, thereby avoiding the occurrence of photoinhibition.

However, there also exists a high light induced increase of thermal dissipation of excitation energy that lasts for several hours and therefore must result from alterations in the thylakoids other than that caused primarily by a high pH gradient. Demmig and Björkman (1987) explained photoinhibition of photosynthesis induced under certain conditions by an increased thermal dissipation of excitation energy at the antenna level, possibly by zeaxanthin formed by the xanthophyll cycle. This conclusion was arrived at after analyses of low temperature (77K) chlorophyll fluorescence kinetics data of photosystem II according to the bipartite model of Butler (1978), and it was primarely based on the observation that both nonvariable (F_0) and maximum (F_m) chlorophyll fluorescence were quenched in photoinhibited leaves.

Table 1. F_0, F_m, F_v, F_v/F_m values of photosystem II
fluorescence kinetics measured at 695 nm
and 77K of needles of Pinus sylvestris after
exposure of detached leaves to 15°C for 2 h in
the dark and in the light of a photon flux
density of 3000 umolm^{-2}s^{-1}. Standard deviation
is given in parentheses for n = 3

Treatment	F_0	F_m	F_v	F_v/F_m
Dark	51 (4)	351 (39)	301 (35)	0.86 (0.01)
Light	56 (6)	115 (7)	60 (4)	0.52 (0.03)

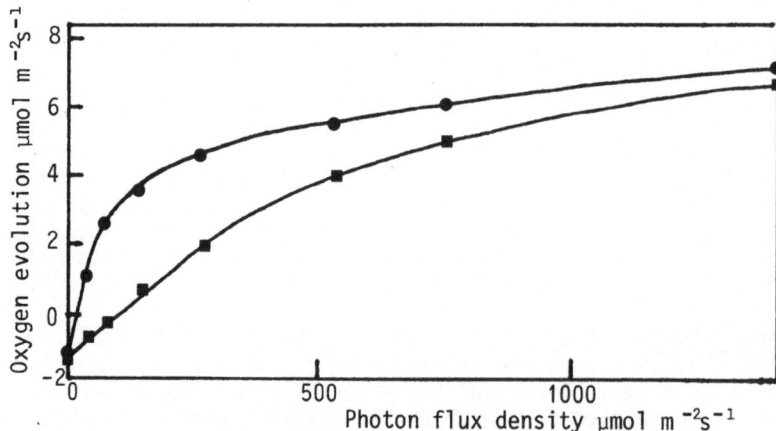

Figure 1. Light response curves of photosynthetic
oxygen evolution in needles of Pinus sylvestris
measured after treatments for 2 h at 5°C in
darkness (●) and in light (■) of 3000 μmolm^{-2}s^{-1}

A similar fluorescence analysis of needles of Pinus sylvestris before and after photoinhibition at 3000 $\mu mol m^{-2} s^{-1}$ and 15°C for 2 h revealed that F_0 was largely unaffected, whereas the variable fluorescence (F_v) was strongly quenched (Table 1). A strong quenching of F_v without effect on F_0 implies that the energy quenching is at the reaction center level; i.e. an increased k_d according to the bipartite model of Butler (1978). The finding that photoinhibition of CO_2-saturated, photosynthetic oxygen evolution of pine was apparent only under light limiting assay conditions, but largely overcome under saturating assay light conditions (Figure 1), suggests that photoinhibited pine had received a decreased photochemical yield of photosystem II because of an increased thermal dissipation competing with photochemistry for photons.

The molecular mechanism for this type of photoinhibition is unknown, but it does apparently not in the first place destroy the reaction centers, which is commonly considered to be the cause of photoinhibition (Ohad et al., 1984). It is hypothesized that the creation of increased thermal dissipation of excitation energy, at antenna as well as reaction ceter levels, is a mean to avoid too severe destruction of the reaction centers due to photodynamic processes. It is suggested that light response curves of photosynthesis can be used to resolve between photoinhibition caused by destruction of reaction centers or by increased heat dissipation. Whether an increased thermal dissipation of excitation energy at the reaction center level (increased k_d; Butler, 1978), supposedly causing a down regulation of the photo-chemical yield of photosystem II, is just an early response in photoinhibition eventually followed by protein breakdown and replacement during recovery, or a mechanism principally different from reaction center destruction, is at present not known.

Energy distribution and photoinhibition. During the light induction phase of photosynthesis phosphorylation of some 20 % of the light-harvesting complex (LHC) of photosystem II occurs resulting in a lateral displacement of phosphorylated LHC away from photosystem II in the grana partition and into the stroma thylakoids where displaced LHC may associate with photosystem I (Williams and Allen, 1987). The decreased antenna size of photosystem II achieved upon phosphorylation of a portion of LHC has by Horton and Lee (1985) been related to an increased tolerance to photoinhibition in isolated pea thylakoid membranes. It is hypothesized that phosphorylation of LHC decreases the excitation pressure of the reaction centers of photosystem II thereby avoiding the occurrence of photoinhibition.

Changes in energy distribution, possibly related to avoiding photo-inhibition of photosystem II, has also been observed in the intertidal red alga Porphyra perforata (Öquist and Fork, 1982). This alga becomes desiccated during low tide but it does not suffer from photodestruction of photosystem II since full photosynthetic competence is regained soon after rewetting. It was found that desiccation of Porphyra resulted in an increased distribution of excitation energy in favour of photosystem I, which is less sensitive to excess light than is photosystem II. It was suggested that this was a mean to protect photosystem II from photo-inhibition in the dry state thereby ensure full photosynthetic competence upon rewetting during periods of high tide.

Tolerance photoinhibition

Besides the avoidance mechanisms discussed above that operate to decrease the excitation density of photosystem II, and thereby avoid photoinhibitory conditions to be created, there appears to be mechanisms related to tolerating photoinhibitory damages rather than to avoiding them. Based on work with the cyanobacterium Anacystis nidulans it was suggested that there is a balance between photoinhibitory and restorative processes

(Samuelsson et al., 1985); the relative rates of these counteracting processes determine the net photoinhibitory damage observed. The restorative process was in the green alga Chlamydomonas reinhardii dependent on protein synthesis occuring on chloroplast encoded mRNA (Ohad et al., 1984; Lidholm et al., 1987). Further support for this hypothesis was obtained in a comparative study of the susceptibility to photoinhibition in high and low light acclimated A. nidulans (Samuelsson et al., 1987). It was shown that high light acclimated A. nidulans had a higher resistance to photoinhibition than low light acclimated A. nidulans. This difference was well matched with the high light acclimated cyanobacteria recovering faster from a photoinhibition damage than did low light acclimated cyanobacteria. This made us suggest that the capacity of a recovery process significantly determine the degree of net photoinhibition observed and contribute to explain the difference in suseptibility to photoinhibition in high and low light acclimated A. nidulans. It is tempting to suggest that the capacity of this recovery process is dependent on the rate of turnover of the protein(s) that is (are) the primary site(s) of photodamage in photosystem II. Support for this hypothesis is also given by others (Ohad, et al., 1984; Greer et al., 1985). The principal difference between this mechanism to protect from photoinhibition and the avoidance mechanisms discussed above is that the photoinhibitory damage occurs but it is repaired, the capacity of the latter process significantly determining the ned photo-inhibitory response observed. There are evidences that a protein involved is the Q_B-protein in photosystem II (Ohad et al., 1984).

CONCLUDING REMARKS

In the case of evergreen conifers, which like Pinus sylvestris is readily photoinhibited both under field (Öquist and Ögren, 1985) and laboratory conditions (Table 1; Figure 1), photoinhibition of photosynthesis may be viewed as a protective mechanism to minimize more severe damages of photodynamic character in the needles during autumn and winter when photosynthesis is limited or fully inhibited by low temperatures. Irrespective of whether the reversible, photoinhibitory damage means physical destruction of reaction centers of photosystem II or only an increased heat dissipation down regulating the photochemical yield of photosystem II, photoinhibition of photosynthesis may be a way by which plants create a mechanism for massive heat dissipation of excessive excitation energy.

The significance of proposed mechanisms to avoid or tolerate photo-inhibition of photosynthesis in vivo is essentially unknown, and there is a need for much more work related to quantitate the involvement of various protective mechanisms in various plants and habitats. The two principle mechanisms proposed for photoinhibition should also be studied in more detail at the molecular levels but they should in addition be considered in relation to ecophysiology and the cost and time of repair of the damage.

REFERENCES

Barényi, B. and Krause, G.H., 1985, Inhibition of photosynthetic reactions by light, Planta, 163:218.
Björkman, O., 1981, Responses to different quantum flux densities, in: Encyclopedia of Plant Physiology, New Series, O.L. Lange, P.S. Nobel, C.B. Osmond and H. Ziegler, eds., Vol. 12A:57, Springer-Verlag, Berlin.
Bongi, G. and Long, S.P., 1987, Light-dependent damage to photosynthesis in olive leaves during chilling and high temperature stress, Plant, Cell Environ., 10:241.
Butler, W.L., 1978, Energy distribution in the photochemical apparatus of photosynthesis, Annu. Rev. Plant Physiol., 29:345.

Deisenhofer, J., Epp, O., Miki, K., Huber, R. and Michel, H., 1985. Structure of the protein subunits in the photosynthetic reaction center of Rhodopseudomonas viridis at 3Å resolution, Nature, 318:618.

Demmig, B. and Björkman, O., 1987, Comparison of the effect of excessive light on chlorophyll fluorescence (77K) on photon yield of O_2-evolution in leaves of higher plants, Planta, 171:171.

Fork, D.C., Öquist, G. and Powles, S.B., 1981, Photoinhibition in bean: a fluorescence analysis, Carnegie Inst. Yearbook, 81:52.

Greer, D.H., Berry, J.A. and Björkman, O., 1986, Photoinhibition of photosynthesis in intact bean leaves: role of light and temperature and requirement for chloroplast protein synthesis during recovery, Planta, 168:253.

Horton, P. and Lee, P., 1985, Phosphorylation of chloroplast membrane proteins partially protects against photoinhibition, Planta, 165:37.

Krause, G.H. and Behrend, U., 1986, DpH-dependent chlorophyll fluorescence quenching indicating a mechanism of protection against photoinhibition of chloroplasts, FEBS Lett., 200:298.

Krause, G.H. and Cornic, G., 1987, CO_2 and O_2 interactions in photoinhibition, in:Topics in Photosynthesis, Photoinhibition, Vol. 9, in press, Elsevier Sci. Publ., Amsterdam.

Krause, G.H. and Laasch, H., 1987, Energy-dependent chlorophyll fluorescence quenching in chloroplasts correlated with yield of photosynthesis, Z. Naturforsch., 42c:581.

Krause, G.H., Köster, S. and Wong, S.C., 1985, Photoinhibition of photosynthesis under anaerobic conditions studied with leaves and chloroplasts of Spinacia oleracea L., Planta, 165:430.

Levitt, J., 1980, Responses of Plants to Environmental Stress, Academic Press, New York.

Lidholm, J., Gustafsson, P. and Öquist, G., 1987, Photoinhibition of photosynthesis and its recovery in the green alga Chlamydomonas reinhardii, Plant Cell Physiol., 28: in press.

Neale, P.J., 1987, Algal photoinhibition and photosynthesis in the aquatic environment, in: Topics in Photosynthesis, Photoinhibition, Vol. 9, in press, Elsevier Sci. Publ., Amsterdam.

Ögren, E. and Öquist, G., 1985, Effects of drought on photosynthesis, chlorophyll fluorescence and photoinhibition susceptibility in intact willow leaves, Planta, 166:380.

Ögren, E., Öquist, G. and Hällgren, J.-E., 1984, Photoinhibition of photosynthesis in Lemna gibba as induced by the interaction between light and temperature. I. Photosynthesis in vivo, Physiol. Plant., 62:181.

Ohad, I., Kyle, D.J. and Arntzen, C.J., 1984, Membrane protein damage and repair: removal and replacement of inactivated 32-kilodalton polypeptides in chloroplast membranes, J. Cell Biol., 99:481.

Öquist, G. and Fork, D.C., 1982, Effects of desiccation on the excitation energy distribution from phycoerythrin to the two photosystems in the red alga Porphyra perforata, Physiol. Plant., 56:56.

Öquist, G. and Ögren, E., 1985, Effects of winter stress on the photosynthetic electron transport and energy distribution between the two photosystems of pine as assayed by chlorophyll fluorescence kinetics, Photosynth. Res., 7:19.

Öquist, G., Greer, D.H. and Ögren, E., 1987, Light stress at low temperature, in: Topics in Photosynthesis, Photoinhibition, Vol.9, in press, Elsevier Sci. Publ., Amsterdam.

Osmond, C.B., 1981, Photorespiration and photoinhibition. Some implications for the energetics of photosynthesis, Biochim. Biophys. Acta, 639:77.

Powles, S.B., 1984, Photoinhibition of photosynthesis induced by visible light, Annu. Rev. Plant Physiol., 35:15.

Samuelsson, G., Lönneborg, A., Rosenqvist, E., Gustafsson, P. and Öquist, G., 1985, Photoinhibition and reactivation of photosynthesis in the cyanobacterium Anacystis nidulans, Plant Physiol., 79:992.

Samuelsson, G., Lönneborg, A., Gustafsson, P. and Öquist, G., 1987, The susceptibility of photosynthesis to photoinhibition and the capacity of recovery in high and low light grown cyanobacteria, Anacystis nidulans, Plant Physiol., 83: 438.

Weis, E. and Berry, J.A., 1987, Quantum efficeincy of photosystem 2 in relation to energy dependent quenching of chlorophyll fluorescence, Biochim. Biophys. Acta, in press.

Williams, W.P. and Allen, J.F., 1987, State 1/State 2 changes in higher plants and algae, Photosynth. Res., 13:19.

THE SUPRAMOLECULAR STRUCTURE OF THE LIGHT-HARVESTING

SYSTEM OF CYANOBACTERIA AND RED ALGAE

Erhard Mörschel

Fachbereich Biologie-Botanik der Philipps-Universität

Karl-von-Frisch-Str., D-3550 Marburg, FRG

Life and development of photosynthetic organisms are strongly dependant on light necessary to drive photosynthesis and morphogenetic processes (Anderson, 1986). Because plants have only limited capabilities to choose optimal light conditions, they had to develope specialized pigments, which adapt the photosynthetic apparatus to the changing light climates, especially to changes in light quantity and quality (Senger and Bauer, 1987). These light collectors, called light harvesting pigments or light harvesting antennae, capture light over a large range of the spectrum and transfer the excitation energy in an energetic cascade finally to the photosynthetic reaction centres where chemical events start. Thereby they increase the photosynthetic efficency of the reaction centres to fully utilize the capacity of the electron-transport chain and that of the carbon dioxide fixation system and thus allow life in ecological niches under limited and unfavourabe conditions.

The major light harvesting pigments of higher plants, green algae, euglenoids and prochlorons are the chlorophylls a and b, absorbing light in the blue and red region of the spectrum. Brown algae, diatoms, dinoflagellates, cryptomonads, red algae and cyanobacteria have developed additional light harvesting pigments as chlorophyll c, fucoxanthin, peridinin and the biliproteins, absorbing light primarily in the green gap of chlorophyll (Glazer, 1983). These pigments are especially important to optimize photosynthesis, because the light window closes in water bodies, depending on depth and water quality, to a gap between 450-600 nm, a spectral range in which chlorophyll does not absorb light efficiently (Larkum and Barrett, 1983). The light harvesting antennae are organized into pigment-protein complexes, which can be i) a part of photosystem I or photosystem II complexes and then called core antennae. ii) They can be associated with the photosystems as the phycobilisomes, or the bound light harvesting chlorophyll a/b complexes, or iii) occur as independant pigment-protein-complexes as the mobile phosphorylated light harvesting chlorophyll a/b complex (LHC). Depending on cell structure and organisation, two different kinds of light harvesting systems have evolved: membrane integrated light harvesting systems and antennae structures, which are extramembraneous and associated with the surface of thylakoids or cytoplasmic membranes (Fig.1). Higher plants, green algae, prochlorons and chromophyta contain the first category of antennae structures which represent mainly light harvesting chlorophyll (chl) a/b, chl a/c and chl-fucoxanthin complexes. Examples of extramembrane light harvesting systems are the chlorosomes of green bacteria (Blankenship and Fuller, 1986) and the phycobilisomes of cyanobacteria and red algae, which are the subject of this review.

In accordance with the extramembraneous localization of the phycobilisomes, cyanobacterial and red algal thylakoids lie singly in the cytoplasm or stroma without any contact to other thylakoids. In most species of cyanobacteria the thylakoids are arranged peripherelly into three or more layers forming an anostomosing network of concentric shells. As the most characteristic feature, the outer surface of the thylakoids is studded with the phycobilisomes, containing the biliproteins as light harvesting pigments (Mörschel and Rhiel, 1987). They built up arrays of short and long rows with a hight regular spacing as shown in Fig. 6b. Three types of phycobilisomes can be distinguished: i) Bundle-shaped phycobilisomes, occuring only in the atypical cyanobacterium <u>Gloeobacter violaceus</u>, which lacks thylakoids. The phycobilisomes are bound to the inner surface of the cytoplasmic membrane (Guglielmi et al., 1981). ii) Hemi-ellipsoidal phycobilisomes. These phycobilisomes are preferably found in red algae and some cyanobacterial <u>Phormidium</u> and <u>Synechococcus</u> strains (Gantt, 1980, Ohki and Fujita, 1987) and iii) Hemidiscoidal phycobilisomes, which are the light harvesting complexes of many cyanobacteria and some red algae as <u>Porphyridium aerugineum</u>. The hemi-discoidal nature of the phycobilisomes and their arrangement into rows is revealed in profile views (Fig. 3a, 6b). Sections running parallel to the plane of the thylakoid show phycobilisomes from above assembled face to face in columns. Sections cut perpendicular to the plane of the thylakoids and parallel to the phycobilisome rows show the phycobilisomes in side view associated with the membrane. The phycobilisomes have a basal length of 36–40 nm, they are 23–33 nm high and about 10 nm thick (Mörschel and Rhiel, 1987).

Phycobilisomes are made up to a major part of the biliproteins and additional linker polypeptides that govern the assembly of phycobilisomes to active light-harvesting complexes. The biliproteins are 1) red phycoerythrins with absorption bands between 500–570 nm, 2) blue phycocyanins with absorption maxima at 610–635 nm, including phycoerythrocyanin absorbing at 575 and 590 nm and 3) allophycocyanin and the allophycocyanin B complex with an absorption at 650 nm and 670 nm respectively. The basic building blocks of the biliproteins are heterodimers composed of two genetically related polypeptides α and ß with molecular weights between 17 and 22 kDa. Occassionally phycoerythrins contain a third γ-subunit (Zuber, 1986).

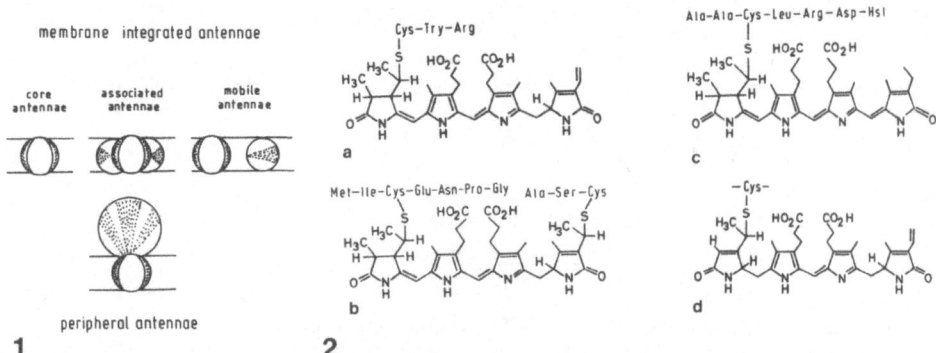

Fig. 1. Types of light-harvesting antennae: membrane integrated chlorophyll-antennae (a) and peripheral antennae such as phycobilisomes (b). RC: reaction centre.
Fig. 2. Structures of bilin chromophores: phycoerythrin linked through ring A (a) and doubly bound through ring A and D (b); phycocyanobilin (c); phycourobilin (d) as reviewed by Glazer (1985).

442

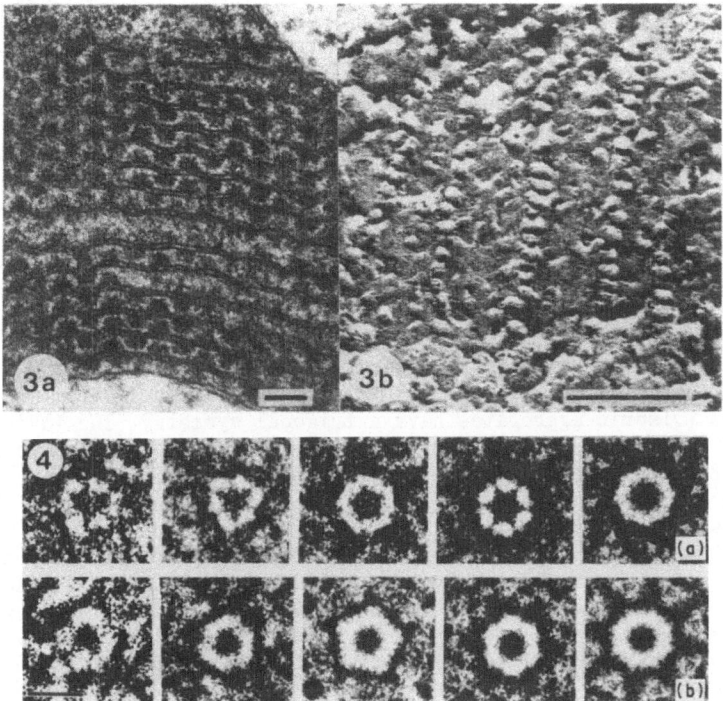

Fig. 3. Section through the chloroplast of the red alga <u>Rhodella</u> <u>violacea</u>
showing thylakoids with hemi-discoidal phycobilisomes displayed in face
views as semicircular aggregates (a); Freeze-fracture through a thylakoid of
the cyanobacterium <u>Synechococcus</u> spec. showing the dimeric 10nm x 20 nm
exoplasmic particles to which phycobilisomes are bound (b). Bars: 100 nm.
Fig. 4. Micrographs of higher magnification and image enhancement by Markham
rotation (3x, 5x, 6x, 8x) show that trimeric phycocyanin (a) and allophycocyanin (b) rings are made up of six globular subunits arranged in a
trigonal or hexagonal pattern. The α- and ß-subunits are arranged in an
alternating series within one plain. Bar: 10 nm.

The spectroscopic properties of the biliproteins are determined by the
covalently bound linear tetrapyrrole chromophores, the phycobilins. Four
different chromophores are found in biliproteins of canobateria and red
algae namely phycoerythrobilin, phycocyanobilin, phycourobilin and a phyco-
biliviolin-like chromophore which is bound to the α-subunit of phycoerythrocyanin. The structures of phycocyanobilin, phycoerythrobilin and phycourobilin are given in Fig. 2. The bilin chromophores are attached to the
polypeptide chains of the α-and ß-subunits by thioether linkages to cysteinyl ·residues. In the majority of the examples the different bilin chromophores appear to be linked to the polypeptide chain through one single bond
at ring A or D to a cysteinyl residue. However, some biliprotein chromophores such as phycoerythrobilin are linked through two cysteinyl-residues
by thioether bonds to ring A and D. The spectral properties of the biliproteins are determined by the types of chromophores described above which
have their basic absorption bands between 300-400 and 500-700 nm. A bathochromic shift occurs with an increasing number of conjugated double bonds
in the tetrapyrrole systems. The intensive visible absorption band of the
native biliproteins is attributed partially to an extended conformation of
the chromophores. In native biliproteins the spectroscopic properties are
modulated by the chromophore-protein interaction, the linkage to the protein and the association of special linker polypeptides to the biliproteins
(Glazer, 1985).

The (αß)-heterodimers have the fundamental property of building up cyclic trimers, which form hexamers and stacks of hexamers as a basic prerequisite for phycobilisome assembly. Figure 4a shows ring-shaped timeric phycocyanin with a diameter of 10-11 nm and a central channel of 3.1-3.7 nm, respectively. The thickness of the rings is about 3.0 nm. Combined biochemical and ultrastructural analyses give a molecular weight of 120 kDa corresponding to a trimeric structure of $(\alpha\beta)_3$. The same basic structure is valid for trimeric allophycocyanin shown in Fig. 4b Hexameric phycocyanin originates through side by side assembly of two trimers (Mörschel et al., 1980a). B-phycoerythrin builts up hexameric double discs made up of two trimers side by side. These two trimers are held together by an additional Y-subunit that penetrates the sandwich ring-structure like an axis (Mörschel et al., 1980b). The central position, which is held by the γ-subunit in B-phycoerythrin is occupied in other biliproteins by linker polypeptides. Recently, the X-ray structures of trimeric and hexameric C-phycocyanin from <u>Mastigocladus laminosus</u> and <u>Agmenellum quadruplicatum</u> were determined. The trimeric aggregates have a trigonal symmetry. The (αß)-subunits form boomerang-shaped heterodimers, which built up the ring-shaped trimers with the form of water-wheels. The contact of α-and ß-subunits within a heterodimer is very intensive while the inter-heterodimer-contact is weeker (Schirmer et al., 1985, 1986).

The structural analysis of phycobilisomes was possible after they were separated from the membranes by the nonionic detergent Triton X-100 and purified by density-gradient centrifugation in 0.65 - 1.0 M phosphate buffers. This high ionic strength is necessary for isolation, because phycobilisomes have to be stabilised by diminishing the activity of water. The electron microscopical investigations showed that isolated hemi-discoidal phycobilisomes had a similar size and semicircular outline compared to phycobilisomes "in situ" (Mörschel and Rhiel, 1987). The hemi-discoidal phycobilisomes consist of two morphologically distinct domains: they are constructed of a core and a periphery built up of short rods (Fig. 5). The core is normally composed of three ring-shaped cylindrical units of about 11 nm diameter arranged in a trigon. Two cylinders are localized at the base, the original attachment site to the membrane "in situ", while the third occupies the groove on top of these structures. From this core usually six peripheral rods radiate in a symmetrical hemi-discoidal array. They are bound to the edge of the core cylinders with their flat faces. These rods consist of stacked hexameric biliprotein double discs corresponding to the formula $(\alpha\beta)_6 L_R$. The discs have diameters of about 10-11 nm and

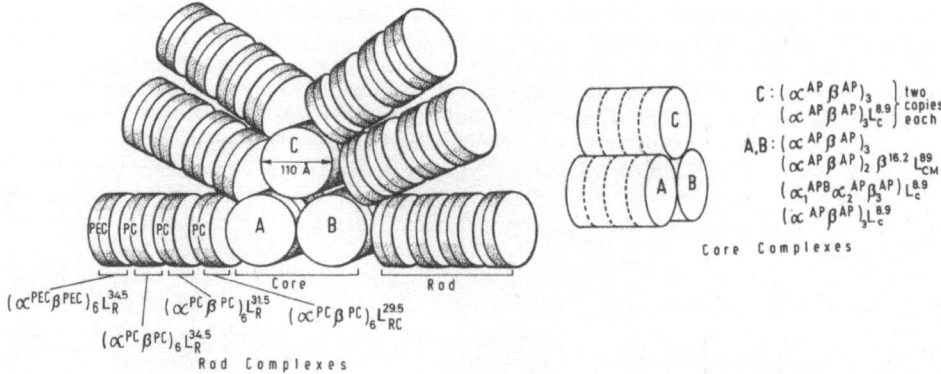

Fig. 5. Phycobilisome model of the cyanobacterium <u>Mastigocladus laminosus</u> with the triangular core and the surrounding biliprotein rods. Allophycocyanin (AP), phycocyanin (PC), phycoerythrocyanin (PEC), linker (L) with indices rod (R), rod-core (RC), core-membrane (CM), (Zuber et al., 1987).

are 5-6 nm thick; they are resolved into two smaller discs of about 2,5-3.0 nm thickness by faint subdivisions parallel to the long axis of the aggregates. Normally 2-6 hexamers make up a rod. The number depends on growth conditions and organism. The principle construction shown in Fig. 5 is valid for nearly all hemi-discoidal phycobilisomes investigated so far.

The sequential and structural arrangement of the biliproteins within the phycobilisomes has been studied extensively by various techniques including dissociation analysis, immune electron microscopy, analysis of mutants and energy transfer experiments. These analyses showed that the peripheral biliprotein rods always contain hexameric phycocyanin aggregates proximal to the core. Depending on the organism, the distal ends are constructed of phycocyanin alone, phycocyanin-phycoerythrin or phycocyanin-phycoerythrocyanin, which are inducible or constitutive chromoproteins. In the case of the red alga Rhodella violacea, the rods contained three biliprotein aggregates, corresponding to two individual hexameric B-phycoerythrin aggregates and one hexameric C-phycocyanin aggregate in a polar arrangement at that site of the rod that was originally bound to the core. The phycoerythrin-phycocyanin aggregates were energetically well coupled and emitted the C-phycocyanin fluorescence when excited in the B-phycoerythrin region. Since allophycocyanin was absent from the peripheral rods, it had to be located in the cylindrical aggregates of the core (Mörschel et al., 1980b). This was directly proven by the isolation of core elements from the cyanobactera Mastigocladus and Synechocystis (Anderson et al., 1984, Wehrmeyer, 1983).

The assembly of such complicated aggregates as phycobilisomes into functional light harvesting aggregates at the thylakoid surface has to proceed in a regulated step by step process. Biliprotein aggregates are very similar in structure and thus additional proteins, the linker polypeptides, are necessary to arrange the biliproteins in the right way. The linker polypeptides are grouped into four categories distinguished by molecular weight ranges of (1) 70-120 kDa, (2) 25-35 kDa, (3) 12-22 kDa and (4) 9-12 kDa. They are genetically related to biliprotein subunits; some of them carry bilin-chromophores (Zuber, 1986). Each of the hexameric biliprotein aggregates making up the phycobilisome rods is associated with a special linker polypeptide of the 25-35 kDa family. These linker polypeptides act as primers for the formation of trimeric and hexameric biliprotein aggregates. They also introduce a vectorial specificity into the hexameric biliprotein complexes that determines location and function within the phycobilisomes. The major domain of the linker polypeptides is buried probably within the central area of the template trimer. However one segment of 4-6 kDa protrudes from the face of the complexes. This exposed portion links the complex directly to the core, thereby initiating rod assembly, or onto other hexameric biliprotein-complexes. The linker polypeptides in addition modulate the spectroscopic properties of hexameric biliproteins by interactions with the ß-84 chromophores. In this way a red shift of the fluorescence maxima from the peripheral biliprotein hexamers of the rods to those proximal to the core is induced, thereby directing the energy flow towards the core and minimizing the random walk of energy within the rod subassemblies (Glazer, 1984). The first phycoerythrin hexamer of Rhodella violacea phycobilisomes fluoresces at 575 nm, the second at 578 nm, phycocyanin at 644 nm. The peripheral phycobilisome rods are an effective system to funnel light energy to the phycobilisome core. The modular design of the rods facilitates modifications due to changed environmental conditions in order to optimize light harvesting.

The core is the domain which attaches the phycobilisomes to the thylakoids and funnels the energy flow from the peripheral rods to the intramembrane chlorophyll antennae of photosystem II. With the exception of the phycobilisomes of Synechococcus 6301 whose cores are of only two cylinders

(Glazer et al., 1983), all other investigated hemidiscoidal phycobilisomes from red algae and cyanobacteria contain three cylinder cores with a triangular symmetry. The cylindrical elements are made up mainly of allophycocyanin. Besides allophycocyanin, the core contains especially the large linker biliprotein of 70-120 kDa. This large polypeptide anchors the phycobilisome on the thylakoid membrane. One domain of this polypeptide is an integral part of the photosystem II particles. It seems reasonable that the part of the polypeptide exposed on the cytoplasmic surface of the thylakoids initiates the self-assembly of the phycobilisomes and fullfills functions in organization and stabilization of the core. It carries one phycocyanobilin chromophore and with a fluorescence maximum of 676 nm acts as a final emitter of energy within the phycobilisome. The second terminal emitter of the phycobilisomes is a special α-allophycocyanin subunit, called α-allophycocyanin B (Glazer, 1984). It differs from α-allophycocyanin in its spectroscopic properties and the amino acid sequence. The isolated subunit absorbs maximally at 645 nm and fluoresces at about 680 nm. A localisation study showed that this α-allophycocyanin subunit is localised in each of the basal core cylinders. The tentative localization of the polypeptides within the core of <u>Mastigocladus</u> <u>laminosusus</u> phycobilisomes is shown in Fig. 5.

Light transduction occurs with an efficiency approaching 100 % in the range of picoseconds (Holzwarth, 1986). Phycobilisomes are constructed so that each domain and each chromophore is qualified to trap light. Light is preferably captured by the rod domain containing most of the biliprotein aggregates and a large numer of chromophores. From the periphery to the final acceptors the absorption cross-section within the phycobilisome is diminished. In <u>Rhodella</u> <u>violacea</u> phycobilisomes 420 phycourobilin and phycoerythrobilin chromophores of B-phycoerythrin channel energy to 108 phycocyanobilin chromophores of C-phycocyanin. From there energy is transferred down the cascade to 68 phycocyanobilin chromophores of allophycocyanin in the core and delivered to only four chromophores of two α-allophycocyanin B and two high molecular weight linker polypeptides (L_{CM}). Phycobilisomes are dynamic structures. Their number and pigment composition is influenced by a number of developmental and environmental factors, such as light intensity and light quality. Generally an inverse correlation between biliprotein content and light intensity is observed. Some cyanobacteria are able to respond to spectral changes in light colour with the synthesis of complementary biliproteins, green light induces synthesis of red pigments, red light that of blue pigments. This phenomenon is called complementary chromatic adaptation (Tandeau de Marsac, 1983). We distinguish three groups:
1) group 1 does not adapt chromatically at all.
2) group 2 regulates phycoerythrin synthesis as a function of light quality, whereas the synthesis of phycocyanin is not affected.
3) group 3 regulates the synthesis of both biliproteins phycoerythrin and phycocyanin by light quality.
When green light adapted cells are exposed to red light, phycoerythrin synthesis is lowered or ceased, whereas phycocyanin synthesis is enhanced. Vice versa, phycoerythrin-synthesis is enhanced and phycocyanin-synthesis lowered, but not completely, so that the proximal phycocyanin parts of the rods as attachment sites to the core are conserved. This is achieved by the occurence of two phycocyanins, the first, phycocyanin-1 is always synthesized, while the second, phycocyanin-2 is inducible. The allophycocyanin level never changes and thus the core domain remains constant as an important prerequisit for energy transfer. Chromatic adaptation as an instrument to optimize the light harvesting capacity is achieved by a simple variation of the distal rod elements in a modular way, by removing, adding or replacing the biliprotein aggregates. The biliprotein variations during complementary chromatic adaptations are controlled by an unidentified photoreceptor pigment that regulates the transcription of specific genes for bilipro-

teins and linker proteins simultaneously. The photoreceptor exhibits two action maxima at 540 and 640 nm; the observations suggest an action scheme comparable to phytochrome. However, unlike phytochrome, which is active only in the red absorbing Pfr-form, this photoreceptor should have two active photoconvertible forms Pg und Pr both triggering the expression of genes. Pg is converted by green light into Pr and Pr into Pg by red light. In cyanobacterial strains where phycocyanin-2 and phycoerythrin are photo-controlled, Pg would induce the expression of the phycocyanin 2α- and ß-genes and that of the corresponding linker polypeptides, and Pr on the other hand would activate the phycoerythrin α-and ß-genes together with the expression of their linker polypeptides.

The well-ordered arrangement of phycobilisomes on the surface of the thylakoids affords an attachment to well aligned intramembrane particles. Thus, the phycobilisome pattern on the membrane should be mirrored by the particle pattern within the thylakoids. In many thylakoid areas we find well aligned particles 1) Broad lanes of densely packed particles of 7.0 - 9.0 nm, the protoplasmic (PF) particles and 2) on the complementary fracture face, rows of well aligned exoplasmic (EF) particles. The minimum centre to centre distance of these EF-particles is normally 45 nm or more and is thus in register with the spacing of phycobilisome rows on the thylakoids (Neushul, 1970). The EF-particles measure 10 x 20 nm and they are aggregated linearly front to back with their longitudinal faces with a periodicity of 10 nm. Most particles reveal a central furrow perpendicular to the long axis dividing them into two side by side domains of 10 nm x 10 nm each, suggesting that each particle represents a dimer (Fig. 3b). The subunits of these dimers are divided an other time parallel to their longitudinal faces. These divisions are complementary to the structure of the two adjacent phycobilisome core cylinders (Mörschel and Schatz, 1987). Freeze-etched thylakoids, exposing simultaneously phycobilisomes and exo-plasmic particles in the same area, show a direct alignment of both systems. The EF-particle rows are oriented towards the centre of the phycobilisome bases and they are directly continued by the lines of phycobilisomes on the outer surface of the thylakoids. On the basis of alignment and periodicity we proposed that hemi-discoidal phycobilisomes are associated peripherally on top of the EF-particles in a 1:1 stoichiometry (Mörschel and Mühlethaler, 1983). Because phycobilisomes transfer the captured light energy mainly to photosystem II, we proposed that the EF-particles correspond to photosystem (PS) II particles, as do those of higher plants. We isolated oxygen developing PS II particles from the cyanobacterium Synecho-

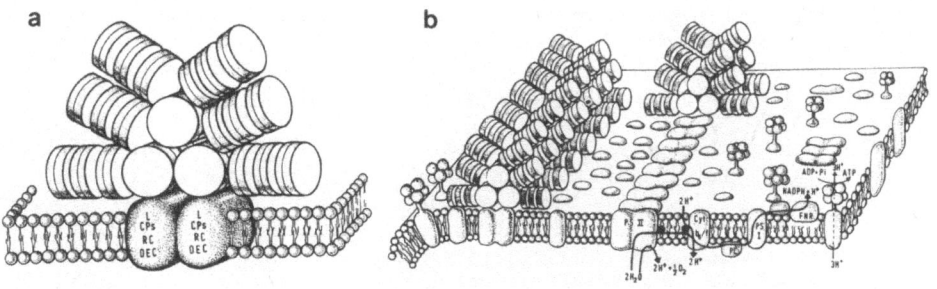

Fig. 6. a: Model of a photosystem II-phycobilisome complex. One hemi-discoidal phycobilisome is associated with a dimeric PSII complex (a). Linker (L), chlorophyll-proteins (CPs), reaction centre (RC), oxygen evolving complex (OEC). b: Model of a thylakoid of cyanobacteria and red algae with PSII-phycobilisome complexes organized into rows. The particles located between the PSII-phycobilisome rows are supposed to correspond to PSI, cytochrome b_6/f, ATP-synthases, and complexes of the respiration chain.

coccus spec. and incorporated these particles into liposomes (Mörschel and Schatz, 1987). The incorporated protein particles have an average diameter of 10 nm and thus correspond to the size of the EF-particles; PF-particles are in a size range of 7-9 nm. A certain fraction is in a dimeric configuration as also observed in negative staining electron microscopy. The PS II particles contain two chlorophyll antennae with apoproteins of 42 and 47 kDa, the reaction centre, the water splitting system and the 70-120 kDa polypeptide that we also found in phycobilisomes. As a component of PS II and phycobilisomes simultaneously this polypeptide may link both systems structurally and couple them energetically. The following scheme explains the energy flow from the phycobilisome to PS II.

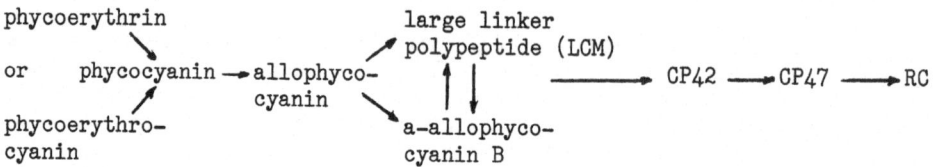

The phycobilisome-PS II complexes are tightly packed within their rows. We assume that energy is transferred not only from phycobilisomes to the underlying PS II units but also between phycobilisomes and photosystem II units along the same row. In this way, phycobilisome-PS II complexes build up energy-conducting tracks that function as an energy distribution system similar to chlorosomes of green bacteria where the baseplate funnels the energy to the reaction centres. Thus the phycobilisome-PS II complexes are not considered as separate units. They work cooperatively as a light-fibre system.

The rowed PSII-phycobilisome particles imply also a spatial separation of PS II and PS I. A lateral heterogenity of PS I and PS II may be postulated, similar to the lateral heterogeneity observed in the thylakoid system of grana containing chloroplasts. The distribution of excitation energy between photosystem II and photosystem I of higher plant chloroplasts is controlled by phosphorylation and migration of the light-harvesting chlorophyll a/b complexes. The mechanisem employed in cyanobateria is not yet clear.

REFERENCES

Anderson, J. M., 1986, Photoregulation of the composition, function, and structure of thylakoid membranes, Ann. Rev. Plant Physiol., 37:93
Anderson, L. K., Rayner, M. C. and Eiserling, F. A., 1984, Ultra-violet mutagenesis of Synechocystis sp. 6701: mutations in chromatic adaptation and phycobilisome assembly. Arch. Microbiol., 138:237
Blankenship, R. E. and Fuller, R. C., 1986, Membrane topology and photochemistry of the green photosynthetic bacterium Chloroflexus aurantiacus, in: " Encyclopedia of Plant Physiology, New Series, Vol. 19", Springer, Berlin, Heidelberg, New York, Tokyo.
Gantt, E., 1980, Structure and function of phycobilisomes: Light harvesting pigment complexes in red and blue-green algae, Int. Rev. Cytol., 66:45.
Glazer, A. N., 1983, Comparative biochemistry of photosynthetic light-harvesting systems, Ann. Rev. Biochem., 52:125
Glazer, A. N., 1984, Phycobilisome. A macromolecular complex optimized for light energy transfer, Biochim. Biophys. Acta, 768:29
Glazer, A. N., 1985, Light harvesting by phycobilisomes, Ann. Rev. Biophys. Biophys. Chem., 14:47
Holzwarth, A. R., 1986, Energy transfer kinetics in phycobilisomes, in: "Antennas and Reaction Centers of Photosynthetic Bacteria," M. E. Michel-Beyerle, ed., Springer, Berlin, Heidelberg, New York, Tokyo.

Larkum, A. W. D. and Barrett, J., 1983, Light-harvesting processes in algae, Adv. Bot. Res., 10:1

Mörschel, E., Koller, K.-P. and Wehrmeyer, W., 1980a, Biliprotein assembly in the disc-shaped phycobilisomes of Rhodella violacea. Electron microscopical and biochemical analyses of C-phycocyanin and allophyco-cyanin aggregates, Arch. Microbiol, 125:43

Mörschel, E. and Mühlethaler, K., 1983, On the linkage of exoplasmic freeze-fracture particles to phycobilisomes, Planta, 158:451

Mörschel, E. and Rhiel, E., 1987, Phycobilisomes and thylakoids: The light-harvesting system of cyanobacteria and red algae, in: "Electron Microscopy of Proteins 6. Membraneous Structures", J.R. Harris and R.W. Horne eds., Academic Press, London, New York, Tokyo.

Mörschel, E. and Schatz, H.-G., 1987, Correlation of photosystem-II com-plexes with exoplasmic freeze-fracture particles of the cyanobacterium Synechococcus sp., Planta, in press

Mörschel, E., Wehrmeyer, W. and Koller, K.-P., 1980b, Biliprotein assembly in the disc-shaped phycobilisomes of Rhodella violacea. Electron microscopical and biochemical analysis of B-phycoerythrin and b-phyco-erythrin-C-phycocyanin aggregates, Eur. J. Cell Biol. 21:319

Neushul, M., 1970, A freeze-etching study of the red alga Porphyridium, Am. J. Bot. 57:1231

Ohki, K. and Fujita, Y., 1987, Non-hemidiscoidal phycobilisomes in cyano-phytes, in:"Progress in Photosynthesis Research," J. Biggins, ed., Martinus Nijhoff, Dortrecht.

Schirmer, T., Bode, W., Huber, R., Sidler, W. and Zuber, H., 1985, X-ray crystallographic structure of the light-harvesting biliprotein C-phy-cocyanin from the thermophilic cyanobacterium Mastigocladus laminosus and its resemblance to globin structures, J. Mol. Biol., 184:257

Schirmer, T., Huber, R., Schneider, M., Bode, W., Miller, M. and Hackert, M. L., 1986, Crystal structure analysis and refinement at 2.5 A of hexameric C-phycocyanin from the cyanobacterium Agmenellum quadru-plicatum. The molecular model and its implications for light-har-vesting, J. Mol. Biol., 188:651

Senger, H. and Bauer, B., 1987, The influence of light quality on adapta-tion and function of the photosynthetic apparatus, Photochem. Photobiol., 45:939.

Tandeau de Marsac, N., 1983, Phycobilisomes and complementary chromatic adapt in cyanobacteria, Bull. L Inst. Pasteur, 81:201

Wehrmeyer, W., 1983, Organization and composition of cyanobacterial and rhodophycean phycobilisomes, in: "Photosynthetic Procaryotes: Cell differentiation and Function," G. C. Papageorgiou and L. Packer, eds., Elsevier, Amsterdam.

Zuber, H., 1986, Primary structure and function of the light-harvesting polypeptides from cyanobacteria, red algae, and purple photosynthetic bacteria, in:"Encyclopedia of Plant Physiology, New Series, Vol. 19", Springer, Berlin, Heidelberg, New York, Tokyo.

Zuber, H., Brunisholz, R. and Sidler, W., 1987, Structure and function of light-harvesting pigment-protein complexes, in: "Photo-synthesis," J. Amesz, ed., Elsevier, Amsterdam.

PHOTOMORPHOGENESIS IN THE NATURAL ENVIRONMENT

M. Geoffrey Holmes

Botany School
University of Cambridge
Downing Street
Cambridge CB2 3EA
U.K.

INTRODUCTION

The survival of plants depends on their ability to sense and exploit their environment. In all phototrophic organisms, the ultimate restriction is the availability of sufficient radiation to synthesize their organic constituents. If this is achieved, the radiation environment can be used by photomorphogenetic pigments to modulate growth and development. The radiation environment therefore acts not only as a source of energy, but also a source of information about the surroundings. The ability of a photoreceptor to influence development depends on both the spectral quality of the radiation in which the plant grows, and on the quantity of radiation present.

Before solar radiation reaches a plant photoreceptor, it has been attenuated both quantitatively and qualitatively by the earth's atmosphere and by the tissue surrounding the photoreceptor. In terrestrial plants solar radiation may also be modified by surrounding vegetation and in aquatic habitats there is additional attenuation by water and by suspended particles. It is impossible to provide an exact description of the quantum and spectral requirements for photomorphogenetic responses in individual species and for individual responses. However, an attempt is made here to describe variations in the natural radiation environment of both terrestrial and aquatic plants in terms of the changes which occur within the known spectral and quantum operating ranges of photomorphogenesis receptors. As plants exploit radiation both as a source of energy and as a source of information, it is convenient to consider natural radiation in terms of the availability of energy for photosynthesis, and also in terms of the information which can be derived from the light environment.

PHOTORECEPTOR SPECTRAL REQUIREMENTS

Photosynthetically active radiation (PAR) has been variously defined as the energy flux within the 400-700nm waveband (Gabrielsen, 1940), the energy flux within the 380-710nm waveband (Ničiporovic, 1961), and the quantum flux within the 400-700nm waveband (Federer and Tanner, 1966; McCree, 1972). This last definition has proven to be an acceptable measure in most laboratory experiments. However, there is evidence that terrestrial plants use wavelengths outside this range for photosynthesis. In *Phaseolus vulgaris*, for example, there is evidence that 28% or more of the energy trapped for photosynthesis in far-red rich environments is derived from wavelengths longer than 700nm (Holmes et al, 1986).

Photomorphogenetically active radiation is more difficult to define. The two main regions of the spectrum which influence photomorphogenetic responses are the ultraviolet/blue (UV/B) waveband and the red/far-red (R/FR) waveband. Responses to UV/B can be categorised into those which can be induced by both the UV and B wavelengths, and those which are induced solely by the B waveband.

The photomorphogenetic photoreceptor phytochrome absorbs throughout the entire 300-800nm waveband (Butler et al, 1960), but the relative quantum efficiency is much greater in the R and FR wavebands than at other wavelengths (Pratt and Briggs, 1966). It is therefore possible to consider only the R (600-700nm) and FR (700-800nm) wavebands as being significant for phytochrome-modulated responses. The approach can be refined further by referring to the R:FR ratio which is the relative photon flux ratio in 10nm bandwidths centred at 660 and 730nm, respectively (Holmes and Smith, 1975). Knowledge of the R:FR ratio provides useful information on the relative extent of photoconversion of the two main component forms (Pr and Pfr) of phytochrome (Smith and Holmes, 1977).

QUANTUM REQUIREMENTS FOR PHOTOMORPHOGENESIS

The threshold quantum requirements for B-induced photomorphogenetic and phototropic responses typically lie within the range 10^{-4} to 10^{-3} μmol m^{-2}, although saturation may require as much as 10^1, or even 10^2 μmol m^{-2} (e.g. Zimmermann and Briggs, 1963; Everett, 1974; Hartmann and Schmidt, 1980).

Threshold requirements in the R waveband for phytochrome-induced reponses have been reported to be as low as 5 x10^{-4} μmol m^{-2} (Withrow, 1959), but 5 x 10^{-2} μmol m^{-2} may be a more realistic figure (Klein et al, 1957). Most phytochrome-induced responses, and measurable phytochrome photoconversion require between 10^{-1} and 103 μmol m^{-2} R light (Holmes and Wagner, 1980).

PHOTOMORPHOGENESIS IN TERRESTRIAL PLANTS

The fundamental properties of the natural radiation environment, such as daylength and the rate of change of daylength are strongly dependent on latitude. In non-mountainous regions at the equator, daylength is constant. But days become longer in summer and shorter in winter with increasing distance from the equator. The rate of daylength change also depends on latitude, but is substantially modified by season, being slowest near the winter and summer solstices and fastest in spring and autumn. The massive daily changes in the amount of radiation reaching plants are attended by only small changes in the spectral quality of natural daylight. At solar elevations of less than 10-15°, however, the air mass through which direct solar radiation has to pass is greatly increased and the increased scattering and absorption in the atmosphere results in a decrease in the R:FR ratio of the direct solar beam. The effect is diluted by the contribution of the scattered sky radiation. Under extreme conditions in temperate regions, the R:FR ratio of total global radiation decreases to about 0.45, which corresponds to a decrease in Pfr/Ptot to about 0.40; a decrease in R:FR ratio from around 1.15 at higher solar elevations to ca. 0.70 is more typical (Holmes and Smith, 1977a).

The most dramatic changes in both the spectral quality and the quantity of radiation received by terrestrial plants are those caused by shading from overlying vegetation. Green leaves selectively absorb, reflect, and transmit solar radiation. Chlorophyll and accessory pigments absorb strongly in the 400-700nm waveband, but transmit most of the incident radiation in the 700-800nm waveband. The result of this selective attenuation is a marked change in the spectral characteristics of the radiation within plant canopies compared to the incident radiation above.

There are at least three spectral changes which appear to be of physiological significance. First, there is a marked reduction in the amount of PAR for plants growing below the canopy. This has obvious consequences for the potential rate of dry mass accumulation. Second, the quantum flux in the B waveband is drastically reduced. This quantity of B is nevertheless sufficient to satisfy the quantum requirements of B-induced responses. It is under these levels of B radiation that directional differences in the quantity of light can play a major role (e.g. phototropism) because the photoreceptor is not saturated.

A third spectral change which is now known to be of major ecological significance is the strong depletion of the R waveband but weak depletion of the FR waveband. There is empirical evidence that the reductions in R:FR ratio found below vegetation canopies cause reductions in phytochrome photoequilibrium *in vivo* (Holmes and Smith, 1975). When etiolated tissue is exposed to a wide variety of natural and artificial radiation sources, a hyperbolic relation exists between the R:FR ratio of the radiation source and the photoequilibrium established in the tissues (Holmes and McCartney, 1976; Smith and Holmes, 1977). The most significant feature of this relation is that phytochrome is most sensitive to changes in R:FR ratio which are characteristic of those found within natural vegetation canopies, i.e between approximately 1.15 in natural daylight and 0.05 in dense canopy shade (Holmes and Smith, 1977b).

PHOTOMORPHOGENESIS IN AQUATIC PLANTS

The UV and PAR wavebands are attenuated only very weakly by pure water and wavelengths longer than about 700nm are more strongly absorbed (Curcio and Petty, 1951), with the result that the R:FR ratio increases with depth. This contrasts with the terrestrial environment where no significant increase in the R:FR ratio have been observed. When dissolved inorganic material is present, in particular fertilizer-derived sulphates, a more marked attenuation of longer wavelengths occurs, resulting in a further increase in R:FR ratio and stronger attenuation of UV and B wavelengths (Kasha, 1948).

In turbid water containing insoluble material, suspended particles influence the attenuation of radiation by both scattering and by absorption. The scattering is largely independent of wavelength and is caused by objects such as silts, clays and phytoplankton. Spectrally selective scattering is caused primarily by small inorganic particles (Spence, 1981). The degree of scattering increases with decreasing wavelength and can reduce severely the penetration of the UV and B wavebands in turbid water containing a large quantity of small inorganic particles. The UV and B wavebands are further absorbed by Gelbstoff which represents about 95% of total organic carbon (Spence, 1981) and is considered to be contained in all natural waters (Hutchinson, 1957; Jerlov, 1968).

The most striking aspect of photomorphogenesis in aquatic habitats is the lack of empirical information. Although a small number of photomorphogenetic responses have been reported, relatively little is known about the underwater light climate and even less is known about about the photoreceptors involved. The over-riding influence of radiation underwater is the limitation set on photosynthesis. This is best defined by determining the depth of the euphotic zone and then determining the quantity of PAR available. Having obtained energy from the radiation environment it would then be of adaptive advantage if the plant could derive information from the same radiation.

Empirical information is not available on the ability of plants to exploit their radiation environment for photomorphogenesis underwater, but it is possible to calculate the availability of radiation for photomorphogenetic responses underwater. The procedures for calculating attenuation with increasing depth in specific water bodies are described in detail elsewhere (Spence, 1981). Using values of 20 μmol m^{-2} s^{-1} and 1 μmol m^{-2} s^{-1} (PAR) for plausible values for the compensation point of aquatic angiosperms and phytoplankton, respectively, the approximate photosynthetic depth limitations in specific water bodies can be calculated if physical characteristics such as the vertical diffuse attenuation coefficient are known. Spence (1981) has carried out comprehensive underwater light measurements in three Scottish lakes which allow such calculations to be made. Using the specific spectral attenuation coefficients for each relevant waveband it is then possible to calculate the quantum flux available (at the various photosynthetically limiting depths) for induction of photomorphogenetic responses caused by the B and R wavebands (Holmes and Klein, 1987). This approach has been used to calculate the data in the Table.

Table (a). Depth in meters at which photosynthesis becomes a limiting factor for growth in three Scottish lakes, assuming compensation points of either 20 μmol m^{-2} s^{-1} or 1.0 μmol m^{-2} s^{-1} and using 10^3 μmol m^{-2} s^{-1} as the photon flux incident at the water surface (all radiation values are for the 400-700nm waveband). (b) Time in seconds required to achieve a B or R-induced photomorphogenetic response at the depths described in Table (a), assuming a quantum requirement of 10^2 μmol m^{-2} for both the B and R wavebands.

(a)

μmol m^{-2} s^{-1}	400-700nm	Depth		
		Loch leven	Black Loch	Loch Borralie
20.0		2.1	3.8	8.5
1.0		3.7	6.7	15.0

(b)

Waveband	400-700nm	Time required for 10^2 μmol m^{-2} s^{-1}		
		Loch leven	Black Loch	Loch Borralie
B	20.0	35	39	2
B	1.0	1250	1470	9
R	20.0	8	4	8
R	1.0	111	36	115

Two generalizations may be made from the Table about the availability of radiation for controlling photomorphogenetic responses at the lower levels of photic zones. First, although B-induced responses may be rapidly saturated in clear lakes (Loch Borralie), the time taken to saturate the responses in lakes containing high levels of Gelbstoff (Loch Levin, Black Loch) may be too long to provide useful information to an organism under any but the highest aerial photon fluxes. Second, the quantum requirements of phytochrome in the R waveband are saturated rapidly in comparison to B, and there is no major dependence on water type for efficient function.

Difficulties arise when we wish to assess the significance of variations in the external radiation environment in terms of phytochrome function. As we have seen above, there appears to be enough energy for phytochrome photoconversion, but the range of R:FR ratios found underwater are too high to cause any major change in phytochrome photoequilibrium, at least in the etiolated test material on which such *in vivo* measurements have been made. However, the most complex – and the most striking – effects of spectral attenuation of natural daylight appear to occur within green tissue, rather than in the surrounding vegetation, water etc.. Experimental measurements of spectral attenuation within plant leaves and buds have been made using a variety of methods (e.g. Seyfried and Fukshansky; Vogelmann and Björn, 1984).

The general trend of the results of these studies is that wavelengths below 700nm and wavelengths above 700nm are distributed differently within plant organs. This is caused by differential absorption and multiple internal scattering within the tissue. The highest levels of radiation below 700nm are situated immediately below the irradiated surface and decrease markedly with increasing distance from the irradiated surface (Vogelmann and Björn, 1984). Whereas the gradient of radiation above 700nm also decreases with increasing distance from the irradiated surface, internal reflectance of these poorly absorbed wavelengths results in a substantially higher photon flux near the irradiated surface than in the incident radiation itself. Studies with *Cucurbita pepo* (Seyfried and Fukshansky, 1983) and with leaves of *Crassula falcata* (Vogelmann and Björn, 1984) indicate an approximately threefold higher level of FR immediately below the irradiated surface than in the incident radiation.

The spectral data of Vogelmann and Björn (1984) can be used to calculate the effects of spectral attenuance within plant leaves on the phytochrome photoequilibrium within green plant leaves. The measured values of R:FR ratio are used to estimate Pfr/Ptot either by reading off the values directly from Fig. 2 in Smith and Holmes (1977) or calculated using the equation:

$$Pfr/Ptot = 0.75/[1 + (0.37/R:FR)]$$

– which is a least squares fit for a rectangular hyperbola applied to the curve in Fig. 2 of Smith and Holmes (1977).

The measurements on *C. falcata* leaves confirm the theoretical predictions (Holmes and Fushansky, 1979) that a gradient in R:FR ratio exists across a green leaf and that the R:FR ratio – and therefore Pfr/Ptot – within the leaf is lower than that in the external radiation environment (Holmes and Klein, 1987). A notable feature is that the thick leaves of *C. falcata* produce a substantially greater decrease in R:FR ratio than that calculated for the thin (ca. 0.25mm) leaves of *P. vulgaris* (for details of those calcul;ations, see Holmes and Fukshansky, 1979).

It is then necessary to determine whether this reduction in R:FR rato is sufficient to bring the ratio into the range in which phytochrome is most sensitive, i.e. R:FR ratios below ca. 1.1. Extending the approach used above, phytochrome photoequilibrium within green leaves can be calculated for various ratios of R:FR in the incident radiation. As no information is available for the localization of phytochrome in green leaves, calculatons are presented for phytochrome situated mid-way between the upper and lower epidermis of a green *P. vulgaris* leaf, and for phytochrome situated 0.5mm and 2.0mm below the irradiated epidermis of a *C. falcata* leaf (see Figure).

It can be seen from the Figure that internal selective screening of
radiation by green tissue provides an extension in the range of incident R:FR
ratios which would bring about a significant change in phytochrome
photoequilibrium. The large shifts in photoequilibrium which have been
measured in etiolated tissue (Holmes and Smith, 1975; Holmes and McCartney,
1976; Smith and Holmes, 1977) occur within the relatively narrow range of
between approximately 0.05 and 1.5 (Fig). In the centre of a green *P.
vulgaris* leaf, the range of R:FR ratios which will produce equivalent shifts
in phytochrome photoequilibrium is extended up to approximately 5. In the
thick leaves of *C. falcata*, a R:FR ratio of 15, or more, will produce the
same Pfr/Ptot value as that produced by a R:FR ratio of 1.5 in etiolated
material.

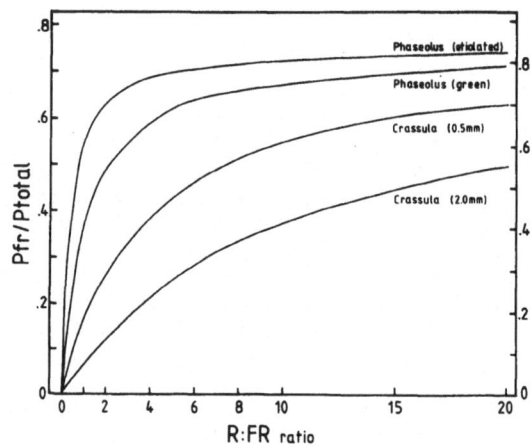

Figure The relation between the R:FR ratio of the incident radiation and the
phytochrome photoequilibrium (Pfr/Ptot) established in dark-grown *Phaseoulus
vulgaris* tissue, or in the middle of a green *P. vulgaris* primary leaf, or at
two distances (0.5mm and 2.0mm) from the irradiated surface of an
approximately 4mm thick *Crassula falcata* leaf.

It must be borne in mind that variables such as the location of the
photoreceptor, direction of propagation of the incident radiation, and
differences between air/tissue and water/tissue interfaces will influence
absolute values. Nevertheless, the inference from the Figure is that
screening by green plant tissue increases the range of spectral sensitivity
to R:FR ratios which are typical of many aquatic habitats. Although very
little experimental work has been done on photomorphogenesis underwater,
theoretical approaches such as these, and the few empirical studies which
have been made (e.g. Spence, 1975, 1981; Bodkin et al, 1980) indicate that
phytochrome is capable of responding to spectral changes underwater. Two
obvious possibilities which would be of adaptive value are the detection of
the increased R:FR ratio associated with increasing depth, and the localized
decreases in R:FR ratio caused by suspended phytoplankton and by macrophytes
competing for nutrients and PAR.

REFERENCES

Beggs, C.J., Holmes, M.G., Jabben, M. and Schäfer, E. (1980c) Action spectra for the inhibition of hypocotyl growth by continuous irradiation in light and dark grown *Sinapis alba* L. seedlings. <u>Plant Physiol</u>. 66, 615-618

Bodkin, P.C., Spence, D.H.N. and Weeks, D.C. (1980) Photoreversible control of heterophylly in *Hippuris vugaris* L. <u>New Phytol</u>. 84, 533-542

Butler, W.L., Hendricks, S.B. and Siegelman, H.W. (1960) *In vivo* and *in vitro* properties of phytochrome. <u>Plant Physiol</u>. 35, supplement 32.

Curcio, J.A and Petty, C.C. (1951) The near-infrared absorption spectrum of liquid water. <u>J. Opt. Soc. Am</u>. 41, 302-304.

Everett, M. (1974) Dose-response curves for radish seedling phototropism. <u>Plant Physiol</u>. 54, 222-225.

Federer, C.A. and Tanner, C.B. (1966) Spectral distribution of light in the forest. <u>Ecology</u> 47, 555-560.

Gabrielsen, E.K. (1940) Einfluss der Lichtfaktoren auf die Kohlensäureassimilation der Laubblätter. <u>Dansk Botanische Archiv</u> 10, 1-177.

Hartmann, E. and Schmidt, K. (1980) Effects of UV and blue light on the bipotential changes in etiolated htpocotyl hooks of dwarf beans. <u>In</u> The Blue Light Syndrome, pp. 221-237, ed. H. Senger. Springer, Berlin.

Holmes, M.G. (1981) Spectral distribution of radiation within plant canopies. <u>In</u> Plants and the Daylight Spectrum, pp. 147-158, ed. H. Smith. Academic Press, London. ISBN 0-12-650980-8

Holmes, M.G. (1983) Perception of shade. <u>Phil. Trans. R. Soc. Lond</u>. B 303, 503-521

Holmes, M.G. and Fukshansky, L. (1979) Phytochrome photoequilibria in green leaves under polychromatic radiation: a theoretical approach. <u>Plant, Cell and Environment</u> 2, 59-65

Holmes, M.G. and Klein, W.H. (1987) Light and temperature variations in aquatic and terrestrial environments. <u>In</u> Plant Life in Aquatic and Amphibious Habitats, pp. 3-22., ed. R.M.M. Crawford. Blackwell, Oxford.

Holmes, M.G. and McCartney, H.A. (1976) Spectral energy distribution in the natural environment and its implications for phytochrome function. <u>In</u> Light and Plant Development, pp. 467-476, ed. H. Smith. Butterworths, London. ISBN 0-408-70719-4

Holmes, M.G. and Smith, H. (1975) The function of phytochrome in plants growing in the natural environment. <u>Nature</u> 254, 512-514

Holmes, M.G. and Smith. H. (1977a) The function of phytochrome in the natural environment - I. Characterization of daylight for studies in photomorphogenesis and photoperiodism. <u>Photochem. Photobiol</u>. 25, 533-538

Holmes, M.G. and Smith. H. (1977b) The function of phytochrome in the natural environment - II. The influence of vegetation canopies on the spectral energy distribution of natural daylight. <u>Photochem. Photobiol</u>. 25, 539-545

Holmes, M.G. and Smith. H. (1977c) The function of phytochrome in the natural environment - IV. Light quality and plant development. Photochem. Photobiol. 25, 551-557

Holmes, M.G., Sager, J.C. and Klein, W.H. (1986) Sensitivity to far-red radiation in stomata of *Phaseolus vulgaris* L.: Rhythmic effects on conductance and photosynthesis. Planta 168, 516-522

Holmes, M.G. and Wagner, E. (1980) A re-evaluation of phytochrome involvement in time measurement in plants. J. Theor. Biol. 83, 255-265

Hutchinson, G.E. (1957) A treatise on limnology. In Geography, Physics and Chemistry. Vol. 1. Wiley, New York.

Jerlov, N.G. (1968) Optical Oceanography. Elsevier, Amsterdam.

Kasha, M. (1948) Transmission filters for the ultraviolet. J. Opt. Soc. Am. 38, 929-934.

Klein, W.H., Withrow, R.B, Elstad, V. and Price, L. (1957) Photocontrol of growth and pigment synthesis in the bean seedling as related to irradiance and wavelength. Am. J. Bot. 44, 15-19

McCree, K.J. (1972) The action spectrum, absorbtance and quantum yield of photosynthesis in crop plants. Agric. Meteorol. 9, 191-216.

Ničiporovic, A.A. (1961) Measurement of visible radiation in plant physiology and ecology, agrometeorology and plant production. Fiziol. Rast. 7, 744-748.

Pratt, L.H. and Briggs, W.R. (1966) Photochemical and non-photochemical reactions of phytochrome *in vivo*. Plant Physiol. 41, 467-474.

Seyfried, M. and Fukshansky, L. (1983) Light gradients in plant tissue. Applied Optics 22, 1402-1408

Smith, H. and Holmes, M.G. (1977) The function of phytochrome in the natural environment - III. Measurement and calculation of phytochrome photoequilibria. Photochem. Photobiol. 25, 547-550

Spence, D.H.N. (1981) Light Quality and plant response underwater. In Plants and the Daylight Spectrum, pp. 245-276, ed. H. Smith. Academic Press, London. ISBN 0-12-650980-8

Vogelmann, T.C. and Björn, L.O. (1984) Measurements of light gradients and spectral regime in plant tissue with a fiber optic probe. Physiol. Plant. 60, 361-368

Withrow, R.B. (1959) A kinetic analysis of photoperiodism. In Photoperiodism and related phenomena in plants and animals. Ed. R.B Withrow, pp. 439-471. American Association for the Advancement of Science, Washington D.C.

Zimmermann, B.K. and Briggs, W.R. (1963) Phototropic dosage-response curves for oat coleoptiles. Plant Physiol. 38, 248-253